RNA Technologies

More information about this series at http://www.springer.com/series/8619

Stefan Jurga • Volker A. Erdmann •
Jan Barciszewski
Editors

Modified Nucleic Acids in Biology and Medicine

Springer

Editors
Stefan Jurga
Nanobiomedical Center
Adam Mickiewicz University
Poznań
Poland

Volker A. Erdmann (Deceased)
Formerly at Institute of Chemistry and Biochemistry
Free University Berlin
Berlin
Germany

Jan Barciszewski
Institute of the Bioorganic Chemistry of the Polish Academy of Sciences
Poznań
Poland

ISSN 2197-9731 ISSN 2197-9758 (electronic)
RNA Technologies
ISBN 978-3-319-34173-6 ISBN 978-3-319-34175-0 (eBook)
DOI 10.1007/978-3-319-34175-0

Library of Congress Control Number: 2016944798

Printed on acid-free paper

This Springer imprint is published by Springer Nature
The registered company is Springer International Publishing AG Switzerland

Preface

The Power of Modified Nucleic Acids

With this volume of the series RNA technologies, we aim to cover various aspects of nucleic acid modifications. This is an interesting issue in the study of macromolecular components of cells.

DNA and RNA are key molecules of the cell. The structure, function, and reactivity of DNA and RNA are central to molecular biology and are crucial for the understanding of complex biological processes. For a long time DNA was considered as the most important molecule of all biology and the key of life. However DNA is not the be all and end all of the living cell, but it appears as an important by-product of the RNA evolution. Deoxyribonucleotides as DNA precursors are synthesized by specific enzymatic modification of ribonucleotides in which the 2′-hydroxyl group of the ribose moiety is replaced by 2′-hydrogen with ribonucleotide reductase. One of these DNA bases, thymine, is produced by methylation of uracil. There are a significant number of adenosines and cytidines in genomic DNA converted by spontaneous or enzymatic deamination to hypoxanthine and uracil, respectively. Cytosine can be methylated to 5-methylcytosine derivative with not only a coding capacity but also a regulatory potential. These data suggest that DNA looks similar to a modified RNA molecule although chemically more stable than RNA. In the eukaryotic cell, all postreplicative modifications of DNA showed a few percent of all bases. Cellular DNAs and RNAs can be chemically modified in more than 100 different ways. Some of these modifications to nucleic acids are random or spontaneous and their formation requires significant energy from the cell. The broad range of chemical modifications to nucleic acids is not restricted to simple nucleophilic substitution but extends to oxidative reactions and C–H activation by various agents. The modifications might occur on DNA as well as different types of RNA such as transfer ribonucleic acids (tRNAs), ribosomal RNA (rRNA), messenger RNA (mRNA), and other noncodings (ncRNAs). Among them, tRNAs represent ca 15 % of the total cellular RNAs and are highly stable. The primary role of tRNA is to deliver amino acids to the polypeptide chain during protein translation. tRNA molecules 73–93 nucleotides long are heavily modified types of ribonucleic acid. tRNA modifications (up to 25 %) are dynamic and adaptive to different environmental changes. Modified nucleosides of tRNAs play an important role in the translation of the genetic code.

The modified nucleosides are utilized to fine tune nucleic acids structure and function. These modifications are dynamic and participate in regulating diverse biological pathways. They can also be used as specific markers of different states of cells and diseases or pathologies. The process of RNAs turnover is directly correlated to their presence in the human. Evaluation of modified nucleosides might become novel markers to facilitate early clinical diagnosis of cancer to improve human cancer risk assessment.

Nucleoside methylation and other nucleic acid modifications are of a great interest, prompted by the discovery of methylation and active demethylation of DNA and RNA. In eukaryotic genomic DNA, 5-methylcytosine is a well-know epigenetic modification and is also known to exist in both rRNA and tRNA.

In response to oxidative stress caused by reactive oxidative species (ROS) as well as nutrient depletion and other growth arrest conditions, modified nucleotides are synthesized in the cell to serve various purposes. Accidental non-enzymatic methylation or oxidation of a base in a DNA and RNA, in addition to the normal enzymatic methylation processes, induces serious problems for living cells, especially for DNA, for which abnormal alkylation can be mutagenic. To remove this type of modification, the cell has developed oxidative mechanisms in an indirect way.

The recent development of high-throughput sequencing technologies has enabled us to identify tRNA-derived RNA fragments. It seems that they are not by-products from random degradation but rather functional molecules that can regulate translation and gene expression.

It takes a large effect to map RNA modifications globally as well as to identify the cellular function as writers, readers, and erasers for each modification. Basic cellular pathways use ubiquitous metabolites and coenzymes to transfer methyl and amino acid groups, isoprenoids, sugars, phosphates, and various metabolite nucleic acid conjugates have been found that affect a functionality of their specific targets.

The turnover of nucleic acids increases when cell proliferation takes place. Any disease or metabolic alteration affecting RNA turnover consequently results in altered nucleoside excretion patterns, leading to the hypothesis that RNA metabolites may be used as early indicators of disease. In addition, increased RNA metabolism with altered nucleoside excretion patterns related to metabolic disorders such as cancer may be suitable markers to facilitate the monitoring of therapeutic intervention.

In this book we have collected work describing modified nucleosides, naturally occurring or chemically synthesized nucleic acids. Their role in cell biology has huge potential for application in medicine. Further research frontiers and new developments are also discussed.

In total there are 18 chapters. Five of them deal with tRNAs and their modifications in relation to biomedical applications. Three discuss modified nucleosides including N^6-methyladenosine and 8-hydroxyguanosine as well as $2'$-O-methylated ribonucleotides. A very interesting chapter describes the role of diadenosine tetraphosphate in health and disease. Similar properties are described for circular RNAs and for modified therapeutic oligonucleotides. Other chapters describe the properties of modified oligonucleotides.

Poznań
Berlin
Poznań
January 2016

Stefan Jurga
Volker A. Erdmann
Jan Barciszewski

RNA Around the Clock: Volker A. Erdmann in Memoriam

On September 11, 2015 we lost our colleague and dear friend Professor Volker A. Erdmann from the Institute of Chemistry–Biochemistry, Freie Universität Berlin, Germany. He was born on February 8, 1941 in Stettin (Germany, now Poland) and later became a U.S. citizen. In 1963 he earned his B.A. in Chemistry and in 1966 an M.Sc. in Biochemistry from the University of New Hampshire, Durham, N. H., USA (advisor: Prof. Dr. E.J. Herbst). From 1966 to 1969 at the Max-Planck-Institut für experimentelle Medizin, Göttingen, Germany, and Technische Universität Braunschweig, Germany, he obtained a Dr. rer. nat. degree in Biochemistry with minors in Chemistry and Microbiology (advisor: Prof. Dr. F. Cramer). After an NIH postdoctoral fellowship with Prof. Dr. M. Nomura at University of Wisconsin, Madison, Wisc., USA, in 1971 he became a research group leader at the Max-Planck-Institut für Molekulare Genetik in Berlin at the Department led by Prof. Dr. H.G. Wittmann. In 1978 Volker did Habilitation in Biochemistry and Molecular Biology at the Freie Universität at Berlin, Germany.

From 1980 he was full professor of Biochemistry in the Department of Chemistry at the Institute of Biochemistry, and from 2009 he was a guest Professor at the Free University of Berlin. In 1987 he received an award for scientific excellence from the German Research Council (DFG), the highest scientific award given in Germany (Förderpreis für deutsche Wissenschaftler im Gottfried Wilhelm Leibniz Programm der Deutschen Forschungsgemeinschaft). He was a member of the Berlin-Brandenburgische Akademie der Wissenschaften (Berlin Brandenburg Academy of Science, former Prussian Academy of Science) and the Polish Academy of Science.

His research interests have always been in the area of gene expression with special emphasis in the structure and function of ribosomes and RNA technologies. His RNA research includes studies on the structure and function of ribozymes, antisense oligonucleotides, siRNAs, micro RNAs, DNAzymes, high affinity RNA molecules (aptamers), enantiomeric catalytic nucleic acids (ribozymes and DNAzymes), and large noncoding RNAs such as the H19 RNA. With his group at Free University he developed methods for the chemical synthesis of RNA molecules, including a large number of modified nucleotides. He has also concentrated on the crystallization of RNA molecules and their protein complexes by X-ray analysis. These crystallization experiments include microgravity experiments (participation in 17 space missions). The results of Volker's research have appeared in more than 450 publications. He created data bases on 5S ribosomal RNA and non-coding RNAs and obtained 14 patents in the area of RNA technologies. In 1998 he and his colleagues founded the Berlin Network for RNA Technologies, with the goal to pursue further the structural and functional potentials of RNA molecules.

Volker's most important discoveries include the first total reconstitution of bacterial 50S and 70S ribosomes, first identification of ribosomal 5S RNA binding proteins and 5S RNA protein complexes, first crystallization of ribosomes, first crystallization of RNA molecules under microgravity conditions, first crystallization and X-ray structural determination of a mirror image RNA structure, first to discover mirror image aptamers and first to discover mirror image L-catalytic nucleic acid as alternatives to siRNAs and microRNAs to cure cancer and viral infections.

These L-form aptamers have a number of advantages when compared with D-aptamers. They are very stable in human sera and cells. Because nature does not make L-nucleic acids, there was no need to develop any enzymes hydrolyzing the L-form of nucleic acids. L-Aptamers can be compared with protein antibodies, and indeed aptamers can assume very similar functions to antibodies. Aptamers are considerably smaller than antibodies and they are easily synthesized by nucleic acid synthesizers. They are not toxic or immunogenic and are therefore most likely ideally suited for the development of new types of pharmaceutical drugs. The development of mirror image catalytic RNA opened new possibilities in basic research and in the area of molecular evolution. Volker A. Erdmann was one of the first Editors-in-Chief and founders of RNA Biology journal published by Landes Bioscience, Georgetown, Texas (USA). He was co-editor of several books in the series "RNA Technologies" published by Springer. In 2013, he established a private-biotech company called Erdmann Technologies based in Berlin.

Married to Hannelore Erdmann, he had two children (Jörn, and Gabriele). Volker was a very quiet and kind person and supportive of young scientists. He directed his students and fellow researchers carefully and with pride. He derived much pleasure from the successes of his scientific offspring—his former students. His influence extended well beyond his scientific contributions to shaping policy on important issues at the interface between science and society. He was a source of inspiration to all around him and will be greatly missed.

Poznań, Poland Jan Barciszewski

Contents

Transfer RNA Modifications: From Biological Functions to Biomedical Applications

Adrian Gabriel Torres and Lluís Ribas de Pouplana

Contents

Abstract Transfer RNAs (tRNAs) are essential components of the protein translation machinery. In order to become fully active, they need to be heavily modified post-transcriptionally. Such modifications affect the structure, stability and functionality of tRNAs; however, their exact roles at the molecular level remain largely elusive. Here we focus on the biological functions of tRNA modifications associated to human diseases and how such information can be used for biomedical applications. We put an emphasis on mitochondrial-linked dysfunctions, metabolic disorders, neurological defects and cancer. We also present methods and approaches currently used in the clinic to detect and monitor different human

A.G. Torres
Institute for Research in Biomedicine (IRB Barcelona), Barcelona Institute for Science and Technology, C/Baldiri Reixac 10, Barcelona 08028, Catalonia, Spain
e-mail: adriangabriel.torres@irbbarcelona.org

L. Ribas de Pouplana (✉)
Institute for Research in Biomedicine (IRB Barcelona), Barcelona Institute for Science and Technology, C/Baldiri Reixac 10, Barcelona 08028, Catalonia, Spain

Catalan Institution for Research and Advanced Studies (ICREA), P/Lluis Companys 23, Barcelona 08010, Catalonia, Spain
e-mail: lluis.ribas@irbbarcelona.org

S. Jurga et al. (eds.), *Modified Nucleic Acids in Biology and Medicine*, RNA Technologies, DOI 10.1007/978-3-319-34175-0_1

pathologies involving tRNA modifications or tRNA modification enzymes, and, additionally, we propose novel tRNA modification-based strategies that could be used for diagnosis, prognosis or treatment of human diseases.

Keywords tRNA • tRNA modification • Protein translation • Human disease • Biomedicine

1 Transfer RNAs Are Post-Transcriptionally Modified

Transfer RNAs (tRNAs) play a key role in the protein translation machinery. They are transcribed as long primary tRNAs that are processed during their biogenesis to yield 70–100 nucleotides long RNA species, that fold into a cloverleaf-shape (2^{ry} structure) (Fig. 1) and L-shape (3^{ry} structure) arrangement (Piñeyro et al. 2014). Following maturation, they are charged at their 3′-end with their cognate amino acid, which will be incorporated into the growing polypeptide chain during protein synthesis. Residues 34, 35 and 36 of the tRNA form the tRNA 'anticodon' (Fig. 1) that pairs specifically with nucleotide triplets on the messenger RNA (mRNA) called 'codons'. Each mRNA codon codifies for a specific amino acid; hence, tRNAs are adaptor molecules that translate specific mRNA codons into specific amino acids during translation (Piñeyro et al. 2014).

During maturation, tRNAs are required to go through a series of post-transcriptional chemical modifications. In general, for a given tRNA, about 10–15 % of the tRNA residues are found modified (Phizicky and Alfonzo 2010). There are more than 50 different chemical modifications described for eukaryotic tRNAs, that include methylations, thiolations, deaminations, acetylations, isomerizations and hydroxylations, among others (Machnicka et al. 2013). Such modifications affect the structure, processing, stability and overall functionality of tRNAs.

Modifications in particular regions of the tRNA affect different aspects of tRNA functionality. In general, modifications in the main body of the tRNA affect the rigidity/flexibility of the molecule. For example, pseudouridines increase the binding affinity of tRNA residues by inducing a C3′-endo sugar conformation, and dihydrouridines make these interactions more flexible by retaining the sugar pucker into a C2′-endo conformation (El Yacoubi et al. 2012). Modifications at the anticodon region of the tRNA have a direct role on codon recognition and prevent frameshifting during protein translation. On the one hand, modifications at position 34 of the tRNA increase (or restrict) the number of codons a tRNA can recognize, by promoting or inhibiting tRNA 'wobbling' (non-Watson-Crick nucleotide pairing) (Crick 1966). Some examples include A34-to-I34 editing (deamination) that allows the modified tRNAs to decode codons ending not only in uridine (U) but also in adenine (A) and cytosine (C) (Torres et al. 2014b) or U34 modifications, such as the one on $tRNA^{Lys}_{(UUU)}$, which allows the tRNA to decode its cognate codons AAA and AAG but restricts the recognition of the near-cognate asparagine

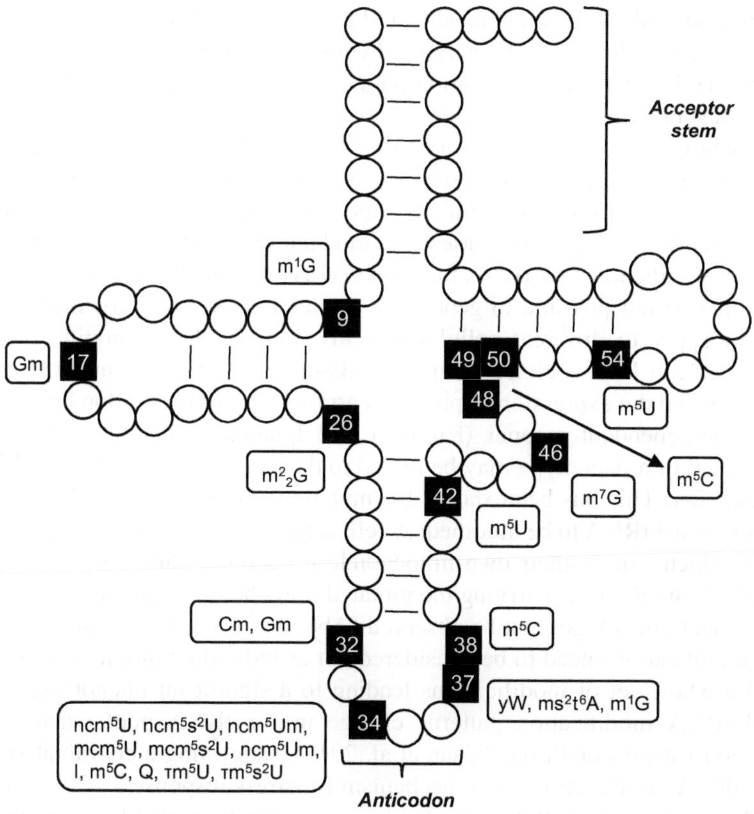

Fig. 1 Representation of the tRNA secondary structure ('cloverleaf'). Post-transcriptionally modified tRNA residues associated to human diseases are shown (*black boxes*). *Abbreviations:* m^1G, 1-methylguanosine; Gm, 2′-*O*-methylguanosine; $m^2{}_2$G, N2,N2-dimethyl guanosine; Cm, 2′-*O*-methylcytidine; ncm^5U, 5-carbamoylmethyluridine; ncm^5s^2U, 5-carbamoylmethyl-2-thiouridine; ncm^5Um, 5-carbamoylmethyl-2′-*O*-methyluridine; mcm^5U, 5-methoxycarbonylmethyluridine; mcm^5s^2U, 5-methoxycarbonylmethyl-2-thiouridine; mcm^5 Um, 5-methoxycarbonylmethyl-2′-*O*-methyluridine; I, inosine; m^5C: 5-methylcytosine; Q, queuosine; τm^5U, 5-taurinomethyluridine; τm^5s^2U, 5-taurinomethyl-2-thiouridine; yW, wybutosine; ms^2t^6A, 2-methylthio-N^6-threonyl carbamoyladenosine; m^5U, 5-methyluridine; m^5 U, 5-methyluridine; m^7G, 7-methylguanosine

codons AAU and AAC (Yarian et al. 2002). On the other hand, modifications at position 37 (adjacent to the anticodon) are usually associated to keeping in-frame translation. These are in general bulk modifications that stabilize codon/anticodon pairing by generating base-stacking interactions. Examples include the wybutosine 37 (yW37) modification that prevents −1 frameshifts (Waas et al. 2007) or the 1-methylguanosine 37 (m^1G37) modification that impedes +1 frameshifting (Urbonavicius et al. 2003). Finally, modifications in the acceptor stem of the tRNA (the stem formed by the 5′- and 3′-ends of the tRNA; Fig. 1) usually serve as identity elements for aminoacyl tRNA synthetases, the enzymes that charge the tRNA with its cognate amino acid. A representative example is the posttranscriptional addition of a G residue at the 5′-end of $tRNA^{His}$ (G_{-1} modification) that is essential for correct charging by histidine-tRNA synthetase (Rudinger et al. 1994). For detailed information on the biological functions of tRNA modifications, several comprehensive reviews are available (Phizicky and Alfonzo 2010; El Yacoubi et al. 2012; Jackman and Alfonzo 2013; Piñeyro et al. 2014).

Interestingly, while tRNA modifications seem very important for tRNA function, the vast majority of them are not essential for cell viability (Piñeyro et al. 2014). In fact, in many cases, the modulation of tRNA modifications does not affect significantly the tRNA function and usually results in subtle phenotypes. However, it is not possible to generalize on this matter, and known observations need not apply to different cellular systems. Indeed, tRNA modification-based phenotypes can be associated to specific tissues. It is well documented that the tRNA pool and the expression of proteins carrying a particular codon bias may vary in a tissue-dependent manner (Kirchner and Ignatova 2015). Additionally, the 'penetrance' of a phenotype may be linked to the degree of the tRNA modification misregulation. This has been shown for mutations on mitochondrial tRNA genes that prevent the tRNA to be modified. A cell has a variable number of mitochondria, each of which carries their own mitochondrial genome. Only when a significant amount of mitochondria carrying the mutated mitochondrial genome variant accumulate, clear phenotypes can be observed (Abbott et al. 2014). Finally, sometimes tRNA modifications need to be considered not as individual modifications but as a part of a whole set of modifications leading to a significant phenotype. In yeast, overall tRNA modification patterns change in the tRNA pool when cells are subjected to stress conditions (Chan et al. 2012), suggesting a coordinated regulation of tRNA modifications as a mechanism for stress response.

In this chapter, we will describe the known connections that have recently been established between post-transcriptional tRNA modifications and human diseases. It will become evident that the roles of such modifications on those diseases are complex and that the molecular mechanisms behind the observed phenotypes remain poorly understood. Importantly, we will not address human diseases caused by mutations on tRNA genes or by defects on tRNA processing and maturation, which have been recently reviewed in depth (Abbott et al. 2014; Kirchner and Ignatova 2015), unless a clear direct effect on the tRNA modification pattern has been observed. Finally, we will present different strategies that are being pursued in the clinic for diagnosis, prognosis and treatment of this type of diseases and will propose potentially novel tRNA modification-based therapeutic approaches to be pushed forward in the future.

2 Links Between tRNA Modifications and Human Diseases

While tRNA modifications and the enzymes that catalyse such modifications have not been studied in depth in metazoans, an association between tRNA modifications and human diseases is starting to emerge (Torres et al. 2014a) (Fig. 1 and Table 1). Several human genetic studies have shown links between mutations in genes that

Table 1 Post-transcriptional tRNA modifications discussed in this chapter, including their role in different tissues and associated human diseases

Tissue	Gene	tRNA modification	Affected tRNAs	tRNA residues	Associated human diseases and phenotypes	References
Brain	FTSJ1	Cm, Gm, $ncm^5U\underline{m}$	Leu, Phe, Trp	32, 34	Non-syndromic X-linked mental retardation	Hamel et al. (1999), Freude et al. (2004), Ramser et al. (2004), Bonnet et al. (2006), Froyen et al. (2007), Dai et al. (2008), Takano et al. (2008), Giorda et al. (2009)
	TRMT1	$m^2{}_2G$	Several	26	Intellectual disability	Najmabadi et al. (2011)
	ADAT3	I	Ala, Pro, Thr, Val, Ser, Arg, Leu, Ile	34	Intellectual disability and strabismus	Alazami et al. (2013)
	Elongator complex (IKAP, ELP2, ELP3, ELP4)	$\underline{\overline{mcm}}^5U$, $\underline{\overline{ncm}}^5U$, and derivatives	Several	34	Familial dysautonomia	Anderson et al. (2001), Slaugenhaupt et al. (2001), Leyne et al. (2003), Karlsborn et al. (2014)
					Intellectual disability	Najmabadi et al. (2011)
					Amyotrophic lateral sclerosis	Simpson et al. (2009)
					Rolandic epilepsy	Strug et al. (2009)
Lung	Elongator complex (IKAP)	$\underline{\overline{mcm}}^5U$, $\underline{\overline{ncm}}^5U$ and derivatives	Several	34	Bronchial asthma in children	Takeoka et al. (2001)
Bladder	HABH8 (ALKBH8)	$\underline{m}cm^5U$	Arg, Glu	34	Bladder cancer	Shimada et al. (2009)
Liver	TARBP1 (TRMT3)	Gm	Ser	17	Morris hepatoma	Randerath et al. (1981)
Breast	TRMT12	yW	Phe	37	Breast cancer	Rodriguez et al. (2007)
	TRMT2A (HTF9C)	m^5U	Several	42, 54		Bartlett et al. (2010)

(continued)

Table 1 (continued)

Tissue	Gene	tRNA modification	Affected tRNAs	tRNA residues	Associated human diseases and phenotypes	References
Multiple tissues	NSUN2	m^5C	Several	34, 48, 49, 50	Autosomal-recessive intellectual disability	Abbasi-Moheb et al. (2012), Khan et al. (2012), Martinez et al. (2012), Fahiminiya et al. (2014), Blanco et al. (2014)
					Dubowitz-like syndrome	Martinez et al. (2012)
					Noonan-like syndrome	Fahiminiya et al. (2014)
					Skin, breast and colorectal cancer	Frye and Watt (2006), Pierga et al. (2007), Frye et al. (2010)
	TRMT10A (HRG9MTD2)	m^1G	Several	9	Intellectual disability, microcephaly, short stature	Igoillo-Esteve et al. (2013), Gillis et al. (2014)
					Colorectal cancer	Berg et al. (2010)
					Hyperinsulinaemic hypoglycaemia	Igoillo-Esteve et al. (2013), Gillis et al. (2014)
	HTRM9L	$\underline{mcm}^5U$, $\underline{mcm}^5s^2U$	Several	34	Different types of cancer	Begley et al. (2013)
	*	Q	Asn, Asp, His, Tyr	34	Different types of cancer	Randerath et al. (1984), Vinayak and Pathak (2010)
	CDKAL1	$\underline{ms}^2t^6A$	Lys	37	Type 2 diabetes	Kirchhoff et al. (2008), Stancakova et al. (2008), Quaranta et al. (2009), Wei et al. (2011), Wei and Tomizawa (2011), Xie et al. (2013)
					Cardiovascular diseases	Saade et al. (2011)
					Crohn's disease	Barrett et al. (2008), Quaranta et al. (2009)
					Psoriasis	Quaranta et al. (2009)
	WDR4	m^7G	Several	46	Down's syndrome	Michaud et al. (2000)
					Microcephalic primordial dwarfism	Shaheen et al. (2015)

	DNMT2 (TRDMT1)	m^5C	Asp	38	Increased red blood cell folate levels	Franke et al. (2009)
					Different types of cancer	Schaefer et al. (2009)
Mitochondrial defects	**	τm^5U	mt-Leu	34	MELAS	Hayashi et al. (1993), Kirino et al. (2004), Suzuki and Nagao (2011)
	**	τm^5s^2U	mt-Lys	34	MERRF	Yasukawa et al. (2001), Suzuki and Nagao (2011)
	MTU1 (TRMU)	s^2U ($\tau m^5\underline{s^2}U$)	mt-Lys, mt-Glu, mt-Gln	34	MERRF	Umeda et al. (2005)
					Acute liver failure in infancy accompanied by lactic acidaemia	Zeharia et al. (2009)
					Deafness associated with mutations in mitochondrial 12S ribosomal RNA	Guan et al. (2006)
	TRMT5	m^1G	mt-Leu	37	Lactic acidosis and multiple mitochondrial respiratory complex deficiencies	Powell et al. (2015)

For complex tRNA modifications, the underlined modification refers to the modification step at which the indicated gene is involved

Cm, 2′-*O*-methylcytidine; Gm, 2′-*O*-methylguanosine; ncm^5U, 5-carbamoylmethyluridine; ncm^5Um, 5-carbamoylmethyl-2′-*O*-methyluridine; m^5C, 5-methylcytosine; mcm^5U, 5-methoxycarbonylmethyluridine; yW, wybutosine; m^5U, 5-methyluridine; m^1G, 1-mehtylguanosine; m^7G, 7-methylguanosine; $m^2{}_2G$, N2,N2-dimethylguanosine; Q, queuosine; ms^2t^6A, 2-methylthio-N6-threonyl carbamoyladenosine; I, inosine; τm^5U, 5-taurinomethyluridine; τm^5s^2U, 5-taurinomethyl-2-thiouridine

*No associated gene

**Unknown enzyme

encode (or are expected to encode) for enzymes that catalyse tRNA modifications and a wide range of complex human pathologies, including neurological disorders, cardiac and respiratory defects, cancer, metabolic dysregulations and mitochondrial-linked dysfunctions (see below). In most of these conditions, an in-depth understanding of the molecular mechanisms of pathology is lacking. It will become clear in this section that unravelling the details of such molecular mechanisms will be very challenging given that some of the observed phenotypes associated to tRNA modifications are tissue-specific and are dependent on the degree to which the tRNA is modified or not (Kirchner and Ignatova 2015).

2.1 Neurological Disorders

The brain is perhaps the most sensitive tissue to defects in tRNA modifications. The FtsJ RNA methyltransferase homolog 1 (FTSJ1) gene is likely the human homolog of the yeast tRNA methyltransferase 7 (TRM7) gene that encodes for the enzyme that methylates positions 32 and 34 on $tRNA^{Leu}$, $tRNA^{Trp}$ and $tRNA^{Phe}$ (Towns and Begley 2012). Mutations in FTSJ1 or a complete deletion of this gene has been linked to non-syndromic X-linked mental retardation (Hamel et al. 1999; Freude et al. 2004; Ramser et al. 2004; Bonnet et al. 2006; Froyen et al. 2007; Dai et al. 2008; Takano et al. 2008) and reported to affect cognitive functions in young males of the Han Chinese population (Gong et al. 2008). The tRNA modification state of these patients was not evaluated, but a mutant FTSJ1 transcript was shown to be very unstable and likely degraded by nonsense-mediated mRNA decay (Freude et al. 2004; Takano et al. 2008). Interestingly, the expression of wild-type FTSJ1 in human tissues was reported to be high in foetal brain (Freude et al. 2004) but low in adult brain (Ramser et al. 2004), consistent with a key role for this protein in the developing brain. Notably, patients bearing a chromosomal duplication of regions involving FTSJ1 and other genes presented mild/moderate mental retardation (Bonnet et al. 2006; Giorda et al. 2009), suggesting that overexpression of wild-type FTSJ1 might also be detrimental. However, a patient with mild mental retardation that presented a smaller chromosomal duplication, also involving FTSJ1, did not show increased levels of FTSJ1 mRNA as measured by quantitative PCR in blood; and instead the phenotype was attributed to the overexpression of three other genes (also located within the duplicated chromosomal region): EBP, WDR13 and ZNF81 (El-Hattab et al. 2011).

Mental retardation has also been reported in human patients with mutations in other genes encoding for tRNA modification enzymes. The tRNA methyltransferase 10A (TRMT10A), also known as human RNA (guanine-9) methyltransferase domain containing 2 (HRG9MTD2), is the human ortholog of the yeast enzyme that catalyses the methylation of guanosine at position 9 of several tRNAs (Towns and Begley 2012). Mutations in the human TRMT10A gene have been associated to microcephaly, short stature and intellectual disability (Igoillo-Esteve et al. 2013; Gillis et al. 2014). TRMT10A was shown to be expressed in human embryonic and foetal brain (Igoillo-Esteve et al. 2013). Furthermore, Gillis and colleagues showed in vitro that the mutant TRMT10A protein could effectively bind to a tRNA substrate but showed a dramatic reduction in methylation activity as compared to the wild-type protein, probably due to impaired ability to bind the methyl donor S-adenosylmethionine (Gillis et al. 2014). As it will be discussed later, defects on this enzyme not only affect brain tissues but also the liver and colon.

Other examples of genes encoding for tRNA modification enzymes and mental retardation include the gene encoding for tRNA methyltransferase 1 (TRMT1), an enzyme that dimethylates guanosines at position 26 of several tRNAs (Liu and Straby 2000), and the gene encoding for ELP2 (a component of the elongator complex; see below). Both enzymes were reported as novel markers for recessive cognitive disorders (Najmabadi et al. 2011). Notably, in mice, a potential homolog of the human TRMT1 named 'TRM1-like' was shown to have a role in motor coordination and exploratory behaviour (Vauti et al. 2007), suggesting a conserved role for this protein in the brain. Finally, the human WD repeat domain 4 (WDR4) gene was found as a potential candidate marker for phenotypes observed in Down's syndrome (Michaud et al. 2000); and mutations in this gene have recently been associated to a distinct form of microcephalic primordial dwarfism (Shaheen et al. 2015). This gene is the human homolog of the yeast TRM82, one of the subunits of the heterodimeric enzyme responsible for the 7-methylguanosine modification at position 46 in several tRNAs (Towns and Begley 2012). Two transcript variants for this gene were described in humans, where the shorter one (of about 1.5 kb) was highly expressed in foetal tissues and the longer one (of about 2.5 kb) showed faint expression in most tissues (Michaud et al. 2000), suggesting a key function for the shorter transcript in developmental processes. Interestingly, the chromosomal region where WDR4 maps has already been associated to several genetic disorders such as maniac-depressive psychosis, autosomal-recessive deafness, Knobloch syndrome and holoprosencephaly (Michaud et al. 2000).

Recently, the human adenosine deaminase acting on tRNA 3 (ADAT3) was validated as one of the subunits of the heterodimeric enzyme responsible for adenosine-to-inosine editing at position 34 of 8 different tRNAs (Torres et al. 2015). A single missense mutation in the human ADAT3 gene was reported to cause intellectual disability and strabismus (Alazami et al. 2013). Interestingly, this protein and its catalytic partner ADAT2 were reported essential in yeast, *Trypanosoma brucei*, *Arabidopsis thaliana* and likely also in human cell lines (Gerber and Keller 1999; Rubio et al. 2007; Zhou et al. 2014; Torres et al. 2015). This suggests that patients carrying mutations in the ADAT3 gene probably still have some residual ADAT activity. However, a more detailed analysis of the functional importance for this enzyme in mammals remains to be addressed (Torres et al. 2014b).

NOP2/Sun RNA methyltransferase family member 2 (NSUN2) is an enzyme that methylates cytosine at positions 34, 48, 49 and 50 on different tRNAs (Brzezicha et al. 2006; Hussain et al. 2013) . Mutations in the NSUN2 gene have been associated to autosomal-recessive intellectual disability (Abbasi-Moheb et al. 2012; Khan et al. 2012; Martinez et al. 2012; Fahiminiya et al. 2014). Khan and colleagues investigated the subcellular localization of wild-type and mutant NSUN2 and found that mutant NSUN2 failed to localize to nucleoli (Khan et al. 2012). Abbasi-Moheb and colleagues further showed that deletion of the fly NSUN2 ortholog resulted in a short-term memory phenotype (Abbasi-Moheb et al. 2012), suggesting an evolutionary conserved function for tRNA methylation in the brain. A recent study has addressed some mechanistic aspects on how NSUN2 defects would be contributing to human disease. The authors showed that lack of 5-methylcytosine (m^5C) on tRNAs results in increased tRNA endonucleolytic cleavage mediated by angiogenin, leading to the accumulation of 5′-tRNA halves that reduce protein translation rates and activate stress pathways in human and mouse cells. Moreover, NSUN2-deficient brains were more sensitive to oxidative stress, and this phenotype could be rescued by inhibiting angiogenin during embryogenesis (Blanco et al. 2014).

Two other reports have associated mutations in NSUN2 to different degrees of mental retardation. Patients with these mutations showed overlapping phenotypes to that of the Dubowitz syndrome (Martinez et al. 2012) and to that of the Noonan syndrome (Fahiminiya et al. 2014). In the first case, the authors showed that patients suffering from this Dubowitz-like syndrome were lacking the m^5C modification at positions 47 and 48 on $tRNA^{Asp}{}_{(GTC)}$, one of the substrates of NSUN2 (Martinez et al. 2012). In the second case, the link to a Noonan-like syndrome suggests a role for NSUN2 beyond the brain, as this is a pathology that affects mainly the cardiac tissue and only about 25 % of affected patients also suffer mental retardation. As it will be discussed later, NSUN2 has also been linked to different forms of cancer (see below) suggesting that this enzyme has key roles not only in the brain and heart but also in different tissue types.

A set of very well studied tRNA modifications associated to neurological disorders are those involving the formation of 5-methoxycarbonylmethyluridine (mcm^5U) and 5-carbamoylmethyluridine (ncm^5U) at position 34 of several tRNAs. These complex tRNA modifications (and derivatives of them) require a methylation step catalysed by the elongator complex. This complex is highly conserved from yeast to humans, where it was shown to be composed of six subunits: IκB kinase complex-associated protein (IKAP/yeast ELP1), Stat3-interacting protein (StIP1/yeast ELP2), elongator protein homolog 3 (ELP3), ELP4 and two unidentified polypeptides (Hawkes et al. 2002).

Mutations in the gene encoding for IKAP (*IKBKAP*) have been linked to familial dysautonomia (FD) (Anderson et al. 2001; Slaugenhaupt et al. 2001; Leyne et al. 2003; Karlsborn et al. 2014). Although some of these mutations are missense mutations, the most prevalent mutation (>99.5 %) was found in homozygosity and involved a point mutation resulting in exon-skipping and aberrant protein truncation (Anderson et al. 2001; Slaugenhaupt et al. 2001). Moreover, levels of mcm^5s^2 U34 were shown to be reduced in brain tissue and fibroblast cell lines derived from FD patients (Karlsborn et al. 2014). Strikingly, even though patients were homozygous for the exon-skipping mutation, they presented variable levels of wild-type IKAP in a tissue-specific manner, where brain cells primarily expressed the mutant

IKBKAP mRNA (Slaugenhaupt et al. 2001). In this regard, *IKBKAP* function may also be important for pulmonary function, as a mutation in this gene has been associated to bronchial asthma in children (Takeoka et al. 2001). Altogether, it seems that a tissue-dependent control of exon skipping of the *IKBKAP* transcript might be central to the symptoms caused by *IKBKAP* mutations.

Mutations in other genes encoding for proteins of the elongator complex have also been linked to neurological dysfunctions (Chen et al. 2009; Simpson et al. 2009; Strug et al. 2009; Najmabadi et al. 2011). As mentioned above, ELP2 has been linked to intellectual disability (Najmabadi et al. 2011). In a study that involved 38 US families, variants of ELP4 have been associated to the electroencephalographic abnormality of centrotemporal sharp waves, a typical clinical trait of Rolandic epilepsy (Strug et al. 2009). Lastly, mutations in ELP3 have been associated to amyotrophic lateral sclerosis (Simpson et al. 2009). Interestingly, in flies and zebrafish, impaired function of ELP3 resulted in neurological phenotypes (Simpson et al. 2009), and in *Caenorhabditis elegans*, mutations in ELPC1 and ELPC3 (homologs of IKAP and ELP3, respectively) also resulted in neurological abnormalities (Chen et al. 2009). This suggests a conserved functional role in metazoan neuronal tissues for these components of the elongator complex.

2.2 Cancer

Mounting evidence suggests that different types of cancer could be attributed to a dysregulation on tRNA modification enzymes (Torres et al. 2014a). In some cases, such dysregulation involves an upregulation of the tRNA modification enzymes, while in other cases, the tRNA modification enzymes (or the modifications itself) are found downregulated. Examples of the first case include upregulation of NSUN2, TRMT12 and human AlkB homolog 8 (HABH8 or HALKBH8), among others. In breast cancer patients and in several breast cancer cell lines, the genomic region encompassing the NSUN2 gene is frequently found duplicated (Pierga et al. 2007; Frye et al. 2010). Indeed, NSUN2 expression was reported to be low in normal tissues but high in different tumour types such as squamous cell carcinoma, breast cancer and colorectal cancer (Frye and Watt 2006). NSUN2 was reported to have a role in keratinocyte proliferation mediated by the proto-oncogene Myc. Further, knockdown of NSUN2 reduced tumour formation in a mice xenograft model for squamous cell carcinoma (Frye and Watt 2006).

TRMT12 is one of the enzymes involved in the formation of wybutosine at position 37 of $tRNA^{Phe}$. This gene was also found amplified in breast cancer cell lines and breast cancer tumours (Rodriguez et al. 2007). Finally, knockdown of HABH8 (the protein likely responsible for the formation of mcm^5U34 on $tRNA^{Arg}$ and $tRNA^{Glu}$) was shown to suppress angiogenesis, invasion and growth of bladder cancers in vivo (Shimada et al. 2009). Many more examples exist of upregulated tRNA modification enzymes and hypermodified tRNAs in cancer. This is especially true for tRNA methyltransferases that, already in the early 1970s, were known to be either upregulated or having higher methyltransferase activity in neoplastic tissues as compared to normal tissues (Kerr and Borek 1973).

Downregulation of tRNA modifications/tRNA modification enzymes has also been found in different types of cancer. Such is the case of another potential human homolog of the yeast TRM9, named tRNA methyltransferase 9-like (HTRM9L) that was reported to be downregulated in bladder, cervix, breast, testicular and colorectal cancer. Moreover, two colorectal carcinoma cell lines showed a dramatic reduction in tumour growth in vivo when HTRM9L expression was recovered (Begley et al. 2013). Similarly, the 2′-*O*-methylguanosine modification at position 17 of $tRNA^{Ser}_{(IGC)}$ has been shown to be absent in Morris hepatoma (Randerath et al. 1981).

Queuosine (Q) is incorporated at position 34 of $tRNA_{GUN}$ (i.e. tRNA Asn, Asp, His and Tyr), replacing G34, in a reaction catalysed by eukaryotic TGTase (Vinayak and Pathak 2010). Eukaryotes are unable to synthesize queuine, the base form of nucleoside Q, and need to obtain it from food intake, the intestinal flora, or by a queuine salvage pathway from tRNA turnover. It is well documented that neoplastic tissues and transformed cells contain hypomodified Q-tRNAs. For example, Morris hepatoma is characterized by a lack of Q on mitochondrial tRNA ^{Asp} (Randerath et al. 1984). Moreover, the degree of Q hypomodification can correlate closely with the degree of malignancy for several types of cancer, making it a promising marker for the prognosis of the disease (Vinayak and Pathak 2010) (see Sect. 3). However, despite strong efforts to understand the biology of Q and Q-tRNAs, a clear picture of the molecular mechanisms linking this tRNA modification to human diseases remains elusive.

2.3 *Metabolic Dysregulations*

Some human diseases associated to tRNA modifications involve complex metabolic dysregulations. DNA (cytosine-5)-methyltransferase-like protein 2 (DNMT2), also known as tRNA aspartic acid methyltransferase 1 (TRDMT1), is responsible for the methylation of cytosine (m^5C) at position 38 of $tRNA^{Asp}$ (Goll et al. 2006). A polymorphism in TRDMT1 was linked to increased red blood cell folate levels that seemed to reduce the risk of developing spina bifida aperta in individuals of Dutch ethnicity (Franke et al. 2009). Interestingly, loss of m^5C at position 34 of $tRNA^{Leu}_{(CAA)}$ by knocking out NSUN2 (see above) resulted in an approximately 30 % weight loss in mice. Moreover, mice also presented alopecia, suggesting that NSUN2 would be playing a role in maintaining skin homeostasis (Blanco et al. 2011). Notably, while single knockout mice for TRDMT1 (DNMT2) or NSUN2 were viable, DNMT2/NSUN2 double knockouts were synthetic lethal and presented an underdeveloped phenotype (Tuorto et al. 2012). These reports highlight a role for the loss of m^5C at different tRNA positions and in different tRNA species with a range of complex phenotypes that could be attributed to metabolic dysfunctions.

Links between tRNA modifications and perturbed glucose metabolism have also been reported. Patients with mutations in TRMT10A (see above) also suffered from hyperinsulinaemic hypoglycaemia (Gillis et al. 2014). It was postulated that the abnormalities in glucose homeostasis triggered by the TRMT10A deficiency may be due to accelerated β-cell apoptosis (Gillis et al. 2014). In line with these results, Igoillo-Esteve and colleagues showed that TRMT10A was enriched in pancreatic islets and that a nonsense mutation on the TRMT10A gene was linked to young-onset diabetes. Moreover, knockdown of TRMT10A in primary rat β-cells and in dispersed human islets induced apoptosis and increased sensitivity to free fatty acids and endoplasmic reticulum stress (Igoillo-Esteve et al. 2013).

Several reports have shown a correlation between mutations in the CDK5 regulatory subunit-associated protein 1-like 1 (CDKAL1) gene and type 2 diabetes (T2D) (Kirchhoff et al. 2008; Stancakova et al. 2008; Wei et al. 2011; Wei and Tomizawa 2011; Xie et al. 2013). CDKAL1 is a methylthiotransferase responsible for the 2-methylthio-N6-threonyl carbamoyladenosine (ms^2t^6A) modification at position 37 of $tRNA^{Lys}_{(UUU)}$ (Wei and Tomizawa 2011). Patients carrying mutations in CDKAL1 show reduced levels of ms^2t^6A37 and impaired proinsulin-to-insulin conversion and insulin secretion (Kirchhoff et al. 2008; Stancakova et al. 2008; Xie et al. 2013). Moreover, misreading of Lys codons in proinsulin was detected in CDKAL1 knockout β-cells from mice (Wei et al. 2011). Notably, polymorphisms on CDKAL1 have also been associated to myocardial infarction and coronary artery disease, probably as a consequence of its influence on T2D (Saade et al. 2011). Mutations in CDKAL1 have also been linked to Crohn's disease and psoriasis (Barrett et al. 2008; Quaranta et al. 2009). These associations were shown to be independent of T2D (Quaranta et al. 2009). Therefore, it is likely that CDKAL1 plays a role in disease in a variety of tissues; and probably, the observed phenotypes are associated to mistranslation of Lys codons on particular tissue-specific Lys-enriched transcripts.

2.4 *Mitochondrial-Linked Dysfunctions*

The mitochondria are responsible for supplying energy to the cell. As such, defects in mitochondrial metabolism can have severe and complex phenotypes especially in tissues that require high amount of energy, such as the brain and muscle. More than 200 different mutations in mitochondrial tRNA (mt-tRNA) genes have been associated to human diseases (Ruiz-Pesini et al. 2007). Here we will focus on two well-characterized mitochondrial-linked human diseases caused by mutations in mt-tRNA genes that result in tRNA hypomodification: mitochondrial encephalomyopathy, lactic acidosis and stroke-like episodes (MELAS) and myoclonic epilepsy with ragged-red fibres (MERRF) (Suzuki and Nagao 2011). Mostly, these diseases arise from the lack of taurine modifications at position 34 of mt-$tRNA^{Leu}_{(UUR)}$ and mt-$tRNA^{Lys}$, respectively (Suzuki and Nagao 2011).

MELAS patients were found containing mutations at different residues of the mt-tRNA^{Leu}$_{(UUR)}$ gene that resulted in a complete lack of 5-taurinomethyluridine (τm^5U34) on this tRNA. However, patients having other mutations in this same gene that resulted in retention of the τm^5U34 modification were diagnosed with mitochondrial diseases other than MELAS such as mitochondrial myopathy (MM), chronic progressive external ophthalmoplegia (CPEO) and maternally inherited mitochondrial myopathy and cardiomyopathy (MMC) (reviewed in Suzuki and Nagao 2011). This suggests that the lack of τm^5U34 on mt-tRNA^{Leu}$_{(UUR)}$ is the causative of MELAS disease. The τm^5U34 modification allows for mt tRNA^{Leu}$_{(UUR)}$ to read both UUR codons: UUA and UUG. However, in the absence of this modification, mt-tRNA^{Leu}$_{(UUR)}$ is only capable of reading its cognate UUA codon (Kirino et al. 2004). Given that in these patients only the decoding of UUG Leu codons is affected, only those transcripts enriched in this codon show translation defects. One such case is that of the mitochondrially encoded NADH dehydrogenase 6 (*ND6*) gene. *ND6* encodes for a component of the respiratory chain complex I. Cybrid cell lines derived from MELAS patients lacking τm^5U34 showed a marked reduction in complex I activity and a reduced rate of translational activity of ND6, while general mitochondrial protein synthesis seemed unaffected (Hayashi et al. 1993). Therefore, the specific translational defects of UUG-enriched mitochondrial transcripts caused by the lack of τm^5U34 on mt-tRNA^{Leu}$_{(UUR)}$ largely explains the molecular mechanism behind MELAS disease.

MERRF patients lack 5-taurinomethyl-2-thiouridine (τm^5s^2U) at position 34 of mt-tRNA^{Lys}. As a result, the unmodified mt-tRNA^{Lys} is unable to efficiently decode any of its Lys codons (AAA and AAG), leading to overall mitochondrial translation defects and a more severe pathological phenotype than that observed in MELAS patients (Yasukawa et al. 2001). In addition to the importance of the taurine modification, this effect may indicate a key role of the s^2U34 modification in efficient tRNA decoding. Knockdown of the mitochondrial tRNA-specific 2-thiouridylase 1 (MTU1, also known as TRMU) in HeLa cells resulted in a reduction of s^2U34 modification on mt-tRNA^{Lys}, reduction on oxygen consumption and defective mitochondrial membrane potentials, all of which are characteristics of MERRF (Umeda et al. 2005). Notably, mutations in MTU1 have been associated to other mitochondrial disorders such as acute liver failure in infancy accompanied by lactic acidaemia (Zeharia et al. 2009) and deafness associated with mutations in mitochondrial 12S ribosomal RNA (Guan et al. 2006).

In this section we have described only two mitochondrial-linked diseases associated to the lack of tRNA modifications. These two pathologies are the ones that have been studied more in depth thus far (see Suzuki and Nagao 2011). However, more reports are emerging with further associations between tRNA modification enzymes and mitochondrial disorders. One such example is the recently published article by Powell and colleagues showing that individuals carrying mutations on the tRNA methyltransferase 5 (*TRMT5*) gene presented lactic acidosis and multiple mitochondrial respiratory complex deficiencies, particularly in the skeletal muscle. The authors also validated that TRMT5 is responsible for m^1G37 modification on mitochondrial tRNA^{Leu} (Powell et al. 2015). We refer the reader to a recent review on mitochondrial tRNA mutations and disease for more information (Abbott et al. 2014).

3 Biomedical Strategies Based on tRNA Modifications

The role of tRNA modifications and human diseases is still an emerging field (Torres et al. 2014a). As such, the full potential of tRNA modification-based approaches to be used in the clinic has not been exploited thus far. In this section we will review the cases where studying and modulating the levels of tRNA modifications or tRNA modification enzymes have proved useful for therapeutic applications; and we will also hypothesize on other tentative approaches that could be worth pursuing in the future to tackle tRNA modification-linked diseases.

3.1 Diagnosis and Prognosis

Several studies have focused on finding typical 'gene signatures' associated to different human diseases. These gene signatures can be in the form of gene expression patterns using 'omics' (e.g. microarrays, RNAseq, proteome profiling, etc.) or as a set of mutations in particular genes obtained by performing population genetic studies in human patients. Interestingly, in many of such type of studies, genes encoding (or expected to encode) for tRNA modification enzymes emerge.

As explained above, there is a strong link between tRNA modifications (and the enzymes that catalyse them) and different forms of cancer. Therefore, tRNA modifications could serve as means of diagnosis and/or prognosis of cancer. One example is Mammostrat, a tool used to stratify patients with early-stage breast cancer into groups having risks of a relapse following treatments with tamoxifen, in order to inform treatment decisions (e.g. if chemotherapy would be additionally recommended or not) (Bartlett et al. 2010). This prognostic tool measures the levels of five genes (i.e. gene signature), one of which is the tRNA methyltransferase homolog 2A (TRMT2A, also known as HTF9C), predicted to be responsible for the formation of 5-methyluridine at positions 42 and 54 of several tRNAs (Towns and Begley 2012).

Another study compared high-resolution genome copy number variation in microsatellite-stable colorectal tumours to identify susceptibility loci (Berg et al. 2010). Ten genomic loci were identified, comprising about 500 protein-coding genes, as markers for susceptibility to early-onset colorectal cancer. Moreover, combined genomics and transcriptomics allowed for the identification of a gene signature of 7 genes that correlates with an increased risk for colorectal cancer. One of these signature genes was TRMT10A (also known as HRG9MTD2; see above) (Berg et al. 2010).

The monitoring of gene signatures involving tRNA modification enzymes for diagnosis or prognosis of disease is not restricted to cancer. Najmabadi and colleagues have performed homozygosity mapping, exon enrichment and next-generation sequencing in 136 consanguineous families and have identified 50 novel genes with potentially pathogenic variants for intellectual disability. Two of these genes encode for tRNA modification enzymes: TRMT1 and ELP2 (discussed above) (Najmabadi et al. 2011). Similarly, a study identified genes related to folate metabolism that are involved in the aetiology of spina bifida. One of the 4 novel genes described in that work was TRDMT1 (DNMT2) (Franke et al. 2009).

Just like genes encoding for tRNA modification enzymes can be monitored to predict or evaluate the development of human diseases, measuring the levels of tRNA modifications themselves can prove useful. The measurement of the levels of Q34 in human lung cancer and normal lung tissue from different patients correlated well with the grade of tumour malignancy and was suggested to be a viable approach to predict patient survival (Huang et al. 1992). Given that other types of cancer also show Q34 hypomodification (see above) (Vinayak and Pathak 2010), it would not be surprising if a similar approach for quantification of Q34 could be applied for prognosis of other cancer types. Likewise, measurement of ms^2 modification on $tRNA^{Lys}{}_{(UUU)}$ by a quantitative PCR technique proved useful to monitor the activity levels of CDKAL1 and assess the risk of developing type 2 diabetes (Xie et al. 2013). Moreover, analysis of total RNA extracted from human peripheral blood samples showed lower levels of ms^2 modification in individuals carrying the CDKAL1 genotype associated to T2D that correlated with lower levels of insulin secretion (Xie et al. 2013).

Even though to date there are few cases reported of successful analyses of tRNA modifications and tRNA modification genes to diagnose and monitor different human diseases, the mounting evidence is strong enough to establish a proof of principle. It is therefore not difficult to imagine that other strategies will emerge soon. These can involve using already available data (e.g. if TRMT1 and ELP2 regulation is associated to intellectual disability, maybe measuring the levels of $m^2{}_2G$ and mcm^5U/ncm^5U could serve as well as molecular markers for this disease) or by performing further 'omics'-based screenings for novel candidate genes in different pathological scenarios. Moreover, as technology progresses, these findings will result in further high-throughput, cost-effective and minimally invasive methods for monitoring disease, as already described for breast cancer (i.e. Mammostrat) or T2D (i.e. quantitative PCR method from peripheral blood samples).

3.2 *Potential Therapeutic Treatments*

Developing drugs for therapeutic applications is often harder than developing tools for diagnosis and prognosis of a disease. However, there is some evidence in the literature of potential approaches that could be helpful for the treatment of complex human diseases based on the biology of tRNA modifications.

For some pathologies, the desired drug effect would be to reduce the levels of tRNA modifications. As discussed above, different types of cancer show high levels of tRNA methylation (Kerr and Borek 1973). Azacytidine, a cytosine analogue, has been developed as a drug for epigenetic cancer therapy through the inhibition of DNA methylation. However, azacytidine, which is commonly used for treatment of myeloid leukaemia, was also shown to specifically inhibit m^5C38 formation on $tRNA^{Asp}$ by DNMT2 (Schaefer et al. 2009). Demethylation of m^5C38 on $tRNA^{Asp}$ in myeloid leukaemia cell lines correlated well with azacytidine, and it has therefore been proposed as a suitable novel molecular marker for azacytidine response (Schaefer et al. 2009). Inhibition of tRNA methyltransferase activity can also be achieved by using different types of adenine analogues, some of which are cytokinins (Wainfan and Borek 1967; Wainfan and Landsberg 1971). In fact, adenine analogues have been proposed as effective compounds for treatment of acute myeloid leukaemia (Honma 2003), while cytokinin ribosides were reported to impair cell viability (Casati et al. 2011).

Another strategy to reduce the levels of tRNA modifications is to directly knockdown tRNA modification enzymes. An example of this approach is the downregulation of NSUN2, shown to inhibit Myc-induced keratinocyte proliferation and to reduce the growth of human squamous cell carcinoma in mice in a dose-dependent manner (Frye and Watt 2006). Likewise, downregulation of HABH8 (ALKBH8) was suggested as a new therapeutic strategy to target urothelial carcinomas (discussed in Sect. 2.2) (Shimada et al. 2009). The same knockdown approaches could be used also for other tRNA modification enzymes such as TRMT12, which is amplified in breast cancer tumours (Rodriguez et al. 2007).

In other cases the desired effect of drugs would be to restore the levels of tRNA modifications. The method to achieve this will be dependent on the type of tRNA modification and the type of tRNA modification enzyme. Enhancing tRNA methyltransferase activity could be useful in patients of neurological disorders caused by mutations in FTSJ1, TRMT10A, TRMT1, WDR4, etc. This could be accomplished by using hormone treatments. As an example, the levels of N2 and N2-dimethylguanine in uteri could be recovered upon addition of estradiol in ovariectomized rats (Sharma et al. 1971).

A completely different mechanism would be the one involving the Q modification. As discussed above, queuine is not readily synthesized by eukaryotes and need to be obtained via different mechanisms in order to produce Q (see above). Queuine has been proposed as an anticarcinogenic agent (Vinayak and Pathak 2010). Externally administered Q reduces cell proliferation in vitro and in vivo probably due to restoration of Q-tRNAs (Pathak et al. 2007). In this scenario, the absence of Q-tRNAs could be due to deficiencies in conversion of queuine into Q; however, the lack of Q-tRNAs could also be due to poor activity of the TGTase, reduced uptake of queuine or dysfunctions in the queuine salvage system (reviewed in Vinayak and Pathak 2010). Therefore, different therapeutic strategies can be devised depending on the affected molecular mechanism for the generation of Q-tRNAs.

An interesting approach to restore tRNA modification is the one currently being pursued for the treatment of FD. It is possible to correct aberrant splicing of *IKBKAP* transcripts by using different types of compounds. Kinetin, a plant cytokinin, is one of the most promising candidates for treating this disease. Patients homozygous for the FD splice mutation were treated with 23.5 mg/kg/day kinetin for 28 days and presented an increase in wild-type *IKBKAP* mRNA expression in leukocytes (Axelrod et al. 2011). Other small molecules are also being investigated as splice-correction agents for treating FD including epigallocatechin gallate (found in green tea), genistein and daidzein (present in soy) and tocotrienols (members of the vitamin E family) (Anderson et al. 2003a, b, 2012; Anderson and Rubin 2005). Importantly, combined treatment using these compounds was shown to have synergistic effects on *IKBKAP* mRNA splicing correction (Anderson et al. 2003a, 2012). The exact molecular mechanism for the efficacy of these small molecules to correct splicing is unclear. A kinetin responsive sequence element has been mapped at the 5′-splice site of *IKBKAP* exon 20; and the actual FD mutation on *IKBKAP* was not required for kinetin activity (Hims et al. 2007). In the case of epigallocatechin gallate, the compound was shown to downregulate the protein HNRNP A2/B1 that promotes the preferential use of the intron distal 5′-splice site that is selected when generating the *IKBKAP* mRNA mutant splice variant (Anderson et al. 2003a). It is conceivable that in the future other small molecules with the ability to correct the FD splicing defect will be described, as well as other splice-correction strategies such as the use of oligonucleotide analogues conjugated to cell-penetrating peptides to mask splicing sites (Betts et al. 2012).

A more dramatic approach to restore tRNA modification levels would be to re-express/overexpress the wild-type versions of tRNA modification enzymes following gene replacement therapies in those cases where patients contain mutations on genes encoding for such tRNA modification enzymes. As mentioned before, re-expression of HTRM9L showed promising potential for controlling tumour growth in a colorectal carcinoma model (see Sect. 2.2) (Begley et al. 2013). Given that gene expression usually needs to be carefully controlled, this kind of approach will probably be more effective when replacing enzymes that are believed to be constitutively active physiologically. This could be applied to the overexpression of wild-type ADAT3 to correct the intellectual disability and strabismus phenotypes (Alazami et al. 2013). It is believed that hetADAT activity is saturated in cells given that apparently $tRNA_{(ANN)}$ seems to be fully modified to I34 (Torres et al. 2014b).

Restoring tRNA modifications associated to mitochondrial-linked diseases opens a whole new set of different therapeutic approaches. MELAS patients are unable to modify mt-$tRNA^{Leu}{}_{(UUR)}$ with τm^5U34 (see Sect. 2.4). Recovery of MELAS phenotypes could be observed in a lung carcinoma cybrid cell line bearing a mutation in the anticodon sequence of the mt-$tRNA^{Leu}{}_{(CUN)}$ gene (the other isoacceptor of mt-$tRNA^{Leu}{}_{(UUR)}$). This mutant resulted in a mt-$tRNA^{Leu}{}_{(CUN)}$ having the anticodon UAA (the same anticodon as mt-$tRNA^{Leu}{}_{(UUR)}$) instead of its wild-type anticodon UAG. Notably, this mutant tRNA was found modified with τm^5U34 and was therefore capable of decoding UUG codons (Kirino et al. 2006). This shows that a potential therapeutic strategy against MELAS is to overexpress a related mt-$tRNA^{Leu}$ isoacceptor bearing the anticodon of mt-$tRNA^{Leu}{}_{(UUR)}$, which can be modified to τm^5U34.

Additionally, the MELAS mutation on $tRNA^{Leu}{}_{(UUR)}$ was also shown to reduce the levels of aminoacylation for this tRNA, and overexpression of the human mitochondrial leucyl-tRNA synthetase (LeuRS) in cells carrying such MELAS mutation recovered respiratory function in a dose-dependent manner (Park et al. 2008; Li and Guan 2010). Park and colleagues showed that, under these conditions, cells showed increased steady-state levels of $tRNA^{Leu}_{(UUR)}$, but the fraction of aminoacylated $tRNA^{Leu}_{(UUR)}$ remained unchanged. Likewise, rates of mitochondrial translation where not increased either in MELAS cells overexpressing LeuRS. The phenotypic recovery observed in these cells was instead attributed to an increase in protein stability (Park et al. 2008). On the contrary, Li and Guang found that overexpression of LeuRS improved aminoacylation efficiency, mt-tRNA stability and mitochondrial translation (Li and Guan 2010). These observations propose yet another therapeutic strategy to treat MELAS, although the molecular mechanism behind the phenotypic recovery is controversial.

Finally, we should also mention mitochondrial gene replacement therapy as means to overcome human diseases associated to mutations in the mitochondrial genome. While this is a sensitive approach due to potential ethical reasons ('three-parent in vitro fertilization'), the technique has already been successfully applied, using the spindle transfer method, in nonhuman primates resulting in healthy offspring (Tachibana et al. 2009). Furthermore, initial studies, using the same mitochondrial gene transfer method, have been reported in human oocytes with promising results (Tachibana et al. 2013). However, even if this approach is taken further into the clinic, it will serve to prevent the transfer of mitochondrial-linked diseases to a child but will not serve to cure existing patients with mitochondrial pathologies.

Clearly, new therapeutic strategies will emerge as our understanding of the molecular and cellular mechanisms triggered by tRNA modification misregulation improves. As exemplified above (see Sect. 2.1), inhibiting angiogenin during embryogenesis could recover some of the phenotypes observed in brain from NSUN2-deficient mice by preventing the accumulation of 5′-tRNA halves derived from tRNAs lacking m^5C (Blanco et al. 2014). It is likely that unexpected or non-canonical mechanisms (e.g. mechanisms beyond aberrant protein translation) play important roles in these different pathologies, and while it will take some time to fully unravel them, they will be key to developing further therapeutic approaches to treat tRNA modification-linked diseases.

4 Conclusions and Perspectives

The knowledge on the roles that tRNA modifications play in human diseases is increasing at a very fast pace. This is opening a whole new and exciting field for research and for development of novel therapeutics. However, in order to exploit its full potential, several key aspects still need to be addressed.

We need to improve our insights into the biological functions of tRNA modifications. For example, a direct role for different types of tRNA modifications on protein translation is often assumed. Not only those potential roles need to be experimentally verified, but also it is important to understand the extent to which protein fidelity and/or efficiency is affected when tRNAs are hypo- or hypermodified. Further, a key question is whether in pathological scenarios general protein synthesis, or only a subset of genes, is affected (Novoa and Ribas de Pouplana 2012). Moreover, tRNA modifications may be important for non-canonical functions of tRNA, such as those performed by tRNA fragments (Durdevic and Schaefer 2013; Anderson and Ivanov 2014). Therefore, we also need to keep an open mind and take into account that some of the observed cellular phenotypes could be due to non-canonical functions of tRNAs driven by the modulation of tRNA modifications.

In addition, there is a need for novel model systems to study tRNA modifications. The vast majority of the published research on tRNA modifications has been carried out in vitro, in bacteria, in lower eukaryotes and, to a less extent, in mammalian cell lines. In many cases, the data obtained using these models cannot be extrapolated to in vivo systems; and in fact our knowledge of tRNA modifications in metazoans is scarce. Therefore, it is critical to develop in vivo models to study tRNA modifications, especially mammalian models. This will not only expand our understanding of the biological functions of tRNA modifications but also will serve as models for the design of tRNA modification-based therapeutics. In this regard, mouse models have been developed to study the function of IKAP (familial dysautonomia) (Dietrich et al. 2012) and NSUN2 (neurological disorders) (Blanco et al. 2014). Other animal models (*Caenorhabditis elegans*, *Drosophila melanogaster* and *Danio rerio*) have also been used to study, for example, other members of the elongator complex (Chen et al. 2009; Simpson et al. 2009).

Lastly, it will also be essential to design new technologies that would allow for proper detection and quantification of tRNA modifications. Fortunately, for some modifications, this is being achieved successfully. Detection of m^5C at transcriptomics scale can be performed using methylation individual-nucleotide resolution cross-linking and immunoprecipitation (miCLIP) or 5-azacytidine-mediated RNA immunoprecipitations (Aza-IP) (Hussain et al. 2013; Khoddami and Cairns 2013). Direct sequencing of tRNAs (tRNAseq) is still very challenging, but novel strategies are being developed that allow for more efficient quantification of tRNA species (Pang et al. 2014) and detection of tRNA modifications (Iida et al. 2009; Torres et al. 2015). For example, the levels of I34 on human tRNAs were successfully monitored, using tRNAseq, in human cell lines upon downregulation of ADAT2 (Torres et al. 2015). Finally, techniques based on HPLC coupled to mass spectrometry are also being developed and have been already proven useful to study the dynamics of a diverse set of tRNA modifications in yeast tRNAs upon cellular stress (Chan et al. 2010). Altogether, it is likely that new technologies will soon allow for strong detection and quantification of tRNA modifications in a cost-efficient and high-throughput manner.

From a strictly therapeutic point of view, the challenges are even higher. On the one hand, the value of tRNA modification-based tools for diagnosis and prognosis will largely depend on the technologies available to detect and quantify tRNA modifications (see above). Moreover, the models that could be used to develop these tools will have to closely mimic pathological scenarios. In this regard, using patient-specific induced pluripotent stem cells (iPSCs) could prove valuable, an approach that has already been tried to model familial dysautonomia (Lee et al. 2009). Likewise, cancer models could be recapitulated using iPSCs and differentiated cells derived from them. On the other hand, the therapeutic strategies based on modulating tRNA modifications will have to be carefully controlled. Many tRNA modification enzymes seem to play different roles in different tissues (Table 1). For example, NSUN2 deficiency in the brain leads to neurological disorders (Abbasi-Moheb et al. 2012; Khan et al. 2012; Martinez et al. 2012; Fahiminiya et al. 2014), while its overexpression in the skin or breast has oncogenic potential (Frye and Watt 2006; Pierga et al. 2007; Frye et al. 2010). Therefore, developing tissue-specific targeting drugs will be key for success.

We have previously predicted that in the future 'epi-tRNAomes' (quantification of the tRNA pool and the evaluation of the modification levels) will be used in personalized medicine (Torres et al. 2014a). We believe that this will become a reality with the advent of cost-effective technologies that combines mass spectrometry, transcriptomics and proteomics and the parallel development of strong bioinformatics that will allow for the rapid and efficient analysis of large datasets.

References

Abbasi-Moheb L, Mertel S, Gonsior M et al (2012) Mutations in NSUN2 cause autosomal-recessive intellectual disability. Am J Hum Genet 90:847–855

Abbott JA, Francklyn CS, Robey-Bond SM (2014) Transfer RNA and human disease. Front Genet 5:158

Alazami AM, Hijazi H, Al-Dosari MS et al (2013) Mutation in ADAT3, encoding adenosine deaminase acting on transfer RNA, causes intellectual disability and strabismus. J Med Genet 50:425–430

Anderson P, Ivanov P (2014) tRNA fragments in human health and disease. FEBS Lett 588:4297–4304

Anderson SL, Rubin BY (2005) Tocotrienols reverse IKAP and monoamine oxidase deficiencies in familial dysautonomia. Biochem Biophys Res Commun 336:150–156

Anderson SL, Coli R, Daly IW et al (2001) Familial dysautonomia is caused by mutations of the IKAP gene. Am J Hum Genet 68:753–758

Anderson SL, Qiu J, Rubin BY (2003a) EGCG corrects aberrant splicing of IKAP mRNA in cells from patients with familial dysautonomia. Biochem Biophys Res Commun 310:627–633

Anderson SL, Qiu J, Rubin BY (2003b) Tocotrienols induce IKBKAP expression: a possible therapy for familial dysautonomia. Biochem Biophys Res Commun 306:303–309

Anderson SL, Liu B, Qiu J et al (2012) Nutraceutical-mediated restoration of wild-type levels of IKBKAP-encoded IKAP protein in familial dysautonomia-derived cells. Mol Nutr Food Res 56:570–579

Axelrod FB, Liebes L, Gold-Von Simson G et al (2011) Kinetin improves IKBKAP mRNA splicing in patients with familial dysautonomia. Pediatr Res 70:480–483

Barrett JC, Hansoul S, Nicolae DL et al (2008) Genome-wide association defines more than 30 distinct susceptibility loci for Crohn's disease. Nat Genet 40:955–962

Bartlett JM, Thomas J, Ross DT et al (2010) Mammostrat as a tool to stratify breast cancer patients at risk of recurrence during endocrine therapy. Breast Cancer Res 12:R47

Begley U, Sosa MS, Avivar-Valderas A et al (2013) A human tRNA methyltransferase 9-like protein prevents tumour growth by regulating LIN9 and HIF1-alpha. EMBO Mol Med 5:366–383

Berg M, Agesen TH, Thiis-Evensen E et al (2010) Distinct high resolution genome profiles of early onset and late onset colorectal cancer integrated with gene expression data identify candidate susceptibility loci. Mol Cancer 9:100

Betts C, Saleh AF, Arzumanov AA et al (2012) Pip6-PMO, a new generation of peptide-oligonucleotide conjugates with improved cardiac exon skipping activity for DMD treatment. Mol Ther Nucleic Acids 1, e38

Blanco S, Kurowski A, Nichols J et al (2011) The RNA-methyltransferase Misu (NSun2) poises epidermal stem cells to differentiate. PLoS Genet 7, e1002403

Blanco S, Dietmann S, Flores JV et al (2014) Aberrant methylation of tRNAs links cellular stress to neuro-developmental disorders. EMBO J 33:2020–2039

Bonnet C, Gregoire MJ, Brochet K et al (2006) Pure de-novo 5 Mb duplication at Xp11.22-p11.23 in a male: phenotypic and molecular characterization. J Hum Genet 51:815–821

Brzezicha B, Schmidt M, Makalowska I et al (2006) Identification of human tRNA:m5C methyltransferase catalysing intron-dependent m5C formation in the first position of the anticodon of the pre-tRNA Leu (CAA). Nucleic Acids Res 34:6034–6043

Casati S, Ottria R, Baldoli E et al (2011) Effects of cytokinins, cytokinin ribosides and their analogs on the viability of normal and neoplastic human cells. Anticancer Res 31:3401–3406

Chan CT, Dyavaiah M, DeMott MS et al (2010) A quantitative systems approach reveals dynamic control of tRNA modifications during cellular stress. PLoS Genet 6, e1001247

Chan CT, Pang YL, Deng W et al (2012) Reprogramming of tRNA modifications controls the oxidative stress response by codon-biased translation of proteins. Nat Commun 3:937

Chen C, Tuck S, Bystrom AS (2009) Defects in tRNA modification associated with neurological and developmental dysfunctions in *Caenorhabditis elegans* elongator mutants. PLoS Genet 5, e1000561

Crick FH (1966) Codon–anticodon pairing: the wobble hypothesis. J Mol Biol 19:548–555

Dai L, Xing L, Gong P et al (2008) Positive association of the FTSJ1 gene polymorphisms with nonsyndromic X-linked mental retardation in young Chinese male subjects. J Hum Genet 53:592–597

Dietrich P, Alli S, Shanmugasundaram R et al (2012) IKAP expression levels modulate disease severity in a mouse model of familial dysautonomia. Hum Mol Genet 21:5078–5090

Durdevic Z, Schaefer M (2013) tRNA modifications: necessary for correct tRNA-derived fragments during the recovery from stress? Bioessays 35:323–327

El Yacoubi B, Bailly M, de Crecy-Lagard V (2012) Biosynthesis and function of posttranscriptional modifications of transfer RNAs. Annu Rev Genet 46:69–95

El-Hattab AW, Bournat J, Eng PA et al (2011) Microduplication of Xp11.23p11.3 with effects on cognition, behavior, and craniofacial development. Clin Genet 79:531–538

Fahiminiya S, Almuriekhi M, Nawaz Z et al (2014) Whole exome sequencing unravels disease-causing genes in consanguineous families in Qatar. Clin Genet 86(2):134–41

Franke B, Vermeulen SH, Steegers-Theunissen RP et al (2009) An association study of 45 folate-related genes in spina bifida: involvement of cubilin (CUBN) and tRNA aspartic acid methyltransferase 1 (TRDMT1). Birth Defects Res A Clin Mol Teratol 85:216–226

Freude K, Hoffmann K, Jensen LR et al (2004) Mutations in the FTSJ1 gene coding for a novel S-adenosylmethionine-binding protein cause nonsyndromic X-linked mental retardation. Am J Hum Genet 75:305–309

Froyen G, Bauters M, Boyle J et al (2007) Loss of SLC38A5 and FTSJ1 at Xp11.23 in three brothers with non-syndromic mental retardation due to a microdeletion in an unstable genomic region. Hum Genet 121:539–547

Frye M, Watt FM (2006) The RNA methyltransferase Misu (NSun2) mediates Myc-induced proliferation and is upregulated in tumors. Curr Biol 16:971–981

Frye M, Dragoni I, Chin SF et al (2010) Genomic gain of 5p15 leads to over-expression of Misu (NSUN2) in breast cancer. Cancer Lett 289:71–80

Gerber AP, Keller W (1999) An adenosine deaminase that generates inosine at the wobble position of tRNAs. Science 286:1146–1149

Gillis D, Krishnamohan A, Yaacov B et al (2014) TRMT10A dysfunction is associated with abnormalities in glucose homeostasis, short stature and microcephaly. J Med Genet 51:581–586

Giorda R, Bonaglia MC, Beri S et al (2009) Complex segmental duplications mediate a recurrent dup(X)(p11.22-p11.23) associated with mental retardation, speech delay, and EEG anomalies in males and females. Am J Hum Genet 85:394–400

Goll MG, Kirpekar F, Maggert KA et al (2006) Methylation of tRNAAsp by the DNA methyltransferase homolog Dnmt2. Science 311:395–398

Gong P, Li J, Dai L et al (2008) Genetic variations in FTSJ1 influence cognitive ability in young males in the Chinese Han population. J Neurogenet 22:277–287

Guan MX, Yan Q, Li X et al (2006) Mutation in TRMU related to transfer RNA modification modulates the phenotypic expression of the deafness-associated mitochondrial 12S ribosomal RNA mutations. Am J Hum Genet 79:291–302

Hamel BC, Smits AP, van den Helm B et al (1999) Four families (MRX43, MRX44, MRX45, MRX52) with nonspecific X-linked mental retardation: clinical and psychometric data and results of linkage analysis. Am J Med Genet 85:290–304

Hawkes NA, Otero G, Winkler GS et al (2002) Purification and characterization of the human elongator complex. J Biol Chem 277:3047–3052

Hayashi J, Ohta S, Takai D et al (1993) Accumulation of mtDNA with a mutation at position 3271 in tRNA(Leu)(UUR) gene introduced from a MELAS patient to HeLa cells lacking mtDNA results in progressive inhibition of mitochondrial respiratory function. Biochem Biophys Res Commun 197:1049–1055

Hims MM, Ibrahim EC, Leyne M et al (2007) Therapeutic potential and mechanism of kinetin as a treatment for the human splicing disease familial dysautonomia. J Mol Med (Berl) 85:149–161

Honma Y (2003) Adenine analogs as potential differentiation therapy agents for acute myeloid leukemia. Drug Dev Res 59:14–22

Huang BS, Wu RT, Chien KY (1992) Relationship of the queuine content of transfer ribonucleic acids to histopathological grading and survival in human lung cancer. Cancer Res 52:4696–4700

Hussain S, Sajini AA, Blanco S et al (2013) NSun2-mediated cytosine-5 methylation of vault noncoding RNA determines its processing into regulatory small RNAs. Cell Rep 4:255–261

Igoillo-Esteve M, Genin A, Lambert N et al (2013) tRNA methyltransferase homolog gene TRMT10A mutation in young onset diabetes and primary microcephaly in humans. PLoS Genet 9, e1003888

Iida K, Jin H, Zhu JK (2009) Bioinformatics analysis suggests base modifications of tRNAs and miRNAs in Arabidopsis thaliana. BMC Genomics 10:155

Jackman JE, Alfonzo JD (2013) Transfer RNA modifications: nature's combinatorial chemistry playground. Wiley Interdiscip Rev RNA 4:35–48

Karlsborn T, Tukenmez H, Chen C, Bystrom AS (2014) Familial dysautonomia (FD) patients have reduced levels of the modified wobble nucleoside mcmsU in tRNA. Biochem Biophys Res Commun 454:441–445

Kerr SJ, Borek E (1973) Regulation of the tRNA methyltransferases in normal and neoplastic tissues. Adv Enzyme Regul 11:63–77

Khan MA, Rafiq MA, Noor A et al (2012) Mutation in NSUN2, which encodes an RNA methyltransferase, causes autosomal-recessive intellectual disability. Am J Hum Genet 90:856–863

Khoddami V, Cairns BR (2013) Identification of direct targets and modified bases of RNA cytosine methyltransferases. Nat Biotechnol 31:458–464

Kirchhoff K, Machicao F, Haupt A et al (2008) Polymorphisms in the TCF7L2, CDKAL1 and SLC30A8 genes are associated with impaired proinsulin conversion. Diabetologia 51:597–601

Kirchner S, Ignatova Z (2015) Emerging roles of tRNA in adaptive translation, signalling dynamics and disease. Nat Rev Genet 16:98–112

Kirino Y, Yasukawa T, Ohta S et al (2004) Codon-specific translational defect caused by a wobble modification deficiency in mutant tRNA from a human mitochondrial disease. Proc Natl Acad Sci U S A 101:15070–15075

Kirino Y, Yasukawa T, Marjavaara SK et al (2006) Acquisition of the wobble modification in mitochondrial tRNALeu(CUN) bearing the G12300A mutation suppresses the MELAS molecular defect. Hum Mol Genet 15:897–904

Lee G, Papapetrou EP, Kim H et al (2009) Modelling pathogenesis and treatment of familial dysautonomia using patient-specific iPSCs. Nature 461:402–406

Leyne M, Mull J, Gill SP et al (2003) Identification of the first non-Jewish mutation in familial Dysautonomia. Am J Med Genet A 118A:305–308

Li R, Guan MX (2010) Human mitochondrial leucyl-tRNA synthetase corrects mitochondrial dysfunctions due to the tRNALeu(UUR) A3243G mutation, associated with mitochondrial encephalomyopathy, lactic acidosis, and stroke-like symptoms and diabetes. Mol Cell Biol 30:2147–2154

Liu J, Straby KB (2000) The human tRNA(m(2)(2)G(26))dimethyltransferase: functional expression and characterization of a cloned hTRM1 gene. Nucleic Acids Res 28:3445–3451

Machnicka MA, Milanowska K, Osman Oglou O et al (2013) MODOMICS: a database of RNA modification pathways–2013 update. Nucleic Acids Res 41:D262–267

Martinez FJ, Lee JH, Lee JE et al (2012) Whole exome sequencing identifies a splicing mutation in NSUN2 as a cause of a Dubowitz-like syndrome. J Med Genet 49:380–385

Michaud J, Kudoh J, Berry A et al (2000) Isolation and characterization of a human chromosome 21q22.3 gene (WDR4) and its mouse homologue that code for a WD-repeat protein. Genomics 68:71–79

Najmabadi H, Hu H, Garshasbi M et al (2011) Deep sequencing reveals 50 novel genes for recessive cognitive disorders. Nature 478:57–63

Novoa EM, Ribas de Pouplana L (2012) Speeding with control: codon usage, tRNAs, and ribosomes. Trends Genet 28:574–581

Pang YL, Abo R, Levine SS, Dedon PC (2014) Diverse cell stresses induce unique patterns of tRNA up- and down-regulation: tRNA-seq for quantifying changes in tRNA copy number. Nucleic Acids Res 42(22), e170

Park H, Davidson E, King MP (2008) Overexpressed mitochondrial leucyl-tRNA synthetase suppresses the A3243G mutation in the mitochondrial tRNA(Leu(UUR)) gene. RNA 14:2407–2416

Pathak C, Jaiswal YK, Vinayak M (2007) Possible involvement of queuine in regulation of cell proliferation. Biofactors 29:159–173

Phizicky EM, Alfonzo JD (2010) Do all modifications benefit all tRNAs? FEBS Lett 584:265–271

Pierga JY, Reis-Filho JS, Cleator SJ et al (2007) Microarray-based comparative genomic hybridisation of breast cancer patients receiving neoadjuvant chemotherapy. Br J Cancer 96:341–351

Piñeyro D, Torres AG, Ribas de Pouplana L (2014) Biogenesis and evolution of functional tRNAs. In: Sesma A, Von der Haar T (eds) Fungal RNA biology. Springer, Switzerland, pp 233–267

Powell CA, Kopajtich R, D'Souza AR et al (2015) TRMT5 mutations cause a defect in post-transcriptional modification of mitochondrial tRNA associated with multiple respiratory-chain deficiencies. Am J Hum Genet 97:319–328

Quaranta M, Burden AD, Griffiths CE et al (2009) Differential contribution of CDKAL1 variants to psoriasis, Crohn's disease and type II diabetes. Genes Immun 10:654–658

Ramser J, Winnepenninckx B, Lenski C et al (2004) A splice site mutation in the methyltransferase gene FTSJ1 in Xp11.23 is associated with non-syndromic mental retardation in a large Belgian family (MRX9). J Med Genet 41:679–683

Randerath E, Gopalakrishnan AS, Gupta RC et al (1981) Lack of a specific ribose methylation at guanosine 17 in Morris hepatoma 5123D tRNASer1IGA. Cancer Res 41:2863–2867

Randerath E, Agrawal HP, Randerath K (1984) Specific lack of the hypermodified nucleoside, queuosine, in hepatoma mitochondrial aspartate transfer RNA and its possible biological significance. Cancer Res 44:1167–1171

Rodriguez V, Chen Y, Elkahloun A et al (2007) Chromosome 8 BAC array comparative genomic hybridization and expression analysis identify amplification and overexpression of TRMT12 in breast cancer. Genes Chromosomes Cancer 46:694–707

Rubio MA, Pastar I, Gaston KW et al (2007) An adenosine-to-inosine tRNA-editing enzyme that can perform C-to-U deamination of DNA. Proc Natl Acad Sci U S A 104:7821–7826

Rudinger J, Florentz C, Giege R (1994) Histidylation by yeast HisRS of tRNA or tRNA-like structure relies on residues -1 and 73 but is dependent on the RNA context. Nucleic Acids Res 22:5031–5037

Ruiz-Pesini E, Lott MT, Procaccio V et al (2007) An enhanced MITOMAP with a global mtDNA mutational phylogeny. Nucleic Acids Res 35:D823–828

Saade S, Cazier JB, Ghassibe-Sabbagh M et al (2011) Large scale association analysis identifies three susceptibility loci for coronary artery disease. PLoS One 6, e29427

Schaefer M, Hagemann S, Hanna K, Lyko F (2009) Azacytidine inhibits RNA methylation at DNMT2 target sites in human cancer cell lines. Cancer Res 69:8127–8132

Shaheen R, Abdel-Salam GM, Guy MP et al (2015) Mutation in WDR4 impairs tRNA m(7)G46 methylation and causes a distinct form of microcephalic primordial dwarfism. Genome Biol 16:210

Sharma OK, Kerr SJ, Lipshitz-Wiesner R et al (1971) Regulation of the tRNA methylases. Fed Proc 30:167–176

Shimada K, Nakamura M, Anai S et al (2009) A novel human AlkB homologue, ALKBH8, contributes to human bladder cancer progression. Cancer Res 69:3157–3164

Simpson CL, Lemmens R, Miskiewicz K et al (2009) Variants of the elongator protein 3 (ELP3) gene are associated with motor neuron degeneration. Hum Mol Genet 18:472–481

Slaugenhaupt SA, Blumenfeld A, Gill SP et al (2001) Tissue-specific expression of a splicing mutation in the IKBKAP gene causes familial dysautonomia. Am J Hum Genet 68:598–605

Stancakova A, Pihlajamaki J, Kuusisto J et al (2008) Single-nucleotide polymorphism rs7754840 of CDKAL1 is associated with impaired insulin secretion in nondiabetic offspring of type 2 diabetic subjects and in a large sample of men with normal glucose tolerance. J Clin Endocrinol Metab 93:1924–1930

Strug LJ, Clarke T, Chiang T et al (2009) Centrotemporal sharp wave EEG trait in rolandic epilepsy maps to Elongator Protein Complex 4 (ELP4). Eur J Hum Genet 17:1171–1181

Suzuki T, Nagao A (2011) Human mitochondrial diseases caused by lack of taurine modification in mitochondrial tRNAs. Wiley Interdiscip Rev RNA 2:376–386

Tachibana M, Sparman M, Sritanaudomchai H et al (2009) Mitochondrial gene replacement in primate offspring and embryonic stem cells. Nature 461:367–372

Tachibana M, Amato P, Sparman M et al (2013) Towards germline gene therapy of inherited mitochondrial diseases. Nature 493:627–631

Takano K, Nakagawa E, Inoue K et al (2008) A loss-of-function mutation in the FTSJ1 gene causes nonsyndromic X-linked mental retardation in a Japanese family. Am J Med Genet B Neuropsychiatr Genet 147B:479–484

Takeoka S, Unoki M, Onouchi Y et al (2001) Amino-acid substitutions in the IKAP gene product significantly increase risk for bronchial asthma in children. J Hum Genet 46:57–63

Torres AG, Batlle E, Ribas de Pouplana L (2014a) Role of tRNA modifications in human diseases. Trends Mol Med 20:306–314

Torres AG, Pineyro D, Filonava L et al (2014b) A-to-I editing on tRNAs: Biochemical, biological and evolutionary implications. FEBS Lett 588:4279–4286

Torres AG, Pineyro D, Rodriguez-Escriba M et al (2015) Inosine modifications in human tRNAs are incorporated at the precursor tRNA level. Nucleic Acids Res 43:5145–5157

Towns WL, Begley TJ (2012) Transfer RNA methytransferases and their corresponding modifications in budding yeast and humans: activities, predications, and potential roles in human health. DNA Cell Biol 31:434–454

Tuorto F, Liebers R, Musch T et al (2012) RNA cytosine methylation by Dnmt2 and NSun2 promotes tRNA stability and protein synthesis. Nat Struct Mol Biol 19:900–905

Umeda N, Suzuki T, Yukawa M et al (2005) Mitochondria-specific RNA-modifying enzymes responsible for the biosynthesis of the wobble base in mitochondrial tRNAs. Implications for the molecular pathogenesis of human mitochondrial diseases. J Biol Chem 280:1613–1624

Urbonavicius J, Stahl G, Durand JM et al (2003) Transfer RNA modifications that alter +1 frameshifting in general fail to affect -1 frameshifting. RNA 9:760–768

Vauti F, Goller T, Beine R et al (2007) The mouse Trm1-like gene is expressed in neural tissues and plays a role in motor coordination and exploratory behaviour. Gene 389:174–185

Vinayak M, Pathak C (2010) Queuosine modification of tRNA: its divergent role in cellular machinery. Biosci Rep 30:135–148

Waas WF, Druzina Z, Hanan M et al (2007) Role of a tRNA base modification and its precursors in frameshifting in eukaryotes. J Biol Chem 282:26026–26034

Wainfan E, Borek E (1967) Differential inhibitors of tRNA methylases. Mol Pharmacol 3:595–598

Wainfan E, Landsberg B (1971) Cytokinins that inhibit transfer RNA methylating enzymes. FEBS Lett 19:144–148

Wei FY, Tomizawa K (2011) Functional loss of Cdkal1, a novel tRNA modification enzyme, causes the development of type 2 diabetes. Endocr J 58:819–825

Wei FY, Suzuki T, Watanabe S et al (2011) Deficit of tRNA(Lys) modification by Cdkal1 causes the development of type 2 diabetes in mice. J Clin Invest 121:3598–3608

Xie P, Wei FY, Hirata S et al (2013) Quantitative PCR measurement of tRNA 2-methylthio modification for assessing type 2 diabetes risk. Clin Chem 59:1604–1612

Yarian C, Townsend H, Czestkowski W et al (2002) Accurate translation of the genetic code depends on tRNA modified nucleosides. J Biol Chem 277:16391–16395

Yasukawa T, Suzuki T, Ishii N et al (2001) Wobble modification defect in tRNA disturbs codon-anticodon interaction in a mitochondrial disease. EMBO J 20:4794–4802

Zeharia A, Shaag A, Pappo O et al (2009) Acute infantile liver failure due to mutations in the TRMU gene. Am J Hum Genet 85:401–407

Zhou W, Karcher D, Bock R (2014) Identification of enzymes for adenosine-to-inosine editing and discovery of cytidine-to-uridine editing in nucleus-encoded transfer RNAs of arabidopsis. Plant Physiol 166:1985–1997

Regulated tRNA Cleavage in Biology and Medicine: Roles of tRNA Modifications

Shawn M. Lyons, Marta M. Fay, and Pavel Ivanov

Contents

Abstract Transfer RNAs (tRNAs) play a key role in translating genomic information and regulating gene expression. tRNA cleavage is an evolutionarily conserved phenomenon serving versatile functions in different organisms. The size distribution and abundance of tRNA-derived fragments suggests that tRNA modifications play important roles in mechanisms that regulate tRNA cleavage and degradation. Here, we discuss the importance of posttranscriptional modifications in controlling processing of tRNAs and describe the functions of tRNA-derived fragments in cell physiology and pathophysiology.

Shawn M. Lyons and Marta M. Fay are equally contributed.

S.M. Lyons • M.M. Fay
Division of Rheumatology, Immunology and Allergy, Brigham and Women's Hospital, Boston, MA 02115, USA

Department of Medicine, Harvard Medical School, Boston, MA 02115, USA

P. Ivanov (✉)
Division of Rheumatology, Immunology and Allergy, Brigham and Women's Hospital, Boston, MA 02115, USA

Department of Medicine, Harvard Medical School, Boston, MA 02115, USA

The Broad Institute of Harvard and M.I.T., Cambridge, MA 02142, USA
e-mail: pivanov@rics.bwh.harvard.edu

S. Jurga et al. (eds.), *Modified Nucleic Acids in Biology and Medicine*, RNA Technologies, DOI 10.1007/978-3-319-34175-0_2

Keywords tRNA • Modifications • Ribonuclease • Angiogenin • tRNA cleavage

1 Introduction

Chemical modifications play an important role in the regulation of biological processes and in the modulation of gene expression. Cellular RNAs can be chemically modified in over a hundred ways (Klungland and Dahl 2014; Wang and He 2014). These modifications play active roles in biological processes and can be either stable (static) or reversible (dynamic). RNA modifications are energy consuming and complex; they require a significant cellular investment (RNA-modifying enzymes account up to 10 % of cellular proteins (Anantharaman et al. 2002)). Transfer RNAs (tRNAs), ancient RNA molecules (Parisien et al. 2013), have the most chemical modifications. Over 90 different structural modifications can be found in an average tRNA with length of 75–90 nucleotides (Machnicka et al. 2014). Importantly, tRNA posttranscriptional modifications are critical for many, if not all, tRNA functions (El Yacoubi et al. 2012).

tRNA modifications are linked to the roles that tRNAs play in protein synthesis (El Yacoubi et al. 2012; Kirchner and Ignatova 2015; Phizicky and Alfonzo 2010; Phizicky and Hopper 2010; Raina and Ibba 2014). As an adaptor molecule that helps the ribosome synthesize proteins by decoding nucleotide triplets thereby linking nucleotide information on the mRNA to amino acid sequence, tRNAs are important for correct and efficient mRNA translation. tRNA modifications stabilize structure, which is critical to determine tRNA quality and, consequently, turnover and its ability to participate in protein synthesis. If unmodified, tRNA can be charged with a near-cognate or non-cognate amino acid leading to mistranslation of the corresponding mRNA codon. Such mistranslation causes introduction of non-coded amino acid into a protein that may, in turn, affect protein stability or create a harmful/toxic protein variant. Similarly, tRNA modifications are also important in the codon–anticodon interactions and regulation of wobble base pairing. These interactions are necessary to prevent mistranslation and frameshifting errors on mRNA (Giege 2008; Phizicky and Hopper 2010).

Beyond the canonical function in protein synthesis, tRNAs are also implicated in a number of other biological processes including cell survival, apoptosis, cell signaling, metabolism of amino acids and porphyrins, and stress response programs (Raina and Ibba 2014; Phizicky and Hopper 2010). The list of such functions continues to grow; the role of tRNA modifications in such processes is also well documented. Interestingly, with the development of high-throughput sequencing technologies, it has become apparent that tRNAs are a rich source of small non-coding RNAs (ncRNAs), and these tRNA-derived ncRNAs are produced by specific and regulated cleavage (Anderson and Ivanov 2014; Gebetsberger and Polacek 2013). Depending on the organism and the ribonuclease, specific RNA modifications can promote or inhibit tRNA cleavage. Here, we review the role of tRNA modifications, discuss tRNA cleavage as a biological phenomenon and the roles of chemical modifications in tRNA cleavage, describe functions of tRNA-derived fragments in different organisms, and summarize the current thinking about the biological and clinical significance of such fragments.

2 Posttranscriptional Processing of tRNAs

2.1 General Remarks on tRNA Processing, Maturation, and Fragmentation

tRNA biogenesis is a complex process (Fig. 1). In all organisms, tRNAs are initially produced as longer transcripts that must undergo 5′- and 3′-end trimming and chemical modifications to form mature molecules. In bacteria, several tRNA cistrons (often as a part of an operon) are first transcribed to generate long polycistronic RNA that are subsequently processed into smaller tRNA-containing RNAs. In contrast, the majority of eukaryotic tRNA genes are transcribed as independent units by RNA polymerase III.

The formation of mature 5′-ends of tRNA in all domains of life (bacteria, archaea, and eukarya) is dependent on the ribonuclease P (RNase P), a unique RNA-based enzyme. The formation/trimming of the 3′-end is more complex and requires either a single endoribonuclease (such as RNase Z in mammalian cells) or a group of endo- and exoribonucleases (such as cooperative actions of RNase E and RNases T and/or D in *E. coli*). Depending on the organism, the universal 3′-CCA end can either be encoded in the tRNA gene (as in *E. coli*) or added posttranscriptionally by a CCA-adding enzyme (such as TRNT1 enzyme in mammalian cells). Finally, some eukaryotic tRNA genes also contain introns, which are spliced out from the precursor tRNA (pre-tRNA). Ligation of the 5′- and 3′-exons can go through two different mechanisms, but the relative contribution of either mechanism is still a matter of debate (Chakravarty et al. 2012; Desai et al. 2014; Mair et al. 2013; Paushkin et al. 2004; Popow et al. 2011; Tanaka et al. 2011) (Fig. 1A). In the first mechanism, RtcB (HSPC17 or C22orf28) works as an RNA 2′,3′-cyclic phosphate and 5′-OH ligase to directly mediate exon ligation (Popow et al. 2014). In the second mechanism, CLP1 (the only identified RNA kinase) associates with the tRNA-splicing endonuclease (TSEN) complex to phosphorylate 3′-tRNA exons and facilitate further ligation. Molecular details of CLP1-facilitated ligation of tRNA exons are unknown (Weitzer and Martinez 2007). Mature tRNAs have a 5′-monophosphate and 3′-CCA charged with the cognate amino acid. All nuclear-encoded eukaryotic tRNAs have four base-paired stems (in 5′-3′-direction: D arm, anticodon arm, T-arm, and acceptor stem) bridging the conserved D-loop, tRNA-specific anticodon loop, variable loop (V-loop), and T-loop (also known as Tψ-loop) (Fig. 1). This linear sequence folds into a cloverleaf-shaped secondary structure and L-shaped tertiary structure (Figs. 1 and 2).

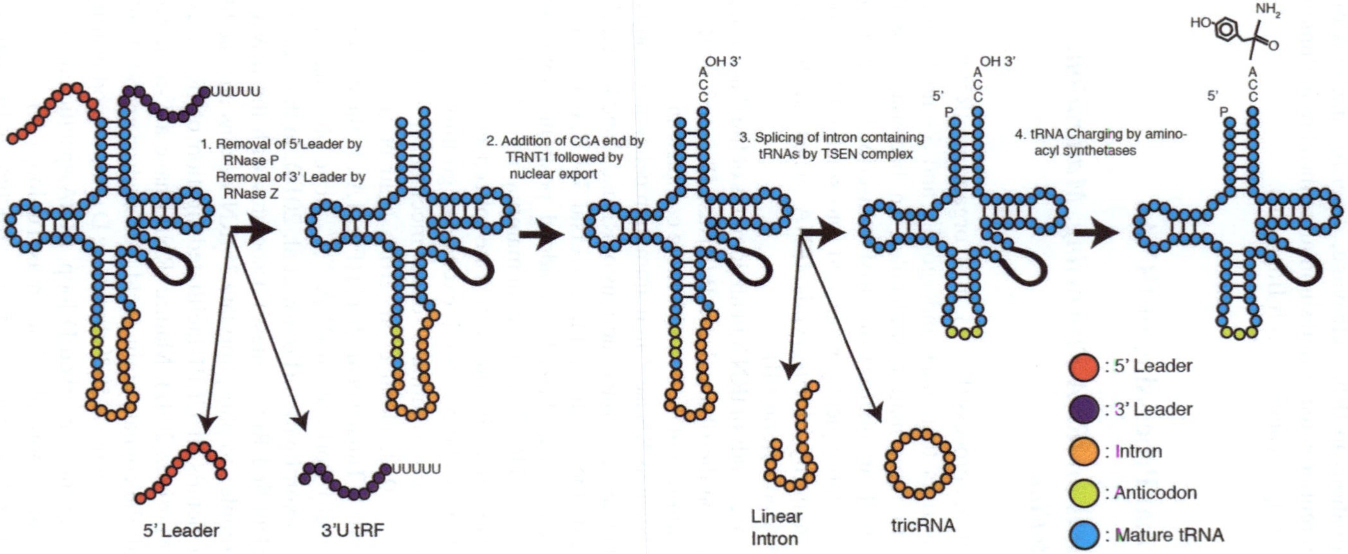

Fig. 1 Eukaryotic tRNA processing. Transcribed pre-tRNA contains distinct 5′- and 3′-extensions (5′-leader (*red*) and 3′-trailer (*purple*)) that have to be removed by RNase P and RNase Z (or other ribonucleases in prokaryotes) during tRNA processing (Step 1). Upon 3′-trailer removal, the CCA-adding enzyme TRNT1 adds CCA to the 3′-ends of tRNA (Step 2). In some cases, 3′-trailers accumulate and are exported to the cytoplasm as 3′U-tRFs. Intron (*orange*)-containing pre-tRNAs are spliced by the TSEN complex (Step 3). Usually, such introns are quickly degraded but in some pathological cases linear and/or circular (tRNA intronic circular RNAs, or tricRNAs) introns can accumulate. 5′-exons and 3′-exons are naturally present in the nucleus as products of intron-containing tRNA splicing. Mature tRNA go through extensive modification reactions, charging with an amino acid and tRNA quality control check (Step 4)

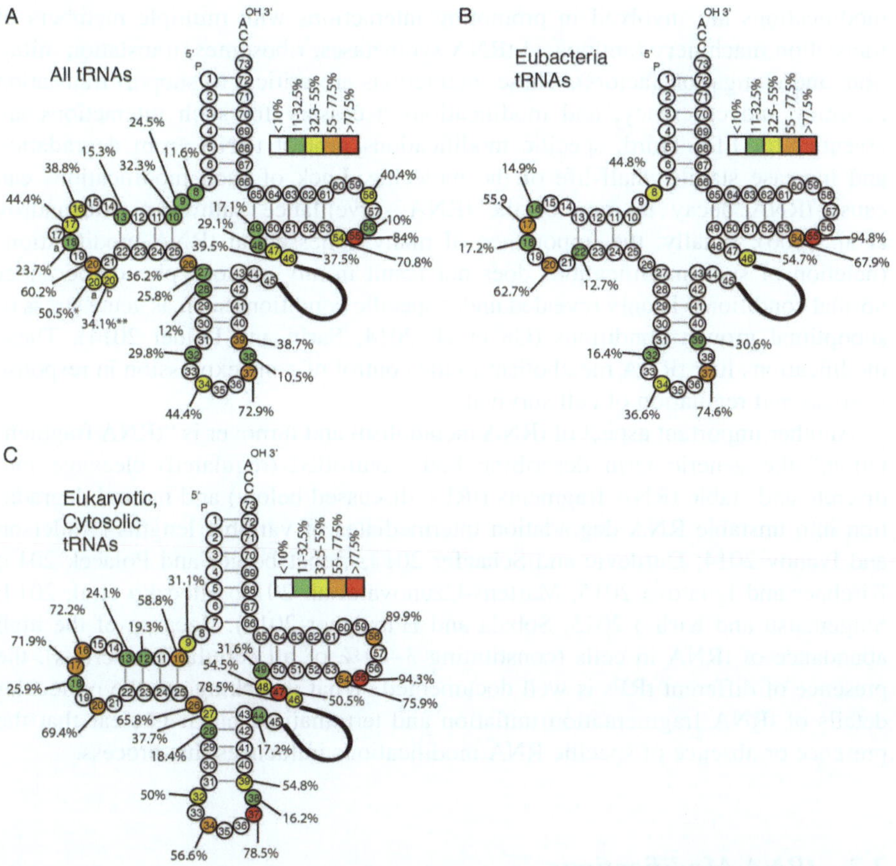

Fig. 2 Distribution of posttranscriptional modifications in tRNA. Comparative analysis of the distribution and frequencies of tRNA modifications (based on the MODOMICS database). (**a**) Consensus tRNA secondary structure, distribution and frequencies of tRNA modifications in an average tRNA (based on the sequences/information of 602 tRNAs). (**b**) Distribution and frequencies of tRNA modifications in bacterial tRNAs. (**c**) Distribution and frequencies of tRNA modifications in eukaryotic tRNAs. The secondary tRNA structure is presented in the cloverleaf form with universal numbering system. Color-coded system is shown to demonstrate the modification frequencies (in %) at a given position on tRNA molecule

In all studied organisms, tRNA molecules are heavily modified (Machnicka et al. 2013) (Fig. 2). Initially, tRNA molecules are transcribed with canonical U, A, G, and C bases. Later, the nucleobases and ribose sugars at selected positions are chemically modified by a system of enzymes to form a different chemical structure. In fact, a large fraction of the genome (up to 10 % of the protein-coding genes in a given genome) (Anantharaman et al. 2002) is dedicated to metabolic and enzymatic pathways that ensure proper posttranscriptional modification of tRNAs, highlighting the importance of tRNA modifications. First, nucleoside modifications have structural roles to support formation of the correct L shape. Second, many tRNA modifications are involved in promoting interactions with multiple members of translation machinery (aminoacyl tRNA synthetases, ribosomes, translation initiation, and elongation factors). These interactions are critical to support translation accuracy and efficiency, and modifications necessary for such interactions are essential for life. Third, specific modifications protect tRNA from degradation and increase stability/half-life of the molecule. Lack of such modifications can cause tRNA decay by one of the tRNA surveillance pathways (Alexandrov et al. 2006). Finally, the importance of many nonessential tRNA modifications (deletion of such modifications does not result in any obvious phenotype under normal conditions) is only revealed under specific conditions such as acute stress or suboptimal growth conditions (Gu et al. 2014; Sarin and Leidel 2014). These modifications link tRNA metabolism to the control of gene expression in response to stress and regulation of cell survival.

Another important aspect of tRNA metabolism and turnover is "tRNA fragmentation," the generic term describing both controlled (regulated) cleavage into discrete and stable tRNA fragments (tRFs, discussed below) and natural degradation into unstable RNA degradation intermediates of variable lengths (Anderson and Ivanov 2014; Durdevic and Schaefer 2013; Gebetsberger and Polacek 2013; Kirchner and Ignatova 2015; Martens-Uzunova et al. 2013; Mleczko et al. 2014; Shigematsu and Kirino 2015; Sobala and Hutvagner 2011). Because of the high abundance of tRNA in cells (constituting 3–10 % of all cellular transcripts), the presence of different tRFs is well documented. What is lacking are the molecular details of tRNA fragmentation initiation and termination, but it is clear that the presence or absence of specific RNA modifications influences this process.

2.2 *tRNA Modifications*

Modifications in tRNA are abundant and universal (Fig. 2). To date, 80–90 modified nucleosides have been identified in tRNAs of different organisms, and new ones are still being discovered. Certain modifications are also conserved in the smallest free-living organisms [such as *Mycoplasma genitalium* (Fraser et al. 1995)] and organelles of eukaryotic organisms [mitochondria and plastids (Machnicka et al. 2014; Suzuki et al. 2011)]. The majority of modifications are in the D stem-loop at the positions 16–18 and 20; the anticodon stem-loop (ASL) at the positions 34 and 37; the T stem-loop at the positions 54, 55, and 58; and the variable loop at the positions 46–47 (Fig. 2A). There are an average of eight modifications per tRNA molecule (derived from viruses, archaea, eubacteria, fungi, plants, animals, as well as mitochondria and chloroplasts) (Phizicky and Alfonzo 2010; Sprinzl and Vassilenko 2005) and 11–13 modifications in the average human tRNA (Saikia et al. 2010) (Fig. 2).

Some modifications are evolutionarily conserved across all three kingdoms of life (Fig. 2A). It is generally believed that modifications in the T- and D-loops are crucial for tRNA structure and stability, whereas modifications in ASL are required for codon–anticodon interactions. The modifications can be broadly subdivided into three categories. The first category represents the group of modified nucleosides that are highly conserved in the majority of tRNA species (from different organisms) for both their position within the tRNA and their chemical identity. These modifications are introduced by evolutionarily conserved enzymes and includes conserved modifications in D-loop (such as dihydrouridine (D) and 2′-*O*-methylguanosine (Gm)) and in T-loop (such as pseudouridine (ψ) and 5-methyluridine (m^5U)). These residues are especially critical for tRNA folding, structure, and stability. The second category includes modifications that are conserved for their position (e.g., many hypermethylated uridines at position 34 or many methylated purines at position 37 of ASL) but not for their chemical structure. Despite their very diverse chemical identities, the functions of the modified nucleosides in these positions are similar and devoted to the accurate decoding of the mRNA codons by the ribosome during protein synthesis. The third category of the modified residues includes all other modified nucleosides scattered throughout tRNA that have no evolutionarily conserved positions, and often represent simple chemical alterations (such as pseudouridines or methylated at base or ribose). The modification enzymes of the second and third categories are diverse and often organism specific.

The most prevalent modification found in tRNA is pseudouridinylation, which is introduced by isomerization of uridine and its selected derivatives (ψ is found in approximately 25 % of all modified positions). Pseudouridine at position 55 is the most common site-specific tRNA modification (found in almost 85 % of all sequenced tRNAs (Machnicka et al. 2013)) (Fig. 2). Other frequent modifications are represented by dihydrouridine and 2′-*O*-methyl derivatives of G, A, C, and U. Interestingly, some positions on tRNA have never been experimentally found to be modified (based on the modification profiles from 602 sequences provided by Modomics RNA modification database (Machnicka et al. 2013)). These are nucleotides at positions 5, 11, 23, 24, 33, 42, 43, 45, 53, 59, 62, 63, and 73. The absence of modifications at these sites may be necessary for correct tRNA folding, and introduction of bulky modifications at these positions may influence conformational dynamics. The universal 3′-CCA is also unmodified, as well as many positions in the V-loop. The number of modified positions in eukaryotes (Fig. 2C) seems to be higher than in bacteria (Fig. 2B), archaea, viruses, and organelles (mitochondria and plastids), although a larger set of tRNA sequences is needed to fully support this conclusion.

Recent data suggest that certain tRNA modifications are dynamic and adaptive to stresses (Gu et al. 2014). Based on the elegant and highly accurate mass-spectrometric approaches to identify the spectra of tRNA modification coupled with genome-wide codon bias analysis (Begley et al. 2007; Cai et al. 2015; Chan et al. 2012; Patil et al. 2012a, b; Rezgui et al. 2013; Su et al. 2014), the Begley and Dedon laboratories introduced an emerging concept on the roles of RNA modification in the regulation of gene expression in response to environmental changes. They provide the evidence that tRNA modifications (specifically tRNA methylation) influence gene expression by modulating mRNA translation. The initial evidence came from a study showing that yeast cells sense stress (such as alkylating stress) and promote selective translation of codon-biased mRNAs by a dynamic change in the levels of a specific modification (mCm^{5}U) at the wobble position of selected tRNAs (tRNA$^{Arg(UCU)}$ and tRNA$^{Glu(UUC)}$) by tRNA methyltransferase 9 (Trm9) (Begley et al. 2007). This modification modulates tRNA–mRNA pairing and allows more efficient binding to the cognate mRNA codon. Such Trm9/mcm^5 U-sensitive mRNAs are enriched with specific arginine and glutamic acid codons and encode DNA damage response proteins; their overexpression leads to the upregulation of DNA damage defense mechanisms against alkylating stress. In subsequent studies, it was shown that such tRNA modification-specific regulation of translation is generally applicable to many other stress conditions (Begley et al. 2007; Cai et al. 2015; Chan et al. 2012; Patil et al. 2012a, b; Rezgui et al. 2013; Su et al. 2014). Moreover, stress-induced tRNA reprogramming (by modulation of diverse tRNA modifications) works in concert with codon usage patterns in specific transcripts (so-called modification tunable transcripts (MoTTs)) to regulate gene expression in a stress-specific manner. These studies suggest unique roles of dynamic tRNA modifications in cell physiology, stress response, and cell survival.

2.3 Role of tRNA Modifications in tRNA Cleavage: Targets or Antidotes?

Cleavage of tRNA is an evolutionarily conserved phenomenon found in most, if not all, organisms. tRNA cleavage parallels the existence of stress-induced tRNA ribonucleases (tRNases), ancient enzymes that are part of the "immune system" of simple unicellular organisms (Ivanov and Anderson 2011). These RNases are used to defend host cells from the invasion of viruses or non-self unicellular species (biotic stresses). Subsequently, such stress-induced tRNases have evolved to participate in a variety of other biological processes. Expression and activities of these tRNases are under tight control under normal optimal conditions; they are induced only under specific stress stimulus. This is achieved by several molecular control mechanisms including physical sequestration of tRNases within membrane-bound organelles (e.g., nuclei), secretion into the extracellular environment, or inactivation by specific RNase inhibitors. Stress-induced activation of tRNases allows rapid alterations in tRNA levels that can profoundly affect cellular physiology due to the high abundance of tRNA in cells.

The relationship between tRNA cleavage and a biological process that is regulated by a tRNA modification is best described in prokaryotic systems. *Escherichia coli* (*E.coli*) tRNA-specific endoribonuclease PrrC (*Eco*PrrC) is activated in response to bacteriophage infection (such as T4 phage) (Kaufmann 2000). In uninfected cells, this RNase is physically associated with an enzyme complex representing the host DNA restriction-modification system *Eco*prrI (encoded by *prrA*, *prrB*, and *prrD* genes of *prr* operon), a classical defense system that limits and/or prevents virus spreading in a bacterial population (Levitz et al. 1990; Tyndall et al. 1994). Upon T4 phage infection, a phage-encoded peptide (Stp) binds to the *Eco*prrI complex to inactivate it (Amitsur et al. 1989, 1992, 2003). The Stp-mediated inactivation of the host restriction-modification system is then sensed by the cell activating the PrrC ribonuclease. In turn, activated PrrC targets bacterial $tRNA^{Lys(UUU)}$ and cleaves directly 5′ of a modified wobble uridine residue (5-methylaminomethyl-2-thiouridine (mN^5s^2U)) within the anticodon loop (Jiang et al. 2001, 2002). By depleting the $tRNA^{Lys}$ pool, PrrC disables both cellular and bacteriophage protein synthesis leading to cell suicide. Bacteria sacrifice individual members of a population, thus preventing the bacteriophage from spreading to adjacent bacteria. Interestingly, T4 phage counteracts this tRNA cleavage-mediated host defense by expressing an RNA repair system (consists of a phage RNA ligase and kinase) that re-ligates cleaved tRNA fragments (Amitsur et al. 1987). Similar antiviral PrrC-like RNases may be a common defense mechanism in diverse microorganisms, since PrrC homologues are widely found in diverse bacterial species.

Additional examples of tRNA cleavage-based strategies to promote survival under adverse conditions are observed in other unicellular organisms. Suboptimal growing conditions stimulate competition between different microorganisms for limited resources. In order to survive, some bacteria and fungi secrete specific toxins (ribotoxins) into the surrounding environment to trigger irreversible growth arrest and/or death of non-self species. For example, certain strains of the dairy yeast *Kluyveromyces lactis* secrete a heterotrimeric protein toxin that inhibits growth of non-self-yeast species such as *Saccharomyces cerevisiae* (Jablonowski and Schaffrath 2007; Keppetipola et al. 2009; Lu et al. 2008). This killer toxin (zymogen) consists of α-, β-, and γ-subunits encoded on cytoplasmic episomes that are subject to cytoplasmic inheritance (Jablonowski and Schaffrath 2007). The α- and β-subunits of the secreted zymogen facilitate transport of the cytotoxic γ-subunit into the cytoplasm of target cells by interaction with the cell wall of susceptible yeast cells. In the cytoplasm, γ-toxin functions as an anticodon tRNase. The cleavage is highly specific to tRNAs possessing a unique 5-methoxycarbonylmethyl-2-thiouridine (mCm^5s^2U) residue at wobble position 34 of the anticodon loop. Recognition of this modification provides a molecular basis for γ-toxin-specific cleavage (Lu et al. 2008), as γ-toxin selectively cleaves $tRNA^{Glu(UUC)}$, $tRNA^{Lys(UUU)}$, and $tRNA^{Gln(UUG)}$ between positions 34 and 35, all of which contain mCm^5s^2U. As a consequence, γ-toxin depletes pools of functional $tRNA^{Glu}$, $tRNA^{Lys}$, and $tRNA^{Gln}$ to promote growth arrest of target cells and eliminate other yeast cells from competing for limited resources.

The bacterium *E. coli* uses a similar γ-toxin-like strategy to limit the propagation of competitor species. Some *E. coli* strains carry *Col* plasmids that encode microbicidal proteins called colicins (Cascales et al. 2007), a functionally diverse group of toxins. Typically, both the toxin and antitoxin ("immunity") genes are encoded in a single gene cluster. The antitoxin binds to colicin and inhibits its activity. Expression of colicin gene clusters is tightly regulated by the SOS response, a major prokaryotic stress response program activated by DNA damage (Cascales et al. 2007). Specific colicins (e.g., colicins D, E3, and E5) are ribonucleases/ribotoxins (Masaki and Ogawa 2002; Ng et al. 2010; Ogawa et al. 2006). Secreted colicins can translocate across the membrane of sensitive *E. coli* strains ("competitors"), which do not possess the immunity gene, resulting in colicin-induced translational repression and growth inhibition. While colicin E3 is a 16S ribosomal RNA-specific RNase (Ng et al. 2010), colicins D and E5 are tRNases targeting anticodon loops (Masaki and Ogawa 2002; Masaki et al. 1997; Ogawa et al. 2006). Colicin E3 binds to the A-site of the 70S ribosome and cleaves between nucleotides A1493 and G1494, causing translational arrest (Ng et al. 2010). Colicin D specifically targets isoacceptors of $tRNA^{Arg}$, while colicin E5 specifically cleaves tRNAs containing the modified nucleotide queuosine in the wobble position of the anticodon loop. Queuosine-containing substrates are $tRNA^{Tyr(QUA)}$, $tRNA^{His(QUG)}$, $tRNA^{Asn(QUU)}$, and $tRNA^{Asp(QUC)}$ (Masaki and Ogawa 2002; Masaki et al. 1997; Ogawa et al. 2006).

The use of selective tRNases that recognize specific modifications in specific tRNAs (Fig. 3) provides an example of a successful strategy used by unicellular organisms to outcompete/eliminate non-self species during unfavorable growth conditions (colicins, γ-toxin). In addition, a similar strategy that targets their own tRNAs is used by other bacteria to defend against the spread of viral infections (e.g., "altruistic suicide" by PrrC).

Other unicellular organisms also utilize tRNA cleavage-based strategies as part of their stress response program to alter cell metabolism and promote cell survival. The protozoan *Giardia lamblia* is an intestinal parasite that survives adverse environments by differentiating from a vegetative trophozoite to a dormant cyst (Adam 2001). During encystation, global tRNA cleavage in the anticodon loop of multiple tRNA species is observed (Li et al. 2008), which causes a limited (less than 20 % of tRNA are cleaved) decrease in global translation. This decrease is, however, sufficient to promote dramatic changes in gene expression and depress metabolism causing cells to enter the dormant stage of the protozoan life cycle. A similar tRNA cleavage phenomenon is described in the protozoa *Tetrahymena thermophila* (Lee and Collins 2005), the bacterium *Streptomyces coelicolor* (Haiser et al. 2008), and the fungi *Saccharomyces cerevisiae* (Thompson et al. 2008; Thompson and Parker 2009a) and *Aspergillus fumigatus* (Jochl et al. 2008) in response to stress and/or during differentiation to a resting state. With the exception of budding yeast, where RNase T2 acts as tRNase (Thompson and Parker 2009a), in all the abovementioned cases, the stress-induced tRNases have not been identified. The roles of chemical modifications in the anticodon loops of tRNA substrates are also unclear and await future investigations.

Although certain ribonucleases such as RNase T2 family members are widely distributed and found in viruses, bacteria, protozoans, fungi, plants, and animals (Deshpande and Shankar 2002; Luhtala and Parker 2010), the RNase A family members are vertebrate specific (Cho et al. 2005; Lander et al. 2001). All RNase A family members are small secreted proteins, with a range of ribonucleolytic activities and cellular functions. Within the RNase A superfamily, RNase 5 or angiogenin (ANG) is the only member found outside of mammalia (Cho et al. 2005)

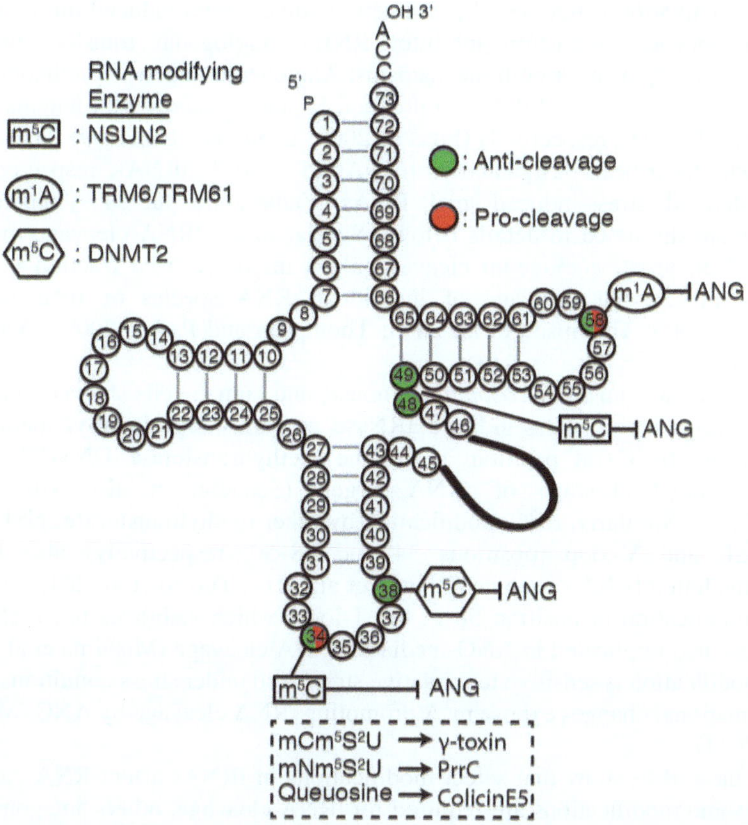

Fig. 3 tRNA modification and tRNA cleavage. Positions of the tRNA modifications known to impact tRNA cleavage are shown either in *red* (attract tRNase, promote tRNA cleavage) or *green* (antagonize actions of tRNase, inhibit tRNA cleavage). Note that the same position on tRNA (e.g., at the wobble base, position 34) can be either "pro-cleavage" (in the case of mCm^5S^2U, mNm^5S^2U, and queuosine modifications; see also *dotted box* for specific tRNases) or "anti-cleavage" (in the case of m^5C modification). The methylation of cytosines (m^5C) at positions 38 (by DNMT2) and/or 34, 48, and 49 (by NSUN2) inhibits tRNA cleavage by ANG. The m^1A modification at the position 58 (modification by TRM6/TRM61 methyltransferases) is required for proper tRNA folding. It is typically buried within the L-shaped structure of a tRNA but becomes exposed during oxidative stress. Exposure of m^1A may then promote a conformational change in tRNA that makes it more susceptible to ANG-mediated tRNA cleavage. Hypothetically, non-oxidized m^1A promotes tRNA stability and antagonizes ANG-mediated tRNA cleavage

and is a stress-induced tRNA-specific ribonuclease with unique cellular functions (Fett et al. 1985).

ANG expression is stress-regulated and increased by hypoxia and acute inflammation (Hartmann et al. 1999; Nakamura et al. 2006; Olson et al. 1998). Different stresses including heat shock, UV irradiation, and oxidative stress activate angiogenin to cleave tRNA (Fu et al. 2009; Yamasaki et al. 2009). The mechanism of ANG activation is not clear but it likely involves stress-induced inactivation of the ribonuclease/angiogenin inhibitor RNH1, angiogenin translocation from nucleus to cytoplasm, or both mechanisms. Angiogenin targets the anticodon loop of mature fully processed tRNA to produce 5′- and 3′-halves with lengths of ~30 and 40 nucleotides, respectively (Fu et al. 2009; Yamasaki et al. 2009). These tRNA fragments have been designated as tiRNAs (5′- and 3′-tiRNAs, respectively) or tRNA-derived, stress-induced small RNAs (Yamasaki et al. 2009; functions of tiRNAs are discussed in details below). Similar to the tRNA cleavage mediated by Rny1 in yeast, angiogenin cleaves only a minor (1–5 %) fraction of tRNA without preferential cleavage of individual tRNA species or their isoforms (Fu et al. 2009; Thompson et al. 2008; Thompson and Parker 2009b; Yamasaki et al. 2009).

Recent data from fly (*Drosophila*), mouse, and human cells showed that tRNA modifications affect stress-induced tRNase activities. Specifically, methylation of cytosine (m^5C) at position 38 by the methyltransferase DNMT2 inhibits ANG-mediated cleavage of tRNA targets (Schaefer et al. 2010; Tuorto et al. 2012). Similarly, m^5C modification by other methyltransferase, NSUN2, in the ASL and V-loop (positions 34 and 48–50, respectively) also inhibits ANG-mediated tRNA cleavage (Blanco et al. 2014; Tuorto et al. 2012). Finally, m^1A modification at position 58 in the T-loop, which stabilizes the L-shape of tRNA, is also implicated in ANG-mediated tRNA cleavage (Mishima et al. 2014). This modification is sensitive to oxidative stress, and under stress conditions, tRNA conformational changes expose m^1A promoting tRNA cleavage by ANG (Mishima et al. 2014).

All these data show that select modifications in tRNAs affect RNA cleavage. While some modifications are required for tRNA cleavage, others antagonize this process (summarized in Fig. 3). Further investigations are required to decipher molecular details connecting specific modifications to the regulation of an endonuclease-induced tRNA cleavage.

3 tRNA-Derived Fragments

Identification of novel ncRNAs derived from tRNAs, known as tRNA-derived fragments, has recently gained significant attention (Anderson and Ivanov 2014; Durdevic and Schaefer 2013; Gebetsberger and Polacek 2013; Kirchner and Ignatova 2015; Martens-Uzunova et al. 2013; Mleczko et al. 2014; Shigematsu and Kirino 2015; Sobala and Hutvagner 2011). Some of these fragments are derived from precursor tRNA molecules; others from mature cytoplasmic tRNAs. Certain fragments are constitutively produced, whereas others are only produced under specific conditions. Here, we will review the biogenesis and functions of tRNA fragments and discuss their potential roles in the pathobiology of human diseases.

3.1 Diversity of tRNA-Derived Fragments

3.1.1 tRNA Halves and tiRNAs

tRNA halves are produced by specific cleavage in the anticodon loop forming 30–35 nucleotide 5′-tRNA and 40–50 nucleotide 3′-tRNA halves (Fig. 4). Under optimal conditions, small quantities of these fragments are present in cells (Kawaji et al. 2008; Saikia et al. 2012; Schutz et al. 2010). It is likely that the majority of these are splicing intermediates from intron-containing tRNAs. Knockout of Clp1, an RNA kinase necessary for tRNA maturation, greatly increases the numbers of these fragments from intron-containing tRNAs (Hanada et al. 2013; Popow et al. 2011). It is unknown if these fragments have biologic activity. In contrast to tRNA halves formed as splicing intermediates, tiRNA expression is induced in response to cellular stress (e.g., oxidative stress, heat shock, or UV irradiation) (Emara et al. 2010; Fu et al. 2009; Ivanov et al. 2011a; Saikia et al. 2012; Yamasaki

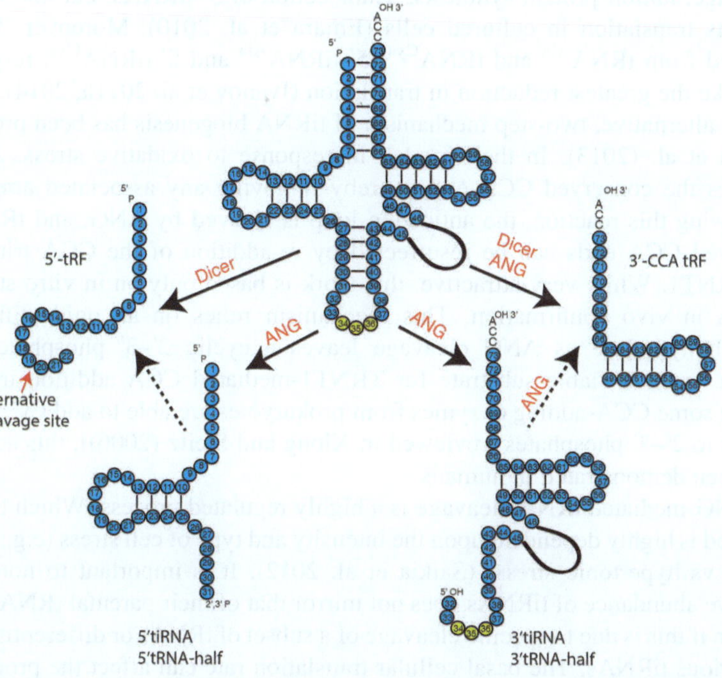

Fig. 4 Processing of mature tRNA into tRNA fragments (tRFs). Processing of mature cytoplasmic tRNA. Mature tRNA can be cleaved in the anticodon loop by angiogenin (ANG) to produce 5′- and 3′-tRNA halves (or 5′- and 3′-tiRNAs, respectively). Dicer-dependent cleavage of tRNA in the D-arm of tRNA results in the production of small tRNA fragments, 5′-tRFs. Similarly, cleavage in the T-arm by Dicer or ANG results in the production of tRFs containing CCA at their 3′-ends (so-called 3′-CCA-tRFs). Both 5′-tRFs and 3′CCA-tRFs may also be processed from 5′- and 3′-tRNA halves, respectively (*dashed arrows*). Alternative cleavage site to produce shorter forms of 5′-tRFs is also shown (*red arrow*)

et al. 2009); they are produced from mature tRNAs by ANG-mediated cleavage within the anticodon loop. Despite the differences in their biogenesis, cytoplasmic tRNA halves and tiRNAs are probably the same molecules and these terms can be used interchangeably.

ANG is a predominantly nuclear protein; however, it is also found in the cytoplasm in complex with RNH1 [reviewed in Li and Hu (2012)]. During the stress response, ANG translocates from the nucleus to the cytoplasm, dissociates from RNH1, and cleaves cytoplasmic tRNA into 5′- and 3′-tiRNAs (Yamasaki et al. 2009) (Figs. 3 and 4). It is tempting to speculate that cleavage of mature tRNAs would have drastic antagonistic effects on the rate of cellular translation. However, cleavage of tRNAs by ANG does not elicit this effect because only a minor fraction (<5 %) of the tRNA pool is cleaved (Fu et al. 2009; Saikia et al. 2012; Yamasaki et al. 2009). When cells are treated with recombinant ANG in the absence of cellular stresses, global translation is reduced by 10–15 %. Strikingly, protein synthesis is similarly reduced when cells are transfected with ANG-induced tiRNAs, suggesting that the tRNA fragments, rather than the tRNA cleavage, inhibit protein synthesis. Transfection of 5′-tiRNAs, but not 3′-tiRNAs, inhibits translation in cultured cells (Emara et al. 2010). Moreover, 5′-tiRNAs derived from $tRNA^{Ala}$ and $tRNA^{Cys}$ (5′-$tiRNA^{Ala}$ and 5′-$tiRNA^{Cys}$, respectively) provoke the greatest reduction in translation (Ivanov et al. 2011a, 2014).

An alternative, two-step mechanism of tiRNA biogenesis has been proposed by Czech et al. (2013). In their model, in response to oxidative stress, ANG first cleaves the conserved CCA-end, thereby removing any associated amino acids. Following this reaction, the anticodon loop is cleaved by ANG, and tRNAs with removed CCA ends can be resurrected by re-addition of the CCA trinucleotide by TRNT1. While very attractive, this work is based only on in vitro studies and awaits in vivo confirmation. This mechanism relies on an unidentified cyclic phosphohydrolase as ANG cleavage leaves a cyclic 2′–3′ phosphate (Fig. 4) which is not a viable substrate for TRNT1-mediated CCA addition in humans. While some CCA-adding enzymes from prokaryotes are able to add CCA trinucleotides to 2′–3′ phosphates [reviewed in Xiong and Steitz (2006)], this activity has not been demonstrated in humans.

ANG-mediated tRNA cleavage is a highly regulated process. Which tRNAs are cleaved is highly dependent upon the intensity and type of cell stress (e.g., oxidative stress vs hypertonic stress) (Saikia et al. 2012). It is important to note that the relative abundance of tiRNAs does not mirror that of their parental tRNAs. It is not known if this is due to specific cleavage of a subset of tRNAs or differential stability of various tiRNAs. The basal cellular translation rate can affect the production of tiRNAs. When cells are treated with inhibitors of translation (e.g., cycloheximide), cleavage of tRNAs is reduced together with translation, suggesting that ANG has better access to tRNA when protein synthesis is active (Saikia et al. 2012). In agreement with this hypothesis, ANG is found in polysomes (the pool of actively translating ribosomes) (Pizzo et al. 2013).

A subset of bioactive tiRNAs inhibits translation by interfering with the assembly of the eIF4F cap-binding complex that is required for canonical translation initiation. eIF4F is a multi-subunit complex composed of eIF4E, eIF4A, and eIF4G (Jackson et al. 2010; Pestova et al. 2001; Sonenberg and Hinnebusch 2009). In addition to eIF4F, eIF4E is also found in a complex with eIF4E-BP. While formation of eIF4F on m^7GTP-capped mRNA promotes translation, assembly of eIF4E:eIF4E-BP on capped mRNA prevents translation. Bioactive tiRNAs specifically target eIF4F but not eIF4E:eIF4E-BP. A terminal oligoguanine (TOG) motif located at the 5′ end of 5′-tiRNA^{Ala} and 5′-tiRNA^{Cys} is required to displace the eIF4F complex and inhibit translation (Ivanov et al. 2011a). The addition of a TOG motif to an inactive 5′-tiRNA, such as 5′-tiRNA^{Met}, endows eIF4F displacing properties. Inhibition of translation initiation by 5′-tiRNAs induces the assembly of stress granules (SGs) (Emara et al. 2010; Ivanov et al. 2011a), dynamic cytoplasmic RNA foci that possess adaptive and pro-survival functions. SGs contain mRNA with stalled 48S pre-initiation complexes bound by specific RNA-binding proteins and signaling molecules (Anderson and Kedersha 2008, 2009; Ivanov et al. 2011b; Kedersha et al. 2013). SGs are typically formed upon activation of stress-sensing serine/threonine kinases that phosphorylate serine-51 of eIF2α (Donnelly et al. 2013), a molecule that regulates the integrated stress response to coordinate cell adaptation and survival under stress conditions [reviewed in (Baird and Wek 2012; Dey et al. 2010; Kroemer et al. 2010)]. Inhibition of translation through phosphorylation of eIF2α results in global translational arrest (Anderson et al. 2015; Ivanov and Anderson 2013; Kedersha et al. 2013). In contrast, tiRNAs assemble SGs in a phospho-eIF2α-independent manner (Emara et al. 2010; Ivanov et al. 2011a) suggesting the existence of an alternative route of signaling and SG assembly. Since ANG-induced tiRNA formation or transfection of tiRNAs only reduces global translation 10–15 %, it is likely that tiRNAs cause translational reprogramming rather than total translation arrest.

ANG-induced tiRNAs have been shown to inhibit hyperosmotic apoptosis in stressed cells by directly binding to cytochrome *c* (Cyt *c*) (Saikia et al. 2014). During stress, Cyt *c* is released from the mitochondria thereby promoting apoptosome formation and subsequent caspase activation. However, upon binding to tiRNAs, Cyt *c* is unable to nucleate the apoptosome and initiate programmed cell death. Importantly, sequencing of tiRNAs bound to Cyt *c* suggests that only a subpopulation of tiRNAs (both 5′-tiRNAs and 3′-tiRNAs, about 20 different species) is highly enriched in these complexes. Previous studies have reported that full-size tRNAs bind Cyt *c* to inhibit apoptosome formation in vitro (Mei et al. 2010; Suryanarayana et al. 2012). However, under stress conditions Cyt *c* preferentially binds to tiRNAs and not tRNAs in vivo (Saikia et al. 2014). Further studies are necessary to characterize the composition of Cyt *c*:tiRNA complexes, sequence/structural determinants of tiRNAs contributing to complex formation, and also the specificity of stresses triggering formation of these complexes.

The tRNA cleavage field is still in its infancy (Anderson and Ivanov 2014). The ability of a subset of tiRNAs to inhibit cap-dependent translation and another subset to inhibit apoptosis reveals the potential diversity of functions that these small RNAs possess. The molecular roles for the majority of 5′- and 3′-tiRNAs as well as their localization, relative abundance, and stability are yet to been determined.

3.1.2 Diverse tRNA Fragments (tRFs)

A second pervasive class of tRNA fragments has been termed tRFs (tRNA fragments), which are fragments of tRNA or pre-tRNA that are typically shorter than tiRNAs/tRNA halves (~12–30 nucleotides) (Fig. 4). Within tRFs, there are several subclasses of tRFs, which are often difficult to classify. Here, we will discuss them in terms of their biogenesis and whether they are derived from mature tRNA or pre-tRNA.

In mammalian cells, mature tRNAs can give rise to 5′-tRFs and 3′-CCA-tRFs that are produced by cleavage of tRNA in the D-loop and in/near the T-loop, respectively (Fig. 4). 5′-tRFs are 19–21 nt fragments that are formed by cleavage in the D-arm (Cole et al. 2009; Lee et al. 2009). Like tiRNAs, the abundance of particular species of 5′-tRFs does not correlate with the parental tRNA. In fact, the abundance of selected 5′-tRF species (e.g., derived from $tRNAs^{Gln/Lys/Val/Arg}$) in HeLa cells is comparable to the levels of abundant microRNAs such as miR-21 or members of the let-7 family (Cole et al. 2009). Similarly, the abundance of selected 5′-tRFs (e.g., derived from $tRNAs^{Glu/Ser/Leu/Gln}$) in prostate cancer cell lines is greater than 90 % of the individual microRNAs in these cells (Lee et al. 2009). It is not clear if these tRNAs are specifically targeted for RNA cleavage or if some cleavage products are preferentially stabilized. In contrast to tiRNAs formed by ANG, Dicer is responsible for the RNA cleavage of 5′-tRFs (Cole et al. 2009). Dicer is a type III endoribonuclease whose best characterized function is in the maturation of miRNAs (Ghildiyal and Zamore 2009). DROSHA first processes primary miRNAs into ~155 nucleotide pre-miRNAs. Dicer acts upon this substrate to produce 20–25 nucleotide mature miRNAs. Structurally, pre-miRNAs fold into a single stem-loop that is required for processing into mature miRNAs. It is unknown how Dicer recognizes tRNAs to produce 5′-tRFs. Mature tRNAs range from 70 to 90 nucleotides, with the majority being 72 nucleotides long, significantly shorter that canonical pre-miRNAs. There is significant base pairing within mature tRNAs, but they contain four short stem-loops rather than the single long stem-loop found in primary miRNAs. Alternatively, 5′-tRFs could be processed from larger tRNA intermediates (e.g., tiRNAs). To further complicate the poorly understood mechanism of 5′-tRF biogenesis, it appears that there also may be a Dicer-independent mechanism that is less understood than the Dicer-dependent mechanism (Li et al. 2012).

5′-tRFs are localized in the cytoplasm and associate, although poorly, with Argonaute (Ago 1–4) proteins that are central to RNA-induced silencing complex (RISC) function (Pratt and MacRae 2009). It is unknown whether 5′-tRFs play a direct role in RNA interference (RNAi), but recently a role in translation inhibition was proposed (Sobala and Hutvagner 2013). It is reported that selected 5′-tRFs (3 out of 4 tested) inhibit translation of mRNA reporters in vitro (Sobala and Hutvagner 2013). The molecular details of this inhibition remain to be determined but they are proposed to affect translation elongation (Sobala and Hutvagner 2013), in contrast to 5′-tiRNAs, which inhibit translation initiation (Emara et al. 2010;

Ivanov et al. 2011a). The universally conserved G18–G19 dinucleotide at the 3′-end of 5′-tRF is absolutely required for translation inhibition. Interestingly, this same mutation in 5′-tiRNA^{Ala} affects 5′-tiRNA-mediated repression of translation approximately two-fold (Ivanov et al. 2011a). Similar 5′-tRFs are found in other domains of life. In archaea, a stress-induced 26-nt 5′-tRF derived from tRNA^{Val} similarly inhibits translation elongation. This fragment directly binds 30S small ribosomal subunit where it inhibits the peptidyl transferase activity of the ribosome (Gebetsberger et al. 2012). Thus, inhibition of translation by 5′-tRNA fragments appears to be an evolutionarily conserved phenomenon.

The 3′-CCA-tRFs are formed by cleavage in the T-loop of mature tRNAs and are characterized by the presence of a universal CCA trinucleotide at their 3′-end (Babiarz et al. 2008; Haussecker et al. 2010; Lee et al. 2009; Li et al. 2012; Maute et al. 2013; Yeung et al. 2009) (Fig. 4). Similar to 5′-tRFs, Lee et al. (2009) reported that the abundance of 3′-CCA-tRFs is comparable to that of microRNAs. The precise mechanism of 3′-CCA-tRF biogenesis is less clear than that of 5′-tRFs; like 5′-tRFs, it has been proposed to proceed through both Dicer-dependent and Dicer-independent mechanisms (Babiarz et al. 2008; Haussecker et al. 2010; Maute et al. 2013; Yeung et al. 2009). For Dicer-independent mechanisms, it has been proposed that ANG cleaves within the T-arm of tRNAs (Li et al. 2012). Strictly speaking, 3′-CCA-tRFs could be formed directly from mature tRNAs or proceed through a tiRNA intermediate. Like 5′-tRFs, selected 3′-CCA-tRFs associate with Argonautes and possess RNAi activities (Haussecker et al. 2010; Maute et al. 2013).

Mammalian pre-tRNAs may also be used as substrates for tRF formation. tRFs generated from these precursors may contain 5′-leader, 3′-trailer, or intron-derived sequences (Fig. 1). The best-studied pre-tRNA-derived fragments belong to so-called 3′U-tRFs (19–25 nt). Pre-tRNAs include a 3′-trailer that includes a stretch of 2–6 uridines at its 3′-end to facilitate RNA polymerase III termination (Haussecker et al. 2010; Lee et al. 2009; Liao et al. 2010). Although biogenesis of 3′U-tRFs is natural and depends on the RNase Z-dependent cleavage of pre-tRNA (Lee et al. 2009), one study proposed Dicer-dependent processing (Babiarz et al. 2008). Surprisingly, 3′U-tRFs are relatively stable molecules that are concentrated in the cytoplasm (Lee et al. 2009; Liao et al. 2010). These fragments have been reported to exist in high levels in human embryonic stem cells and embryoid bodies (Morin et al. 2008). Moreover, the biogenesis of 3′U-tRFs is proposed to occur in the cytoplasm from pre-tRNA by a cytoplasmic RNase Z homologue ELAC2 and might be regulated by cell proliferation (Lee et al. 2009). It is not clear how pre-tRNAs are exported from the nucleus to the cytoplasm to produce 3′U-tRFs. In the cytoplasm, 3′U-tRFs bind to Argonaute proteins (preferentially Ago3 and Ago4) and compete with cellular microRNAs and siRNAs for incorporation into RISCs (Haussecker et al. 2010).

Mammalian intron-containing pre-tRNAs also produce a distinct subset of relatively stable tRFs (Hanada et al. 2013; Karaca et al. 2014; Schaffer et al. 2014) (Fig. 1). Some of these tRFs are splicing products of pre-tRNAs that lack leader and trailer sequences. Although these fragments resemble tiRNAs, they are concentrated in the nucleus rather than the cytoplasm. Much larger tRFs containing 5′-leader sequences followed by the entire 5′-exon of pre-tRNA (5′-leader-exon-tRF, e.g., produced from pre-tRNA$^{Tyr/Arg}$) are predominantly nuclear and induced by oxidative stress in an ANG-independent manner (Hanada et al. 2013). In some pathological conditions (discussed below), accumulation of relatively abundant linear introns derived from pre-tRNA is also observed (Karaca et al. 2014). Splicing of intron-containing tRNAs can generate stable circular RNA species called tricRNAs (tRNA intronic circular RNAs) (Lu et al. 2015). Circular RNAs are generated from the introns of pre-mRNAs. However, this process utilizes Alu repeats and the nuclear spliceosome. As tRNAs are spliced in the cytoplasm by the TSEN complex, an alternative mechanism must be utilized. Circular RNAs derived from pre-mRNA introns have been proposed to act as a sink for miRNAs. It is unknown if tricRNAs are able to function in a similar manner. Additional tRFs (e.g., derived from pre-tRNA$^{Ile/Tyr}$) containing 5′-trailer sequences followed by a partial exon (5′-leader-partial-tRF) or 3′-exon followed by 3′-trailer sequences (3′-partial-exon-trailer) have been reported (Karaca et al. 2014). However, these fragments are less abundant than intron-containing and 5′-leader-exon-tRFs, and their functions are unknown (Anderson and Ivanov 2014).

3.2 *tRNA-Derived Fragments in Human Health and Disease*

3.2.1 tRNA Fragments and Cellular Stress

We and other groups identified that when cells are stressed, ANG cleaves tRNA to produce tiRNAs (Fu et al. 2009; Mishima et al. 2014; Yamasaki et al. 2009). Mechanistically, oxidative stress alters tRNA conformation allowing angiogenin accessibility. Taking advantage of this conformational shift, Mishima et al. generated an antibody specific to a modified nucleotide, 1-methyladenosine (m^1A) (Fig. 3), which is typically buried within the L-shaped structure of a tRNA but becomes exposed during oxidative stress. This modification-specific antibody detects unfolded tRNA, 3′-tiRNAs, smaller tRFs, and free m^1A and can be used to quantify tRNA fragments by ELISA and to visualize oxidative stress by immunofluorescence. The m^1A antibody may prove to be a clinically useful tool to detect oxidative damage by sensing changes in tRNA biology as shown in human and mouse models of renal ischemia/reperfusion (I/R) injury and cisplatin-mediated nephrotoxicity. It will be interesting to see whether this antibody will be clinically applicable.

While Mishima et al. have shown that detection is possible, whether it is feasible to accurately quantify circulating tRFs using this antibody is unknown. It is known that tRNA-derived fragments can be found in circulating extracellular vesicles known as exosomes (Mishima et al. 2014). Additionally, 5′-tiRNAs were found to associate with 100–300 kDa protein–RNA complexes in serum. Levels of

5′-tiRNAs from specific tRNA isoacceptors are altered in response to physiological changes such as aging and caloric intake (Dhahbi et al. 2013).

3.2.2 tRNA Fragments and Cancer

Cancer is characterized by abnormal and generally uncontrolled cell growth. This aberrant growth is often marked by increases in translation and the corresponding machinery. Early studies indicated that tumor tissues turn over tRNA quicker than noncancerous tissue and that cancer patients and tumor-burdened animals excrete modified nucleosides (Borek et al. 1977), which are highly enriched in tRNA. tRNA levels are increased in several cancers or cancer cell lines compared to corresponding normal controls, including breast and multiple myeloma (Pavon-Eternod et al. 2009; Zhou et al. 2009). Interestingly, genome-wide tRNA analysis indicates increases are specific to certain isoacceptors (Pavon-Eternod et al. 2009). Moreover, overexpression of the initiator tRNA is sufficient to promote cell proliferation and alter cell metabolism by reprogramming gene expression (Pavon-Eternod et al. 2013). It is unknown whether isoacceptor-specific overexpression represents a typical cancer pattern or whether such increases are cancer subtype specific. Whether tRNA overexpression or isoacceptor-specific overexpression is required for cancer growth is also unclear.

ANG was initially identified based on its ability to stimulate blood vessel formation in the chicken embryo chorioallantoic membrane angiogenesis assay (Fett et al. 1985). Since tumors are known to exist in a hypoxic microenvironment, the increase in both expression and secretion of angiogenin helps tumors adapt to such conditions by stimulating new blood vessel formation to promote further tumor growth. ANG is directly involved in tumor growth, and its expression and secretion are upregulated in many types of cancers (Olson et al. 1994, 1995; Tello-Montoliu et al. 2006), and its ability to stimulate angiogenesis requires its RNase activity (Tsuji et al. 2005). Secreted ANG is readily taken up by target cells where it stimulates cell migration, triggers second messenger production, promotes cell adhesion, and induces cell proliferation (Gao and Xu 2008). In a striking contrast to the cancer-associated ANG overexpression, ANG "loss-of-function" mutations that affect its RNase activity are found in patients with amyotrophic lateral sclerosis (ALS), a fatal neurodegenerative disease (Greenway et al. 2004, 2006).

While ANG is currently thought to promote proliferation and tumor angiogenesis via stimulation of rRNA transcription (Xu et al. 2002), the only known RNA cleavage target of ANG is tRNAs. Pharmacological targeting of ANG RNase activity could be clinically useful as it prevents tumor formation in mouse xenograft tumor models (Kao et al. 2002). While the role of ANG is currently thought to be due to stimulating rRNA transcription, it is unknown whether ANG-produced tiRNAs also play a role in promoting tumor growth.

Both tRNAs and ANG are often overexpressed in cancers yet it is unknown whether this leads to an increase in tRFs or tiRNAs. Several reports indicate that tRFs are observed in cancers, including prostate cancer and B-cell lymphoma

(Li et al. 2012; Martens-Uzunova et al. 2012; Maute et al. 2013). Small RNA transcriptome analysis of prostate cancers indicates potentially differential processing of tRNA between nonmetastatic and metastatic samples, as tRFs that are 18-nt and 27-nt in length, respectively, were identified (Martens-Uzunova et al. 2012). As this is an expanding field, it will be interesting to determine whether all metastatic tumors have a similar tRF profile and whether tRFs can be used as biomarkers.

Due to the vast number of potential tRFs, much work is needed to decipher the functions of the fragments. The few cancer-related studies that have identified functions of specific tRFs are summarized below:

1. 3′-CCA-tRF (or CU1276), derived from $tRNA^{Gly}$, is produced in a Dicer-dependent manner and acts as a miRNA in mature B cells. It is downregulated in B-cell lymphomas to promote malignant cell growth by altering cell proliferation and DNA damage by regulating levels of the DNA replication and repair protein RPA1 (Maute et al. 2013).
2. 3′U-tRF (or tRF-1001), derived from pre-$tRNA^{Ser}$, is produced by cleavage of cytoplasmic pre-tRNA by ELAC2, a prostate cancer susceptibility gene (Tavtigian et al. 2001). tRF-1001 is highly expressed in different cancer cell lines and its levels correlate with cell proliferation and viability of prostate cancer cells (Lee et al. 2009), yet molecular details are unknown.
3. 3′-CCA-tRFs (derived from $tRNA^{His(GTG)}$ and $tRNA^{Leu(CAG)}$) were identified in B-cell lymphomas and are produced in a Dicer-independent manner. Mechanistically, these 3′-CCA-tRFs associate with Ago2 and can direct Ago2-mediated cleavage of an mRNA reporter (Li et al. 2012). This could potentially have a role in downregulating retroviral movement within the genome as retroviral elements and retroviruses (such as HIV) utilize tRNAs to initiate retroviral genome replication (Das et al. 1995). This appears to be likely for HIV as the 3′-CCA-tRF 18-nt fragment derived from $tRNA^{Lys}$ is enriched in HIV-infected T cells and is complimentary to the site necessary to initiate retroviral replication (Yeung et al. 2009).

These studies have found that tRFs may have completely different functions indicating a clear need for further investigation.

3.2.3 tRNA Fragments and Neurological Disorders

Although not exclusively expressed in neuronal tissues and thought to have an essential function within all cells, mutations in tRNA metabolism enzymes is a reoccurring theme in neuro-disorders. The exact mechanistic details remain unknown but the studies thus far provide an intriguing basis for further investigation.

Mutations in the ribonuclease ANG have been linked to the neurodegenerative disorders, Parkinson's disease and ALS (Greenway et al. 2006; van Es et al. 2011). Many of the identified mutations decrease ribonuclease activity (Wu et al. 2007), and while little work has been done to characterize the function of ANG in neurons, the loss of ANG activity could promote motor neuron death. Indeed, data from our lab suggests that ANG and selected 5′-tiRNAs assemble unique G-quadruplex structures (Ivanov et al. 2014) that are neuroprotective molecules. Delivery of G-quadruplex-forming 5′-tiRNAs and their DNA analogues protect motor neuron from stress-induced injuries and promote their survival (Ivanov et al. 2014).

Mutations in the tRNA methyltransferase NSUN2 are linked to intellectual disorders, syndromic form of intellectual disability (ID), and a Dubowitz-like syndrome in humans (Abbasi-Moheb et al. 2012; Khan et al. 2012; Martinez et al. 2012). NSUN2 methylates a subset of tRNAs (Asp, Glu, Gly, His, Lys, and Val) at cytosine C34 within the anticodon loop and C48/49/50 where the variable loop meets the T-arm. NSUN2 methylation blocks ANG-mediated cleavage and therefore loss of NSUN2 promotes accumulation of 5′-tRNA fragments. Phenotypically, NSUN2 loss leads to decreased cell size and apoptosis of cortical, hippocampal, and striatal neurons (Blanco et al. 2014). Why a mutation in an enzyme that causes a tRNA modification displays a specific neuronal phenotype is unknown, yet there appears to be a clear connection between NSUN2, ANG-mediated cleavage, and neuronal cell death. We hypothesize that the loss of NSUN2-mediated methylation in target tRNAs leads to the excessive ANG-mediated cleavage and increased accumulation of tiRNAs that may affect neuron survival.

A mutation (R140A) in the RNA kinase CLP1 is linked to pontocerebellar hypoplasia (PCH), a group of inherited neurodegenerative disorders. Wild-type CLP1 plays a role in tRNA splicing and the R140A mutation causes dysregulation of tRNA levels, accumulation of unspliced pre-tRNAs (Schaffer et al. 2014), and linear introns (Karaca et al. 2014). Animals (zebrafish and mice) lacking CLP1 or CLP1 kinase activity display similar phenotypes (Hanada et al. 2013; Karaca et al. 2014; Schaffer et al. 2014) and accumulate 5′-leader-exon-tRFs (Hanada et al. 2013). Both CLP1 kinase-dead mice and transfection of a natural CLP1 substrate, 3′-exon of pre-tRNA^{Tyr}, cause increased sensitivity to oxidative stress and cell death (Hanada et al. 2013; Schaffer et al. 2014).

4 Concluding Remarks and Perspectives

In contrast to regulatory or structural information encoded by the primary sequence, RNA modifications expand this information to new horizons. These modifications can greatly affect RNA stability, conformation, and localization; modulate RNA–protein interactions; and change/fine-tune RNA functions. Because RNA modifications require significant cellular energy, their biosynthesis and introduction into RNA must be responsive to changes in the environment and the nutritional/metabolic cellular status. Since many genetic alterations in RNA modification enzymes are associated with diverse pathological conditions and disease states, we suggest that such alterations may lead to aberrant RNA modification changes contributing to the development of human disease.

tRNAs are the most heavily modified types of cellular RNA. Emerging concepts such as tRNA modification-dependent translation reprogramming and/or tRNA fragment-specific regulation of gene expression suggest that the canonical role of tRNA, as an adaptor molecule that helps ribosome to synthesize proteins, is oversimplified. Taking into consideration the large number of tRNA genes [the human genome contains over 500 genes (Abe et al. 2014; Chan and Lowe 2009; Parisien et al. 2013)] and the number of possible modifications (80–90 modified nucleosides are found in tRNAs), we predict that additional functions of tRNAs will be discovered. We propose that tRNA diversity provides substrates for diverse classes of tRNA-derived fragments that have important effects on cell physiology and pathophysiology.

Moving forward, it is clear that we have to overcome a few obstacles to achieve further progress in this area. A major limitation in the field of tRNA modifications is the technical challenges inherent to sequencing the highly modified and structured tRNA molecules to identify the residues, which are actually modified. While several thousand tRNA genes are available from various genomes, only approximately 600–700 tRNAs were directly sequenced. The same is true for tRNA-derived fragments; their cellular repertoire is only partially determined. The recent advances in tRNA sequencing (tRNA-seq) will be instrumental to overcome such difficulties (Cozen et al. 2015; Zheng et al. 2015). We also need to generate a wide set of the high-affinity reagents (antibodies, small chemical molecules) to monitor and capture specific tRNA modifications.

In conclusion, we are looking forward to discovering the functions of tRNA-derived fragments, the function of different tRNA modifications, and the roles of tRNA modifications in regulating tRNA cleavage. Progress in this area of biology and medicine will provide valuable insights into human physiology and pathophysiology.

References

Abbasi-Moheb L, Mertel S, Gonsior M et al (2012) Mutations in NSUN2 cause autosomal-recessive intellectual disability. Am J Hum Genet 90:847–855

Abe T, Inokuchi H, Yamada Y et al (2014) tRNADB-CE: tRNA gene database well-timed in the era of big sequence data. Front Genet 5:114

Adam RD (2001) Biology of *Giardia lamblia*. Clin Microbiol Rev 14:447–475

Alexandrov A, Chernyakov I, Gu W et al (2006) Rapid tRNA decay can result from lack of nonessential modifications. Mol Cell 21:87–96

Amitsur M, Levitz R, Kaufmann G (1987) Bacteriophage T4 anticodon nuclease, polynucleotide kinase and RNA ligase reprocess the host lysine tRNA. EMBO J 6:2499–2503

Amitsur M, Morad I, Kaufmann G (1989) In vitro reconstitution of anticodon nuclease from components encoded by phage T4 and Escherichia coli CTr5X. EMBO J 8:2411–2415

Amitsur M, Morad I, Chapman-Shimshoni D et al (1992) HSD restriction-modification proteins partake in latent anticodon nuclease. EMBO J 11:3129–3134

Amitsur M, Benjamin S, Rosner R et al (2003) Bacteriophage T4-encoded Stp can be replaced as activator of anticodon nuclease by a normal host cell metabolite. Mol Microbiol 50:129–143

Anantharaman V, Koonin EV, Aravind L (2002) Comparative genomics and evolution of proteins involved in RNA metabolism. Nucleic Acids Res 30:1427–1464

Anderson P, Ivanov P (2014) tRNA fragments in human health and disease. FEBS Lett 588:4297–4304

Anderson P, Kedersha N (2008) Stress granules: the Tao of RNA triage. Trends Biochem Sci 33:141–150

Anderson P, Kedersha N (2009) Stress granules. Curr Biol 19:R397–398

Anderson P, Kedersha N, Ivanov P (2015) Stress granules, P-bodies and cancer. Biochim Biophys Acta 1849:861–870

Babiarz JE, Ruby JG, Wang Y et al (2008) Mouse ES cells express endogenous shRNAs, siRNAs, and other microprocessor-independent, Dicer-dependent small RNAs. Genes Dev 22:2773–2785

Baird TD, Wek RC (2012) Eukaryotic initiation factor 2 phosphorylation and translational control in metabolism. Adv Nutr 3:307–321

Begley U, Dyavaiah M, Patil A et al (2007) Trm9-catalyzed tRNA modifications link translation to the DNA damage response. Mol Cell 28:860–870

Blanco S, Dietmann S, Flores JV et al (2014) Aberrant methylation of tRNAs links cellular stress to neuro-developmental disorders. EMBO J 33:202–2039

Borek E, Baliga BS, Gehrke CW et al (1977) High turnover rate of transfer RNA in tumor tissue. Cancer Res 37:3362–3366

Cai WM, Chionh YH, Hia F et al (2015) A platform for discovery and quantification of modified ribonucleosides in RNA: application to stress-induced reprogramming of tRNA modifications. Methods Enzymol 560:29–71

Cascales E, Buchanan SK, Duche D et al (2007) Colicin biology. Microbiol Mol Biol Rev 71:158–229

Chakravarty AK, Subbotin R, Chait BT et al (2012) RNA ligase RtcB splices 3′-phosphate and 5′-OH ends via covalent RtcB-(histidinyl)-GMP and polynucleotide-(3′)pp(5′)G intermediates. Proc Natl Acad Sci U S A 109:6072–6077

Chan PP, Lowe TM (2009) GtRNAdb: a database of transfer RNA genes detected in genomic sequence. Nucleic Acids Res 37:D93–97

Chan CT, Pang YL, Deng W et al (2012) Reprogramming of tRNA modifications controls the oxidative stress response by codon-biased translation of proteins. Nat Commun 3:37

Cho S, Beintema JJ, Zhang J (2005) The ribonuclease A superfamily of mammals and birds: identifying new members and tracing evolutionary histories. Genomics 85:08–220

Cole C, Sobala A, Lu C et al (2009) Filtering of deep sequencing data reveals the existence of abundant Dicer-dependent small RNAs derived from tRNAs. RNA 15:2147–2160

Cozen AE, Quartley E, Holmes AD et al (2015) ARM-seq: AlkB-facilitated RNA methylation sequencing reveals a complex landscape of modified tRNA fragments. Nat Methods 12:879–884

Czech A, Wende S, Morl M et al (2013) Reversible and rapid transfer-RNA deactivation as a mechanism of translational repression in stress. PLoS Genet 9, e1003767

Das AT, Klaver B, Berkhout B (1995) Reduced replication of human immunodeficiency virus type 1 mutants that use reverse transcription primers other than the natural tRNA(3Lys). J Virol 69:3090–3097

Desai KK, Cheng CL, Bingman CA et al (2014) A tRNA splicing operon: Archease endows RtcB with dual GTP/ATP cofactor specificity and accelerates RNA ligation. Nucleic Acids Res 42:3931–3942

Deshpande RA, Shankar V (2002) Ribonucleases from T2 family. Crit Rev Microbiol 28:79–122

Dey S, Baird TD, Zhou D et al (2010) Both transcriptional regulation and translational control of ATF4 are central to the integrated stress response. J Biol Chem 285:33165–33174

Dhahbi JM, Spindler SR, Atamna H et al (2013) 5′ tRNA halves are present as abundant complexes in serum, concentrated in blood cells, and modulated by aging and calorie restriction. BMC Genomics 14:298

Donnelly N, Gorman AM, Gupta S et al (2013) The eIF2alpha kinases: their structures and functions. Cell Mol Life Sci 70:3493–3511
Durdevic Z, Schaefer M (2013) tRNA modifications: necessary for correct tRNA-derived fragments during the recovery from stress? Bioessays 35:323–327
El Yacoubi B, Bailly M, de Crecy-Lagard V (2012) Biosynthesis and function of posttranscriptional modifications of transfer RNAs. Ann Rev Genet 46:69–95
Emara MM, Ivanov P, Hickman T et al (2010) Angiogenin-induced tRNA-derived stress-induced RNAs promote stress-induced stress granule assembly. J Biol Chem 285:10959–10968
Fett JW, Strydom DJ, Lobb RR et al (1985) Isolation and characterization of angiogenin, an angiogenic protein from human carcinoma cells. Biochemistry 24:5480–5486
Fraser CM, Gocayne JD, White O et al (1995) The minimal gene complement of *Mycoplasma genitalium*. Science 270:397–403
Fu H, Feng J, Liu Q et al (2009) Stress induces tRNA cleavage by angiogenin in mammalian cells. FEBS Lett 583:437–442
Gao X, Xu Z (2008) Mechanisms of action of angiogenin. Acta Biochim Biophys Sinica 40:619–624
Gebetsberger J, Polacek N (2013) Slicing tRNAs to boost functional ncRNA diversity. RNA Biol 10:1798–1806
Gebetsberger J, Zywicki M, Kunzi A et al (2012) tRNA-derived fragments target the ribosome and function as regulatory non-coding RNA in Haloferax volcanii. Archaea 2012:260909
Ghildiyal M, Zamore PD (2009) Small silencing RNAs: an expanding universe. Nat Rev Genet 10:94–108
Giege R (2008) Toward a more complete view of tRNA biology. Nat Struct Mol Biol 15:1007–1014
Greenway MJ, Alexander MD, Ennis S et al (2004) A novel candidate region for ALS on chromosome 14q11.2. Neurology 63:1936–1938
Greenway MJ, Andersen PM, Russ C et al (2006) ANG mutations segregate with familial and 'sporadic' amyotrophic lateral sclerosis. Nat Genet 38:411–413
Gu C, Begley TJ, Dedon PC (2014) tRNA modifications regulate translation during cellular stress. FEBS Lett 588:4287–4296
Haiser HJ, Karginov FV, Hannon GJ et al (2008) Developmentally regulated cleavage of tRNAs in the bacterium *Streptomyces coelicolor*. Nucleic Acids Res 36:732–741
Hanada T, Weitzer S, Mair B et al (2013) CLP1 links tRNA metabolism to progressive motor-neuron loss. Nature 495:474–480
Hartmann A, Kunz M, Kostlin S et al (1999) Hypoxia-induced up-regulation of angiogenin in human malignant melanoma. Cancer Res 59:1578–1583
Haussecker D, Huang Y, Lau A et al (2010) Human tRNA-derived small RNAs in the global regulation of RNA silencing. RNA 16:673–695
Ivanov P, Anderson P (2011) Stress-induced ribonucleases. In: Nicholson AW (ed) Ribonucleases, nucleic acids and molecular biology. Springer, Berlin, pp 115–119
Ivanov P, Anderson P (2013) Post-transcriptional regulatory networks in immunity. Immunol Rev 253:253–272
Ivanov P, Emara MM, Villen J et al (2011a) Angiogenin-induced tRNA fragments inhibit translation initiation. Mol Cell 43:613–623
Ivanov P, Kedersha N, Anderson P (2011b) Stress puts TIA on TOP. Genes Dev 25:2119–2124
Ivanov P, O'Day E, Emara MM et al (2014) G-quadruplex structures contribute to the neuroprotective effects of angiogenin-induced tRNA fragments. Proc Natl Acad Sci U S A 111:18201–18206
Jablonowski D, Schaffrath R (2007) Zymocin, a composite chitinase and tRNase killer toxin from yeast. Biochem Soc Trans 35:1533–1537
Jackson RJ, Hellen CU, Pestova TV (2010) The mechanism of eukaryotic translation initiation and principles of its regulation. Nat Rev Mol Cell Biol 11:113–127

Jiang Y, Meidler R, Amitsur M et al (2001) Specific interaction between anticodon nuclease and the tRNA(Lys) wobble base. J Mol Biol 305:377–388
Jiang Y, Blanga S, Amitsur M et al (2002) Structural features of tRNALys favored by anticodon nuclease as inferred from reactivities of anticodon stem and loop substrate analogs. J Biol Chem 277:3836–3841
Jochl C, Rederstorff M, Hertel J et al (2008) Small ncRNA transcriptome analysis from *Aspergillus fumigatus* suggests a novel mechanism for regulation of protein synthesis. Nucleic Acids Res 36:2677–2689
Kao RY, Jenkins JL, Olson KA et al (2002) A small-molecule inhibitor of the ribonucleolytic activity of human angiogenin that possesses antitumor activity. Proc Natl Acad Sci U S A 99:10066–10071
Karaca E, Weitzer S, Pehlivan D et al (2014) Human CLP1 mutations alter tRNA biogenesis, affecting both peripheral and central nervous system function. Cell 157:636–650
Kaufmann G (2000) Anticodon nucleases. Trends Biochem Sci 25:70–74
Kawaji H, Nakamura M, Takahashi Y et al (2008) Hidden layers of human small RNAs. BMC Genomics 9:157
Kedersha N, Ivanov P, Anderson P (2013) Stress granules and cell signaling: more than just a passing phase? Trends Biochem Sci 38:494–506
Keppetipola N, Jain R, Meineke B et al (2009) Structure-activity relationships in Kluyveromyces lactis gamma-toxin, a eukaryal tRNA anticodon nuclease. RNA 15:1036–1044
Khan MA, Rafiq MA, Noor A et al (2012) Mutation in NSUN2, which encodes an RNA methyltransferase, causes autosomal-recessive intellectual disability. Am J Hum Genet 90:856–863
Kirchner S, Ignatova Z (2015) Emerging roles of tRNA in adaptive translation, signalling dynamics and disease. Nat Rev Genet 16:98–112
Klungland A, Dahl JA (2014) Dynamic RNA modifications in disease. Curr Opin Genet Dev 26:47–52
Kroemer G, Marino G, Levine B (2010) Autophagy and the integrated stress response. Mol Cell 40:280–293
Lander ES, Linton LM, Birren B et al (2001) Initial sequencing and analysis of the human genome. Nature 409:860–921
Lee SR, Collins K (2005) Starvation-induced cleavage of the tRNA anticodon loop in *Tetrahymena thermophila*. J Biol Chem 280:42744–42749
Lee YS, Shibata Y, Malhotra A et al (2009) A novel class of small RNAs: tRNA-derived RNA fragments (tRFs). Genes Dev 23:2639–2649
Levitz R, Chapman D, Amitsur M et al (1990) The optional *E. coli* prr locus encodes a latent form of phage T4-induced anticodon nuclease. EMBO J 9:1383–1389
Li S, Hu GF (2012) Emerging role of angiogenin in stress response and cell survival under adverse conditions. J Cell Physiol 227:2822–2826
Li Y, Luo J, Zhou H et al (2008) Stress-induced tRNA-derived RNAs: a novel class of small RNAs in the primitive eukaryote *Giardia lamblia*. Nucleic Acids Res 36:6048–6055
Li Z, Ender C, Meister G et al (2012) Extensive terminal and asymmetric processing of small RNAs from rRNAs, snoRNAs, snRNAs, and tRNAs. Nucleic Acids Res 40:6787–6799
Liao JY, Ma LM, Guo YH et al (2010) Deep sequencing of human nuclear and cytoplasmic small RNAs reveals an unexpectedly complex subcellular distribution of miRNAs and tRNA 3′ trailers. PLoS One 5, e10563
Lu J, Esberg A, Huang B et al (2008) Kluyveromyces lactis gamma-toxin, a ribonuclease that recognizes the anticodon stem loop of tRNA. Nucleic Acids Res 36:1072–1080
Lu Z, Filonov GS, Noto JJ et al (2015) Metazoan tRNA introns generate stable circular RNAs in vivo. RNA 21:1554–1565
Luhtala N, Parker R (2010) T2 Family ribonucleases: ancient enzymes with diverse roles. Trends Biochem Sci 35:253–259

Machnicka MA, Milanowska K, Osman Oglou O et al (2013) MODOMICS: a database of RNA modification pathways–2013 update. Nucleic Acids Res 41:D262–267
Machnicka MA, Olchowik A, Grosjean H et al (2014) Distribution and frequencies of post-transcriptional modifications in tRNAs. RNA Biol 11:1619–1629
Mair B, Popow J, Mechtler K et al (2013) Intron excision from precursor tRNA molecules in mammalian cells requires ATP hydrolysis and phosphorylation of tRNA-splicing endonuclease components. Biochem Soc Trans 41:831–837
Martens-Uzunova ES, Jalava SE, Dits NF et al (2012) Diagnostic and prognostic signatures from the small non-coding RNA transcriptome in prostate cancer. Oncogene 31:978–991
Martens-Uzunova ES, Olvedy M, Jenster G (2013) Beyond microRNA–novel RNAs derived from small non-coding RNA and their implication in cancer. Cancer Lett 340:201–211
Martinez FJ, Lee JH, Lee JE et al (2012) Whole exome sequencing identifies a splicing mutation in NSUN2 as a cause of a Dubowitz-like syndrome. J Med Genet 49:380–385
Masaki H, Ogawa T (2002) The modes of action of colicins E5 and D, and related cytotoxic tRNases. Biochimie 84:433–438
Masaki H, Ogawa T, Tomita K et al (1997) Colicin E5 as a new type of cytotoxin, which cleaves a specific group of tRNAs. Nucleic Acids Symp Ser 37:287–288
Maute RL, Schneider C, Sumazin P et al (2013) tRNA-derived microRNA modulates proliferation and the DNA damage response and is down-regulated in B cell lymphoma. Proc Natl Acad Sci U S A 110:1404–1409
Mei Y, Yong J, Liu H et al (2010) tRNA binds to cytochrome c and inhibits caspase activation. Mol Cell 37:668–678
Mishima E, Inoue C, Saigusa D et al (2014) Conformational change in transfer RNA is an early indicator of acute cellular damage. J Am Soc Nephrol 25:2316–2326
Mleczko AM, Celichowski P, Bakowska-Zywicka K (2014) Ex-translational function of tRNAs and their fragments in cancer. Acta Biochim Pol 61:211–216
Morin RD, O'Connor MD, Griffith M et al (2008) Application of massively parallel sequencing to microRNA profiling and discovery in human embryonic stem cells. Genome Res 18:610–621
Nakamura M, Yamabe H, Osawa H et al (2006) Hypoxic conditions stimulate the production of angiogenin and vascular endothelial growth factor by human renal proximal tubular epithelial cells in culture. Nephrol Dial Transplant 21:1489–1495
Ng CL, Lang K, Meenan NA et al (2010) Structural basis for 16S ribosomal RNA cleavage by the cytotoxic domain of colicin E3. Nat Struct Mol Biol 17:1241–1246
Ogawa T, Inoue S, Yajima S et al (2006) Sequence-specific recognition of colicin E5, a tRNA-targeting ribonuclease. Nucleic Acids Res 34:6065–6073
Olson KA, French TC, Vallee BL et al (1994) A monoclonal antibody to human angiogenin suppresses tumor growth in athymic mice. Cancer Res 54:4576–4579
Olson KA, Fett JW, French TC et al (1995) Angiogenin antagonists prevent tumor growth in vivo. Proc Natl Acad Sci U S A 92:442–446
Olson KA, Verselis SJ, Fett JW (1998) Angiogenin is regulated in vivo as an acute phase protein. Biochem Biophys Res Commun 242:480–483
Parisien M, Wang X, Pan T (2013) Diversity of human tRNA genes from the 1000-genomes project. RNA Biol 10:1853–1867
Patil A, Chan CT, Dyavaiah M et al (2012a) Translational infidelity-induced protein stress results from a deficiency in Trm9-catalyzed tRNA modifications. RNA Biol 9:990–1001
Patil A, Dyavaiah M, Joseph F et al (2012b) Increased tRNA modification and gene-specific codon usage regulate cell cycle progression during the DNA damage response. Cell Cycle 11:3656–3665
Paushkin SV, Patel M, Furia BS et al (2004) Identification of a human endonuclease complex reveals a link between tRNA splicing and pre-mRNA 3′ end formation. Cell 117:311–321
Pavon-Eternod M, Gomes S, Geslain R et al (2009) tRNA over-expression in breast cancer and functional consequences. Nucleic Acids Res 37:7268–7280

Pavon-Eternod M, Gomes S, Rosner MR et al (2013) Overexpression of initiator methionine tRNA leads to global reprogramming of tRNA expression and increased proliferation in human epithelial cells. RNA 19:461–466
Pestova TV, Kolupaeva VG, Lomakin IB et al (2001) Molecular mechanisms of translation initiation in eukaryotes. Proc Natl Acad Sci U S A 98:7029–7036
Phizicky EM, Alfonzo JD (2010) Do all modifications benefit all tRNAs? FEBS Lett 584:265–271
Phizicky EM, Hopper AK (2010) tRNA biology charges to the front. Genes Dev 24:1832–1860
Pizzo E, Sarcinelli C, Sheng J et al (2013) Ribonuclease/angiogenin inhibitor 1 regulates stress-induced subcellular localization of angiogenin to control growth and survival. J Cell Sci 126:4308–4319
Popow J, Englert M, Weitzer S et al (2011) HSPC117 is the essential subunit of a human tRNA splicing ligase complex. Science 331:760–764
Popow J, Jurkin J, Schleiffer A et al (2014) Analysis of orthologous groups reveals archease and DDX1 as tRNA splicing factors. Nature 511:104–107
Pratt AJ, MacRae IJ (2009) The RNA-induced silencing complex: a versatile gene-silencing machine. J Biol Chem 284:17897–17901
Raina M, Ibba M (2014) tRNAs as regulators of biological processes. Front Genet 5:171
Rezgui VA, Tyagi K, Ranjan N et al (2013) tRNA tKUUU, tQUUG, and tEUUC wobble position modifications fine-tune protein translation by promoting ribosome A-site binding. Proc Natl Acad Sci U S A 110:12289–12294
Saikia M, Fu Y, Pavon-Eternod M et al (2010) Genome-wide analysis of N1-methyl-adenosine modification in human tRNAs. RNA 16:1317–1327
Saikia M, Krokowski D, Guan BJ et al (2012) Genome-wide identification and quantitative analysis of cleaved tRNA fragments induced by cellular stress. J Biol Chem 287:42708–42725
Saikia M, Jobava R, Parisien M et al (2014) Angiogenin-cleaved tRNA halves interact with cytochrome c, protecting cells from apoptosis during osmotic stress. Mol Cell Biol 34:2450–2463
Sarin LP, Leidel SA (2014) Modify or die?—RNA modification defects in metazoans. RNA Biol 11:1555–1567
Schaefer M, Pollex T, Hanna K et al (2010) RNA methylation by Dnmt2 protects transfer RNAs against stress-induced cleavage. Genes Dev 24:1590–1595
Schaffer AE, Eggens VR, Caglayan AO et al (2014) CLP1 founder mutation links tRNA splicing and maturation to cerebellar development and neurodegeneration. Cell 157:651–663
Schutz K, Hesselberth JR, Fields S (2010) Capture and sequence analysis of RNAs with terminal 2′,3′-cyclic phosphates. RNA 16:621–631
Shigematsu M, Kirino Y (2015) tRNA-derived short non-coding RNA as interacting partners of argonaute proteins. Gene Regul Syst Biol 9:27–33
Sobala A, Hutvagner G (2011) Transfer RNA-derived fragments: origins, processing, and functions. Wiley Interdiscip Rev RNA 2:853–862
Sobala A, Hutvagner G (2013) Small RNAs derived from the 5′ end of tRNA can inhibit protein translation in human cells. RNA Biol 10:553–563
Sonenberg N, Hinnebusch AG (2009) Regulation of translation initiation in eukaryotes: mechanisms and biological targets. Cell 136:731–745
Sprinzl M, Vassilenko KS (2005) Compilation of tRNA sequences and sequences of tRNA genes. Nucleic Acids Res 33:D139–140
Su D, Chan CT, Gu C et al (2014) Quantitative analysis of ribonucleoside modifications in tRNA by HPLC-coupled mass spectrometry. Nat Protoc 9:828–841
Suryanarayana T, Uppala JK, Garapati UK (2012) Interaction of cytochrome c with tRNA and other polynucleotides. Mol Biol Rep 39:9187–9191
Suzuki T, Nagao A, Suzuki T (2011) Human mitochondrial tRNAs: biogenesis, function, structural aspects, and diseases. Annu Rev Genet 45:299–329
Tanaka N, Meineke B, Shuman S (2011) RtcB, a novel RNA ligase, can catalyze tRNA splicing and HAC1 mRNA splicing in vivo. J Biol Chem 286:30253–30257

Tavtigian SV, Simard J, Teng DH et al (2001) A candidate prostate cancer susceptibility gene at chromosome 17p. Nat Genet 27:172–180

Tello-Montoliu A, Patel JV, Lip GY (2006) Angiogenin: a review of the pathophysiology and potential clinical applications. J Thromb Haemost 4:1864–1874

Thompson DM, Parker R (2009a) The RNase Rny1p cleaves tRNAs and promotes cell death during oxidative stress in Saccharomyces cerevisiae. J Cell Biol 185:43–50

Thompson DM, Parker R (2009b) Stressing out over tRNA cleavage. Cell 138:215–219

Thompson DM, Lu C, Green PJ et al (2008) tRNA cleavage is a conserved response to oxidative stress in eukaryotes. RNA 14:2095–2103

Tsuji T, Sun Y, Kishimoto K et al (2005) Angiogenin is translocated to the nucleus of HeLa cells and is involved in ribosomal RNA transcription and cell proliferation. Cancer Res 65:1352–1360

Tuorto F, Liebers R, Musch T et al (2012) RNA cytosine methylation by Dnmt2 and NSun2 promotes tRNA stability and protein synthesis. Nat Struct Mol Biol 19:900–905

Tyndall C, Meister J, Bickle TA (1994) The *Escherichia coli* prr region encodes a functional type IC DNA restriction system closely integrated with an anticodon nuclease gene. J Mol Biol 237:266–274

van Es MA, Schelhaas HJ, van Vught PW et al (2011) Angiogenin variants in Parkinson disease and amyotrophic lateral sclerosis. Ann Neurol 70:964–973

Wang X, He C (2014) Dynamic RNA modifications in posttranscriptional regulation. Mol Cell 56:5–12

Weitzer S, Martinez J (2007) The human RNA kinase hClp1 is active on 3′ transfer RNA exons and short interfering RNAs. Nature 447:222–226

Wu D, Yu W, Kishikawa H et al (2007) Angiogenin loss-of-function mutations in amyotrophic lateral sclerosis. Ann Neurol 62:609–617

Xiong Y, Steitz TA (2006) A story with a good ending: tRNA 3′-end maturation by CCA-adding enzymes. Curr Opin Struct Biol 16:12–17

Xu ZP, Tsuji T, Riordan JF, Hu GF (2002) The nuclear function of angiogenin in endothelial cells is related to rRNA production. Biochem Biophys Res Commun 294:287–292

Yamasaki S, Ivanov P, Hu GF et al (2009) Angiogenin cleaves tRNA and promotes stress-induced translational repression. J Cell Biol 185:35–42

Yeung ML, Bennasser Y, Watashi K et al (2009) Pyrosequencing of small non-coding RNAs in HIV-1 infected cells: evidence for the processing of a viral-cellular double-stranded RNA hybrid. Nucleic Acids Res 37:6575–6586

Zheng G, Qin Y, Clark WC et al (2015) Efficient and quantitative high-throughput tRNA sequencing. Nat Meth 12:835–837

Zhou Y, Goodenbour JM, Godley LA et al (2009) High levels of tRNA abundance and alteration of tRNA charging by bortezomib in multiple myeloma. Biochem Biophys Res Commun 385:160–164

Sulfur Modifications in tRNA: Function and Implications for Human Disease

Naoki Shigi

Contents

Abstract Transfer RNA is an adaptor molecule that links amino acids to codons on messenger RNA. Functional tRNA molecules are produced by posttranscriptional processing events, such as splicing, end maturation, and chemical modifications of bases and sugars. More than one hundred types of naturally occurring chemical modifications of RNA are currently known. This chapter will summarize the recent advances in our understanding of the sulfur modifications of tRNA and their roles in cellular functions. The biosynthesis of tRNA sulfur modifications involves unique sulfur trafficking systems and modification enzymes that eventually result in the incorporation of a sulfur atom into tRNA. tRNA thionucleosides have been known for some time to be important for accurate and efficient translation, but more recently, these modifications and the codon usage bias of genes have been proposed to control the translation efficiency of specific groups of genes, allowing the organism to adapt to specific environments. Sulfur modifications of tRNA have also far-reaching implications for the molecular pathogenesis of human diseases, and this chapter provides a comprehensive and up-to-date overview of advances in our knowledge of the mechanisms involved.

N. Shigi (✉)
National Institute of Advanced Industrial Science and Technology (AIST), 2-4-7 Aomi, Koto-ku, Tokyo 135-0064, Japan
e-mail: naoki-shigi@aist.go.jp

S. Jurga et al. (eds.), *Modified Nucleic Acids in Biology and Medicine*, RNA Technologies, DOI 10.1007/978-3-319-34175-0_3

Keywords Codon bias • Mitochondria • Posttranscriptional modification • Translation

1 Introduction

One of the characteristic features of RNA molecules is their posttranscriptional chemical modification. In tRNA, more than 90 forms of naturally occurring chemical modification, whose primary role is the fine-tuning of protein synthesis, have been found. Besides the importance of tRNA modifications for the maturation and structural stability of tRNA itself, these modifications are also critical for translation, such as for precise codon recognition and reading frame maintenance, and providing recognition signals for the translation apparatus (Bjork 1995). The distribution of RNA modifications and an overview of their biosynthesis pathways are available from an online database maintained at http://mods.rna.albany.edu/ and http://modomics.genesilico.pl/.

Among the many modified nucleosides, four kinds of thionucleoside derivatives are found in tRNAs: 4-thiouridine (s^4U) at position 8, 2-thiocytidine (s^2C) at position 32, 2-thiouridine derivatives (xm^5s^2U) at positions 34 and 54, and 2-methylthioadenosine derivatives (ms^2x^6A) at position 37 ("x" means several functional groups found in each species) (Fig. 1). The wobble base (position 34)

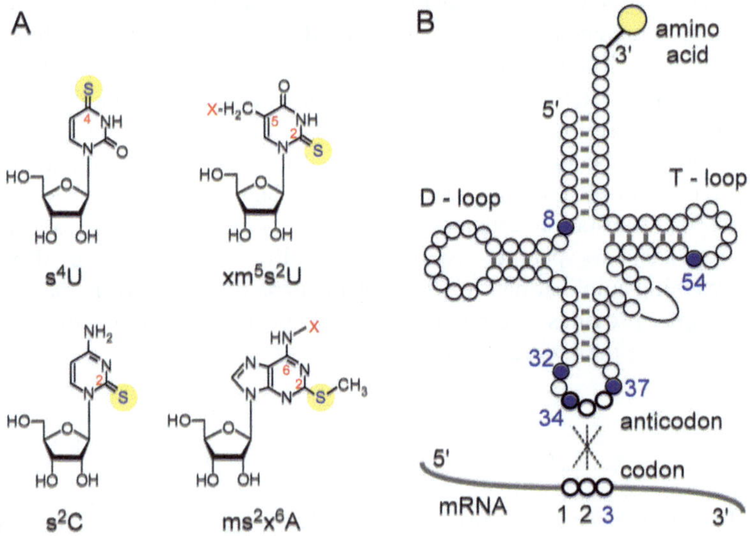

Fig. 1 Sulfur modifications in tRNA. (**a**) Chemical structures of sulfur modifications. s^4U (4-thiouridine), s^2C (2-thiocytidine), xm^5s^2U (5-methyl-2-thiouridine derivatives), and ms^2x^6A (2-methylthioadenosine derivatives) are shown. (**b**) Positions of sulfur modifications in tRNA

of tRNAs for Lys, Glu, and Gln is universally modified to xm^5s^2U, although modifications at 5-carbon vary with species. The distributions of thio modifications at positions other than position 34 vary with species. Modifications in the anticodon loop (positions 32, 34, and 37) are important for precise codon recognition, while s^4 U8 and s^2U54 are responsible for ultraviolet (UV) light sensing in bacteria and thermal adaptation in thermophilic microorganisms, respectively.

In this chapter, I will focus on the recent advances in the functional characterization of sulfur modifications in tRNA. For a detailed description of the reaction mechanisms of modification enzymes, please refer to my previous review (Shigi 2014). Genome-wide analyses in model organisms have revealed the importance of 2-thiouridine in cellular homeostasis, and other studies have deepened our understanding of the molecular basis of human diseases whose pathogenesis has been attributed to deficiencies in tRNA modifications.

2 Distribution and Functional Properties of Sulfur Modifications

Modifications of the wobble base in anticodon loop of tRNA have a critical role in precise codon recognition at the ribosome A site. The wobble bases of tRNAs for Lys, Glu, and Gln are universally modified to xm^5s^2U (Fig. 1). C5 modifications vary depending on the species; 5-methylaminomethyl (mnm^5) and 5-carboxymethylaminomethyl ($cmnm^5$) are found in bacterial tRNAs, 5-methoxycarbonylmethyl (mcm^5) in eukaryotic cytosolic tRNAs, $cmnm^5$ in yeast mitochondrial tRNA, and 5-taurinomethyl (τm^5) in mammalian mitochondrial tRNAs (Suzuki 2005). The ribose of xm^5s^2U tends to adopt the C3′-*endo* conformation because of steric hindrance between the bulky 2-thio group and the 2′-OH group of ribose (Yokoyama et al. 1985; Agris et al. 1992). This conformation of xm^5s^2U stabilizes base paring with NNA and NNG codons that specify Lys, Glu, and Gln, while it prevents unwanted base paring with NNC and NNU codons that specify near-cognate amino acids (Yokoyama et al. 1985; Murphy et al. 2004; Durant et al. 2005; Johansson et al. 2008; Vendeix et al. 2012; Rodriguez-Hernandez et al. 2013). xm^5s^2U34 also has a role in preventing frame shift (Urbonavicius et al. 2001; Atkins and Bjork 2009; Isak and Ryden-Aulin 2009; Jager et al. 2013; Tukenmez et al. 2015).

Modifications near the anticodon triplet also have important roles in translation. 2-Methylthio-N^6-isopentenyladenosine (ms^2i^6A) and 2-methylthio-N^6-threonylcarbamoyladenosine (ms^2t^6A) are found at position 37 in most bacterial tRNAs that read UNN codons and ANN codons, respectively (Fig. 1). (Recently, cyclic forms of t^6A (ct^6A) were reported to be bona fide modifications in many species of bacteria, fungi, and plants (Miyauchi et al. 2013).) These modifications may stabilize codon-anticodon interactions (Vacher et al. 1984; Wilson and Roe 1989; Wei et al. 2011) and prevent frame shifts (Urbonavicius et al. 2001). The structure of $tRNA^{Phe}$ in complex with mRNA in the ribosome clearly showed that the 2-methylthio group of ms^2i^6A stabilizes the codon-anticodon interaction by enhancing stacking interactions with the base of the first nucleoside of the codon in mRNA (Jenner et al. 2010). In mammals, ms^2t^6A is found only in cytosolic tRNA $^{Lys(UUU)}$ (Wei et al. 2011), whereas ms^2i^6A is found on several mitochondrial tRNAs (mt-tRNAs), such as mt-$tRNA^{Phe}$, mt-$tRNA^{Trp}$, mt-$tRNA^{Tyr}$, and mt-tRNA $^{Ser(UCN)}$ (Wei et al. 2015). s^2C is found at position 32 in some bacterial tRNAs (Fig. 1). In *E. coli*, s^2C32 is found on three $tRNA^{Arg}$ and one $tRNA^{Ser(GCU)}$. The structural roles of s^2C in translation differ depending on the tRNA species (Jager et al. 2004).

The modified nucleoside s^4U is found at position 8 (and 9) of several bacterial and archaeal tRNAs (Fig. 1). s^4U is unique in that its UV adsorption spectrum has a maximum around 330 nm, unlike other bases whose maxima are around 260–280 nm. Therefore, upon irradiation with near-UV light, the s^4U at position 8 cross-links with the structurally close cytidine at position 13, which results in translation arrest (Carre et al. 1974; Ryals et al. 1982). In this way bacteria can withstand near-UV light irradiation and survive the harmful effects of this type of irradiation.

The modified nucleoside 5-methyl-2-thiouridine (m^5s^2U, also known as 2-thioribothymidine, s^2T) (Fig. 1) is found at position 54 in the T-loop of almost all tRNAs in several thermophilic bacteria (Watanabe et al. 1974) and archaea (Kowalak et al. 1994). The rigid C3′-*endo* conformation of m^5s^2U54 stabilizes the A-form duplex formed at the interaction site of the D-loop and T-loop. The 2-thio modification of m^5s^2U54 is a partial modification, and the ratio of 2-thiolated tRNA to non-thiolated tRNA increases with cultivation temperature (Watanabe et al. 1976; Kowalak et al. 1994). The melting temperature of the tRNA increases concomitantly with 2-thiolation (Watanabe et al. 1976; Horie et al. 1985), and 2-thiolated tRNAs work more efficiently in translation at higher temperatures (Yokoyama et al. 1987). Mutants defective in the 2-thio modification of m^5s^2U54 cannot grow at high temperatures (Shigi et al. 2006a).

3 Biosynthesis Pathways of Sulfur Modification

In many tRNA thionucleoside biosynthesis pathways, the sulfur flows from free L-cysteine (Fig. 2). A pyridoxal-5′-phosphate (PLP)-dependent enzyme, cysteine desulfurase, transfers the sulfur atom of L-cysteine to its catalytic cysteine residue in the form of a persulfide (R-SSH) (Flint 1996; Lauhon and Kambampati 2000; Lauhon 2002; Nilsson et al. 2002). The persulfides formed on the enzyme are transferred to downstream sulfur-carrier proteins or domains, before being finally used by tRNA modification enzymes as sulfur donors. One of the unusual features of this pathway is the existence of many sulfur-carrier proteins (or domains) that act as molecular cargos of activated sulfur species, such as persulfide (R-SSH) and thiocarboxylate (R-COSH). Modification enzymes bind substrate tRNAs and

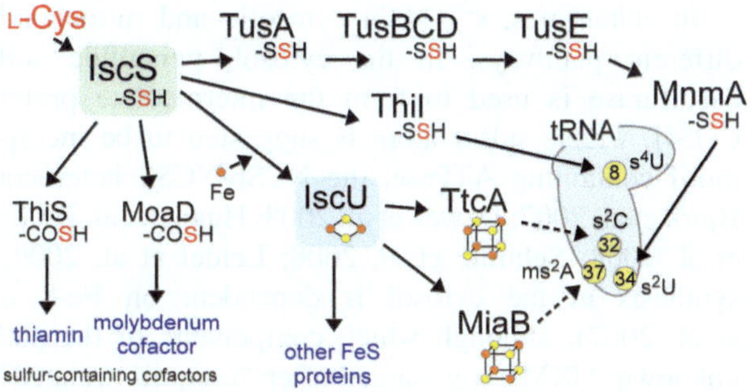

Fig. 2 Biosynthesis pathways of sulfur-containing molecules in *E. coli*. The sulfur atom of cysteine is delivered to sulfur-containing molecules by many sulfur-carrier proteins in each pathway

activate and transfer the activated sulfur atoms to the target nucleosides (Numata et al. 2006). Remarkably, these biosynthesis pathways are part of a larger picture involving other sulfur metabolites, such as the iron-sulfur (Fe-S) cluster, thiamin, and the molybdenum cofactor (Moco) (Fig. 2) (Schindelin et al. 2001; Settembre et al. 2003; Hidese et al. 2011). Therefore, each pathway is probably mutually influenced by other pathways involved in sulfur trafficking as part of a network (Maynard et al. 2012; Dahl et al. 2013). The detailed mechanisms of sulfur transfer have been revealed by structure-based analyses of cysteine desulfurase complexed with sulfur-carrier proteins (Shi et al. 2010) and those of modification enzymes (Numata et al. 2006; Forouhar et al. 2013; Nakagawa et al. 2013; Neumann et al. 2014).

The biosynthesis of these thionucleosides can be divided into two groups in terms of their dependency on iron-sulfur (Fe-S) cluster biosynthesis (Fig. 2). The biosynthesis of s^4U8 and s^2U34 in bacteria is not dependent on Fe-S cluster biogenesis, while the biosynthesis of s^2C32 and ms^2A37 is dependent (Lauhon et al. 2004; Leipuviene et al. 2004). s^4U8 and s^2U34 formation is catalyzed by PP motif containing ATPases, ThiI (Ryals et al. 1982; Mueller and Palenchar 1999; Kambampati and Lauhon 2000; Mueller et al. 2001) and MnmA (Sullivan et al. 1985; Kambampati and Lauhon 2003), respectively. Persulfide sulfurs generated by IscS cysteine desulfurase are relayed by persulfide carriers and are finally incorporated into target nucleosides by these ATPases. In *E. coli* and closely related species, the persulfide is relayed to MnmA by a number of proteins, such as TusA, TusBCD, and TusE (Ikeuchi et al. 2006), although in *Bacillus subtilis*, there is no need for such intermediate carrier proteins (Black and Dos Santos 2015). s^2C32 is formed by TtcA, an Fe-S protein with a PP motif (Jager et al. 2004; Bouvier et al. 2014). ms^2i^6A37 and ms^2t^6A37 are formed by Fe-S proteins with the radical SAM domains, MiaB (Pierrel et al. 2004) and MtaB (Arragain et al. 2010), respectively. The Fe-S clusters are built on a scaffold protein such as IscU, with the sulfur atoms derived from cysteine desulfurases (Hidese et al. 2011). The Fe-S clusters are subsequently transferred to these modification enzymes, where they are known to be oxygen labile. The sulfur atom of s^2C32 is derived from cysteine desulfurase (Bouvier et al. 2014), whereas the sulfur donor for ms^2A37 remains to be identified (Forouhar et al. 2013).

In eukaryotes, s^2U34 in cytosolic and mitochondrial tRNAs are formed by different pathways. In the cytosol, persulfide sulfur generated by cysteine desulfurase is used to form the intermediate protein thiocarboxylate, URM1-COSH, whose sulfur atom is suggested to be incorporated into s^2U34 by a PP motif containing ATPase, the NCS6/NCS2 heterocomplex (Esberg et al. 2006; Bjork et al. 2007; Dewez et al. 2008; Huang et al. 2008; Nakai et al. 2008; Schlieker et al. 2008; Schmitz et al. 2008; Leidel et al. 2009; Noma et al. 2009). s^2U34 synthesis in the cytosol is dependent on Fe-S cluster biosynthesis (Nakai et al. 2007), although which components of the pathways are Fe-S proteins is unknown. URM1 was named after "ubiquitin-related modifier 1" and also works as a posttranslational protein modifier of many cellular proteins (Furukawa et al. 2000; Goehring et al. 2003), in addition to its role as a thiocarboxylate carrier for s^2U synthesis. The components of the biosynthesis pathway, such as NCS6/NCS2, are covalently modified by URM1 (Van der Veen et al. 2011), although the function of urmylation in s^2U34 synthesis is unknown. In mitochondria, s^2U34 formation is catalyzed by MTU1 (Umeda et al. 2005), a eukaryotic homolog of bacterial MnmA. The sulfur-carrier protein has not been identified yet, if it exists.

Intriguingly, in a thermophilic bacterium, *Thermus thermophilus*, s^2U54 is synthesized by a very similar pathway to that of s^2U34 in the eukaryotic cytosol. Thiocarboxylate formed on TtuB (a bacterial homolog of URM1) is used as a sulfur donor by the modification enzyme, TtuA (a bacterial homolog of NCS6) (Shigi et al. 2002, 2006a, b, 2008). TtuB is also a posttranslational protein modifier and the active site of TtuA is covalently modified by TtuB, suggesting that the thiolation activity of TtuA is regulated by TtuB conjugation (Shigi 2012; Nakagawa et al. 2013).

4 Wobble Modification Affects Global Translation by Modulating Specific Groups of Genes

The loss of the mcm^5s^2U34 modification from cytosolic $tRNA^{Lys}$, $tRNA^{Glu}$, and $tRNA^{Gln}$ reduces cellular resistance to high temperature, oxidative stress, and rapamycin, a Tor-signaling inhibitor, in yeast (Furukawa et al. 2000; Goehring et al. 2003) and worm (Dewez et al. 2008). The phenotypes can be suppressed by overexpression of unmodified tRNAs (Leidel et al. 2009), implying that defects in the translation of Lys, Glu, and Gln codons are responsible for these phenotypes. Indeed, the mcm^5s^2U34 modification improves the expression of a set of genes involved in the DNA damage response, telomeric gene silencing, and cell division in yeast and worm (Begley et al. 2007; Chen et al. 2011; Bauer et al. 2012; Fernandez-Vazquez et al. 2013; Rezgui et al. 2013). The mRNAs of these genes have a lot more cognate codons for Lys, Glu, and Gln than other mRNAs (biased codon usage), indicating that they can be effectively read with mcm^5s^2U34 modified anticodons (Fig. 3A).

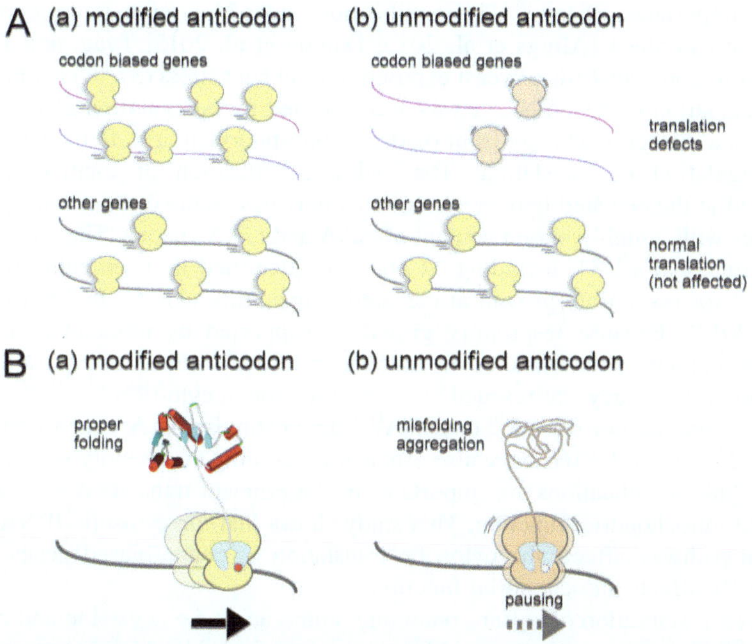

Fig. 3 Functions of anticodon modifications in translation. (A) Anticodon modifications in tRNAs affect the translation of specific groups of genes whose cognate codons are enriched. (**a**) The mRNAs which have a lot more cognate codons for Lys, Glu, and Gln (codon-biased genes) can be effectively read with modified anticodons. (**b**) If there are no modifications in the anticodons, only the translation of codon-biased genes is affected. (B) (**a**) Anticodon modifications enable efficient translation, which leads to proper folding of the proteome. (**b**) If there is no modifications in the anticodons, slow translation causes misfolding and aggregation of the proteome

Genome-wide ribosome distributions on mRNAs at single-codon resolution were analyzed in yeast strains lacking 5 or 2 modification of mcm^5s^2U34 (Zinshteyn and Gilbert 2013) by using the ribosome profiling technique, which involves deep sequencing of mRNA fragments protected from RNase digestion by translating ribosomes. Moderate accumulation of ribosomes occurred at AAA (Lys), CAA (Gln), and GAA (Glu) codons, suggesting that the translation of these codons is moderately slower in the absence of U34 modifications. Recently, a similar analysis was also conducted in yeast and worm, which revealed more pronounced ribosome accumulation at AAA and CAA codons in yeast lacking the 5-mcm and/or 2-thio modification of mcm^5s^2U34 and ribosome accumulation at AAA, CAA, and GAA codons in worm mutants lacking the 2-thio modification of mcm^5s^2U (Nedialkova and Leidel 2015). Intriguingly, they also found that pausing of the ribosome in a yeast strain lacking modifications of mcm^5s^2U34 caused protein misfolding and aggregation and that overexpression of unmodified tRNAs reversed the translational pausing and restored the conformational integrity of the proteome (Fig. 3B). This study clearly shows that optimal codon translation mediated by tRNA wobble modifications is critical for maintaining proteome integrity.

The importance of the s^2U34 modification under high temperature conditions has been examined (Alings et al. 2015; Damon et al. 2015; Tyagi and Pedrioli 2015). Proteome analysis of yeast exposed to mild heat stress (37 °C) revealed that downregulation of NCS6/NCS2 was accompanied by upregulation of heat shock proteins, a metabolic change from oxidative phosphorylation to fermentation, and downregulation of translation. The 2-thio modification of mcm^5s^2U34 was impaired at the elevated temperature, which may have caused the downregulation of genes with a highly biased content of AAA and GAA codons. The biosynthesis genes for mcm^5s^2U34, including *NCS6*, were identified in a screening of genes required for respiratory growth at elevated temperature (37 °C) in yeast (Tigano et al. 2015). Because respiratory growth is supported by mitochondria, *NCS6* deletion mutants were impaired in mitochondrial translation, as expected. Their respiratory deficiency was rescued by overexpression of cyto-$tRNA^{Lys(UUU)}$ but not by overexpression of $tRNA^{Glu}$ or $tRNA^{Gln}$. Interestingly, AAA codon-rich genes, such as *BCK1* and *HFM1*, were also efficient in rescuing the phenotype, suggesting that wobble modifications are important for the efficient translation of genes that regulate mitochondrial function. This study shows that the cytosolic tRNA modification pathway affects the cytosolic translation of codon-biased genes, which eventually affects mitochondrial function.

In yeast, starvation of sulfur-containing amino acids, i.e., cysteine and methionine, leads to a decrease in the modification status of s^2U34, which in turn results in a global reduction in translation and cell growth (Laxman et al. 2013). Lys, Glu, and Gln codons are overrepresented in genes essential for translation and growth. Cells respond to availability of sulfur-containing amino acid by using sulfur modification of tRNAs and their function at cognate codons. Overexpression of these tRNAs did not rescue the phenotypes of s^2U34-deficient mutants in this case.

5 Sulfur Modification and Molecular Pathogenesis of Human Disease

A number of human diseases have been linked to sulfur modifications in tRNAs, probably because these modifications are indispensable for translation. Abnormal translation, both in mitochondria and the cytosol, is linked to human pathogenesis. Mitochondria disease is a symptom of mitochondria dysfunction and mainly affects tissues in which energy demands are high, such as the brain and skeletal muscle. More than 100 mutations in mt-tRNA genes on the mt-DNA genome are reported, but mutations in the nuclear genome resulting in mitochondria disease have also been reported (Viscomi et al. 2015). MELAS (mitochondrial myopathy, encephalopathy, lactic acidosis, and stroke-like episodes) and MEERF (myoclonus epilepsy associated with ragged-red fibers) are two representative mitochondria diseases whose pathogenic mutations reside in the mt-$tRNA^{Leu(UUR)}$ gene and mt-$tRNA^{Lys}$ gene, respectively. MELAS and MEERF patients have modification-deficient mt-tRNAs lacking the normal 5-taurinomethyl-(2-thio)uridine ($\tau m^5(s^2)U$) modification at position 34, which results in inefficient mitochondrial translation (Yasukawa et al. 2001; Kirino et al. 2004) and eventually clinical manifestations (Kirino and Suzuki 2005). It is speculated that point mutations outside of position 34 in tRNA may affect the recognition of mutant tRNAs by U34 modifying enzymes, leading to the loss of the U34 modification.

A number of mutations in the *MTU1* gene are associated with mitochondria diseases (Guan et al. 2006; Zeharia et al. 2009; Schara et al. 2011; Uusimaa et al. 2011; Gaignard et al. 2013). MTU1 (also known as TRMU) is a mitochondria-specific thiouridylase responsible for the formation of s^2U34 on human mt-tRNA^{Lys}, mt-tRNA^{Glu}, and mt-tRNA^{Gln} (Umeda et al. 2005). Disruption of the *MTU1* gene in yeast results in a mitochondrial translation defect as well as reduced oxygen consumption. Knockdown of *MTU1* in HeLa cells also results in reduced oxygen consumption and lower mitochondrial membrane potentials. It is of note that, in addition to the lack of s^2U modification (Umeda et al. 2005), lower amounts of mt-tRNAs may also contribute to these defects (Wang et al. 2007).

MTU1 mutations (Y77H, exon-3 skip, or G14S) have been mapped in infants with acute liver failure associated with lactic acidemia (also known as transient infantile liver failure (LFIT)) (Zeharia et al. 2009). In fibroblasts from these patients, s^2U levels were markedly lower, and these lower levels were associated with reduced levels of some mt-tRNAs and mitochondrial translation activity. In patients with another mutation in the *MTU1* gene (compound heterozygote with 9-bp in-frame and 1-bp insertion in each allele), although s^2U levels were markedly lower, the amounts of mt-tRNAs and mitochondrial translation were normal (Sasarman et al. 2011). Another *MTU1* mutation (A10S) associated with mutations in mitochondrial 12S ribosomal RNA (A1555G) led to lower levels of s^2U, reduced amounts of mt-tRNAs, and lower mitochondrial translation activity (Guan et al. 2006). *MTU1* knockdown in myoblasts from a patient with reversible infantile respiratory chain deficiency (RIRCD), who had a mutation in mt-tRNA^{Glu}, resulted in similar phenotypes (Boczonadi et al. 2013). Therefore, lower levels of the s^2U modification caused by the *MTU1* mutation and other unknown mechanisms may contribute to the pathogenic phenotypes associated with *MTU1* mutations, although further analysis is required (Armengod et al. 2014).

A characteristic feature of LFIT patients with the *MTU1* mutation is that those who survive the initial acute episode can recover and develop normally (Zeharia et al. 2009). LFIT develops in the neonatal period, in which the availability of L-cysteine, a sulfur donor for s^2U synthesis, is limited. The endogenous synthesis of cysteine is markedly attenuated, because the rate-limiting enzyme in the transsulfuration pathway of cysteine synthesis, cystathionase, is very low in this period (Zlotkin and Cherian 1988). In addition, another source of cysteine, metallothionein, is also low in this period. Therefore, the authors proposed that dietary- or metallothionein-derived cysteine in the neonatal period might provide some protection against LFIT development in people with the *MTU1* mutation (Zeharia et al. 2009). This possibility was further tested in in vitro experiments. Supplementation of MTU1-deficient fibroblasts with cysteine increased the amount of the mitochondrial respiratory complex, although the recovery of 2-thio levels in s ^{2}U of mt-tRNAs was not investigated (Boczonadi et al. 2013).

Oxidative stress in MELAS and MEERF cybrids can induce the upregulation of microRNA9/9*, which in turn suppress the expression of mitochondrial U34 modification enzymes including MTU1. Expectedly, s^{2}U34 modification in mt-tRNA^{Glu} and mt-tRNA^{Lys} is downregulated significantly in the MELAS cybrid, which possesses a mutant mt-tRNA$^{Leu(UUR)}$ gene (Meseguer et al. 2015). The authors speculated that, in addition to the loss of τm^{5}U34 modification in mutant tRNA$^{Leu\ (UUR)}$, downregulation of these modification enzymes contributes to the MELAS phenotype (Kirino et al. 2004).

CDKAL1 and CDK5RAP1 are eukaryotic 2-methylthio transferases responsible for the synthesis of ms^{2}t^{6}A37 in cytosolic tRNA^{Lys} (Arragain et al. 2010; Wei et al. 2011) and ms^{2}i^{6}A37 in mt-tRNA^{Phe}, mt-tRNA^{Trp}, mt-tRNA^{Tyr}, and mt-tRNA $^{Ser(UCN)}$ (Reiter et al. 2012; Wei et al. 2015), respectively. CDKAL1 and CDK5RAP1 are homologous to bacterial MiaB (Pierrel et al. 2004) and possess ER- and mitochondria-targeting signals, respectively. Type 2 diabetes (T2D) is the most frequent form of diabetes and affects more than 350 million people worldwide. High levels of blood sugar are a characteristic feature of T2D and are caused by defects in insulin production and insulin resistance. Genome-wide analysis has revealed that genetic variations of the *CDKAL1* gene are associated with T2D (Diabetes Genetics Initiative of Broad Institute et al. 2007). Detailed investigations have been made using *CDKAL1* knockout mice, which recapitulate T2D-associated phenotypes, such as pancreatic islet hypertrophy, a decrease in insulin secretion, and impaired blood glucose control (Wei et al. 2011). It is suggested that misreading of a critical Lys codon in proinsulin results in a reduction in glucose-stimulated proinsulin synthesis. Low insulin secretion was found to be associated with low levels of ms^{2}t^{6}A37 caused by CDKAL1 mutation in a human blood sample (Xie et al. 2013), providing additional evidence that ms^{2}t^{6}A functions in cytosolic translation and that a lack of this modification increases the risk of T2D.

A recent study revealed a correlation between mitochondrial activity and the amount of ms^{2}i^{6}A synthesized by CDK5RAP1 in pork tissues (Reiter et al. 2012). In the muscle and brain, higher cytochrome C oxidase activity, an indication of increased mitochondrial activity, and higher amounts of ms^{2}i^{6}A amounts were observed. The importance of ms^{2}i^{6}A37 in mt-tRNAs for mitochondrial function has also been revealed by an analysis of *CDK5RAP1* KO mice (Wei et al. 2015). The 2-methylthio modification in mt-tRNAs was absent and mitochondrial translation was lower in KO MEF cells and tissues. Importantly, under stress conditions, the KO mice exhibited symptoms such as myopathy and cardiac dysfunction. Methylthio modification levels were lower in blood cells from MELAS patients and H_2O_2-treated HeLa cells than in those from normal cells. The authors also observed a correlation between heteroplasmy of mt-DNA and the amount of methylthio modification in human blood samples from MELAS patients. Because the mutant cells exhibited heightened oxidative stress (Crimi et al. 2005), the authors speculated that CDK5RAP1 was inactivated in these cells via disruption of the oxidation-liable Fe-S clusters (Pierrel et al. 2004; Arragain et al. 2010).

Reduced mitochondrial translation may be caused by lower levels of the methylthio modification in MELAS patients, in addition to the loss of τm^5U34 modification of mutant $tRNA^{Leu\ (UUR)}$ (Kirino et al. 2004).

6 Concluding Remarks

Considering the importance of sulfur modifications for tRNA function, it is no wonder that many human diseases have their origins in abnormalities in these modifications and/or in biosynthesis genes responsible for these modifications. Intensive characterization of mitochondrial diseases and diabetes whose pathogenesis involves alterations in tRNA sulfur modification has led to a deeper understanding of the pathogenic mechanisms and opened up the prospect, in some instances, of medical treatment. In addition, it was also reported that abnormal expression of *NCS6* mRNA is associated with prostate and breast cancer (Yousef et al. 2004).

Codon usage bias in mRNAs and amounts of each tRNA species differ significantly between organisms (Maraia and Iben 2014; Quax et al. 2015). Furthermore, even among individuals, copy-number variations in tRNA genes exist (Iben and Maraia 2014). These variations may influence the functional importance of modified nucleosides in each organism; therefore, one should keep this possibility in mind when analyzing the pathogenic mechanism of the diseases in model systems (Maraia and Iben 2014). It is also important to investigate whether dysregulation of gene expression by codon bias and sulfur modifications in higher eukaryotes is associated with human diseases. There also exists considerable variation in the biosynthesis pathways of sulfur modifications between species, and there may exist sulfur-carrier proteins that have not yet been discovered. Mutations or environmental conditions that affect the sulfur-trafficking network may also be involved in the development of human diseases.

Although the biosynthesis of sulfur modifications has been extensively studied over the last decade, the mechanisms involved in the removal of these modifications have received less attention. Active removal of s^2U34 in tRNAs has not been observed in yeast (Alings et al. 2015). Modification of m^6A on mRNA is suggested to be widely used for posttranscriptional regulation of mRNA with a set of enzymes: m^6A (methyltransferases (writer)), m^6A (demethylating enzymes (eraser)), and m^6A (binding proteins (reader)) (Meyer and Jaffrey 2014). In addition, a recent report describing the in vitro oxidation of s^2U has shown that under oxidative conditions s^2U is mainly transformed to 4-pyrimidinone riboside (H_2U) and not to U, and this conversion results in an altered base-pairing mode (Sochacka et al. 2015). As thionucleosides are considered to be susceptible to oxidation, the removal or repair of oxidized nucleosides may be important.

Acknowledgment I would like to thank all our collaborators, including Drs. Tsutomu Suzuki and Kimitsuna Watanabe (University of Tokyo), and Dr. Shigeyuki Yokoyama (RIKEN), as well as members of their laboratories. I would also like to thank Dr. Kenjyo Miyauchi (University of Tokyo) for comments on the manuscript. This work was supported in part by KAKENHI Grant (24570173) of the Ministry of Education, Culture, Sports, Science, and Technology of Japan and the Takeda Science Foundation.

References

Agris PF, Sierzputowska-Gracz H, Smith W et al (1992) Thiolation of uridine carbon-2 restricts the motional dynamics of the transfer RNA wobble position nucleoside. J Am Chem Soc 114:2652–2656

Alings F, Sarin LP, Fufezan C et al (2015) An evolutionary approach uncovers a diverse response of tRNA 2-thiolation to elevated temperatures in yeast. RNA 21:202–212

Armengod ME, Meseguer S, Villarroya M et al (2014) Modification of the wobble uridine in bacterial and mitochondrial tRNAs reading NNA/NNG triplets of 2-codon boxes. RNA Biol 11:1495–1507

Arragain S, Handelman SK, Forouhar F et al (2010) Identification of eukaryotic and prokaryotic methylthiotransferase for biosynthesis of 2-methylthio-N6-threonylcarbamoyladenosine in tRNA. J Biol Chem 285:28425–28433

Atkins JF, Bjork GR (2009) A gripping tale of ribosomal frameshifting: extragenic suppressors of frameshift mutations spotlight P-site realignment. Microbiol Mol Biol Rev 73:178–210

Bauer F, Matsuyama A, Candiracci J et al (2012) Translational control of cell division by Elongator. Cell Rep 1:424–433

Begley U, Dyavaiah M, Patil A et al (2007) Trm9-catalyzed tRNA modifications link translation to the DNA damage response. Mol Cell 28:860–870

Bjork GR (1995) Biosynthesis and function of modified nucleosides. In: Soll D, RajBhabdary UL (eds) tRNA: structure, biosynthesis, and function. ASM Press, Washington, DC, pp 165–205

Bjork GR, Huang B, Persson OP et al (2007) A conserved modified wobble nucleoside (mcm5s2U) in lysyl-tRNA is required for viability in yeast. RNA 13:1245–1255

Black KA, Dos Santos PC (2015) Abbreviated pathway for biosynthesis of 2-thiouridine in *Bacillus subtilis*. J Bacteriol 197:1952–1962

Boczonadi V, Smith PM, Pyle A et al (2013) Altered 2-thiouridylation impairs mitochondrial translation in reversible infantile respiratory chain deficiency. Hum Mol Genet 22:4602–4615

Bouvier D, Labessan N, Clemancey M et al (2014) TtcA a new tRNA-thioltransferase with an Fe-S cluster. Nucleic Acids Res 42:7960–7970

Carre DS, Thomas G, Favre A (1974) Conformation and functioning of tRNAs: cross-linked tRNAs as substrate for tRNA nucleotidyl-transferase and aminoacyl synthetases. Biochimie 56:1089–1101

Chen C, Huang B, Eliasson M et al (2011) Elongator complex influences telomeric gene silencing and DNA damage response by its role in wobble uridine tRNA modification. PLoS Genet 7, e1002258

Crimi M, Bordoni A, Menozzi G et al (2005) Skeletal muscle gene expression profiling in mitochondrial disorders. FASEB J 19:866–868

Dahl JU, Radon C, Buhning M et al (2013) The sulfur carrier protein TusA has a pleiotropic role in Escherichia coli that also affects molybdenum cofactor biosynthesis. J Biol Chem 288:5426–5442

Damon JR, Pincus D, Ploegh HL (2015) tRNA thiolation links translation to stress responses in Saccharomyces cerevisiae. Mol Biol Cell 26:270–282

Dewez M, Bauer F, Dieu M et al (2008) The conserved Wobble uridine tRNA thiolase Ctu1-Ctu2 is required to maintain genome integrity. Proc Natl Acad Sci U S A 105:5459–5464

Diabetes Genetics Initiative of Broad Institute LU, and Novartis Institutes of BioMedical Research, Saxena R, Voight BF, Lyssenko V et al (2007) Genome-wide association analysis identifies loci for type 2 diabetes and triglyceride levels. Science 316:1331–1336

Durant PC, Bajji AC, Sundaram M et al (2005) Structural effects of hypermodified nucleosides in the Escherichia coli and human tRNALys anticodon loop: the effect of nucleosides s2U, mcm5U, mcm5s2U, mnm5s2U, t6A, and ms2t6A. Biochemistry 44:8078–8089

Esberg A, Huang B, Johansson MJ et al (2006) Elevated levels of two tRNA species bypass the requirement for elongator complex in transcription and exocytosis. Mol Cell 24:139–148

Fernandez-Vazquez J, Vargas-Perez I, Sanso M et al (2013) Modification of tRNA(Lys) UUU by elongator is essential for efficient translation of stress mRNAs. PLoS Genet 9, e1003647

Flint DH (1996) *Escherichia coli* contains a protein that is homologous in function and N-terminal sequence to the protein encoded by the nifS gene of Azotobacter vinelandii and that can participate in the synthesis of the Fe-S cluster of dihydroxy-acid dehydratase. J Biol Chem 271:16068–16074

Forouhar F, Arragain S, Atta M et al (2013) Two Fe-S clusters catalyze sulfur insertion by radical-SAM methylthiotransferases. Nat Chem Biol 9:333–338

Furukawa K, Mizushima N, Noda T et al (2000) A protein conjugation system in yeast with homology to biosynthetic enzyme reaction of prokaryotes. J Biol Chem 275:7462–7465

Gaignard P, Gonzales E, Ackermann O et al (2013) Mitochondrial infantile liver disease due to TRMU gene mutations: three new cases. JIMD Rep 11:117–123

Goehring AS, Rivers DM, Sprague GF Jr (2003) Attachment of the ubiquitin-related protein Urm1p to the antioxidant protein Ahp1p. Eukaryot Cell 2:930–936

Guan MX, Yan Q, Li X et al (2006) Mutation in TRMU related to transfer RNA modification modulates the phenotypic expression of the deafness-associated mitochondrial 12S ribosomal RNA mutations. Am J Hum Genet 79:291–302

Hidese R, Mihara H, Esaki N (2011) Bacterial cysteine desulfurases: versatile key players in biosynthetic pathways of sulfur-containing biofactors. Appl Microbiol Biotechnol 91:47–61

Horie N, Hara-Yokoyama M, Yokoyama S et al (1985) Two tRNAIle1 species from an extreme thermophile, *Thermus thermophilus* HB8: effect of 2-thiolation of ribothymidine on the thermostability of tRNA. Biochemistry 24:5711–5715

Huang B, Lu J, Bystrom AS (2008) A genome-wide screen identifies genes required for formation of the wobble nucleoside 5-methoxycarbonylmethyl-2-thiouridine in Saccharomyces cerevisiae. RNA 14:2183–2194

Iben JR, Maraia RJ (2014) tRNA gene copy number variation in humans. Gene 536:376–384

Ikeuchi Y, Shigi N, Kato J et al (2006) Mechanistic insights into sulfur relay by multiple sulfur mediators involved in thiouridine biosynthesis at tRNA wobble positions. Mol Cell 21:97–108

Isak G, Ryden-Aulin M (2009) Hypomodification of the wobble base in tRNAGlu, tRNALys, and tRNAGln suppresses the temperature-sensitive phenotype caused by mutant release factor 1. J Bacteriol 191:1604–1609

Jager G, Leipuviene R, Pollard MG et al (2004) The conserved Cys-X1-*X*2-Cys motif present in the TtcA protein is required for the thiolation of cytidine in position 32 of tRNA from Salmonella enterica serovar Typhimurium. J Bacteriol 186:750–757

Jager G, Nilsson K, Bjork GR (2013) The phenotype of many independently isolated +1 frameshift suppressor mutants supports a pivotal role of the P-site in reading frame maintenance. PLoS One 8, e60246

Jenner LB, Demeshkina N, Yusupova G et al (2010) Structural aspects of messenger RNA reading frame maintenance by the ribosome. Nat Struct Mol Biol 17:555–560

Johansson MJ, Esberg A, Huang B et al (2008) Eukaryotic wobble uridine modifications promote a functionally redundant decoding system. Mol Cell Biol 28:3301–3312

Kambampati R, Lauhon CT (2000) Evidence for the transfer of sulfane sulfur from IscS to ThiI during the in vitro biosynthesis of 4-thiouridine in *Escherichia coli* tRNA. J Biol Chem 275:10727–10730

Kambampati R, Lauhon CT (2003) MnmA and IscS are required for in vitro 2-thiouridine biosynthesis in *Escherichia coli*. Biochemistry 42:1109–1117

Kirino Y, Suzuki T (2005) Human mitochondrial diseases associated with tRNA wobble modification deficiency. RNA Biol 2:41–44

Kirino Y, Yasukawa T, Ohta S et al (2004) Codon-specific translational defect caused by a wobble modification deficiency in mutant tRNA from a human mitochondrial disease. Proc Natl Acad Sci U S A 101:15070–15075

Kowalak JA, Dalluge JJ, McCloskey JA et al (1994) The role of posttranscriptional modification in stabilization of transfer RNA from hyperthermophiles. Biochemistry 33:7869–7876

Lauhon CT (2002) Requirement for IscS in biosynthesis of all thionucleosides in *Escherichia coli*. J Bacteriol 184:6820–6829

Lauhon CT, Kambampati R (2000) The iscS gene in *Escherichia coli* is required for the biosynthesis of 4-thiouridine, thiamin, and NAD. J Biol Chem 275:20096–20103

Lauhon CT, Skovran E, Urbina HD et al (2004) Substitutions in an active site loop of *Escherichia coli* IscS result in specific defects in Fe-S cluster and thionucleoside biosynthesis in vivo. J Biol Chem 279:19551–19558

Laxman S, Sutter BM, Wu X et al (2013) Sulfur amino acids regulate translational capacity and metabolic homeostasis through modulation of tRNA thiolation. Cell 154:416–429

Leidel S, Pedrioli PG, Bucher T et al (2009) Ubiquitin-related modifier Urm1 acts as a sulphur carrier in thiolation of eukaryotic transfer RNA. Nature 458:228–232

Leipuviene R, Qian Q, Bjork GR (2004) Formation of thiolated nucleosides present in tRNA from Salmonella enterica serovar Typhimurium occurs in two principally distinct pathways. J Bacteriol 186:758–766

Maraia RJ, Iben JR (2014) Different types of secondary information in the genetic code. RNA 20:977–984

Maynard ND, Macklin DN, Kirkegaard K et al (2012) Competing pathways control host resistance to virus via tRNA modification and programmed ribosomal frameshifting. Mol Syst Biol 8:567

Meseguer S, Martinez-Zamora A, Garcia-Arumi E et al (2015) The ROS-sensitive microRNA-9/9*controls the expression of mitochondrial tRNA-modifying enzymes and is involved in the molecular mechanism of MELAS syndrome. Hum Mol Genet 24:167–184

Meyer KD, Jaffrey SR (2014) The dynamic epitranscriptome: N6-methyladenosine and gene expression control. Nat Rev Mol Cell Biol 15:313–326

Miyauchi K, Kimura S, Suzuki T (2013) A cyclic form of N6-threonylcarbamoyladenosine as a widely distributed tRNA hypermodification. Nat Chem Biol 9:105–111

Mueller EG, Palenchar PM (1999) Using genomic information to investigate the function of ThiI, an enzyme shared between thiamin and 4-thiouridine biosynthesis. Protein Sci 8:2424–2427

Mueller EG, Palenchar PM, Buck CJ (2001) The role of the cysteine residues of ThiI in the generation of 4-thiouridine in tRNA. J Biol Chem 276:33588–33595

Murphy FV, Ramakrishnan V, Malkiewicz A et al (2004) The role of modifications in codon discrimination by tRNA(Lys)UUU. Nat Struct Mol Biol 11:1186–1191

Nakagawa H, Kuratani M, Goto-Ito S et al (2013) Crystallographic and mutational studies on the tRNA thiouridine synthetase TtuA. Proteins 81:1232–1244

Nakai Y, Nakai M, Lill R et al (2007) Thio modification of yeast cytosolic tRNA is an iron-sulfur protein-dependent pathway. Mol Cell Biol 27:2841–2847

Nakai Y, Nakai M, Hayashi H (2008) Thio-modification of yeast cytosolic tRNA requires a ubiquitin-related system that resembles bacterial sulfur transfer systems. J Biol Chem 283:27469–27476

Nedialkova DD, Leidel SA (2015) Optimization of codon translation rates via tRNA modifications maintains proteome integrity. Cell 161:1606–1618

Neumann P, Lakomek K, Naumann PT et al (2014) Crystal structure of a 4-thiouridine synthetase-RNA complex reveals specificity of tRNA U8 modification. Nucleic Acids Res 42:6673–6685

Nilsson K, Lundgren HK, Hagervall TG et al (2002) The cysteine desulfurase IscS is required for synthesis of all five thiolated nucleosides present in tRNA from Salmonella enterica serovar typhimurium. J Bacteriol 184:6830–6835

Noma A, Sakaguchi Y, Suzuki T (2009) Mechanistic characterization of the sulfur-relay system for eukaryotic 2-thiouridine biogenesis at tRNA wobble positions. Nucleic Acids Res 37:1335–1352

Numata T, Ikeuchi Y, Fukai S et al (2006) Snapshots of tRNA sulphuration via an adenylated intermediate. Nature 442:419–424

Pierrel F, Douki T, Fontecave M et al (2004) MiaB protein is a bifunctional radical-S-adenosylmethionine enzyme involved in thiolation and methylation of tRNA. J Biol Chem 279:47555–47563

Quax TE, Claassens NJ, Soll D et al (2015) Codon Bias as a Means to Fine-Tune Gene Expression. Mol Cell 59:149–161

Reiter V, Matschkal DM, Wagner M et al (2012) The CDK5 repressor CDK5RAP1 is a methylthiotransferase acting on nuclear and mitochondrial RNA. Nucleic Acids Res 40:6235–6240

Rezgui VA, Tyagi K, Ranjan N et al (2013) tRNA tKUUU, tQUUG, and tEUUC wobble position modifications fine-tune protein translation by promoting ribosome A-site binding. Proc Natl Acad Sci U S A 110:12289–12294

Rodriguez-Hernandez A, Spears JL, Gaston KW et al (2013) Structural and mechanistic basis for enhanced translational efficiency by 2-thiouridine at the tRNA anticodon wobble position. J Mol Biol 425:3888–3906

Ryals J, Hsu RY, Lipsett MN et al (1982) Isolation of single-site *Escherichia coli* mutants deficient in thiamine and 4-thiouridine syntheses: identification of a nuvC mutant. J Bacteriol 151:899–904

Sasarman F, Antonicka H, Horvath R et al (2011) The 2-thiouridylase function of the human MTU1 (TRMU) enzyme is dispensable for mitochondrial translation. Hum Mol Genet 20:4634–4643

Schara U, von Kleist-Retzow JC, Lainka E et al (2011) Acute liver failure with subsequent cirrhosis as the primary manifestation of TRMU mutations. J Inherit Metab Dis 34:197–201

Schindelin H, Kisker C, Rajagopalan KV (2001) Molybdopterin from molybdenum and tungsten enzymes. Adv Protein Chem 58:47–94

Schlieker CD, Van der Veen AG, Damon JR et al (2008) A functional proteomics approach links the ubiquitin-related modifier Urm1 to a tRNA modification pathway. Proc Natl Acad Sci U S A 105:18255–18260

Schmitz J, Chowdhury MM, Hanzelmann P et al (2008) The sulfurtransferase activity of Uba4 presents a link between ubiquitin-like protein conjugation and activation of sulfur carrier proteins. Biochemistry 47:6479–6489

Settembre E, Begley TP, Ealick SE (2003) Structural biology of enzymes of the thiamin biosynthesis pathway. Curr Opin Struct Biol 13:739–747

Shi R, Proteau A, Villarroya M et al (2010) Structural basis for Fe-S cluster assembly and tRNA thiolation mediated by IscS protein-protein interactions. PLoS Biol 8, e1000354

Shigi N (2012) Posttranslational modification of cellular proteins by a ubiquitin-like protein in bacteria. J Biol Chem 287:17568–17577

Shigi N (2014) Biosynthesis and functions of sulfur modifications in tRNA. Front Genet 5:67

Shigi N, Suzuki T, Tamakoshi M et al (2002) Conserved bases in the TPsi C loop of tRNA are determinants for thermophile-specific 2-thiouridylation at position 54. J Biol Chem 277:39128–39135

Shigi N, Sakaguchi Y, Suzuki T et al (2006a) Identification of two tRNA thiolation genes required for cell growth at extremely high temperatures. J Biol Chem 281:14296–14306

Shigi N, Suzuki T, Terada T et al (2006b) Temperature-dependent biosynthesis of 2-thioribothymidine of *Thermus thermophilus* tRNA. J Biol Chem 281:2104–2113

Shigi N, Sakaguchi Y, Asai S et al (2008) Common thiolation mechanism in the biosynthesis of tRNA thiouridine and sulphur-containing cofactors. EMBO J 27:3267–3278

Sochacka E, Szczepanowski RH, Cypryk M et al (2015) 2-Thiouracil deprived of thiocarbonyl function preferentially base pairs with guanine rather than adenine in RNA and DNA duplexes. Nucleic Acids Res 43:2499–2512

Sullivan MA, Cannon JF, Webb FH et al (1985) Antisuppressor mutation in *Escherichia coli* defective in biosynthesis of 5-methylaminomethyl-2-thiouridine. J Bacteriol 161:368–376

Suzuki T (2005) Biosynthesis and function of tRNA wobble modifications. In: Grosjean H (ed) - Fine-tuning of RNA functions by modification and editing. Springer, Heidelberg, Germany, pp 23–69

Tigano M, Ruotolo R, Dallabona C et al (2015) Elongator-dependent modification of cytoplasmic tRNALysUUU is required for mitochondrial function under stress conditions. Nucleic Acids Res 43:8368–8380

Tukenmez H, Xu H, Esberg A et al (2015) The role of wobble uridine modifications in +1 translational frameshifting in eukaryotes. Nucleic Acids Res. doi:10.1093/nar/gkv832

Tyagi K, Pedrioli PG (2015) Protein degradation and dynamic tRNA thiolation fine-tune translation at elevated temperatures. Nucleic Acids Res 43:4701–4712

Umeda N, Suzuki T, Yukawa M et al (2005) Mitochondria-specific RNA-modifying enzymes responsible for the biosynthesis of the wobble base in mitochondrial tRNAs. Implications for the molecular pathogenesis of human mitochondrial diseases. J Biol Chem 280:1613–1624

Urbonavicius J, Qian Q, Durand JM et al (2001) Improvement of reading frame maintenance is a common function for several tRNA modifications. EMBO J 20:4863–4873

Uusimaa J, Jungbluth H, Fratter C et al (2011) Reversible infantile respiratory chain deficiency is a unique, genetically heterogenous mitochondrial disease. J Med Genet 48:660–668

Vacher J, Grosjean H, Houssier C et al (1984) The effect of point mutations affecting Escherichia coli tryptophan tRNA on anticodon-anticodon interactions and on UGA suppression. J Mol Biol 177:329–342

Van der Veen AG, Schorpp K, Schlieker C et al (2011) Role of the ubiquitin-like protein Urm1 as a noncanonical lysine-directed protein modifier. Proc Natl Acad Sci U S A 108:1763–1770

Vendeix FA, Murphy FV, Cantara WA et al (2012) Human tRNA(Lys3)(UUU) is pre-structured by natural modifications for cognate and wobble codon binding through keto-enol tautomerism. J Mol Biol 416:467–485

Viscomi C, Bottani E, Zeviani M (2015) Emerging concepts in the therapy of mitochondrial disease. BBA-Bioenergetics 1847:544–557

Wang X, Yan Q, Guan MX (2007) Deletion of the MTO2 gene related to tRNA modification causes a failure in mitochondrial RNA metabolism in the yeast *Saccharomyces cerevisiae*. FEBS Lett 581:4228–4234

Watanabe K, Oshima T, Saneyoshi M et al (1974) Replacement of ribothymidine by 5-methyl-2-thiouridine in sequence GT psi C in tRNA of an extreme thermophile. FEBS Lett 43:59–63

Watanabe K, Shinma M, Oshima T et al (1976) Heat-induced stability of tRNA from an extreme thermophile, *Thermus thermophilus*. Biochem Biophys Res Commun 72:1137–1144

Wei FY, Suzuki T, Watanabe S et al (2011) Deficit of tRNA(Lys) modification by Cdkal1 causes the development of type 2 diabetes in mice. J Clin Invest 121:3598–3608

Wei FY, Zhou B, Suzuki T et al (2015) Cdk5rap1-mediated 2-methylthio modification of mitochondrial tRNAs governs protein translation and contributes to myopathy in mice and humans. Cell Metab 21:428–442

Wilson RK, Roe BA (1989) Presence of the hypermodified nucleotide N6-(delta 2-isopentenyl)-2-methylthioadenosine prevents codon misreading by *Escherichia coli* phenylalanyl-transfer RNA. Proc Natl Acad Sci U S A 86:409–413

Xie P, Wei FY, Hirata S et al (2013) Quantitative PCR measurement of tRNA 2-methylthio modification for assessing type 2 diabetes risk. Clin Chem 59:1604–1612

Yasukawa T, Suzuki T, Ishii N et al (2001) Wobble modification defect in tRNA disturbs codon-anticodon interaction in a mitochondrial disease. EMBO J 20:4794–4802

Yokoyama S, Watanabe T, Murao K et al (1985) Molecular mechanism of codon recognition by tRNA species with modified uridine in the first position of the anticodon. Proc Natl Acad Sci U S A 82:4905–4909
Yokoyama S, Watanabe K, Miyazawa T (1987) Dynamic structures and functions of transfer ribonucleic acids from extreme thermophiles. Adv Biophys 23:115–147
Yousef GM, Borgono CA, Michael IP et al (2004) Molecular cloning of a new gene which is differentially expressed in breast and prostate cancers. Tumour Biol 25:122–133
Zeharia A, Shaag A, Pappo O et al (2009) Acute infantile liver failure due to mutations in the TRMU gene. Am J Hum Genet 85:401–407
Zinshteyn B, Gilbert WV (2013) Loss of a conserved tRNA anticodon modification perturbs cellular signaling. PLoS Genet 9, e1003675
Zlotkin SH, Cherian MG (1988) Hepatic metallothionein as a source of zinc and cysteine during the first year of life. Pediatr Res 24:326–329

Regulation of Protein Synthesis via the Network Between Modified Nucleotides in tRNA and tRNA Modification Enzymes in *Thermus thermophilus*, a Thermophilic Eubacterium

Hiroyuki Hori, Ryota Yamagami, and Chie Tomikawa

Contents

Abstract *Thermus thermophilus* is a living organism that utilizes changes in the structural rigidity (flexibility) of the tRNA via multiple nucleoside modifications for the adaptation of protein synthesis process depending on environmental changes. *Thermus thermophilus* grows at a wide range of temperatures (50~83 °C) and can synthesize proteins in response to changes in the temperature. Three distinct modified nucleosides (Gm18, m^5s^2U54, and m^1A58) are found in *T. thermophilus* tRNA, and the combination of these modifications increased the melting temperature of tRNA by nearly 10 °C. The extents of these modifications

H. Hori (✉) • R. Yamagami • C. Tomikawa
Department of Materials Sciences and Biotechnology, Graduate School of Science and Engineering, Ehime University, Bunkyo 3, Matsuyama, Ehime 790-8577, Japan
e-mail: hori@eng.ehime-u.ac.jp

S. Jurga et al. (eds.), *Modified Nucleic Acids in Biology and Medicine*, RNA Technologies, DOI 10.1007/978-3-319-34175-0_4

were relatively low in tRNA from cells that were cultured at 50 °C but increased with the rise in temperature, indicating that the regulation of protein synthesis was temperature dependent. Several common modifications (e.g., m^7G46 and Ψ55) exist in *T. thermophilus* tRNA in addition to Gm18, m^5s^2U54, and m^1A58. These nucleosides and tRNA modification enzymes form a network, and this network regulates protein synthesis by controlling tRNA flexibility under a wide range of temperatures. In this chapter, we describe the regulation of protein synthesis through the network between modified nucleotides in tRNA and tRNA modification enzymes in *T. thermophilus* and that the studies on eubacterial tRNA modification enzymes contribute to the medical science.

Keywords Dihydrouridine • *N*^{7}-Methylguanosine • 5-Methyluridine • Pseudouridine • tRNA modification enzyme

1 Introduction

Transfer RNA (tRNA) is an adaptor molecule that enables the genetic code of nucleic acids to be converted to amino acids in protein. In general, tRNA is composed of 55–95 nucleotides, and the secondary structure can be depicted as the cloverleaf structure (Fig. 1a). Each nucleotide is numbered from the 5′-terminus (Sprinzl et al. 1998). Three nucleotides at positions 34, 35, and 36 form an anticodon (in the case of Fig. 1a, G34A35A36), which decodes the corresponding codon in mRNA on ribosome. An amino acid is attached to the 3′-end (CCA terminus) by the corresponding aminoacyl tRNA synthetase. The tertiary structure of tRNA is represented as "L-shaped" (Fig. 1b and Robertus et al. 1974; Kim et al. 1974; Shi and Moore 2000). Many non-Watson-Crick base pairs are formed in the L-shaped tRNA structure, and these base pairs are called tertiary base pairs. Tertiary base pairs are essential for the maintenance of L-shaped tRNA structure and arrangement of the spatial locations between anticodon and CCA terminus. As shown in Fig. 1a, tRNA contains numerous modified nucleosides. Indeed, more than 80 % of modified nucleosides were identified from tRNA (Machnicka et al. 2013). The primary functions of individual tRNA modifications are linked to the different steps of protein synthesis such as regulation of codon-anticodon interaction (Takai and Yokoyama 2003), maintenance of reading frame during protein synthesis (Björk et al. 1989; Urbonavicius et al. 2001), stabilization of L-shaped tRNA structure (Motorin and Helm 2010), precise recognition by aminoacyl tRNA synthetases (Muramatsu et al. 1988; Perret et al. 1990), and so on.

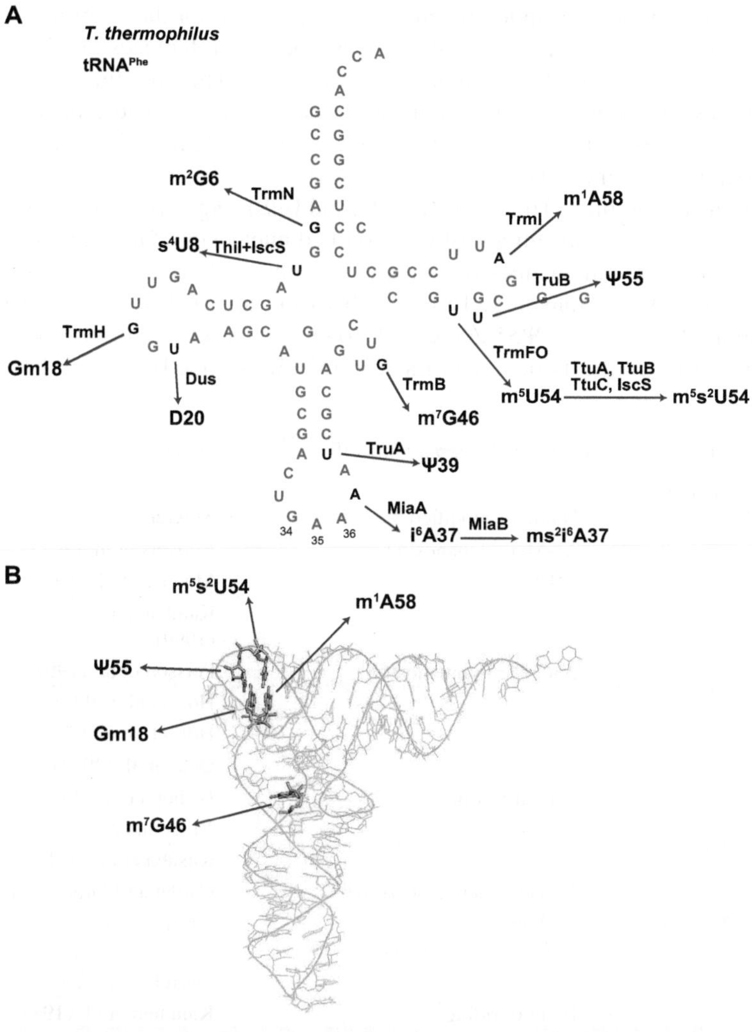

Fig. 1 (**a**) Modified nucleosides and responsible enzymes in *T. thermophilus* $tRNA^{Phe}$ are depicted on the cloverleaf structure. Full names of modified nucleosides and references are shown in Table 1. Three nucleotides in anticodon are numbered. (**b**) The locations of five modified nucleosides, Gm18, m^7G46, m^5s^2U54, Ψ55, and m^1A58, are highlighted by stick models on the L-shaped tRNA structure

2 Modified Nucleosides in *Thermus thermophilus* tRNA and Enzymes Responsible for Such Modifications

Thermus thermophilus is an extreme thermophilic eubacterium that was originally isolated from Mine Hot Spring in Japan and can grow at a wide range of temperatures (50–83 °C) (Oshima and Imahori 1974). Figure 1a shows the modified nucleosides in $tRNA^{Phe}$ from *T. thermophilus* (Grawunder et al. 1992; Tomikawa et al. 2010) and the corresponding enzymes responsible for the modifications. The formal names of the modified nucleosides and the references of these tRNA modification enzymes are summarized in Table 1. The structures of modified nucleosides are shown in Fig. 2. In some cases, the biosynthesis pathway of modified nucleoside includes multiple steps. For example, the biosynthesis of m^5s^2U54 requires the methylation by TrmFO and the sulfur transfer reaction by the sulfur transfer complex (TtuA, TtuB, TtuC, and IscS) (Fig. 1a). Transfer RNA from *T. thermophilus* contains three distinct modified nucleosides Gm18, m^5s^2U54, and m^1A58. These modified nucleosides are assembled at the elbow region in the L-shaped tRNA structure (Fig. 1b and Robertus et al. 1974; Kim et al. 1974; Shi and Moore 2000). Gm18-Ψ55 and G19-C56 tertiary bases are formed between the D- and T-arms. In the T-loop, the m^5s^2U54-m^1A58 reverse Hoogsteen base pair is

Table 1 Modified nucleosides in *T. thermophilus* $tRNA^{Phe}$ and references

Abbreviation in Fig. 1A	Name of modified nucleoside	Reference
m^2G6	N^2-Methylguanosine	Roovers et al. (2012)
s^4U8	4-Thiouridine	Mueller et al. (1998)
		Kambampati and Lauhon (1999)
Gm18	2′-*O*-Methylguanosine	Persson et al. (1997)
		Hori et al. (1998)
		Hori et al. (2002)
		Ochi et al. (2013)
D20	Dihydrouridine	Bishop et al. (2002)
		Yu et al. (2011)
		Kusuba et al. (2015)
i^6A37	N^6-Isopentenyladenosine	Caillet and Droogmans (1988)
ms^2i^6A37	2-Methylthio-N^6-isopentenyladenosine	Esberg et al. (1999)
		Pierrel et al. (2004)
Ψ39	Pseudouridine	Kammen et al. (1988)
m^7G46	N^7-Methylguanosine	De Bie et al. (2003)
		Tomikawa et al. (2010)
m^5U54	5-Methyluridine	Urbonavicius et al. (2005)
		Nishimasu et al. (2009)
		Yamagami et al. (2012)
m^5s^2U54	5-Methyl-2-thiouridine	Watanabe et al. (1976)
		Shigi et al. (2006a)
		Shigi (2012)
Ψ55	Pseudouridine	Nurse et al. (1995)
		Ishida et al. (2011)
m^1A58	N^1-Methyladenosine	Droogmans et al. (2003)
		Takuma et al. (2015)

The abbreviation in Fig. 1A includes the position of each modification. For example, m^2G6 means N^2-methylguanosine at position 6

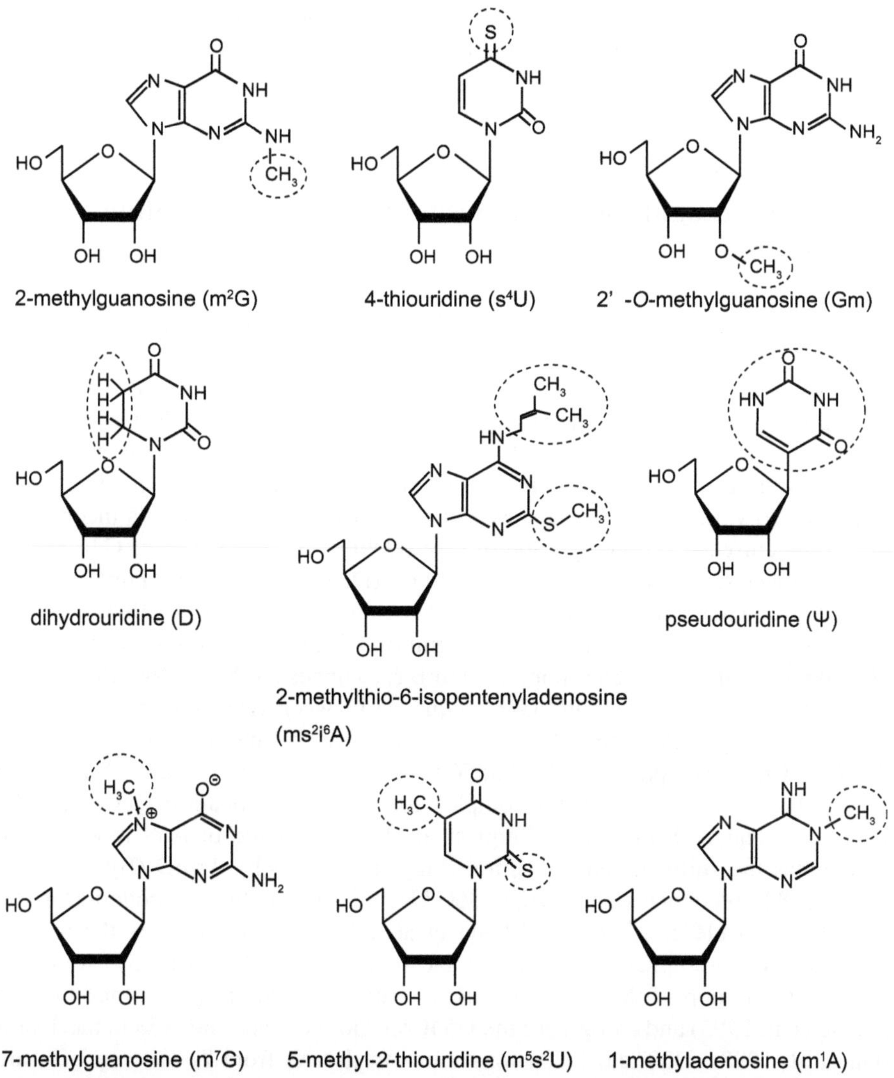

Fig. 2 Structures of modified nucleosides in *T. thermophilus* $tRNA^{Phe}$. The modifications are enclosed by broken circles. Pseudouridine is synthesized by the isomerization of uridine

formed, and this base pair stacks with the other tertiary base pairs. The m^7G46 forms a tertiary base pair with the C13-G22 Watson-Crick base pair in the D-stem. These tertiary interactions of the modified nucleosides stabilize the L-shaped tRNA structure. Indeed, the existence of these modified nucleosides increases the melting temperature of $tRNA^{Phe}$ by nearly 10 °C relative to the unmodified tRNA transcript (Tomikawa et al. 2010). A high-field proton NMR study suggested that m^5s^2U54 in *T. thermophilus* tRNA mainly contributes to the increase in the melting temperature

(Davanloo et al. 1979), consistent with the comparison of melting temperatures of tRNA^{Met}f species from *T. thermophilus* and *Escherichia coli* (Watanabe et al. 1980).

3 tRNA Recognition Mechanism by tRNA Modification Enzymes

In many cases, tRNA modification enzymes recognize the local structure of tRNA, in which a structural element(s) such as a stem-loop structure(s) is contained. For example, TrmH methylates the 2′-OH of ribose of G18 (Persson et al. 1997; Hori et al. 2002) and recognizes the D-arm structure with two bulges (Hori et al. 2003). The L-shaped tRNA structure with conserved sequences enhances the activity of TrmH (Hori et al. 1998). The site where TrmH methylates tRNA is changeable according to variations in sequences and sizes of the D-arm, and this mechanism brings the correct 2′-*O*-methylation of G18 (Ochi et al. 2010). The m^{7}G46 formation in eubacterial tRNA is catalyzed by TrmB (De Bie et al. 2003). TrmB proteins from thermophiles have a long C-terminal region (Okamoto et al. 2004), and this region is required for the accurate recognition of the methylation site by TrmB (Tomikawa et al. 2008). Thermophilic TrmB recognizes the NNG*NN (the asterisk represents the methylation site, G46) sequence between two stem-loop structures (Okamoto et al. 2004). TrmFO, which brings the m^{5}U54 modification, recognizes the G53-C61 base pair and U54U55C56 sequence in the T-arm (Yamagami et al. 2012). The sulfur transfer complex for s^{2}U54 modification recognizes the U55, C56, and A58 in T-arm (Shigi et al. 2002), and the m^{1}A58 modification accelerates the sulfur transfer reaction (Shigi et al. 2006b). TrmD (Byström and Björk 1982) recognizes the purine 36G37 sequence in the anticodon-arm-like microhelix (Brulé et al. 2004; Takeda et al. 2006). TrmI catalyzes the m^{1}A58 modification (Droogmans et al. 2003) and recognizes C56, purine 57, A58, and U60 in the T-loop (Takuma et al. 2015). TruB catalyzes the Ψ55 modification (Nurse et al. 1995) and recognizes the U54U55C56 sequence and A58 in the T-arm (Gu et al. 1998). D20 and D20a formations in tRNAs from *T. thermophilus* are catalyzed by Dus (Kusuba et al. 2015). Although the tRNA recognition mechanism by *T. thermophilus* Dus is not completely clarified, a crystal structure of the Dus-tRNA complex suggests that Dus requires the tertiary interaction between the D- and T-arms for tRNA recognition (Yu et al. 2011). In the case of eukaryotic tRNA modification enzymes, several enzymes such as Trm5 [eukaryotic tRNA (m^1 G37) methyltransferase] require the tertiary interaction between the D- and T-arms (Christian and Hou 2007; Goto-Ito et al. 2009). However, the requirement of T- and D-arm interaction by eubacterial Dus is a rare case in the tRNA modification enzymes from eubacteria.

The target site for modification is often embedded in the L-shaped tRNA structure (Fig. 1b). Consequently, in many cases, tRNA recognition by tRNA modification enzymes seems to involve multiple steps (initial binding and then induced fit processes). Indeed, in the crystal structure of archaeosine tRNA-guanine transglycosylase and tRNA complex, the L-shaped tRNA structure was changed to the λ form (Ishitani et al. 2003), demonstrating the existence of an induced fit process. Although it is very difficult to prepare intermediate complexes, we recently analyzed the initial binding and changes in the structure of TrmH by stopped-flow presteady-state kinetic analysis (Ochi et al. 2010, 2013). TrmH binds to tRNA within 10 ms in the initial binding process, in which the substrate and non-substrate (methylated) tRNAs are not distinguished. Methylated tRNA is then excluded from the complex due to steric hindrance between the methyl groups in the tRNA and *S*-adenosyl-L-methionine (methyl group donor) before the induced-fit process occurs. The advantage of this mechanism is that methylated tRNA does not severely inhibit the methyl transfer reaction as a competitive inhibitor. Subsequently, in the induced-fit process, which takes more than 50 ms, G18 is recognized and the ribose is introduced into the catalytic pocket. Therefore, in the reactions by tRNA modification enzymes from *T. thermophilus* (Fig. 1a and Table 1), the L-shaped tRNA structure seems to be often disrupted.

4 Network Between Modified Nucleotides in tRNA and tRNA Modification Enzymes in *T. thermophilus*

4.1 Temperature-Dependent Modifications and Their Regulation of the Poly(U)-Dependent Poly(Phe) Synthesis

Figure 3a shows the modification levels of Gm18, m^7G46, m^5s^2U54, and m^1A58 in class I tRNAs, which contain a regular size variable region, from *T. thermophilus* cells that were cultured at various temperatures (Watanabe et al. 1976, 1984; Shigi et al. 2006b; Tomikawa et al. 2010; Yamagami et al. unpublished results). The modification level of m^7G46 is nearly 100 %, irrespective of the culture temperatures. In contrast, the modification levels of Gm18, m^5s^2U54, and m^1A58 increase according to the increase of culture temperatures. Since Ψ modifications are present in multiple sites in tRNA, we could not measure the level of Ψ55 in class I tRNAs precisely. However, the levels of Ψ55 in $tRNA^{Met}f$, in which Ψ is present only at position 55, from the cells cultured at 50 and 80 °C, were near 100 % (Ishida et al. 2011), suggesting that the modification level of Ψ55 in tRNA is near 100 % irrespective of the culture temperatures, like m^7G46.

The differences of modification levels in tRNAs reflect on protein synthesis activity (Fig. 3b and Yokoyama et al. 1987). Transfer RNA^{Phe} from cells cultured at 80 °C efficiently synthesizes poly(U) at high temperatures (above 65 °C) (Fig. 3b filled circles). In contrast, $tRNA^{Phe}$ from cells cultured at 50 °C, in which the levels of the three modifications are low, works efficiently at low temperatures (Fig. 3b

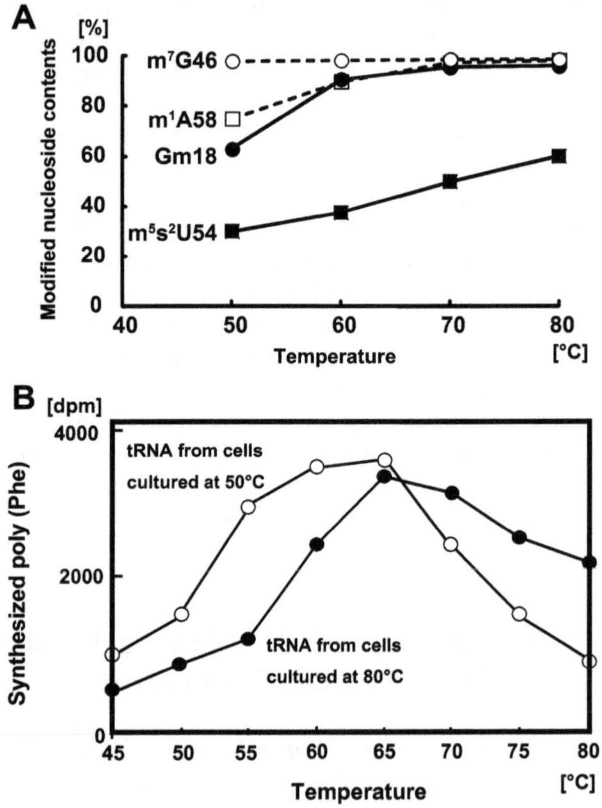

Fig. 3 (**a**) Contents of m^7G46, m^1A58, Gm18, and m^5s^2U54 in class I tRNAs at various temperatures. The contents of m^7G46, m^1A58, and Gm18 were calculated from the methyl group acceptance activities of class I tRNAs. The contents of m^5s^2U54 were according to reference previous study (Watanabe et al. 1976). (**b**) Transfer RNA^{Phe} from cells cultured at 80 °C (*filled circles*) can efficiently synthesize poly(U) at high temperatures. In contrast, at low temperatures, $tRNA^{Phe}$ from cells cultured at 50 °C (*open circles*) can work more efficiently than tRNA from cells cultured at 80 °C

open circles). Thus, the levels of three modified nucleosides, Gm18, m^5s^2U54, and m^1A58, in tRNA, control the elongation of translation via the flexibility of the tRNA. Under natural conditions, the temperature of hot springs can change dramatically due to several factors such as the overflow of hot spring water, snowfall, and the influx of river water. Therefore, the adaptation of protein synthesis to temperature change through the regulation of modification levels in tRNA is required for survival of *T. thermophilus*.

4.2 Growth Phenotype of the trmB Gene Disruption Strain

At the beginning of the twenty-first century, the mechanisms of regulation of these modifications (Gm18, m^5s^2U54, and m^1A58) remained unknown. Initially, we assumed that transcriptional and/or translational regulation of the expression of tRNA modification enzymes was involved in the regulation of the three modifications. However, unexpectedly, we have observed that the phenomenon can be simply explained by the RNA recognition mechanisms of tRNA modification enzymes (Tomikawa et al. 2010; Ishida et al. 2011). There are common modified nucleosides such as m^7G46 and Ψ55 in $tRNA^{Phe}$ as well as Gm18, m^5s^2U54, and m^1 A58 (Fig. 1a). The m^7G46 and Ψ55 modifications are catalyzed by TrmB and TruB, respectively. When the *trmB* gene was disrupted, this gene disruption (*ΔtrmB*) strain could not grow at high temperatures (80 °C) (Tomikawa et al. 2010). To investigate what happened in the *ΔtrmB* strain at high temperatures, we tested growth phenotypes of this strain under several conditions. We found that the *ΔtrmB* strain could survive at 80 °C when the strain was cultured at 70 °C until the middle log phase. Under this condition, the modification levels of several modified nucleosides (Gm18, m^1G37, and m^5s^2U54) in class I tRNAs from the *ΔtrmB* strain were clearly low compared to those from the wild-type strain. This observation was confirmed by in vitro experiments with the tRNA transcript and purified enzymes. These experimental results provided a first clue to clarify the network between modified nucleosides in tRNA and tRNA modification enzymes in *T. thermophilus*.

4.3 Network Between Modified Nucleosides in tRNA and tRNA Modification Enzymes at High Temperatures

Until now, we have investigated seven gene disrupt ant strains of *T. thermophilus*, in which the modified nucleoside(s) is absent. Figure 4 summarizes the network between modified nucleosides in tRNA and tRNA modification enzymes in *T. thermophilus*. At high temperature (80 °C), the m^7G46 modification functions as a key factor. Although the m^7G46 modification increases the melting temperature of $tRNA^{Phe}$ transcript only by 0.1 °C, the presence of this modification functions as a marker of precursor tRNA and increases the activity of TrmD (for m^1G37), TrmH (for Gm18), and TrmI (for m^1A58) (Tomikawa et al. 2010). The m^1 A58 modification is a positive determinant for the sulfur transfer complex (TtuA, TtuB, TtuC, and IscS), which catalyzes the 2-thio modification of m^5U54 (Shigi et al. 2002; Shigi 2012, 2014). As a result, the increased m^1A58 modification increases the s^2U54 modification. Therefore, the presence of m^7G46 increases the modification levels of m^1G37, Gm18, m^1A58, and 2-thiolation of m^5U54. The presence of m^5U54 accelerates the modification speed of m^1A58 by TrmI (Yamagami et al., unpublished results). These effects coordinately contribute to the stabilization of L-shaped tRNA structure at high temperatures. The stabilized D- and T-arm interaction is required for effective D20 formation in $tRNA^{Phe}$ by Dus at high temperatures (Kusuba et al. 2015).

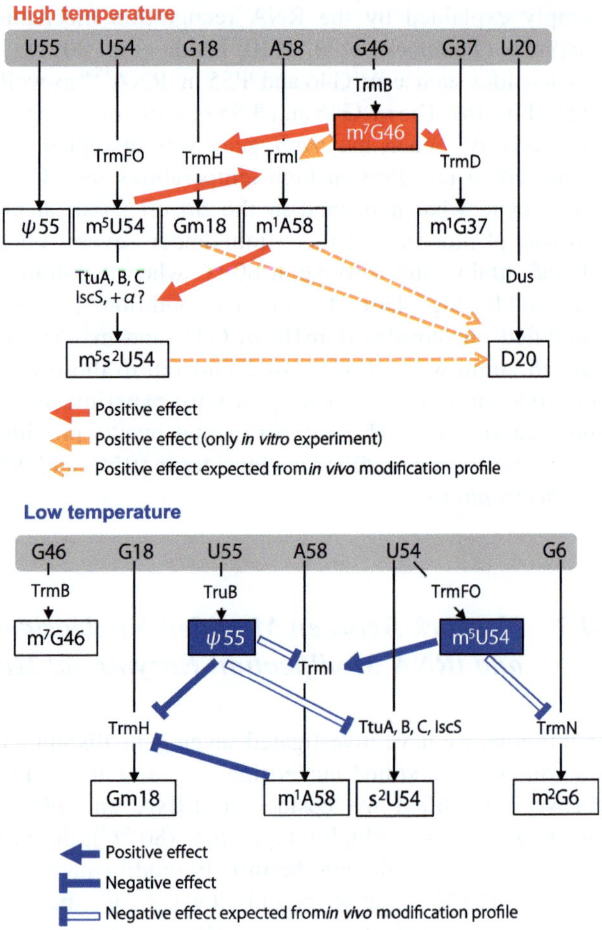

Fig. 4 Networks between modified nucleosides in tRNA and tRNA modification enzymes in *T. thermophilus*. This figure summarizes our recent studies (Tomikawa et al. 2010; Ishida et al. 2011; Yamagami et al. 2012; Kusuba et al. 2015; Yamagami et al., unpublished results 2015)

4.4 Network Between Modified Nucleosides in tRNA and tRNA Modification Enzymes at Low Temperatures

The presence of Ψ55 modification rigidifies the local structure of tRNA. At a low temperature (50 °C), this rigidity slows down the formation speed of other modified nucleosides around Ψ55, i.e., Gm18 by TrmH, s^2U54 by the sulfur transfer complex, and m^1A by TrmI (Fig. 4 lower panel and Ishida et al. 2011). At 65 °C, this effect is weakened, and the presence of Ψ55 does not affect the formation speed of other modified nucleosides. The m^5U54 modification causes similar effects on other modified nucleosides; however, the effect by m^5U54 is considerably weaker compared to that of Ψ55 (Yamagami et al. unpublished results). Furthermore, the analysis of $tRNA^{Phe}$ from the *ΔtrmFO* strain, in which m^5U54 is absent, suggested that m^5U54 slows down the m^2G6 modification by TrmN. However, this phenomenon has not been shown by in vitro experiments. In sum, Ψ55 and m^5U54 coordinately regulate the formation of other modified nucleosides at low temperature, with Ψ55 playing the main role. Therefore, in the case of *T. thermophilus*, TruB is essential for survival at 50 °C (Ishida et al. 2011).

4.5 Degradation of Hypo-modified tRNA at High Temperatures

The lack of m^7G46 modification in tRNA causes hypo-modifications in the tRNA through the network. The decrease of modification levels in tRNA decreases the melting temperature of tRNA and results in the degradation of $tRNA^{Phe}$ and $tRNA^{Lys}$ at 80 °C (Tomikawa et al. 2010). The degradation decreases protein synthesis activity and impairs growth (Fig. 5). Until now, although the degradation of some

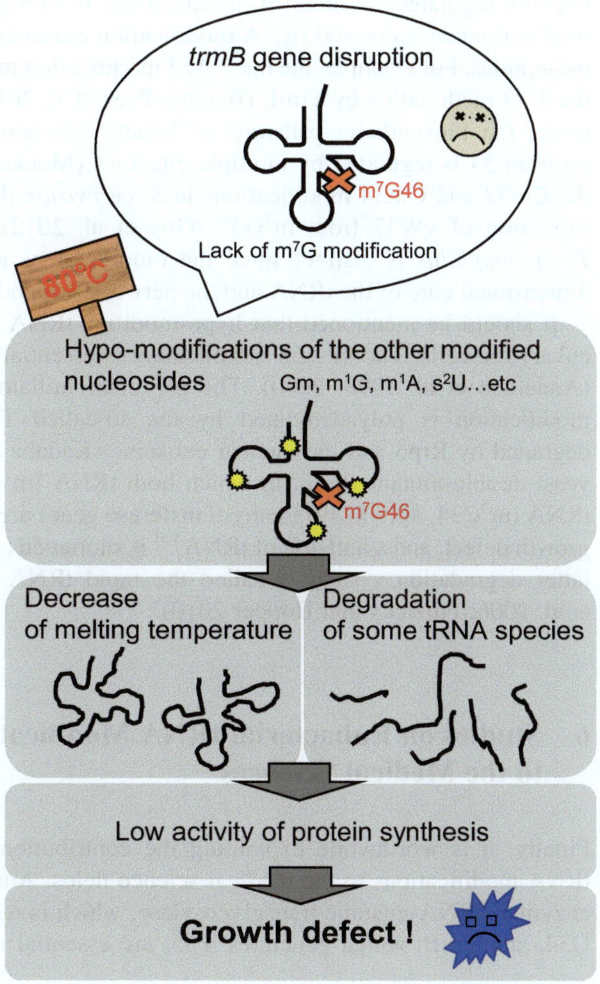

Fig. 5 Schematics of growth impairment caused by the absence of m^7G46 modification

tRNA species is only observed in the *ΔtrmB* strain, the absence of sulfur transfer complex for s^2U54 modification and/or TrmI for m^1A58 may cause the degradation of tRNA that is observed in the *ΔtrmB* strain. These modification enzymes were reported to be essential for survival at high temperatures (Shigi et al. 2006b; Droogmans et al. 2003).

5 Comparison of the *T. thermophilus* Network and Degradation of Hypo-modified tRNA with Those in Mesophiles

In the case of *T. thermophilus*, the network is a mechanism to maintain the rigidity (flexibility) of tRNA at a wide range of temperatures. However, consequently, the network regulates the order of modifications in tRNA. Similar networks between modified nucleosides and tRNA modification enzymes have also been reported in mesophiles. For example, the ms^2i^6A37 modification in *E. coli* tRNA is required for the 2′-*O*-methylation by TrmL (Benítez-Páez et al. 2010; Liu et al. 2013). Furthermore, the biosynthetic pathway of 5-methylaminomethyluridine derivatives at position 34 is regulated by multiple enzymes (Moukadiri et al. 2014). Moreover, the Cm32 and Gm34 modifications in *S. cerevisiae* $tRNA^{Phe}$ are required for the formation of yW37 from m^1G37 (Guy et al. 2012). However, the network in *T. thermophilus* is distinct since the modifications are almost all in the three-dimensional core of the tRNA and the network responds to environmental changes.

It should be mentioned that hypo-modified tRNA is aggressively degraded in eukaryotic cells. The m^1A58 modification is essentially required for cell viability (Anderson et al. 1998, 2000). The precursor initiator $tRNA^{Met}$ without m^1A58 modification is polyadenylated by the so-called TRAMP complex and then degraded by Rrp6 and the nuclear exosome (Kadaba et al. 2004). Furthermore, a yeast double-mutant strain, in which both tRNA (m^7G46) methyltransferase and tRNA (m^5C34, 40, 48, 49) methyltransferase genes are disrupted, exhibits a severe growth defect, and a half-life of $tRNA^{Val}$ is shortened (Alexandrov et al. 2006). The latter degradation system is called the rapid tRNA decay system (Alexandrov et al. 2006; Phizicky and Hopper 2010).

6 Studies on Eubacterial tRNA Modifications Contribute to the Medical Sciences

Finally, it is worthwhile explaining the contributions of studies on eubacterial tRNA modifications to the medical science fields. Among the tRNA modification enzymes, tRNA-guanine transglycosylase, which is required for the production of Q34, and TruB, which generates Ψ55, are essential infection factors in *Shigella flexneri* (Durand et al. 1994) and *Pseudomonas aeruginosa* (Saga et al. 1997), respectively. Therefore, the studies on eubacterial tRNA modifications are required for understanding these mechanisms of infection. Furthermore, several modified nucleosides in tRNA are generated by different enzymes between eubacteria and eukaryotes. For example, the m^{5}U54 modification in eukaryote is formed by an *S*-adenosyl-L-methionine-dependent tRNA methyltransferase, Trm2 (Nordlund et al. 2000). In contrast, in almost all gram-positive and some gram-negative eubacteria, a FAD/folate-dependent tRNA methyltransferase, TrmFO, generates the m^{5}U54 modification (Urbonavicius et al. 2005). Similarly, the m^{1}A58 modification in eukaryotic and eubacterial tRNAs is catalyzed by the Trm6-Trm62 complex (Anderson et al. 2000) and TrmI (Droogmans et al. 2003), respectively. Furthermore, m^{1}G37 modification in eukaryotic and eubacterial tRNAs is catalyzed by Trm5 (Björk et al. 2001; Brulé et al. 2004; Christian et al. 2004) and TrmD (Byström and Björk 1982), respectively. Moreover, m^{7}G46 modification in eukaryotic and eubacterial tRNAs is formed by the Trm8-Trm82 complex (Alexandrov et al. 2002) and TrmB (De Bie et al. 2003), respectively. Given that the m^{7}G46 modification in tRNA is essentially required for infection by the phytopathogenic fungus *Colletotrichum lagenarium* (Takano et al. 2006), tRNA (m^{7}G46) methyltransferase can be the target for drug design. Considering that TrmFO (Nishimasu et al. 2009; Yamagami et al. 2012), TrmI (Barraud et al. 2008; Takuma et al. 2015; Dégut et al. 2016), TrmD (Ahn et al. 2003; Elkins et al. 2003; Liu et al. 2003; Takeda et al. 2006; Christian and Hou 2007; Ito et al. 2015),and TrmB (Okamoto et al. 2004; Zegers et al. 2006) are structurally and enzymatically different from their counterparts in eukaryotes, these enzymes are convincing targets for drug design of antibiotics. Moreover, the Gm18 modification by TrmH suppresses immunostimulation through toll-like receptor 7 (Gehrig et al. 2012; Jöckel et al. 2012). Therefore, enterobacteria exploit the Gm18 modification in tRNA to avoid the host immune system. In addition, given that the Gm18-modified tRNA acts as an antagonist of toll-like receptor 7 (Jöckel et al. 2012), Gm18-modified tRNA might be an effective anti-inflammatory drug.

Therefore, studies on eubacterial tRNA modification systems can contribute to medical sciences.

Acknowledgments This work was supported by Grant-in-Aid for JSPS Fellows 20-4827 (to C. T.) and 26-8015 (to R. Y.), Grant-in-Aid for Scientific Research 19350087 and 23350081 (to H. H.), and a Grant-in-Aid for Exploratory Research 24655156 (to H. H.) from the Japan Society for the Promotion of Science (JSPS).

References

Ahn HJ, Kim HW, Yoon HJ et al (2003) Crystal structure of tRNA (m^{1}G37) methyltransferase: insights into tRNA recognition. EMBO J 22:2593–2603

Alexandrov A, Martzen MR, Phizicky EM (2002) Two proteins that form a complex are required for 7-methylguanosine modification of yeast tRNA. RNA 8:1253–1266

Alexandrov A, Chernyakov I, Gu W et al (2006) Rapid tRNA decay can result from lack of nonessential modifications. Mol Cell 21:87–96
Anderson J, Phan L, Cuesta R et al (1998) The essential Gcd10p-Gcd14p nuclear complex is required for 1-methyladenosine modification and maturation of initiator methionyl-tRNA. Genes Dev 12:3650–3652
Anderson J, Phan L, Hinnebusch AG (2000) The Gcd10p/Gcd14p complex is the essential two-subunit tRNA(1-methyladenosine) methyltransferase of *Saccharomyces cerevisiae*. Proc Natl Acad Sci U S A 97:5173–5178
Barraud P, Golinelli-Pimpaneau B, Atmanene C et al (2008) Crystal structure of *Thermus thermophilus* tRNA m^1A58 methyltransferase and biophysical characterization of its interaction with tRNA. J Mol Biol 377:535–550
Benítez-Páez A, Villarroya M, Douthwaite S et al (2010) YibK is the 2′-*O*-methyltransferase TrmL that modifies the wobble nucleotide in *Escherichia coli* tRNA(Leu) isoacceptors. RNA 16:2131–2143
Bishop AC, Xu J, Johnson RC et al (2002) Identification of the tRNA-dihydrouridine synthase family. J Biol Chem 277:25090–25095
Björk GR, Wikstrom PM, Byström AS (1989) Prevention of translational frameshifting by the modified nucleoside 1-methylguanosine. Science 244:986–989
Björk GR, Jacobsson K, Nilsson K et al (2001) A primordial tRNA modification required for the evolution of life? EMBO J 20:231–239
Brulé H, Elliott M, Redlak M et al (2004) Isolation and characterization of the human tRNA-(N^1 G37) methyltransferase (TRM5) and comparison to the *Escherichia coli* TrmD protein. Biochemistry 43:9243–9255
Byström AS, Björk GR (1982) Chromosomal location and cloning of the gene (*trmD*) responsible for the synthesis of tRNA (m^1G) methyltransferase in *Escherichia coli* K-12. Mol Gen Genet 188:440–446
Caillet J, Droogmans L (1988) Molecular cloning of the *Escherichia coli miaA* gene involved in the formation of delta 2-isopentenyl adenosine in tRNA. J Bacteriol 170:4147–4152
Christian T, Hou YM (2007) Distinct determinants of tRNA recognition by the TrmD and Trm5 methyl transferases. J Mol Biol 373:623–632
Christian T, Evilia C, Williams S et al (2004) Distinct origins of tRNA (m^1G37) methyltransferase. J Mol Biol 339:707–719
Davanloo P, Sprinzl M, Watanabe K et al (1979) Role of ribothymidine in the thermal stability of transfer RNA as monitored by proton magnetic resonance. Nucleic Acids Res 6:1571–1581
De Bie LG, Roovers M, Oudjama Y et al (2003) The *yggH* gene of *Escherichia coli* encodes a tRNA (m^7G46) methyltransferase. J Bacteriol 185:3238–3243
Dégut C, Ponchon L, Folly-Klan M et al (2016) The m^1A58 modification in eubacterial tRNA: an overview of tRNA recognition and mechanism of catalysis by TrmI. Biophys Chem 210:27–34
Droogmans L, Roovers M, Bujnicki JM et al (2003) Cloning and characterization of tRNA (m^1 A58) methyltransferase (TrmI) from *Thermus thermophilus* HB27, a protein required for cell growth at extreme temperatures. Nucleic Acids Res 31:2148–2156
Durand JM, Okada N, Tobe T et al (1994) *vacC*, a virulence-associated chromosomal locus of *Shigella flexneri*, is homologous to *tgt*, a gene encoding tRNA-guanine transglycosylase (Tgt) of *Escherichia coli* K-12. J Bacteriol 176:4627–4634
Elkins PA, Watts JM, Zalacain M et al (2003) Insights into catalysis by a knotted TrmD tRNA methyltransferase. J Mol Biol 333:931–949
Esberg B, Leung HC, Tsui HC et al (1999) Identification of the *miaB* gene, involved in methylthiolation of isopentenylated A37 derivatives in the tRNA of *Salmonella typhimurium* and *Escherichia coli*. J Bacteriol 181:7256–7265
Gehrig S, Eberle ME, Botschen F et al (2012) Identification of modifications in microbial, native tRNA that suppress immunostimulatory activity. J Exp Med 209:225–233
Goto-Ito S, Ito T, Kuratani M et al (2009) Tertiary structure checkpoint at anticodon loop modification in tRNA functional maturation. Nat Struct Mol Biol 16:1109–1115

Grawunder U, Schön A, Sprintzl M (1992) Sequence and base modifications of two phenylalanine-tRNAs from *Thermus thermophilus* HB8. Nucleic Acids Res 20:137

Gu X, Yu M, Ivanetich KM et al (1998) Molecular recognition of tRNA by tRNA pseudouridine 55 synthase. Biochemistry 37:339–343

Guy MP, Podyma BM, Preston MA et al (2012) Yeast Trm7 interacts with distinct proteins for critical modifications of the $tRNA^{Phe}$ anticodon loop. RNA 18:1921–1933

Hori H, Yamazaki N, Matsumoto T et al (1998) Substrate recognition of tRNA (Guanosine-2′-)-methyltransferase from *Thermus thermophilus* HB27. J Biol Chem 273:25721–25727

Hori H, Suzuki T, Sugawara K et al (2002) Identification and characterization of tRNA (Gm18) methyltransferase from *Thermus thermophilus* HB8: domain structure and conserved amino acid sequence motifs. Genes Cells 7:259–272

Hori H, Kubota S, Watanabe K et al (2003) *Aquifex aeolicus* tRNA (Gm18) methyltransferase has unique substrate specificity: tRNA recognition mechanism of the enzyme. J Biol Chem 278:25081–25090

Ishida K, Kunibayashi T, Tomikawa C et al (2011) Pseudouridine at position 55 in tRNA controls the contents of other modified nucleotides for low-temperature adaptation in the extreme-thermophilic eubacterium *Thermus thermophilus*. Nucleic Acids Res 39:2304–2318

Ishitani R, Nureki O, Nameki N et al (2003) Alternative tertiary structure of tRNA for recognition by a posttranscriptional modification enzyme. Cell 113:383–394

Ito T, Masuda I, Yoshida K et al (2015) Structural basis for methyl-donor-dependent and sequence-specific binding to tRNA substrates by knotted methyltransferase TrmD. Proc Natl Acad Sci U S A 112:E4197–4205

Jöckel S, Nees G, Sommer R et al (2012) The 2′-*O*-methylation status of a single guanosine controls transfer RNA-mediated Toll-like receptor 7 activation or inhibition. J Exp Med 209:235–241

Kadaba S, Krueger A, Trice T et al (2004) Nuclear surveillance and degradation of hypomodified initiator $tRNA^{Met}$ in *S. cerevisiae*. Genes Dev 18:1227–1240

Kambampati R, Lauhon CT (1999) IscS is a sulfurtransferase for the in vitro biosynthesis of 4-thiouridine in *Escherichia coli* tRNA. Biochemistry 38:16561–16568

Kammen HO, Marvel CC, Hardy L et al (1988) Purification, structure, and properties of *Escherichia coli* tRNA pseudouridine synthase I. J Biol Chem 263:2255–2263

Kim SH, Suddath FL, Quigley GJ et al (1974) Three-dimensional tertiary structure of yeast phenylalanine transfer RNA. Science 185:435–440

Kusuba H, Yoshida T, Iwasaki E et al (2015) In vitro dihydrouridine formation by tRNA dihydrouridine synthase from *Thermus thermophilus*, an extreme-thermophilic eubacterium. J Biochem 158(6):513–21

Liu J, Wang W, Shin DH et al (2003) Crystal structure of tRNA (m^1G37) methyltransferase from *Aquifex aeolicus* at 2.6 A resolution: a novel methyltransferase fold. Proteins 53:326–328

Liu RJ, Zhou M, Fang ZP et al (2013) The tRNA recognition mechanism of the minimalist SPOUT methyltransferase, TrmL. Nucleic Acids Res 41:7828–7842

Machnicka MA, Milanowska K, Osman Oglou O et al (2013) MODOMICS: a database of RNA modification pathways—2013 update. Nucleic Acids Res 41:D262–267

Motorin Y, Helm M (2010) tRNA stabilization by modified nucleotides. Biochemistry 49:4934–4944

Moukadiri I, Garzón MJ, Björk GR et al (2014) The output of the tRNA modification pathways controlled by the *Escherichia coli* MnmEG and MnmC enzymes depends on the growth conditions and the tRNA species. Nucleic Acids Res 42:2602–2623

Mueller EG, Buck CJ, Palenchar PM et al (1998) Identification of a gene involved in the generation of 4-thiouridine in tRNA. Nucleic Acids Res 26:2606–2610

Muramatsu T, Nishikawa K, Nemoto F et al (1988) Codon and amino-acid specificities of a transfer RNA are both converted by a single post-transcriptional modification. Nature 336:79–181

Nishimasu H, Ishitani R, Yamashita K et al (2009) Atomic structure of a folate/FAD-dependent tRNA T54 methyltransferase. Proc Natl Acad Sci U S A 106:8180–8185

Nordlund ME, Johansson JO, von Pawel-Rammingen U et al (2000) Identification of the TRM2 gene encoding the tRNA(m^5U54)methyltransferase of *Saccharomyces cerevisiae*. RNA 6:844–860

Nurse K, Wrzesinski J, Bakin A et al (1995) Purification, cloning, and properties of the tRNA Ψ55 synthase from *Escherichia coli*. RNA 1:102–112

Ochi A, Makabe K, Kuwajima K et al (2010) Flexible recognition of the tRNA G18 methylation target site by TrmH methyltransferase through first binding and induced fit processes. J Biol Chem 285:9018–9029

Ochi A, Makabe K, Yamagami R et al (2013) The catalytic domain of topological knot tRNA methyltransferase (TrmH) discriminates between substrate tRNA and nonsubstrate tRNA *via* an induced-fit process. J Biol Chem 288:25562–25574

Okamoto H, Watanabe K, Ikeuchi Y et al (2004) Substrate tRNA recognition mechanism of tRNA (m7G46) methyltransferase from *Aquifex aeolicus*. J Biol Chem 279:49151–49159

Oshima T, Imahori K (1974) Description of *Thermus thermophilus* (Yoshida and Oshima) comb. nov., a Nonsporulating Thermophilic Bacterium form a Japanese Thermal Spa. Int J Syst Bacteriol 24:102–112

Perret V, Garcia A, Grosjean H et al (1990) Relaxation of a transfer RNA specificity by removal of modified nucleotides. Nature 344:787–789

Persson BC, Jager G, Gustafsson C (1997) The spoU gene of *Escherichia coli*, the fourth gene of the spoT operon, is essential for tRNA (Gm18) 2′-*O*-methyltransferase activity. Nucleic Acids Res 25:4093–4097

Phizicky EM, Hopper AK (2010) tRNA biology charges to the front. Genes Dev 24:1832–1860

Pierrel F, Douki T, Fontecave M et al (2004) MiaB protein is a bifunctional radical-S-adenosylmethionine enzyme involved in thiolation and methylation of tRNA. J Biol Chem 279:47555–47563

Robertus JD, Ladner JE, Finch JT et al (1974) Structure of yeast phenylalanine tRNA at 3 Å resolution. Nature 250:546–551

Saga AE, Vasil AI, Vasil ML (1997) Molecular characterization of mutants affected in the osmoprotectant-dependent induction of phospholipase C in *Pseudomonas aeruginosa* PAO1. Mol Microbiol 23:43–56

Shi H, Moore PB (2000) The crystal structure of yeast phenylalanine tRNA at 1.93 A resolution: a classic structure revisited. RNA 6:1091–1105

Shigi N (2012) Posttranslational modification of cellular proteins by a ubiquitin-like protein in bacteria. J Biol Chem 287:17568–17577

Shigi N (2014) Biosynthesis and functions of sulfur modifications in tRNA. Front Genet 5:67

Shigi N, Suzuki T, Tamakoshi M et al (2002) Conserved bases in the TΨC loop of tRNA are determinants for thermophile-specific 2-thiouridylation at position 54. J Biol Chem 277:39128–39135

Shigi N, Sakaguchi Y, Suzuki T et al (2006a) Identification of two tRNA thiolation genes required for cell growth at extremely high temperatures. J Biol Chem 281:14296–14306

Shigi N, Suzuki T, Terada T et al (2006b) Temperature-dependent biosynthesis of 2-thioribothymidine of *Thermus thermophilus* tRNA. J Biol Chem 281:2104–2113

Sprinzl M, Horn C, Brown M et al (1998) Compilation of tRNA sequences and sequences of tRNA genes. Nucleic Acids Res 26:148–153

Takai K, Yokoyama S (2003) Roles of 5-substituents of tRNA wobble uridines in the recognition of purine-ending codons. Nucleic Acids Res 31:6383–6391

Takano Y, Takayanagi N, Hori H et al (2006) A gene involved in modifying transfer RNA is required for fungal pathogenicity and stress tolerance of *Colletotrichum lagenarium*. Mol Microbiol 60:81–92

Takeda H, Toyooka T, Ikeuchi Y et al (2006) The substrate specificity of tRNA (m^1G37) methyltransferase (TrmD) from *Aquifex aeolicus*. Genes Cells 11:1353–1365

Takuma H, Ushio N, Minoji M et al (2015) Substrate tRNA recognition mechanism of eubacterial tRNA (m^1A58) methyltransferase (TrmI). J Biol Chem 290:5912–5925

Tomikawa C, Ochi A, Hori H (2008) The C-terminal region of thermophilic tRNA (m7G46) methyltransferase (TrmB) stabilizes the dimer structure and enhances fidelity of methylation. Proteins 71:1400–1408

Tomikawa C, Yokogawa T, Kanai T et al (2010) N^7-Methylguanine at position 46 (m^7G46) in tRNA from *Thermus thermophilus* is required for cell viability through a tRNA modification network. Nucleic Acids Res 38:942–957

Urbonavicius J, Qian Q, Durand JM et al (2001) Improvement of reading frame maintenance is a common function for several tRNA modifications. EMBO J 20:4863–4873

Urbonavicius J, Skouloubris S, Myllykallio H et al (2005) Identification of a novel gene encoding a flavin-dependent tRNA: m5U methyltransferase in bacteria—evolutionary implications. Nucleic Acids Res 33:3955–3964

Watanabe K, Shinma M, Oshima T et al (1976) Heat-induced stability of tRNA from an extreme thermophile, *Thermus thermophilus*. Biochem Biophys Res Commun 72:1137–1144

Watanabe K, Oshima T, Iijima K et al (1980) Purification and thermal stability of several amino acid-specific tRNAs from an extreme thermophile, *Thermus thermophilus* HB8. J Biochem 87:1–13

Watanabe K, Himeno H, Ohta T (1984) Selective utilization of 2-thioribothymidine- and ribothymidine-containing tRNAs by the protein synthetic systems of *Thermus thermophilus* HB 8 depending on the environmental temperature. J Biochem 96:1625–1632

Yamagami R, Yamashita K, Nishimasu H et al (2012) The tRNA recognition mechanism of folate/FAD-dependent tRNA methyltransferase (TrmFO). J Biol Chem 287:42480–42494

Yokoyama S, Watanabe K, Miyazawa T (1987) Dynamic structures and functions of transfer ribonucleic acids from extreme thermophiles. Adv Biophys 23:115–147

Yu F, Tanaka Y, Yamashita K et al (2011) Molecular basis of dihydrouridine formation on tRNA. Proc Natl Acad Sci U S A 108:19593–19598

Zegers I, Gigot D, van Vliet F et al (2006) Crystal structure of *Bacillus subtilis* TrmB, the tRNA (m^7G46) methyltransferase. Nucleic Acids Res 34:1925–1934

Post-Transcriptional Modifications of RNA: Impact on RNA Function and Human Health

Kyla M. Frohlich, Kathryn L. Sarachan, Gabrielle C. Todd, Maria Basanta-Sanchez, Ville Y.P. Väre, and Paul F. Agris

Contents

Abstract Epitranscriptomics is the study of global modification patterns to both coding and noncoding RNA. Understanding the epitranscriptomic profile of disease states or individual patients is imperative to understanding human health and molecular disease pathology. Modifications have long been established as

K.M. Frohlich • K.L. Sarachan • G.C. Todd • M. Basanta-Sanchez • V.Y.P. Väre • P.F. Agris (✉)
The RNA Institute, University at Albany-SUNY, 1400 Washington Ave., Albany, NY 12222, USA
e-mail: PAgris@albany.edu

S. Jurga et al. (eds.), *Modified Nucleic Acids in Biology and Medicine*, RNA Technologies, DOI 10.1007/978-3-319-34175-0_5

important determinants of tRNA stability, dynamics, and ribosome binding and of maintenance of the translational reading frame. These modifications also serve as biomarkers for several human diseases, including type 2 diabetes, cardiac dysfunction, intellectual disability, and skin, breast, and colorectal cancers. Of particular note, several mitochondrial disorders trace their molecular pathogenesis to deficiencies in specific tRNA modifications. Pathology can also be attributed to mutations affecting protein recognition of tRNA substrates. However, protein recognition of RNA modification is at present an underdeveloped field and the subject of increasing attention. Epitranscriptomic profiling will be readily achievable with new advances in the detection of RNA modifications by peptides and mass spectrometry at the attomole level. These technologies will allow for single-cell analysis of modifications and will serve as a platform for increased sensitivity for biomarker identification. Thus, RNA modifications are a real-time code to RNA structure and function that has yet to be deciphered.

Keywords RNA modification • Epitranscriptomics • tRNA modification disease biomarkers • Emerging RNA modification technology

1 Introduction

Both DNA and proteins undergo dynamic chemical modifications that do not change the primary polymer sequence, but greatly impact its structural and functional properties (Dominissini 2014; Meyer and Jaffrey 2014). DNA methylation patterns, as well as methylation, acetylation, phosphorylation, and ubiquitinylation of histone proteins, play critical roles in determining which genes are expressed in a cell at any given time (Franckenberg et al. 2012; Paska and Hudler 2015; Zhang et al. 2015). Dynamic and sometimes reversible modifications to the ribonucleosides A, G, C, and U add an additional layer of control to RNAs to regulate cellular processes and allow for rapid response to environmental signals. In addition, the RNA backbone has six dihedral angles allowing for the flexibility to adapt an extraordinary array of conformations and flexibility for the many roles RNA play in controlling gene expression (Moore 1999; Ulyanov and James 2010; Reiter et al. 2011; Lilley 2012; Muto and Yokoyama 2012; Halder and Bhattacharyya 2013).

Although less well studied than DNA, it is clear that some RNA molecules are also reversibly modified posttranscriptionally to alter their structural and functional properties. The most common modifications are methylation and pseudouridylation. However, there are over 100 different modification chemistries associated with RNAs; tRNAs can have up to 35 % of their nucleosides modified (Cantara et al. 2011; Machnicka et al. 2013). RNA modifications have the same chemistries as those of amino acid side chains (Agris 1996) and have been classified as being hydrophobic (aromatic and aliphatic) or hydrophilic (polar, charged, and transiently charged). The chemical attributes these modifications impart to RNA include hydrogen bonding and negation of hydrogen bonding, enhanced stacking and reordering of water, metal ion binding, restraining and enhancing dynamics, and altering backbone conformation (Agris 1996). The ten proposed chemical attributes for RNA modifications have been verified, and in general, modifications affect RNA thermal stability (Ashraf et al. 2000); restrict dynamics (Stuart et al. 2003; Agris 2008); serve as protein recognition determinants (Sylvers et al. 1993; Putz et al. 1994; Madore et al. 1999; Agris et al. 2007; Agris 2008; Hashimura et al. 2009); in tRNAs are requisite for codon binding affinity (Agris et al. 2007; Agris 2008), codon discrimination (Agris et al. 2007; Agris 2008), speed of translation (Kruger et al. 1998), and maintenance of the translational reading frame (Bjork et al. 1999; Urbonavicius et al. 2001; Gustilo et al. 2008; Maehigashi et al. 2014); and are recognition determinants for aminoacyl-tRNA synthetases (Freist et al. 1997; Agris et al. 2007; Agris 2008; Gustilo et al. 2008).

In this chapter we will discuss well-known examples of modified RNAs, their potential biological functions, and the molecular mechanisms by which aberrant modifications, especially those in highly modified tRNAs, can lead to severe human diseases. Furthermore, we will discuss the development of increasingly sophisticated technologies with which to detect and study the functional relevance of RNA posttranscriptional modifications (Fig. 1).

2 Types of Modified RNA

2.1 Messenger RNA Methylation

The study of RNA posttranscriptional modification patterns on a global scale in cells or tissues is called epitranscriptomics. This term most commonly refers to analysis of messenger RNAs (mRNAs) and long noncoding RNAs (lncRNAs) for methylation at the N^6 position of adenosine (m^6A) (Wang et al. 2014; Liu and Pan 2015; Yue et al. 2015) and (about fourfold less common) methylation at position 5 of cytosine (m^5C) (Meyer and Jaffrey 2014). Although m^6A modifications were originally identified in the 1970s in polyadenylated RNAs (Desrosiers et al. 1974; Yue et al. 2015), global detection of these modifications remained challenging for decades because normal sequencing techniques first convert RNA into DNA by reverse transcription, and the m^6A modification does not alter A-T Watson-Crick binding (Dominissini 2014; Meyer and Jaffrey 2014; Liu and Pan 2015; O'Connell 2015; Yue et al. 2015). Furthermore, methyl groups are not chemically reactive and therefore not easy to label with more readily detectable chemistries (Meyer and Jaffrey 2014; Yue et al. 2015). More recently, high-throughput sequencing methodologies such as MeRIP-seq (Meyer et al. 2012) or m^6A-seq (Dominissini et al. 2012) to detect methylation sites in mRNAs and noncoding mRNAs have

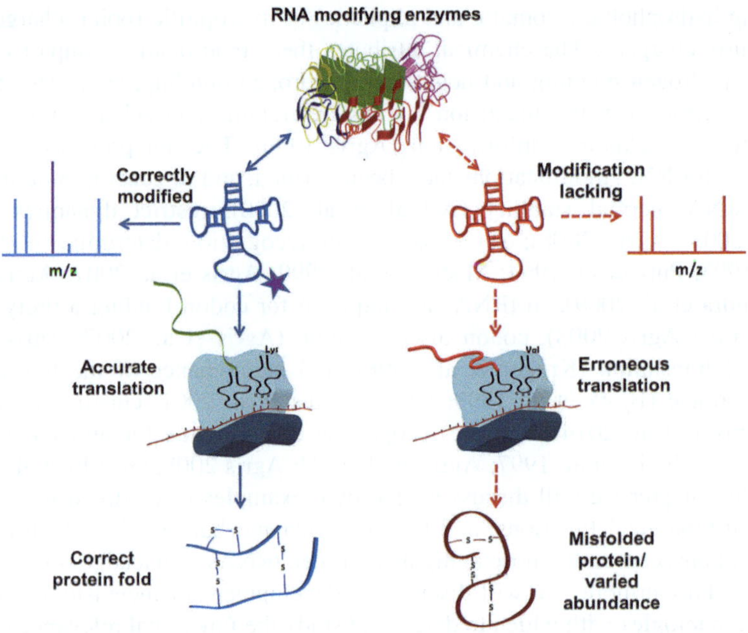

Fig. 1 RNA modifications serve as contributors to and biomarkers of disease. Specific proteins or RNA-protein complexes catalyze chemical modifications of RNA, which are essential for proper RNA folding, dynamics, binding partner recognition, and cellular function. Exciting new technologies enhancing the detection and analysis of modified RNAs will pave the way for improved biomarker screening and therapeutic development. tRNA is the representative example pictured here. Improper modification, as detected by MS, can result in mistranslation and improperly folded protein

permitted detailed and robust studies of the role of RNA methylation in gene expression; these emerging technologies will be discussed in more detail at the end of this chapter.

For chemical modifications to affect dynamic regulation of cellular processes, there must be "writers" and "erasers" to alter methylation states in response to environmental changes, as well as "readers" to recognize modifications and alter downstream events (Fig. 2) (Liu and Pan 2015). While likely not accounting for all m^6A modifications, a methylation complex composed of methyltransferase like 3 (METTL3), methyltransferase like 14 (METTL14), and Wilms' tumor 1-associating protein (WTAP) is thought to be a major writer of m^6A methylation (Yue et al. 2015). METTL3 and METTL14 have been shown in vitro to be methyltransferases, using S-adenosylmethionine as a methyl donor (Liu et al. 2014). WTAP (not a methyltransferase) is thought to aid in localization of the METTL3-METTL14 complex to nuclear speckles (Ping et al. 2014). Knockdown of METTL3, METTL14, or WTAP decreases m^6A modification in mammalian cells (Liu et al. 2014; Meyer and Jaffrey 2014).

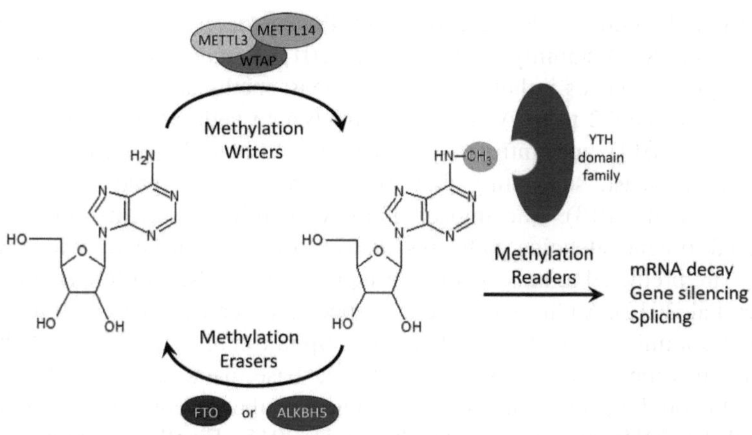

Fig. 2 Dynamic regulation of m^{6}A modification of mRNA in humans. The N^6 position of adenosine can be methylated by the METTL3-METTL14-WTAP writing complex and demethylated by the erasers FTO or ALKBH5. Proteins in the YTH domain family can read the status of methylation sites in mRNA to regulate mRNA decay, gene silencing, and alternative splicing

Removal of m^{6}A methylation can be catalyzed by two known demethylases, or erasers, from the AlkB family of nonheme Fe(II)/α-ketoglutarate (α-KG)-dependent dioxygenases: fat mass and obesity-associated protein (FTO) and AlkB family member 5 (ALKBH5) (Zheng et al. 2013; Yue et al. 2015). FTO removal of methylation sites by oxidation has been demonstrated in vitro, and levels of FTO in human cells are inversely correlated with m^{6}A levels: upregulation of FTO decreases methylation, while knockdown of FTO enhances methylation (Jia et al. 2011). ALKBH5 exhibits similar behavior both in vitro and in vivo (Zheng et al. 2013). Furthermore, both FTO and ALKBH5 localize to nuclear speckles in human cells (Jia et al. 2011; Zheng et al. 2013), which is the suspected site of dynamic methylation and processing of transcripts (Yue et al. 2015).

The functional consequence of dynamic methylation/demethylation of mRNAs is interpreted via methylation reader proteins (Meyer and Jaffrey 2014). Although methylation is a very small chemical change, a single methyl group can enhance binding affinity of a protein by >20-fold (Meyer and Jaffrey 2014). Human m^{6}A reader proteins in the YTH domain family (YTHDF), including YTHDF1 (Xu et al. 2014), YTHDF2 (Dominissini et al. 2012), and YTHDF3 (Dominissini et al. 2012), have affinities for m^{6}A-modified RNA between 180 and 520 nM (Wang et al. 2014). The specific functional roles played by most mRNA methylation remain elusive (Yue et al. 2015). Thus far, all tissues examined contain m^{6}A-modified mRNAs, but the extent of modification appears to differ among tissue types as well as temporally during development (Meyer and Jaffrey 2014). m^{6}A modifications occur in conserved regions of mRNA (Meyer and Jaffrey 2014), and methylation patterns in specific genes are well conserved between mice and humans (Batista et al. 2014). These results strongly support the hypothesis that methylation plays critical functional roles in regulating cellular processes. m^6A modification has been linked to RNA stability (Berulava et al. 2015; Spitale et al. 2015), as the most highly expressed genes in humans were the least methylated (Dominissini 2014). Binding of YTHDF2 to m^6A can recruit mRNA to cellular sites for RNA decay (Wang et al. 2014), and, intriguingly, there are no homologues for demethylases expressed in yeast, suggesting that m^6A is a passive signal for degradation (Schwartz et al. 2013). The impact of m^6A on mRNA stability might also be connected to gene silencing pathways, as 67 % of 3′ untranslated regions with m^6A peaks are predicted to have a non-overlapping microRNA (miRNA) binding site (Meyer et al. 2012). Additional evidence supports a specific role of m^6A in alternate splicing (Dominissini 2014; Ping et al. 2014; Spitale et al. 2015; Yue et al. 2015) as this modification destabilizes Watson-Crick base pairs that would otherwise occlude the binding site of heterogeneous nuclear ribonucleoprotein C (HNRNPC) (Liu and Pan 2015; Yue et al. 2015; Zhou et al. 2015). Finally, aberrant activity of the demethylase FTO has been linked to obesity (Jia et al. 2011; Liu and Pan 2015) and diabetes (Liu and Pan 2015).

Known modifications to mRNA are confined to m^6A, m^5C, or 2′-*O*-methylation (Nm) by present technology and methods (Liu and Pan 2015). However, as detection methods become more sensitive, it is highly likely that a broad range of chemical modifications that play important functional roles in mRNA and lncRNA will be discovered. For instance, the detection and localization of m^5C during deep sequencing is made possible by prior bisulfite deamination of cytosines (Schaefer et al. 2009a, b). The epitranscriptomic detection of pseudouridine (ψ) is now possible through more than one method (Cao and Limbach 2015; Carlile et al. 2015; Li et al. 2015).

2.2 Ribosomal RNA Posttranscriptional Modifications

Ribosomal RNAs (rRNAs) are heavily modified with Nm and pseudouridine (Ψ) in all kingdoms of life (Sharma and Lafontaine 2015). The number of modified rRNA nucleotides increases with the complexity of the organism, with 35 ribosomal modifications identified in *Escherichia coli* (Fischer et al. 2015), 100 in yeast, and >200 in humans (Liu and Pan 2015). In rRNA, modifications consist mostly of ribose 2′-hydroxyl and nucleobase methylation as well as pseudouridylation (Chow et al. 2007; Liu and Pan 2015). Modifications occur predominantly in regions of critical functional importance (Chow et al. 2007; Fischer et al. 2015; Sharma and Lafontaine 2015) that are highly conserved (Chow et al. 2007) and lack protein binding partners (Chow et al. 2007). They are particularly prevalent in the peptidyl-transferase center, at subunit interfaces, and at the decoding region (Chow et al. 2007; Fischer et al. 2015). These modifications likely play a role in altering the available Watson-Crick or non-canonical base-pairing interactions, driving rRNA folding toward its optimal functional structure (Chow et al. 2007). Interestingly, the location of Ψ, and Ψ derivatives such as 3-(3-amino-3-carboxypropyl)uridine, is the most conserved of modifications. For example, the Ψs of the ribosome's large subunit, helix 69 of *E. coli* and its analogues in other organisms, are found within the terminal loop (Ψs 1911, 1915, 1917, *E. coli* numbering) at the interface of the two subunits and the location of ribosome release factor interaction (Ofengand 2002).

The method of rRNA modification is dependent upon the complexity of the organism. Modifications are introduced to *E. coli* rRNA by site-specific methyltransferases (MTases) or pseudouridine synthases (Chow et al. 2007). In eukaryotes, however, modification is directed by small nucleolar RNA-protein complexes (snoRNPs) (Chow et al. 2007; Sharma and Lafontaine 2015). snoRNAs act as guide RNAs and form 10–20 nucleotide long regions of complementarity to rRNA, recruiting associated proteins to the appropriate site for modification. There are two types of snoRNAs: C/D box snoRNAs, which catalyze 2′-*O*-methylation, and H/ACA box snoRNAs, which catalyze the isomerization of uridine to pseudouridine (Reichow et al. 2007). The timing of rRNA modification is also critical for proper ribosome biogenesis. Most 2′-*O*-methylation and pseudouridylation of rRNA occurs co-transcriptionally, while base methylation typically takes place in later stages of assembly. The addition of modifying proteins and/or snoRNPs to assembling ribosome complexes, as well as the timing of their catalysis, serves as structural and temporal cues to accurately choreograph the complex folding of rRNA and incorporation of ribosomal protein components (Sharma and Lafontaine 2015).

Importantly, rRNA modification in bacteria often overlaps with the critical functional sites to which antibiotics bind (Chow et al. 2007; Fischer et al. 2015). For example, A1518 and A1519 (*E. coli* rRNA numbering) are converted to two dimethyladenosine (m_2^6A) residues in every domain of life. Loss of these modifications in *E. coli* leads to reduced growth and translational fidelity, but confers resistance to the antibiotic kasugamycin (van Buul et al. 1984; van Buul and van Knippenberg 1985). Similarly, 2′-*O*-methylation of C1409 in helix 44 and C1920 in helix 69 is carried out by the TlyA MTase in *Mycobacterium tuberculosis* and aids in function of the intersubunit bridge B2a. Inactivation of this methylase renders the bacteria resistant to cyclic peptide antibiotics capreomycin and viomycin (Johansen et al. 2006). While individual modifications often may not be essential for survival of the organism, their absence can lead to significant fitness costs, and in the absence of an antibiotic stress, these organisms are readily outcompeted by wild-type strains (Chow et al. 2007).

2.3 *tRNA Posttranscriptional Modifications*

tRNA molecules are by far the most heavily modified RNAs in the cell, with more than 90 known modifications (Agris 2008). The wobble position on tRNAs (nucleotide 34) and the universal purine at position 37 (directly 3′ of the anticodon) are the most commonly modified nucleotides in tRNAs and carry the widest variety of chemical modifications (Agris 1996). Modifications at the wobble position are used to expand or restrict the number of codons that can be read by a particular tRNA. For example, an inosine at the wobble position expands decoding to include codons with any nucleotide in the third position (Crick 1966; Lim 1995). Modifications at position 37 also are critical for codon recognition. For codons beginning with a U, almost all decoding tRNAs contain an N^6-isopentenyladenosine (i^6A_{37}) and codons beginning with an A often contain an N^6-threonylcarbamoyladenosine (t^6A_{37}) (Agris et al. 2007). These chemical modifications in the anticodon loop of tRNAs alter its conformation and reduce flexibility, decreasing the entropic penalty of proper codon-anticodon minihelix formation in the decoding center of the ribosome (Agris 2004, 2008). Proper geometry serves as a proofreading mechanism, which, if correctly aligned, enables rapid transition to the closed conformation of the ribosome (Ogle et al. 2002). In addition, an optimal rate of protein synthesis is achieved at 10–20 amino acids per second (Andersson et al. 1982) with a low error rate of 1 in 10^3–10^4 amino acids (Rodnina and Wintermeyer 2001). Modifications also maintain proper reading frame (Bjork et al. 1999; Urbonavicius et al. 2001; Gustilo et al. 2008; Maehigashi et al. 2014). For example, expansion of the anticodon loop, or lack of the posttranscriptional modification N^1-methylguanosine at position 37 (m^1G_{37}) of $tRNA^{Pro}$, results in a +1 frameshift along an mRNA (Maehigashi et al. 2014). Hypomodified tRNAs slow the rate of decoding, giving the P-site tRNA a greater chance to slip into the +1 or -1 reading frame (Bjork et al. 1999). Finally, modified nucleotides in tRNAs also serve as important components of protein recognition elements, especially by tRNA synthetases that need to aminoacylate the proper tRNA with its cognate amino acid (Beuning and Musier-Forsyth 1999). Recognition of modified nucleosides by specific proteins is discussed in detail below.

2.4 Other Noncoding RNAs That Are Modified

Functionally relevant modifications are now also being discovered in other types of noncoding RNAs (ncRNAs). All of the U-rich spliceosomal RNAs, for example, appear to contain pseudouridine modifications. U2 RNA contains 14 pseudouridine residues, which represent 7 % of its nucleotides. Pseudouridines are added to spliceosomal snRNAs via either H/ACA box snoRNP complexes or by individual proteins. Four of U2's pseudouridine residues are clustered around the branch site recognition region and are critical for splicing activity (Adachi and Yu 2014). In addition, miRNA has been observed with 2′-*O*-methylation and m^6A modifications (Berulava et al. 2015). Knockdown of FTO in mammalian cells altered the steady-state levels of several miRNA species (either up or down), suggesting that methylation likely plays a role in miRNA biogenesis and/or stability, although the exact mechanism of action remains unclear (Berulava et al. 2015).

3 Effects of Aberrant tRNA Modification on Human Health

Altered protein translation as a result of improperly modified tRNAs can cause several human disorders, including various metabolic, mitochondrial, neurological, and respiratory diseases as well as multiple types of cancer (Fig. 3) (Torres et al. 2014; Kirchner and Ignatova 2015). Mutations affecting the modification of tRNA can occur in either the tRNA genes or, more commonly, genes coding for proteins responsible for tRNA processing and modification. Disease-associated alterations in tRNA modifications have been found in the D-stem, anticodon stem and loop, variable region between the anticodon and T-stems, and the T-stem and loop (Fig. 3).

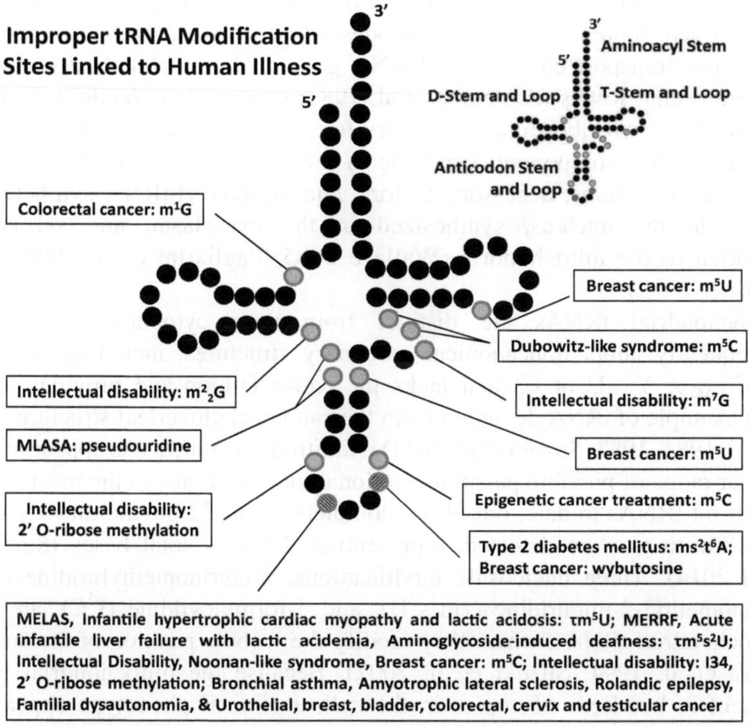

Fig. 3 Disease states linked to improperly modified tRNA. tRNA nucleosides are represented by circles. *Gray circles* indicate a position at which an improper modification leads to disease. *Gray-* and *black-lined circles* indicate a position at which multiple diseases and/or improper modifications are linked. N^1-methylguanosine, m^1G; 2-dimethylguanosine, $m^2{}_2G$; 5-methyluridine, ribothymidine, m^5U; 5-methylcytidine, m^5C; N^7-methylguanosine, m^7G; 2-methylthio-N^6-threonylcarbamoyladenosine, ms^2t^6A; 5-taurinomethyluridine, τm^5U; 5-taurinomethyl-2-thiouridine, τm^5s^2U; inosine, I; MELAS, mitochondrial encephalomyopathy, lactic acidosis, and stroke-like episodes; MERRF, myoclonus epilepsy associated with ragged red fibers; MLASA, mitochondrial myopathy and sideroblastic anemia

3.1 tRNA Modifications and Inherited Mitochondrial Disease

Many human diseases attributed to improperly modified tRNAs are associated primarily with mutations in modification enzymes. However, modifications can also be altered by mutations in tRNA genes themselves, which interestingly are found exclusively in the maternally inherited mitochondrial tRNAs (Abbott et al. 2014). No cytoplasmic tRNA gene mutations associated with disease have been identified to date. Accordingly, mitochondrial diseases account for a significant portion of disorders related to aberrant modification of tRNAs (Florentz et al. 2003; Kirino et al. 2005; Umeda et al. 2005; Yasukawa et al. 2005).

Mitochondria (mt) employ oxidative phosphorylation to generate the majority of the cell's energy in the form of adenosine triphosphate (ATP) (Ernster and Schatz 1981; Hatefi 1985; Wallace 2005). The maternally inherited human mitochondrial genome (mt DNA) encodes 13 genes for protein components of the electron transport chain and 24 RNA genes: 22 mitochondrial (mt) tRNAs and 2 ribosomal RNAs (Anderson et al. 1981; Gray 2013). While mitochondria maintain their own ribosomal machinery to translate those 13 proteins, only the rRNA and tRNA components originate in the human mitochondria themselves; all ribosomal proteins, accessory factors, and aminoacyl-tRNA synthetases are encoded in the nucleus, synthesized in the cytoplasm, and subsequently transported to the mitochondria (Wallace 2005; Pagliarini et al. 2008; Ahmed and Fisher 2009).

Mitochondrial tRNAs are distinct from their cytoplasmic counterparts and frequently adopt noncanonical secondary structures, including one variety ($tRNA^{Ser}{}_{AGY}$, Y = U or C) that lacks the entire D-loop and provides the only known example of tRNA deviation from the canonical cloverleaf structure (Anderson et al. 1981, 1982; Suzuki et al. 2011). Additionally, mt tRNAs appear to exhibit a smaller range of possible posttranscriptional modifications; in the most complete study of mt tRNAs to date, out of the complete set of 22 bovine mt tRNAs, only 15 modifications were detected, representing 7.5 % of total bases (Suzuki and Suzuki 2014). Three nucleoside modifications, 5-taurinomethyluridine (τm^5U), 5-taurinomethyl-2-thiouridine (τm^5s^2U), and 5-formylcytidine (f^5C), are unique to mitochondrial tRNAs, where they modify the wobble position of the anticodon (Moriya et al. 1994; Suzuki et al. 2002). Because the mitochondrial genome is present in hundreds of copies in the cell, a clinical phenotype appears only when the percentage of mtDNA copies containing the mutation exceeds a certain tolerance, typically >80 % (Dimauro and Davidzon 2005; Elliott et al. 2008; Stewart and Chinnery 2015). A severe disruption of tRNA function resulting from prevalence of a mutation is likely to be embryonic lethal (Brandon et al. 2005).

3.2 MELAS and MERFF

The best-characterized mitochondrial diseases linked to abnormal tRNA modification patterns are MELAS (mitochondrial encephalomyopathy, lactic acidosis, and stroke-like episodes) and MERFF (myoclonus epilepsy associated with ragged red fibers) syndromes. MELAS, a progressive mitochondrial cytopathy with an estimated prevalence of 60:100,000, is one of the more frequently diagnosed mitochondrial disorders (Goto et al. 1990, 1992; El-Hattab et al. 2014). MELAS stems from a respiratory defect in complex I or IV and is associated with a range of symptoms, including multiple stroke-like episodes with secondary neurological deficits, lactic acidosis, and progressive muscle weakness (Koo et al. 1993; El-Hattab et al. 2014). Symptoms of MERRF present more variability in terms of severity and age at onset and include myoclonus, epilepsy, ragged red fibers, progressive ataxia, deafness, and dementia (Luft 1994). Neither has a cure (Luft 1994; Scaglia and Northrop 2006).

MELAS and MERRF most commonly arise from point mutations to the mttRNA$^{Leu(UUR)}$ and mttRNA^{Lys} genes, respectively; an A-to-G mutation at position 3243 (A3243G) and a T-to-C mutation at position 3271 (T3271C) are the most common disease mutations in MELAS (Goto et al. 1990, 1991; Kobayashi et al. 1990), while the most significant subgroup of MERRF patients carries an A-to-G mutation at position 8344 (A8344G) (Yasukawa et al. 2000a, b). mttRNA$^{Leu(UUR)}$ from patients displaying these and several other MELAS mutations lacked the expected posttranscriptional modification τm^5U at the anticodon wobble position uridine (Yasukawa et al. 2000a, b). Similarly, τm^5s^2U resulting from thiolation of wobble position τm^5U to τm^5s^2U by an additional 2-thiouridylase (Mtu1) (Umeda et al. 2005) was absent from mttRNA^{Lys} transcripts of MERRF patients (Yasukawa et al. 2000a, b). These results strongly suggested that MELAS and MERRF disease mutations prevent τm^5U biogenesis.

Crystal structures of an anticodon stem-loop containing τm^5U interacting with a UUA or UUG codon at the ribosomal A-site suggest that τm^5U and τm^5s^2U at the wobble position of mttRNA$^{Leu(UUR)}$ and mttRNA^{Lys} facilitate ribosome binding by stabilizing the U:G wobble base pair (Kurata et al. 2008). Correspondingly, the τm^5 U-deficient mttRNA$^{Leu(UUR)}$ exhibits deficient UUG (though not UUA) decoding, while mutant mttRNA^{Lys} loses translational activity for both of its cognate codons due to its wobble modification defect (Yasukawa et al. 2001). In one subsequent hypothesis, reduced expression of the ND6 subunit of mitochondrial respiratory complex I, which alone of 13 mt proteins contains a disproportionate number of relatively rare UUG codons (Goto et al. 1992), underpins MELAS pathology, consistent with the observation of a reduction in complex I activity in patients (Koga et al. 1988).

While most cases of MELAS and MERRF trace their molecular pathogenesis to mutations in mitochondrial DNA, an increasing number have also been attributed to nuclear-encoded modification factors (Villarroya et al. 2008; Ghezzi et al. 2012; Baruffini et al. 2013; Kopajtich et al. 2014; Taylor et al. 2014). GTPBP3 and MTO1 are the nuclear-encoded human enzymes predicted to catalyze τm^5U biogenesis in mttRNA$^{Leu(UUR)}$ and mttRNA^{Lys} (the latter receiving a subsequent 2-thiolation to form τm^5s^2U) (Li and Guan 2002; Li et al. 2002; Suzuki et al. 2011). In analogy to their bacterial orthologues MnmE and MnmG, GTPBP3 and MTO1 are proposed to form a protein complex that takes as substrates taurine, an unknown tetrahydrofolate derivative, FAD and GTP to synthesize τm^5U (Moukadiri et al. 2009). Mutations in patients with infantile hypertrophic cardiomyopathy and lactic acidosis, related to mitochondrial respiratory chain deficiency, include frame shift mutations to MTO1 leading to a nonfunctional truncated protein product and missense mutations predicted to interfere with FAD binding and tRNA substrate recognition (Ghezzi et al. 2012; Baruffini et al. 2013).

Mutations in the *MTU1* (*TRMU*) gene for the 2-thiouridylase by which the τm^5U is further modified to τm^5s^2U (Umeda et al. 2005) were indicated in homozygosity studies of families of children with acute infantile liver failure accompanied by lactic acidemia, a life-threatening transient disorder of hepatic function (Zeharia et al. 2009). In total cellular RNA from affected patients, overall thiouridylation levels were shown to be reduced, confirming the role of Mtu1-mediated tRNA modification in the pathogenesis of the disease (Zeharia et al. 2009). Mtu1 is also implicated in modulating the phenotypical appearance of both aminoglycoside-induced and nonsymptomatic deafness related to mutations to the 12S ribosomal RNA (rRNA), particularly A1555G. The relatively conservative missense Mtu1 mutation of a conserved alanine to a serine at position 10, when combined with known 12 rRNA disease mutations, was found to be associated with profound deafness (Guan et al. 2006). The A10S mutation, which was also one of the several Mtu1 mutations highlighted in acute infantile liver failure, did not prevent the mitochondrial localization and import of Mtu1. However, it did affect levels of tRNA 2-thiouridylation and significantly decreased the steady-state levels of several substrate and non-substrate tRNAs when appearing in conjunction with A1555G (Guan et al. 2006). The exact molecular pathogenesis of the profound deafness observed in these patients remains to be explored.

Finally, a mutation in the nuclear-encoded *PUS1* gene, encoding pseudouridine synthase 1 (Pus1p), has been linked to mitochondrial myopathy and sideroblastic anemia (MLASA) (Patton et al. 2005). MLASA, a rare mitochondrial autosomal-recessive disorder of oxidative phosphorylation and iron metabolism, affects skeletal muscle and bone marrow (Bergmann et al. 2010). Genetic analysis of affected individuals in two families with MLASA identified two disease mutations in Pus1p: a nonsense mutation and a nonconservative missense mutation in which a tryptophan replaces a highly conserved arginine at position 116, located within the active site near a catalytically critical aspartate (Bykhovskaya et al. 2004; Numata et al. 2006; Fernandez-Vizarra et al. 2007). Both cytoplasmic and mitochondrial tRNAs from MLASA patients showed an absence of Ψ at position 28 and positions 27 and 28, respectively, and whole cell extracts from patients exhibited impaired Pus1p enzymatic activity (Patton et al. 2005). Pseudouridines are known to stabilize tRNA secondary structure (Davis 1995; Auffinger and Westhof 1998), and their absence in one or more tRNA species owing to Pus1p deficiency may result in disorders of tRNA aminoacylation or protein translation. While both mitochondrial and cytoplasmic localizations of Pus1p have been observed (Patton et al. 2005), its effect on mitochondrial tRNAs has been suggested to be more pronounced. The observation that combined absence of Ψ_{27} and Ψ_{55} in cytoplasmic tRNAs is known to be lethal (Simos et al. 1996), together with the fact that several species of mitochondrial tRNAs lack Ψ_{55}, may render mt tRNAs less able to tolerate the lack of Ψ_{27} (Patton et al. 2005). Additional cases of MLASA have also been traced to mutations in the gene for YARS2, the mitochondrial tyrosyl-tRNA synthetase, suggesting that the molecular pathogenesis of MLASA may stem from a disorder in mitochondrial protein translation (Riley et al. 2010).

3.3 tRNA Modifications and Non-mitochondrial Disease

Mutations affecting proper modification of positions 34 and 37 are linked to approximately half of the known non-mitochondrial as well as mitochondrial diseases associated with improper tRNA modification (Fig. 3) (Torres et al. 2014; Kirchner and Ignatova 2015). Improper modification at these positions is often caused by mutations in genes encoding tRNA modification enzymes, suggesting modification genes are a possible target of intervention to treat many human health problems (Fig. 3) (Ishiwata et al. 2004; Kato et al. 2005; Kirino et al. 2005; Yasukawa et al. 2005; Begley et al. 2013; Slotkin and Nishikura 2013; Abbott et al. 2014; Torres et al. 2014; Kirchner and Ignatova 2015). However, in order to develop these genes as therapeutic targets of intervention for personalized medicine, the fundamental biochemistry of RNA modification enzymes needs to be elucidated (Agris 1996). For example, though studied extensively in bacterial tRNAs (Pierrel et al. 2002; Atta et al. 2012), the mammalian 2-methylthio- (ms^2) modification is poorly understood in the context of human disease. Several bioinformatics studies link this modification to disease states, but the mechanism of disease at a biochemical level is lacking.

3.4 tRNA Modification and Type 2 Diabetes

Recently, a strong association was made between the tRNA modification enzyme Cdkal1 and type 2 diabetes (T2D) (Diabetes Genetics Initiative of Broad Institute of et al. 2007; Scott et al. 2007; Steinthorsdottir et al. 2007; Zeggini et al. 2007; Iwata et al. 2012; Chen et al. 2013; Horikoshi et al. 2013; Ng et al. 2013). Single nucleotide polymorphisms (SNPs) of *cdkal1* are recognized as the most reproducible risk factor for T2D. This significance was observed in genome-wide association studies that have been independently replicated by different groups (Diabetes Genetics Initiative of Broad Institute of et al. 2007; Scott et al. 2007; Steinthorsdottir et al. 2007; Zeggini et al. 2007; Iwata et al. 2012; Chen et al. 2013; Horikoshi et al. 2013; Ng et al. 2013). The mutations are recessive, and only homozygous carriers present with significant diabetic symptoms (Pascoe et al. 2007; Steinthorsdottir et al. 2007; Omori et al. 2008). Cdkal1 is responsible for adding a unique, essential modification to human $tRNA^{Lys3}$, which is needed to decode lysine AAA/G codons (Wei et al. 2011; Wei and Tomizawa 2011). Thus, an understanding of the relationship between Cdkal1 SNPs and modification of $tRNA^{Lys3}$ provides a unique treatment opportunity for T2D outside of the traditional focus on insulin.

Cdkal1 is a member of a superfamily of radical *S*-adenosyl-*L*-methionine (SAM) enzymes and is a tRNA modification enzyme for the rare methylthio- (ms^2-) modification found on $tRNA^{Lys3}$. It transfers the ms^2-moiety from SAM to tRNA ^{Lys3}, completing modification of N^6-threonylcarbamoyladenosine-37 (t^6A_{37}) to 2-methylthio-t^6A_{37} ($ms^2t^6A_{37}$) (Fig. 4) (Arragain et al. 2010). Under normal conditions, alternative splicing events located on intron 5 generate three isoforms of the Cdkal1 protein. However, only isoform 1 (~61 KDa with 579 amino acids) is expressed in human pancreatic islets where it is anchored to the endoplasmic reticulum (Brambillasca et al. 2012). SNPs in the noncoding region of *cdkal1* result in mRNA transcripts that are not spliced, or not spliced properly (Diabetes Genetics Initiative of Broad Institute of et al. 2007; Steinthorsdottir et al. 2007), resulting in a lack of functional enzyme to add the ms^2-modification to $tRNA^{Lys3}$. Studies conducted with bacterial homologues of Cdkal1 suggest that the ms^2t^6A modification of $tRNA^{Lys3}$ is important to prevent misreading (Arragain et al. 2010;

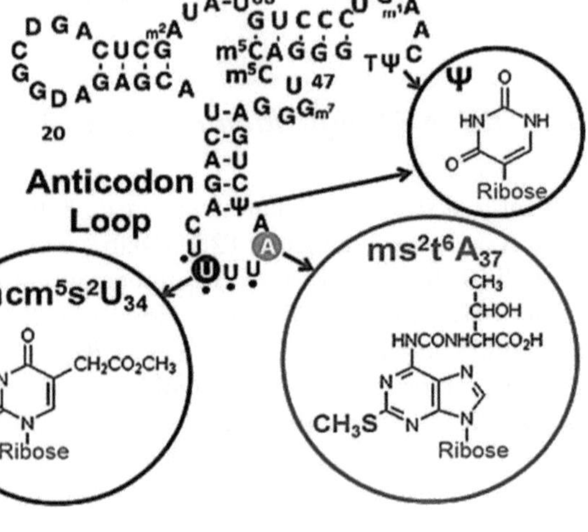

Fig. 4 Secondary structure of human $tRNA^{Lys3}$. Positions and chemical structures of the three modified nucleosides in the anticodon domain are circled. The ms^2-modification by Cdkal1 of $ms^2t^6A_{37}$ is *gray*. The anticodon, UUU, and the adjacent 5′-invariant U and 3′-adenosine are denoted. The missing ms^2- from t^6A_{37} has been linked to aberrant synthesis of pre-proinsulin

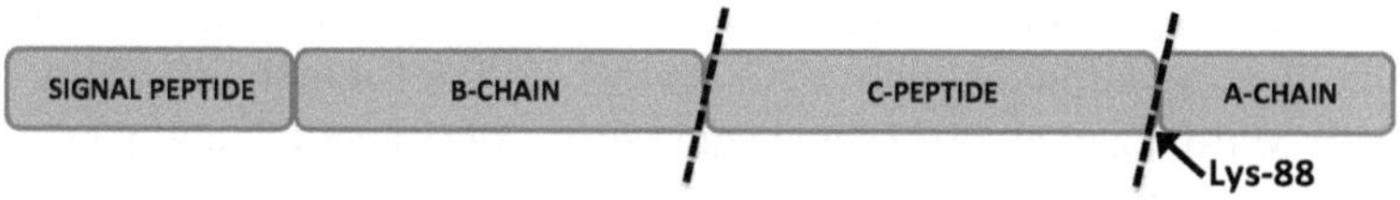

Fig. 5 Schematic of human pre-proinsulin. Protease cleavages (processing) occur at native residues (*dashed lines*) including the critical Lys-88. Translation of Lys-88 is dependent on properly modified $tRNA^{Lys3}$

Wei et al. 2011). Together, these results suggest that SNPs in *cdkal1* lead to incorrect amino acid incorporation at lysine codons during translation.

The pre-proinsulin mRNA contains lysine codons at positions 53 and 88, the latter of which is positioned at a critical protease cleavage site separating the A-chain from the C-peptide (Fig. 5) (Kaufman 2011). Proper cleavage of proinsulin at Lys88 is required for correct insulin processing and functioning. Cdkal1-deficient cells have shown aberrant insulin synthesis and secretion, decreased glucose tolerance, and increased ER stress (Wei et al. 2011; Wei and Tomizawa 2012). These results imply a strong link between the proper modification of $tRNA^{Lys3}$ by Cdkal1 and T2D, perhaps by the need to accurately incorporate Lys88 at the protease cleavage recognition site.

3.5 tRNA Modifications in Neurological Disease and Cancer

Several neurological diseases are also linked to improper modification of the tRNA anticodon stem and loop domain (ASL), often at position 34 (Fig. 3). Intellectual disability has been associated with mutations in the modification enzymes for five different ASL modifications (Torres et al. 2014). Most notable is the 2′-*O*-methyl ribose of wobble position 34. Several studies have linked mutations of FTSJ1, a homologue of a yeast methyltransferase 7 (TRM7), to mental retardation (Freude et al. 2004; Ramser et al. 2004; Froyen et al. 2007; Dai et al. 2008; Gong et al. 2008; Takano et al. 2008; Guy et al. 2015). TRM7 acts upon positions 32 and 34 of three tRNAs: $tRNA^{Trp}$, $tRNA^{Leu}$, and $tRNA^{Phe}$ (Guy et al. 2012; Towns and Begley 2012). Although the reported location of TRM7 expression is variable, it is thought that proper expression of this methyltransferase is critical in fetal brain development (Freude et al. 2004; Ramser et al. 2004). Additional disease states linked to improper modification of the ASL at position 34 are bronchial asthma (Takeoka et al. 2001) and Noonan-like syndrome, which presents with a variety of symptoms that can include cardiac defects, deafness, and cognitive deficits (Fahiminiya et al. 2014; Satterlee et al. 2014). Single nucleotide polymorphisms of IKAP, a member of the elongator complex involved in the early steps of 5-methoxycarbonylmethyl-2-thiouridine synthesis, are strongly correlated with childhood bronchial asthma (Takeoka et al. 2001). Interestingly, mutations in NSUN2, the methyltransferase responsible for the 5-methylcytosine modification (Squires et al. 2012), have been reported in a wide range of diseases including cases of Noonan-like syndrome (Fahiminiya et al. 2014), neurological disorders, Dubowitz-like syndrome, autosomal-recessive intellectual disability (Khan et al. 2012; Martinez et al. 2012), and some forms of cancer including skin, breast, and colorectal (Fig. 3) (Frye and Watt 2006; Vachon et al. 2007).

Cancer is one of the most diverse disease states, arising from a multitude of factors, arguing for a personalized medicine approach to its treatment. Understanding the variety of etiologies among distinct types of cancer and between patients, including those involving tRNA modification abnormalities, will be paramount in devising personalized strategies for treatment. Modifications linked to various types of cancers are found equally within and outside the ASL (Fig. 3) (Torres et al. 2014; Kirchner and Ignatova 2015). Interestingly, both up- and downregulation of enzymes involving tRNA modifications are associated with cancer, emphasizing the need for detailed biochemical dissection of the involvement of modification enzymes. For example, TRMT12 (Rodriguez et al. 2007) and HTRM9L (Begley et al. 2013), involved in the hypermodification of guanosine 3′-adjacent to the anticodon and 5-methoxycarbonylmethyluridine modification of position 34, respectively, are linked to a variety of cancers including breast and colorectal carcinomas (Towns and Begley 2012). TRMT12 is found to be commonly overexpressed in several breast tumors (Rodriguez et al. 2007). Conversely, HTRM9L is downregulated in breast bladder, colorectal, cervical, and testicular cancer, and the restoration of HTRM9L methyltransferase activity abrogated many of the tumorlike qualities (Begley et al. 2013). Already, exploration of the mechanisms of epigenetic cancer therapies on the modification of tRNA, in addition to DNA, is providing valuable information suggesting that treatment targeted to RNA is a viable source of new cancer therapeutics (Schaefer et al. 2009a, b). Thus, identification followed by exploration of the biological effects of altered tRNA modification may uncover new and novel treatment options for not only a variety of cancers but also for a diverse range of human diseases.

3.6 *Impact of Cellular tRNA Levels on Posttranscriptional Modifications*

Alterations to the tRNA pool (both overall quantity and relative abundance of each species) have been observed in a variety of infectious and noninfectious diseases, although little is known about what causes this variation. Examples include metabolic diseases (Krokowski et al. 2013), several types of cancer including breast and myeloma (Pavon-Eternod et al. 2009; Zhou et al. 2009), neurological diseases such as Huntington's disease (Girstmair et al. 2013), and a variety of viral infections (influenza A, vaccinia, West Nile virus, Japanese encephalitis virus, and HIV) (Pavon-Eternod et al. 2010, 2013; Clarke et al. 2014). Variations in tRNA pools can range from increased aminoacyl-tRNA synthetase expression and aminoacylation of a particular tRNA species to drastic alterations in translationally active tRNA pools. In general, noninfectious diseases in which tRNA pools are altered have not been linked to specific gene mutations and are likely a byproduct of the primary disease-state factors (Pavon-Eternod et al. 2009; Zhou et al. 2009; Girstmair et al. 2013; Krokowski et al. 2013). As a logical outcome of altered tRNA levels, one would expect discrepancies in tRNA modification patterns, especially if a particular tRNA is drastically upregulated without the simultaneous upregulation of its cognate modification enzymes. The modification profile of altered tRNA pools associated with various disease states has not yet been explored and is a subject to which new technologies for tRNA modification detection, discussed later in this chapter, can be applied.

4 RNA Modifications Impact Protein Recognition

Now more than ever, the contribution of RNA modifications to disease and the importance of epitranscriptomic profiling are recognized as critical fields to human disease identification and treatment. Success in these disciplines will be dependent on both emerging technologies for the detection and study of RNA modifications and the study of proteins that recognize modifications. In many instances, the primary function of an RNA modification is to serve as a protein recognition determinant. Thus, identifying, characterizing, and utilizing protein motifs that recognize modified RNA will further our understanding of the complex interactions necessary for regulating normal cellular function. Peptides provide tools to both detect RNA modifications and study the proteins that rely on modified RNA recognition determinants.

4.1 Relationships Between Modification Chemistry, Structure, and Function

More than 100 RNA modified nucleoside chemistries affect tertiary structure and dynamics important to function (Agris 1996). Some strengthen hydrogen bonding, whereas others negate Watson-Crick hydrogen bonding. More hydrophobic modifications enhance stacking and reorder water locally. Larger, more chemically complex modifications such as $ms^2t^6A_{37}$ have been shown to produce a platform on which nucleosides upstream and downstream are stacked (Vendeix et al. 2012). Interestingly, modifications that are small or lack chemical complexity are able to drastically change the character of RNAs. For instance, 2-thiouridine, s^2U, a substitution of sulfur for oxygen, dramatically alters the nucleoside's sugar conformation and function in decoding when located at the wobble position 34 of tRNAs (Sierzputowska-Gracz et al. 1987; Ashraf et al. 1999). Dihydrouridine (D), the only fully saturated naturally occurring modification, acts similarly to a proline in proteins, altering the backbone of the macromolecule both locally and downstream (Stuart et al. 1996). Thus, the uniqueness of chemistry, structure, and dynamics contributed by modifications creates a distinctiveness to RNAs that is recognized by proteins.

4.2 Protein Recognition of RNA

The physicochemical contributions of RNA modifications are important recognition determinants for proteins. Yet, much more is known about the contributions of modifications to RNA function and the mechanisms of some of the enzymes responsible for the modifications than is known about how proteins recognize RNA modifications (Freist et al. 1997; Agris et al. 2007; Agris 2008; Gustilo et al. 2008). Protein/RNA interactions regulate gene expression at the transcriptional and translational levels and create superstructures for the most important cellular processes such as the ribosome for translation and the spliceosome for mRNA processing. RNAs have chemical and structural elements, RNA recognition determinants, that are recognized by proteins in vitro and in vivo (Anko and Neugebauer 2012). These structural elements comprise duplex stems, internal loops, tetraloops, hairpins, U-turns, K-turns, and cross-strand stacking structures (Moore 1999). The proteins that bind these RNA determinants have corresponding recognition domains, the so-called RNA recognition motifs (RRMs) (Daubner et al. 2013) and RNA-binding domains (RBDs) (Thapar et al. 2014). Proteins that bind RNAs to form highly stable complexes, such as the ribosome and mRNA-processing machinery, have come to be designated ribonucleoproteins, or RNPs. Transient complexes composed of enzymes and factors with RNA substrates can also be formed, e.g., ribonucleases, aminoacyl-tRNA synthetases (aaRS), modification enzymes, and translation factors (Draper 1999; Hsieh et al. 2004).

Bioinformatic databanks report that some 15 % of all known proteins are RNA-binding proteins (UniProt 2009); there are likely many more non-annotated RNA-binding proteins whose functions are unknown and whose RNA-binding residues are difficult to predict (Spriggs et al. 2009). The increasing number of protein-RNA complexes whose structures have been determined is making it easier for investigators to predict RRMs and RBDs from sequence information (Spriggs et al. 2009). RNA-binding proteins have been found for almost all of the RNA structural motifs created by both canonical and noncanonical base pairings (Moore 1999; Ulyanov and James 2010; Reiter et al. 2011; Lilley 2012; Muto and Yokoyama 2012; Halder and Bhattacharyya 2013). Broadly speaking, RNA-binding proteins are divided between those that bind single-stranded (ss) RNA (Auweter et al. 2006) and those that bind double-stranded (ds) RNA (Masliah et al. 2013). The array of protein secondary structures and their combinations yields a diverse set of RNA-binding domains of which the RNA binding is studied mostly in vitro sans modified nucleosides (Ellis et al. 2007; Lunde et al. 2007; Mackereth and Sattler 2012). Simple electrostatic interactions often place arginines and lysines in contact with the RNA's polyanionic backbone and 2′-OH moieties. In contrast to backbone interactions, hydrogen bonding and hydrophobic interactions with RNA bases contribute to sequence and structural specificity that is attributed to some proteins. Hydrophobic and van der Waals interactions appear to be more prevalently involved in nucleobase recognition than hydrogen bonds (Ellis et al. 2007). Tyrosines, phenylalanines, and tryptophans are the three amino acids most often involved in base recognition, with the two latter amino acids interacting with adenine and guanine, respectively (Ellis et al. 2007).

To date, the chemical and physical properties contributed by posttranscriptional RNA modifications, and the corresponding character of modification-dependent protein binding sites, have not been considered in studies of RRM, RBD, and even RNP recognition of RNA. Thus, very little is known about the basic biochemical principles of modification-dependent binding of RNA by proteins, including binding to tRNAs. Complementary information is also lacking about modification chemistry and structure that imparts distinct conformation and dynamics that constitute identity determinants for RRMs and RBDs.

Of the too-few clear examples of proteins that distinguish RNAs by virtue of their modified nucleosides, the first that come to mind are the aminoacyl-tRNA synthetases (aaRS). aaRS recognition, binding, and aminoacylation of tRNA are often affected by alteration of, or lack of, one or more tRNA modifications. The first evidence of aaRS recognition of a modification was reported 40 years ago: the dependency of glnRS and gluRS on the single atom substitution of a U to a 2-thioU (s^2U_{34}) at the tRNA's anticodon wobble position 34 (Seno et al. 1974). Other aaRSs require different anticodon modifications for recognition and effective kinetics of aminoacylation of their cognate tRNAs (Fig. 6) (Kern and Lapointe 1979; Muramatsu et al. 1988; Sylvers et al. 1993; Cusack et al. 1996; Commans et al. 1998; Madore et al. 1999; Ikeuchi et al. 2005; Nakanishi et al. 2009). Some modifications have been shown to be negative determinants, guarding against mis-acylations, e.g., arginine vs. asparagine (Putz et al. 1994). A single modification is also the recognition determinant for the bacterial $tRNA^{Lys}$-specific PrrC-anticodon nuclease (Jiang et al. 2001). Enzymes involved in completing the syntheses of hypermodified nucleosides also require that the precursor of the modification be present. For example, the bacterial and mammalian 2-methylthiotransferases that complete the hypermodification of N^6-threonylcarbamoyladenosine and N^6-isopentenyladenosine (t^6A_{37} and i^6A_{37}, respectively) at position 37, 3′-adjacent to the anticodons of specific tRNAs, require that the N^6-modification be present (Pierrel et al. 2002, 2004; Hernandez et al. 2007; Arragain et al. 2010; Kaufman 2011). The N1-methyltransferase that methylates the already-modified pseudouridine (Ψ_{1915}) of the 23(28)S rRNA helix 69 (H69) is required for maturation of the eukaryotic ribosome's small subunit rRNA (Thomas et al. 2011). Ribosome release factor (RF2) interacts with H69 and is influenced by that RNA's modification state (Fig. 6) (Kipper et al. 2011). The mRNA-processing enzymes are dependent on the guide RNAs, such as U2 snRNA, having been modified for effective, accurate splicing (Yu et al. 1998, 2011). Two of the 13 Ψs grouped near the 5′-terminal sequence of U2 snRNA are required for efficient pre-mRNA splicing (Donmez

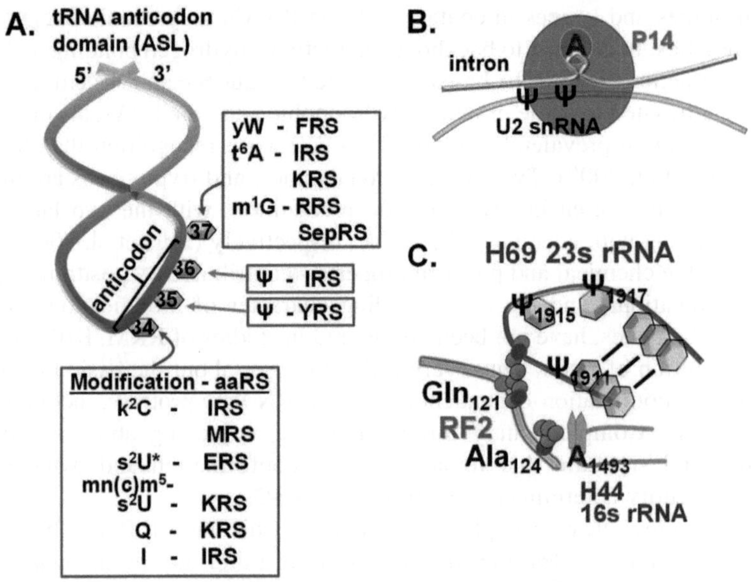

Fig. 6 RNA's modified nucleosides are identity determinants and important to RNA function. (**a**) Aminoacyl-tRNA synthetase (aaRS) recognition elements include modification in tRNA's anticodon domain, particularly anticodon wobble position 34 and purine-37 3′-adjacent to the anticodon. (**b**) U2 snRNA Ψ-facilitated recognition of mRNA branch point A of the intron. Branch point A is within a pocket of the spliceosome protein P14. (**c**) Ribosome release factor RF2 may recognize Ψ of helix 69 of the 23S rRNA, large subunit. Adapted (Dominissini 2014; Meyer and Jaffrey 2014; Ulyanov and James 2010)

et al. 2004). The intron at the branch point A is annealed to U2 snRNA opposite the Ψs, and the A is bound in a pocket of the spliceosomal protein p14(SF3b) (Fig. 6) (Schellenberg et al. 2011). In contrast, some Ψs in snRNA appear to be negative effectors of U2AF binding (Chen et al. 2010), thus regulating pre-mRNA cleavage. Proteins that bind m^6A-modified RNAs have been identified by using a viral RNA for affinity capture (Dominissini et al. 2012).

4.3 Peptide Tools Mimic Modification-Dependent RNA-Binding Proteins

The functional interactions between macromolecules in living systems are elegant and complex. Determining a protein's binding to recognition determinants is a considerable task. Protein domains recognize other proteins, DNAs or RNAs, and bind with specificity and high affinity; peptides mimic this binding and provide information about the amino acids involved in the ligand recognition. Although peptides that mimic protein-protein and protein-DNA interactions are more common, a small number of studies exploring peptides that bind unmodified RNA have generated important information (Grate and Wilson 1997; Weiss and Narayana 1998; Long and Crothers 1999; Ye et al. 1999; Austin et al. 2002; Barkan et al. 2012). In one study, the swapping of peptide sequences has demonstrated that as few as 10 amino acids confer protein specificity for a tRNA's anticodon domain (Auld and Schimmel 1995). Peptides were designed to discriminate a single base pair in a helix (Frugier and Schimmel 1997; Schimmel et al. 1997), and natural peptides derived from aminoacyl-tRNA sequences have been used to inhibit tumor angiogenesis (Tzima and Schimmel 2006). High-affinity peptides have been designed by taking advantage of arginine's propensity to bind RNA's polyanionic nature (Long and Crothers 1999; Ye et al. 1999). Experimentation with both RNA and peptide sequences has produced information about the RNA's identity elements that are recognized by amino acids as being crucial for binding. Peptides can mimic and interfere with the recognition of RNA by the parent protein. Studies of peptides that mimic functionally important domains of proteins that interact with viral RNAs have been particularly helpful in elucidating both the amino acids involved and the RNA nucleoside ligand of the native protein. The TAT-TAR interaction of HIV has been an exemplar system (Calnan et al. 1991; Tao and Frankel 1992; Long and Crothers 1999; Athanassiou et al. 2007; Niu et al. 2011; Li et al. 2013; Sudrik et al. 2013). The success has been transferred to studies of not only HIV but also hepatitis C and dengue viral RNAs (Freire et al. 2013; Manna et al. 2013).

The combination of nucleoside modification chemistries in the anticodon domains of yeast $tRNA^{Phe}$ and human $tRNA^{Lys3}{}_{UUU}$ are unique, and specific to these two tRNAs, allowing for the selection of peptides of 15 and 16 amino acids that mimic the binding of the modified anticodon domain by proteins and providing excellent tools to investigate these protein-tRNA interactions (Mucha et al. 2003; Graham et al. 2011). Over the last 15 years, phage display selection of peptides demonstrating modification-dependent RNA binding has been developed and implemented toward this goal (Agris et al. 1999; Mucha et al. 2002, 2003, 2004; Eshete et al. 2007; Graham et al. 2011).

The ability to design specific multifunctional proteins computationally has improved dramatically in recent years as computational design algorithms have matured and the protein database has expanded. Computational design is used to systematically evaluate the merits of different candidate sequences and analyze the consequences of sequence perturbation when experimental validation is difficult or time consuming. Computationally based strategies (Halperin et al. 2002; Lippow and Tidor 2007; Samish et al. 2011) are used to search the broad sequence space to design proteins for particular purposes, e.g., novel enzymes (Jiang et al. 2008), protein-DNA interactions (Ashworth et al. 2006), and immune epitopes (Ofek et al. 2010). Most protein design strategies are based on assuming a fixed backbone (Dahiyat and Mayo 1997; Voigt et al. 2000). Accommodating backbone flexibility in protein design is challenging although several successes have been achieved in protein docking (Wang et al. 2007; Chaudhury and Gray 2008), conformation prediction (Georgiev et al. 2008; Mandell and Kortemme 2009), and interface modeling (Correia et al. 2011; Karanicolas et al. 2011). Therefore, to improve upon the peptide selection process, new algorithms leading to a signature amino acid sequence are being developed (Spears et al. 2014; Xiao et al. 2014, 2015).

The addition of computational tools to peptide selection has aided in achieving the highest affinity and selectivity (Spears et al. 2014; Xiao et al. 2014, 2015). For example, the selection of peptides with chemistry and structure that mimic the yeast pheRS and HIV replication proteins in their abilities to bind $tRNA^{Phe}$ and $tRNA^{Lys}$, respectively (Mucha et al. 2001; Graham et al. 2011), has shed new light on the most distinctive features of these two tRNAs: the combination of modified nucleoside chemistries and conformational dynamics in their anticodon stem and loop domains ($ASL^{Phe}$and $ASL^{Lys3}{}_{UUU}$) critical to aminoacylation of and mRNA decoding by the tRNAs (Agris et al. 1986; Schmidt et al. 1987; Dao et al. 1992; Chen et al. 1993; Ashraf et al. 2000; Stuart et al. 2000, 2003; Yarian et al. 2000, 2002; Murphy et al. 2004; Bilbille et al. 2009; Vendeix et al. 2012).

The use of peptides to mimic protein-RNA interactions and detect modifications has proven to be a powerful tool. However, to fully explore RNA modifications in the context of epitranscriptomic profiling, more sensitive technologies, such as mass spectrometry and advanced deep sequencing, need to be harnessed. Further discovery of RNA modifications coupled to an increased understanding of their roles by newly developed technologies will feed peptide technologies, thereby extending the biological understanding of modifications and their contributions toward molecular mechanisms of disease.

5 Emerging Methodologies for Detection of Modified RNA

Although previously the study of RNA modifications was thought to be tangential to the study of RNA function, recent technological advances in global RNA modification detection and identification have linked modifications to myriad cellular functions. This emerging field is known as epitransciptomics. RNA modifications can occur independently from standard regulatory processes such as DNA synthesis and cell division, providing alternative mechanisms of cell response to environmental change (Agris 2015). Temporal changes in the epitranscriptome that result from mutations or as a response to cellular signaling and/or external insults can also serve as biomarkers of disease (Begley et al. 2007; Hsu et al. 2009; Satterlee et al. 2014). Thus further improvement of these technologies is imperative for the development of new diagnostics and therapeutics.

5.1 New Technological Advances in Epitranscriptomic Profiling

The variety of cellular roles for RNA coupled with the large number of possible chemical modifications present a complicated picture of the roles modifications

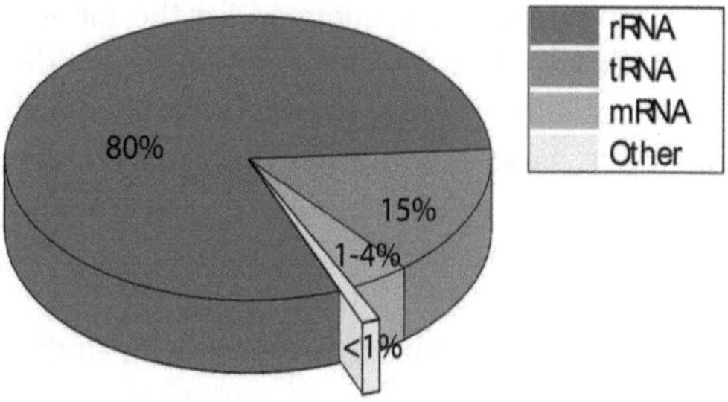

Fig. 7 Total RNA content per cell

play in gene regulation. RNA populations are complex, and the levels at which various RNAs are found in the cell vary considerably by type. In general, rRNAs and tRNAs are relatively abundant, while mRNAs and noncoding RNAs exist at much lower levels (Fig. 7) and require laboratory amplification to be studied (Brown 2002). However, standard amplification protocols are broadly incompatible with preserving RNA modifications, with a few exceptions: Ψ, inosine (I), m^6A, 6-methyl-2′*O*-methyladenosine (m^6Am), and m^5C (Meyer et al. 2012; Edelheit et al. 2013; Li and Church 2013; Carlile et al. 2014; Linder et al. 2015). Developments in RNA sequencing technologies for these most tractable of modifications opened a window onto the important functional roles of RNA modifications and their temporal changes in response to cellular signals.

However, for many reasons, not the least of which being low abundance, most modified RNAs and the 100+ modifications remain difficult to study in their native states with their native modifications. As interest in the epitranscriptome accelerates, there is a pressing need to develop highly selective, sensitive, and rapid technologies to identify all currently known modifications and analyze their dynamic nature in a reproducible, accurate, and quantifiable manner.

5.2 *High-Throughput Sequence Detection of RNA Modifications*

The field of epitranscriptomics exploded with the development of technologies like inosine chemical erasing (Sakurai and Suzuki 2011; Cattenoz et al. 2013), bisulfite-seq (Song et al. 2012), pseudo-seq (Zhao and He 2015), MeRIP-seq (Meyer et al. 2012), and m^6A-seq (Dominissini et al. 2012), which enable the detection of modification sites in mRNAs and noncoding mRNAs by high-throughput sequencing techniques. The first three methodologies use chemical cocktails to alter the structure of modified nucleobases, which generate detectable differences by reverse transcription. Inosine cyanoethylation blocks reverse transcription creating distinct patterns between treated and untreated cDNA (Sakurai and Suzuki 2011). Glyoxal is known to form a stable adduct with guanosine but not with inosine; therefore, digestion with RNAse T1 yields RNA fragments that can be further sequenced for the identification of A > I editing sites (Cattenoz et al. 2013). The distinct reaction of cytidine, but not m^5C, with sodium bisulfite to deaminate cytidine to uracil allows for the detection of m^5C modification sites at single base resolution (Song et al. 2012; Booth et al. 2013). Additionally, pseudouridine reacts with *N*-cyclohexyl-*N′*-(2-morpholinoethyl)carbodiimidemetho-p-toluenesulphonate (CMC) to form N3-CMC-Ψ, which blocks reverse transcription (Carlile et al. 2014; Schwartz et al. 2014).

Alternative methods for modification detection employ antibodies against modifications such as m^6A to immunoprecipitate methylated RNAs and subsequently identify these fragments by deep sequencing (Dominissini 2014; Yue et al. 2015). Methylation sites can be detected from resolution of ~100 nucleotides (Liu and Pan 2015). In regard to sequence specificity, initial data suggests m^6A modifications generally fall within the consensus sequence RRACU (R = purine), although not every instance of this sequence is methylated (Dominissini et al. 2012). In mammalian cells, mRNAs are methylated an average of three times (O'Connell 2015; Yue et al. 2015), and modifications tend to cluster near stop codons (Dominissini 2014; Meyer and Jaffrey 2014; Yue et al. 2015), within 5′ and 3′ untranslated regions (Meyer and Jaffrey 2014; Yue et al. 2015) and in long internal exons (Dominissini 2014; Yue et al. 2015). Methylated regions often lack significant secondary structure (Meyer and Jaffrey 2014; Spitale et al. 2015; Yue et al. 2015), likely because m^6A destabilizes traditional Watson-Crick A-T base pairs (Meyer and Jaffrey 2014; Spitale et al. 2015) and prevents the adenosine from participating in Hoogsteen interactions or forming base triples (Meyer and Jaffrey 2014). High-throughput sequencing has revolutionized epitranscriptomics research; however, these techniques are limited in that they are specific for a single modification and are difficult to implement for low-abundance RNAs. Thus additional technologies are needed to evaluate modification levels and changes throughout the epitranscriptome.

5.3 Mass Spectrometric-Based Methodologies for Simultaneous Detection of Many Types of RNA Modifications

Ultra-high performance liquid chromatography (UHPLC) coupled with tandem mass spectrometry (MS/MS) has emerged as an ideal tool to provide high levels of accuracy, selectivity, and sensitivity in the rigorous quantitative study of RNA nucleoside modifications (Meng and Limbach 2006; Chan et al. 2010; Addepalli and Limbach 2011; Gaston and Limbach 2014; Laourdakis et al. 2014). In this technique, modified RNA is first completely digested enzymatically into its constituent nucleosides under physiological conditions. This preserves the most complex of modification chemistries, given the susceptibility of a number of them to alkaline, acidic, free radical, and oxidative conditions (Davis et al. 1979; Gehrke et al. 1982). This enzymatic approach was first pioneered by Agris and Gehrke, who coupled initial enzymatic digestions of RNA to HPLC separation and UV-diode array identification to study many of the modified nucleosides known 30 years ago (Gehrke et al. 1982, 1984; Gehrke and Kuo 1990). The development of UHPLC-MS/MS methods may be considered a direct descendant of this method (Crain 1990; Pomerantz and McCloskey 1990), evincing higher sensitivities that permit the direct analysis of modified RNAs without subjecting them to possible degradation under laboratory RNA amplification schemes (Begley et al. 2007; Kellner et al. 2010; Gaston and Limbach 2014; Wang et al. 2014; Rose et al. 2015). Recently, a UHPLC-MS/MS-based study demonstrated the separation of RNA major nucleosides and 28 of their modifications in a relatively short analysis time, twenty minutes compared to 40–60 minutes reported previously (Chan et al. 2010; Russell and Limbach 2013; Gaston and Limbach 2014; Su et al. 2014), and with limits of detection to the femtogram, attomole level (Basanta-Sanchez et al. 2015). The use of an oligonucleotide-based exclusion list composed of unmodified sequences provides a platform of information to which all modified RNAs can be compared. All modified nucleosides, with the exception of the different methylations of the same parent nucleoside and Ψ, will increase the mass above that of the exclusion list. This approach begins with high-resolution modified nucleoside sequence analyses (Cao and Limbach 2015).

Accurate quantification of individual nucleosides is dependent on the individual nucleoside extinction coefficients and UV spectra under experimental pH and buffer conditions at a given concentration. Extinction coefficients of major RNA nucleosides have been well characterized (Cavaluzzi and Borer 2004), yet limited information is available for modified RNA nucleosides. A recent study has calculated the extinction coefficient at λ_{max} and 260 nm of major RNA and 28 RNA modifications under specific UHPLC-MS/MS experimental conditions to allow simultaneous and absolute quantification of RNA modifications. Drastically different UV profiles were observed among modifications, demonstrating the importance of extinction coefficient calculation for each distinctive chemistry (Fig. 8) (Basanta-Sanchez et al. 2015). Alternative studies have reported quantities of modified nucleosides by calculating relative peak areas to that of their parental nucleosides, A, G, C, and U (Luo et al. 2014; Wang et al. 2014; Geula et al. 2015). However, without individual extinction coefficients for the specific modified nucleosides, absolute quantification of the modified nucleosides is not accurate.

UHPLC-MS/MS enables the study of the epitranscriptome by the identification and quantification of RNA modifications directly and simultaneously in low-abundance RNAs extracted from prokaryotic and eukaryotic cells, as well as tissues. A recent study on m^6A demonstrated the ability of amplification and mass spectrometric technologies to decipher its involvement in regulating pluripotency and other developmental processes (Zheng et al. 2013; Geula et al. 2015). The analysis and absolute quantification of modifications has been extended to 28 modified nucleosides from as little as 100 ng of RNA from human pluripotent stem cell-derived neural cells with concentrations detected as low as 23.0 femtograms or 64.09 attomoles (Basanta-Sanchez et al. 2015). These studies uncovered the

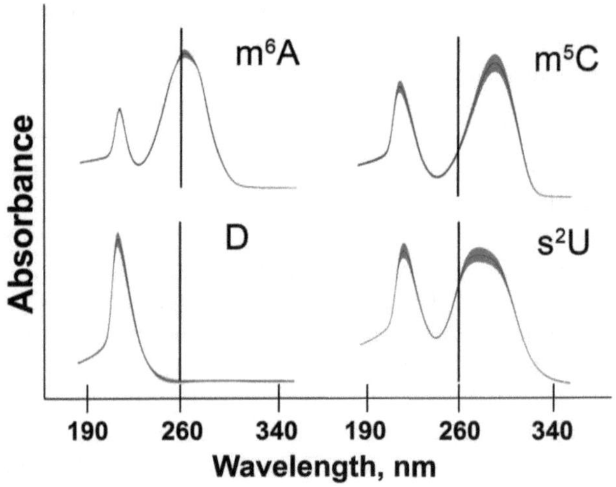

Fig. 8 Modified nucleoside UV spectra. The *X* axis represents the wavelength range from 190 to 340 nm. The *Y* axis shows absorbance units. The 260 nm absorbance is represented by the vertical line and standard deviation represented by gray shadow (Adapted from Basanta-Sanchez et al. 2015)

potential for mass spectrometry-based technologies to reveal functional relationships of the epitranscriptome by investigating temporal changes in modifications during biological processes. Furthermore, advancements aimed toward high-throughput methods and expansion of our analysis capabilities to 50 or more modifications will be critical for detailed exploration of novel epigenetic regulatory functions. The discovery of these fundamental molecular processes will accelerate the pace of current research, creating new avenues of research and discovery.

6 Conclusion: Role of RNA Modifications in Cellular Regulation

The universal genetic code is the basis for the biological dogma of DNA to RNA to protein; with the unveiling of this code came the understanding that RNA was central to all of life. Modified nucleosides found within tRNAs, rRNAs and mRNAs were long viewed as providing a fitness advantage without being strictly necessary to enabling the pathway from transcription through translation. We understand now, however, that posttranscriptional modifications to RNA are required for myriad essential functions, including accurate and efficient processing of RNAs (Agris et al. 1983) and the translation of mRNAs (Agris et al. 2007; Agris 2008), with implications as diverse as modulating type 2 diabetes and guiding stem cell differentiation (Wei et al. 2011; Wei and Tomizawa 2011; Batista et al. 2014; Geula et al. 2015). Thus, the modified nucleosides of tRNA, rRNA, mRNA, and snRNAs, and those yet to be discovered in miRNAs and lncRNAs constitute an additional layer to the genetic code (Agris 2004, 2015) that, when deciphered, will provide a window through which we will gain additional insight into the control of gene expression. This code will allow us to fashion new tools with which to detect normal and abnormal states of human health (Fig. 1).

We underestimate the capacity of cells, particularly mammalian cells, to respond to normal cues and abnormal insults in the absence of replication and new transcription. Yet, we are fully cognizant that each cell possesses a library of already-transcribed RNAs that, for unknown reasons, are not functionalized, or for which functions are yet to be discovered. Naturally occurring modified nucleosides functionalize mRNA for translation and tRNA to be able to read specific codons accurately. Modifications of miRNAs and lncRNAs may be equally or even more important in the control of gene expression and, therefore, of critical importance to understanding states of disease and infection. Experimental approaches and techniques that assess individual modifications link these specific modifications to particular RNA functions, but are apt to miss a more global view of cellular RNA responses. We are well aware that RNA function in control of gene expression comprises a network of interactions in response to small molecule metabolites, chemical insults, and large molecule or protein factors. Many of the numerous RNA modification enzymes are constitutively expressed, while others are upregulated at specific times. Thus, we should understand and study the modification of RNA as a systems biology response that provides cells with temporal flexibility in the absence of new transcription or in the control of new transcription. Our ability to decode the array of modifications that prompts change during normal or abnormal biological phenomena is key to understanding the function that is encoded in modification. New technologies and methods will enable us to "connect the dots" of individual modifications, providing insights not otherwise possible.

References

Abbott JA, Francklyn CS, Robey-Bond SM (2014) Transfer RNA and human disease. Front Genet 5:158

Adachi H, Yu YT (2014) Insight into the mechanisms and functions of spliceosomal snRNA pseudouridylation. World J Biol Chem 5:398–408

Addepalli B, Limbach PA (2011) Mass spectrometry-based quantification of pseudouridine in RNA. J Am Soc Mass Spectrom 22:1363–1372

Agris PF (1996) The importance of being modified: roles of modified nucleosides and Mg^{2+} in RNA structure and function. Prog Nucleic Acid Res Mol Biol 53:79–129

Agris PF (2004) Decoding the genome: a modified view. Nucleic Acids Res 32:223–238

Agris PF (2008) Bringing order to translation: the contributions of transfer RNA anticodon-domain modifications. EMBO Rep 9:629–635

Agris PF (2015) The importance of being modified: an unrealized code to RNA structure and function. RNA 21:552–554

Agris PF, Playl T, Goldman L et al (1983) Processing of tRNA is accomplished by a high-molecular-weight enzyme complex. Recent Results Cancer Res 84:237–254

Agris PF, Sierzputowska-Gracz H, Smith C (1986) Transfer RNA contains sites of localized positive charge: carbon NMR studies of [13C]methyl-enriched *Escherichia coli* and yeast tRNAPhe. Biochemistry 25:5126–5131

Agris PF, Marchbank MT, Newman W et al (1999) Experimental models of protein-RNA interaction: isolation and analyses of tRNA(Phe) and U1 snRNA-binding peptides from bacteriophage display libraries. J Protein Chem 18:425–435

Agris PF, Vendeix FA, Graham WD (2007) tRNA's wobble decoding of the genome: 40 years of modification. J Mol Biol 366:1–13

Ahmed AU, Fisher PR (2009) Import of nuclear-encoded mitochondrial proteins: a cotranslational perspective. Int Rev Cell Mol Biol 273:49–68

Anderson S, Bankier AT, Barrell BG et al (1981) Sequence and organization of the human mitochondrial genome. Nature 290:457–465

Anderson S, de Bruijn MH, Coulson AR et al (1982) Complete sequence of bovine mitochondrial DNA. Conserved features of the mammalian mitochondrial genome. J Mol Biol 156:683–717

Andersson DI, Bohman K, Isaksson LA et al (1982) Translation rates and misreading characteristics of rpsD mutants in *Escherichia coli*. Mol Gen Genet 187:467–472

Anko ML, Neugebauer KM (2012) RNA-protein interactions in vivo: global gets specific. Trends Biochem Sci 37:255–262

Arragain S, Handelman SK, Forouhar F et al (2010) Identification of eukaryotic and prokaryotic methylthiotransferase for biosynthesis of 2-methylthio-N-6-threonylcarbamoyladenosine in tRNA. J Biol Chem 285:28425–28433

Ashraf SS, Sochacka E, Cain R et al (1999) Single atom modification (O→S) of tRNA confers ribosome binding. RNA 5:188–194

Ashraf SS, Guenther RH, Ansari G et al (2000) Role of modified nucleosides of yeast tRNA(Phe) in ribosomal binding. Cell Biochem Biophys 33:241–252

Ashworth J, Havranek JJ, Duarte CM et al (2006) Computational redesign of endonuclease DNA binding and cleavage specificity. Nature 441:656–659

Athanassiou Z, Patora K, Dias RL et al (2007) Structure-guided peptidomimetic design leads to nanomolar beta-hairpin inhibitors of the Tat-TAR interaction of bovine immunodeficiency virus. Biochemistry 46:741–751

Atta M, Arragain S, Fontecave M et al (2012) The methylthiolation reaction mediated by the Radical-SAM enzymes. Biochim Biophys Acta 1824:1223–1230

Auffinger P, Westhof E (1998) Effects of pseudouridylation on tRNA hydration and dynamics: a theoretical approach. In: Grosjean H, Benne R (eds) Modification and Editing of RNA. ASM Press, Washington, DC, pp 103–112

Auld DS, Schimmel P (1995) Switching recognition of two tRNA synthetases with an amino acid swap in a designed peptide. Science 267:994–1996

Austin RJ, Xia T, Ren J et al (2002) Designed arginine-rich RNA-binding peptides with picomolar affinity. J Am Chem Soc 124:10966–10967

Auweter SD, Oberstrass FC, Allain FH (2006) Sequence-specific binding of single-stranded RNA: is there a code for recognition? Nucleic Acids Res 34:4943–4959

Barkan A, Rojas M, Fujii S et al (2012) A combinatorial amino acid code for RNA recognition by pentatricopeptide repeat proteins. PLoS Genet 8, e1002910

Baruffini E, Dallabona C, Invernizzi F et al (2013) MTO1 mutations are associated with hypertrophic cardiomyopathy and lactic acidosis and cause respiratory chain deficiency in humans and yeast. Hum Mutat 34:1501–1509

Basanta-Sanchez M, Temple S, Ansari SA et al (2015) Attomole quantification and global profile of RNA modifications: epitranscriptome of human neural stem cells. Nucleic Acids Res. doi:10.1093/nar/gkv971

Batista PJ, Molinie B, Wang J et al (2014) m(6)A RNA modification controls cell fate transition in mammalian embryonic stem cells. Cell Stem Cell 15:707–719

Begley U, Dyavaiah M, Patil A et al (2007) Trm9-catalyzed tRNA modifications link translation to the DNA damage response. Mol Cell 28:860–870

Begley U, Sosa MS, Avivar-Valderas A et al (2013) A human tRNA methyltransferase 9-like protein prevents tumour growth by regulating LIN9 and HIF1-alpha. EMBO Mol Med 5:366–383

Bergmann AK, Campagna DR, McLoughlin EM et al (2010) Systematic molecular genetic analysis of congenital sideroblastic anemia: evidence for genetic heterogeneity and identification of novel mutations. Pediatr Blood Cancer 54:273–278
Berulava T, Rahmann S, Rademacher K et al (2015) N6-adenosine methylation in MiRNAs. PLoS One 10, e0118438
Beuning PJ, Musier-Forsyth K (1999) Transfer RNA recognition by aminoacyl-tRNA synthetases. Biopolymers 52:1–28
Bilbille Y, Vendeix FA, Guenther R et al (2009) The structure of the human tRNALys3 anticodon bound to the HIV genome is stabilized by modified nucleosides and adjacent mismatch base pairs. Nucleic Acids Res 37:3342–3353
Bjork GR, Durand JM, Hagervall TG et al (1999) Transfer RNA modification: influence on translational frameshifting and metabolism. FEBS Lett 452:47–51
Booth MJ, Ost TW, Beraldi D et al (2013) Oxidative bisulfite sequencing of 5-methylcytosine and 5-hydroxymethylcytosine. Nat Protoc 8:1841–1851
Brambillasca S, Altkrueger A, Colombo SF et al (2012) CDK5 Regulatory Subunit-associated Protein 1-Like 1 (CDKAL1) Is a Tail-anchored Protein in the Endoplasmic Reticulum (ER) of Insulinoma Cells. J Biol Chem 287:41808–41819
Brandon MC, Lott MT, Nguyen KC et al (2005) MITOMAP: a human mitochondrial genome database--2004 update. Nucleic Acids Res 33(Database issue):D611–613
Brown TA (2002) Genomes. Wiley-Liss, Oxford
Bykhovskaya Y, Casas K, Mengesha E et al (2004) Missense mutation in pseudouridine synthase 1 (PUS1) causes mitochondrial myopathy and sideroblastic anemia (MLASA). Am J Hum Genet 74:1303–1308
Calnan BJ, Tidor B, Biancalana S et al (1991) Arginine-mediated RNA recognition: the arginine fork. Science 252:1167–1171
Cantara WA, Crain PF, Rozenski J et al (2011) The RNA Modification Database, RNAMDB: 2011 update. Nucleic Acids Res 39(Database issue):D195–201
Cao X, Limbach PA (2015) Enhanced detection of post-transcriptional modifications using a mass-exclusion list strategy for RNA modification mapping by LC-MS/MS. Anal Chem 87:8433–8440
Carlile TM, Rojas-Duran MF, Zinshteyn B et al (2014) Pseudouridine profiling reveals regulated mRNA pseudouridylation in yeast and human cells. Nature 515:143–146
Carlile TM, Rojas-Duran MF, Gilbert WV (2015) Pseudo-Seq: genome-wide detection of pseudouridine modifications in RNA. Methods Enzymol 560:219–245
Cattenoz PB, Taft RJ, Westhof E et al (2013) Transcriptome-wide identification of A > I RNA editing sites by inosine specific cleavage. RNA 19:257–270
Cavaluzzi MJ, Borer PN (2004) Revised UV extinction coefficients for nucleoside-5-′-monophosphates and unpaired DNA and RNA. Nucleic Acids Res 32, e13
Chan CT, Dyavaiah M, DeMott MS et al (2010) A quantitative systems approach reveals dynamic control of tRNA modifications during cellular stress. PLoS Genet 6, e1001247
Chaudhury S, Gray JJ (2008) Conformer selection and induced fit in flexible backbone protein-protein docking using computational and NMR ensembles. J Mol Biol 381:1068–1087
Chen Y, Sierzputowska-Gracz H, Guenther R et al (1993) 5-Methylcytidine is required for cooperative binding of Mg^{2+} and a conformational transition at the anticodon stem-loop of yeast phenylalanine tRNA. Biochemistry 32:10249–10253
Chen C, Zhao X, Kierzek R et al (2010) A flexible RNA backbone within the polypyrimidine tract is required for U2AF65 binding and pre-mRNA splicing in vivo. Mol Cell Biol 30:4108–4119
Chen G, Xu Y, Lin Y et al (2013) Association study of genetic variants of 17 diabetes-related genes/loci and cardiovascular risk and diabetic nephropathy in the Chinese She population. J Diabetes 5:136–145
Chow CS, Lamichhane TN, Mahto SK (2007) Expanding the nucleotide repertoire of the ribosome with post-transcriptional modifications. ACS Chem Biol 2:610–619

Clarke P, Leser JS, Bowen RA et al (2014) Virus-induced transcriptional changes in the brain include the differential expression of genes associated with interferon, apoptosis, interleukin 17 receptor A, and glutamate signaling as well as flavivirus-specific upregulation of tRNA synthetases. MBio 5:e00902–00914

Commans S, Lazard M, Delort F et al (1998) tRNA anticodon recognition and specification within subclass IIb aminoacyl-tRNA synthetases. J Mol Biol 278:801–813

Correia BE, Ban YE, Friend DJ et al (2011) Computational protein design using flexible backbone remodeling and resurfacing: case studies in structure-based antigen design. J Mol Biol 405:284–297

Crain PF (1990) Preparation and enzymatic hydrolysis of DNA and RNA for mass spectrometry. Methods Enzymol 193:782–790

Crick FH (1966) Codon–anticodon pairing: the wobble hypothesis. J Mol Biol 19:548–555

Cusack S, Yaremchuk A, Tukalo M (1996) The crystal structures of *T. thermophilus* lysyl-tRNA synthetase complexed with *E. coli* tRNA(Lys) and a *T. thermophilus* tRNA(Lys) transcript: anticodon recognition and conformational changes upon binding of a lysyl-adenylate analogue. EMBO J 15:6321–6334

Dahiyat BI, Mayo SL (1997) De novo protein design: fully automated sequence selection. Science 278:82–87

Dai L, Xing L, Gong P et al (2008) Positive association of the FTSJ1 gene polymorphisms with nonsyndromic X-linked mental retardation in young Chinese male subjects. J Hum Genet 53:592–597

Dao V, Guenther RH, Agris PF (1992) The role of 5-methylcytidine in the anticodon arm of yeast tRNA(Phe): site-specific Mg2+ binding and coupled conformational transition in DNA analogs. Biochemistry 31:11012–11019

Daubner GM, Clery A, Allain FH (2013) RRM-RNA recognition: NMR or crystallography and new findings. Curr Opin Struct Biol 23:100–108

Davis DR (1995) Stabilization of RNA stacking by pseudouridine. Nucleic Acids Res 23:5020–5026

Davis GE, Gehrke CW, Kuo KC et al (1979) Major and modified nucleosides in tRNA hydrolysates by high-performance liquid chromatography. J Chromatogr 173:281–298

Desrosiers R, Friderici K, Rottman F (1974) Identification of methylated nucleosides in messenger RNA from Novikoff hepatoma cells. Proc Natl Acad Sci U S A 71:3971–3975

Diabetes Genetics Initiative of Broad Institute of H, Mit LU, Novartis Institutes of BioMedical R et al (2007) Genome-wide association analysis identifies loci for type 2 diabetes and triglyceride levels. Science 316:1331–1336

Dimauro S, Davidzon G (2005) Mitochondrial DNA and disease. Ann Med 37:222–232

Dominissini D (2014) Genomics and proteomics. Roadmap to the epitranscriptome. Science 346:1192

Dominissini D, Moshitch-Moshkovitz S, Schwartz S et al (2012) Topology of the human and mouse m6A RNA methylomes revealed by m6A-seq. Nature 485:201–206

Donmez G, Hartmuth K, Luhrmann R (2004) Modified nucleotides at the 5′ end of human U2 snRNA are required for spliceosomal E-complex formation. RNA 10:1925–1933

Draper DE (1999) Themes in RNA-protein recognition. J Mol Biol 293:255–270

Edelheit S, Schwartz S, Mumbach MR et al (2013) Transcriptome-wide mapping of 5-methylcytidine RNA modifications in bacteria, archaea, and yeast reveals m5C within archaeal mRNAs. PLoS Genet 9, e1003602

El-Hattab AW, Emrick LT, Chanprasert S et al (2014) Mitochondria: role of citrulline and arginine supplementation in MELAS syndrome. Int J Biochem Cell Biol 48:85–91

Elliott HR, Samuels DC, Eden JA et al (2008) Pathogenic mitochondrial DNA mutations are common in the general population. Am J Hum Genet 83:254–260

Ellis JJ, Broom M, Jones S (2007) Protein-RNA interactions: structural analysis and functional classes. Proteins 66:903–911

Ernster L, Schatz G (1981) Mitochondria: a historical review. J Cell Biol 91:227s–255s

Eshete M, Marchbank MT, Deutscher SL et al (2007) Specificity of phage display selected peptides for modified anticodon stem and loop domains of tRNA. Protein J 26:61–73
Fahiminiya S, Almuriekhi M, Nawaz Z et al (2014) Whole exome sequencing unravels disease-causing genes in consanguineous families in Qatar. Clin Genet 86:134–141
Fernandez-Vizarra E, Berardinelli A, Valente L et al (2007) Nonsense mutation in pseudouridylate synthase 1 (PUS1) in two brothers affected by myopathy, lactic acidosis and sideroblastic anaemia (MLASA). J Med Genet 44:173–180
Fischer N, Neumann P, Konevega AL et al (2015) Structure of the *E. coli* ribosome-EF-Tu complex at <3 A resolution by Cs-corrected cryo-EM. Nature 520:567–570
Florentz C, Sohm B, Tryoen-Toth P et al (2003) Human mitochondrial tRNAs in health and disease. Cell Mol Life Sci 60:1356–1375
Franckenberg S, Becker T, Beckmann R (2012) Structural view on recycling of archaeal and eukaryotic ribosomes after canonical termination and ribosome rescue. Curr Opin Struct Biol 22:786–796
Freire JM, Veiga AS, de la Torre BG et al (2013) Peptides as models for the structure and function of viral capsid proteins: Insights on dengue virus capsid. Biopolymers 100:325–336
Freist W, Gauss DH, Soll D et al (1997) Glutamyl-tRNA synthetase. Biol Chem 378:1313–1329
Freude K, Hoffmann K, Jensen LR et al (2004) Mutations in the FTSJ1 gene coding for a novel S-adenosylmethionine-binding protein cause nonsyndromic X-linked mental retardation. Am J Hum Genet 75:305–309
Froyen G, Bauters M, Boyle J et al (2007) Loss of SLC38A5 and FTSJ1 at Xp11.23 in three brothers with non-syndromic mental retardation due to a microdeletion in an unstable genomic region. Hum Genet 121:539–547
Frugier M, Schimmel P (1997) Subtle atomic group discrimination in the RNA minor groove. Proc Natl Acad Sci U S A 94:11291–11294
Frye M, Watt FM (2006) The RNA methyltransferase Misu (NSun2) mediates Myc-induced proliferation and is upregulated in tumors. Curr Biol 16:971–981
Gaston KW, Limbach PA (2014) The identification and characterization of non-coding and coding RNAs and their modified nucleosides by mass spectrometry. RNA Biol 11:1568–1585
Gehrke CW, Kuo KCT (1990) Chromatography and modification of nucleosides. Amsterdam; New York, NY, U.S.A., Elsevier; Distributors for the U.S. and Canada Elsevier Science Pub. Co.
Gehrke CW, Kuo KC, McCune RA et al (1982) Quantitative enzymatic hydrolysis of tRNAs: reversed-phase high-performance liquid chromatography of tRNA nucleosides. J Chromatogr 230:297–308
Gehrke CW, McCune RA, Gama-Sosa MA et al (1984) Quantitative reversed-phase high-performance liquid chromatography of major and modified nucleosides in DNA. J Chromatogr 301:199–219
Georgiev I, Keedy D, Richardson JS et al (2008) Algorithm for backrub motions in protein design. Bioinformatics 24:i196–204
Geula S, Moshitch-Moshkovitz S, Dominissini D et al (2015) Stem cells. m6A mRNA methylation facilitates resolution of naive pluripotency toward differentiation. Science 347:1002–1006
Ghezzi D, Baruffini E, Haack TB et al (2012) Mutations of the mitochondrial-tRNA modifier MTO1 cause hypertrophic cardiomyopathy and lactic acidosis. Am J Hum Genet 90:1079–1087
Girstmair H, Saffert P, Rode S et al (2013) Depletion of cognate charged transfer RNA causes translational frameshifting within the expanded CAG stretch in huntingtin. Cell Rep 3:148–159
Gong P, Li J, Dai L et al (2008) Genetic variations in FTSJ1 influence cognitive ability in young males in the Chinese Han population. J Neurogenet 22:277–287
Goto Y, Nonaka I, Horai S (1990) A mutation in the tRNA(Leu)(UUR) gene associated with the MELAS subgroup of mitochondrial encephalomyopathies. Nature 348:651–653

Goto Y, Nonaka I, Horai S (1991) A new mtDNA mutation associated with mitochondrial myopathy, encephalopathy, lactic acidosis and stroke-like episodes (MELAS). Biochim Biophys Acta 1097:238–240

Goto Y, Horai S, Matsuoka T et al (1992) Mitochondrial myopathy, encephalopathy, lactic acidosis, and stroke-like episodes (MELAS): a correlative study of the clinical features and mitochondrial DNA mutation. Neurology 42:545–550

Graham WD, Barley-Maloney L, Stark CJ et al (2011) Functional recognition of the modified human tRNALys3(UUU) anticodon domain by HIV's nucleocapsid protein and a peptide mimic. J Mol Biol 410:698–715

Grate D, Wilson C (1997) Role REVersal: understanding how RRE RNA binds its peptide ligand. Structure 5:7–11

Gray MW (2013) Mitochondrial genome. In: Maloy S, Hughes K (eds) Brenner's encyclopedia of genetics. Academic Press, San Diego, CA, pp 441–442

Guan MX, Yan Q, Li X et al (2006) Mutation in TRMU related to transfer RNA modification modulates the phenotypic expression of the deafness-associated mitochondrial 12S ribosomal RNA mutations. Am J Hum Genet 79:291–302

Gustilo EM, Vendeix FA, Agris PF (2008) tRNA's modifications bring order to gene expression. Curr Opin Microbiol 11:134–140

Guy MP, Podyma BM, Preston MA et al (2012) Yeast Trm7 interacts with distinct proteins for critical modifications of the tRNAPhe anticodon loop. RNA 18:1921–1933

Guy MP, Shaw M, Weiner CL et al (2015) Defects in tRNA anticodon loop 2′-O-methylation are implicated in nonsyndromic X-linked intellectual disability due to mutations in FTSJ1. Hum Mutat 36:1176–1187

Halder S, Bhattacharyya D (2013) RNA structure and dynamics: a base pairing perspective. Prog Biophys Mol Biol 113:264–283

Halperin I, Ma B, Wolfson H et al (2002) Principles of docking: an overview of search algorithms and a guide to scoring functions. Proteins 47:409–443

Hashimura Y, Nozu K, Kanegane H et al (2009) Minimal change nephrotic syndrome associated with immune dysregulation, polyendocrinopathy, enteropathy, X-linked syndrome. Pediatr Nephrol 24:1181–1186

Hatefi Y (1985) The mitochondrial electron transport and oxidative phosphorylation system. Annu Rev Biochem 54:1015–1069

Hernandez HL, Pierrel F, Elleingand E et al (2007) MiaB, a bifunctional radical-S-adenosylmethionine enzyme involved in the thiolation and methylation of tRNA, contains two essential [4Fe-4S] clusters. Biochemistry 46:5140–5147

Horikoshi M, Yaghootkar H, Mook-Kanamori DO et al (2013) New loci associated with birth weight identify genetic links between intrauterine growth and adult height and metabolism. Nat Genet 45:76–82

Hsieh J, Andrews AJ, Fierke CA (2004) Roles of protein subunits in RNA-protein complexes: lessons from ribonuclease P. Biopolymers 73:79–89

Hsu WY, Chen WT, Lin WD et al (2009) Analysis of urinary nucleosides as potential tumor markers in human colorectal cancer by high performance liquid chromatography/electrospray ionization tandem mass spectrometry. Clin Chim Acta 402:31–37

Ikeuchi Y, Soma A, Ote T et al (2005) Molecular mechanism of lysidine synthesis that determines tRNA identity and codon recognition. Mol Cell 19:235–246

Ishiwata S, Ozawa Y, Katayama J et al (2004) Elevated expression level of 60-kDa subunit of tRNA-guanine transglycosylase in colon cancer. Cancer Lett 212:113–119

Iwata M, Maeda S, Kamura Y et al (2012) Genetic risk score constructed using 14 susceptibility alleles for type 2 diabetes is associated with the early onset of diabetes and may predict the future requirement of insulin injections among Japanese individuals. Diabetes Care 35:1763–1770

Jia G, Fu Y, Zhao X et al (2011) N6-methyladenosine in nuclear RNA is a major substrate of the obesity-associated FTO. Nat Chem Biol 7:885–887

Jiang Y, Meidler R, Amitsur M et al (2001) Specific interaction between anticodon nuclease and the tRNA(Lys) wobble base. J Mol Biol 305:377–388
Jiang L, Althoff EA, Clemente FR et al (2008) De novo computational design of retro-aldol enzymes. Science 319:1387–1391
Johansen SK, Maus CE, Plikaytis BB et al (2006) Capreomycin binds across the ribosomal subunit interface using tlyA-encoded 2′-O-methylations in 16S and 23S rRNAs. Mol Cell 23:173–182
Karanicolas J, Corn JE, Chen I et al (2011) A de novo protein binding pair by computational design and directed evolution. Mol Cell 42:250–260
Kato T, Daigo Y, Hayama S et al (2005) A novel human tRNA-dihydrouridine synthase involved in pulmonary carcinogenesis. Cancer Res 65:5638–5646
Kaufman RJ (2011) Beta-cell failure, stress, and type 2 diabetes. N Engl J Med 365:1931–1933
Kellner S, Burhenne J, Helm M (2010) Detection of RNA modifications. RNA Biol 7:237–247
Kern D, Lapointe J (1979) Glutamyl transfer ribonucleic acid synthetase of *Escherichia coli*. Effect of alteration of the 5-(methylaminomethyl)-2-thiouridine in the anticodon of glutamic acid transfer ribonucleic acid on the catalytic mechanism. Biochemistry 18:5819–5826
Khan MA, Rafiq MA, Noor A et al (2012) Mutation in NSUN2, which encodes an RNA methyltransferase, causes autosomal-recessive intellectual disability. Am J Hum Genet 90:856–863
Kipper K, Sild S, Hetenyi C et al (2011) Pseudouridylation of 23S rRNA helix 69 promotes peptide release by release factor RF2 but not by release factor RF1. Biochimie 93:834–844
Kirchner S, Ignatova Z (2015) Emerging roles of tRNA in adaptive translation, signalling dynamics and disease. Nat Rev Genet 16:98–112
Kirino Y, Goto Y, Campos Y et al (2005) Specific correlation between the wobble modification deficiency in mutant tRNAs and the clinical features of a human mitochondrial disease. Proc Natl Acad Sci U S A 102:7127–7132
Kobayashi Y, Momoi MY, Tominaga K et al (1990) A point mutation in the mitochondrial tRNA (Leu)(UUR) gene in MELAS (mitochondrial myopathy, encephalopathy, lactic acidosis and stroke-like episodes). Biochem Biophys Res Commun 173:816–822
Koga Y, Nonaka I, Kobayashi M et al (1988) Findings in muscle in complex I (NADH coenzyme Q reductase) deficiency. Ann Neurol 24:749–756
Koo B, Becker LE, Chuang S et al (1993) Mitochondrial encephalomyopathy, lactic acidosis, stroke-like episodes (MELAS): clinical, radiological, pathological, and genetic observations. Ann Neurol 34:25–32
Kopajtich R, Nicholls TJ, Rorbach J et al (2014) Mutations in GTPBP3 cause a mitochondrial translation defect associated with hypertrophic cardiomyopathy, lactic acidosis, and encephalopathy. Am J Hum Genet 95:708–720
Krokowski D, Han J, Saikia M et al (2013) A self-defeating anabolic program leads to beta-cell apoptosis in endoplasmic reticulum stress-induced diabetes via regulation of amino acid flux. J Biol Chem 288:17202–17213
Kruger MK, Pedersen S, Hagervall TG et al (1998) The modification of the wobble base of tRNAGlu modulates the translation rate of glutamic acid codons in vivo. J Mol Biol 284:621–631
Kurata S, Weixlbaumer A, Ohtsuki T et al (2008) Modified uridines with C5-methylene substituents at the first position of the tRNA anticodon stabilize U.G wobble pairing during decoding. J Biol Chem 283:18801–18811
Laourdakis CD, Merino EF, Neilson AP et al (2014) Comprehensive quantitative analysis of purines and pyrimidines in the human malaria parasite using ion-pairing ultra-performance liquid chromatography-mass spectrometry. J Chromatogr B Analyt Technol Biomed Life Sci 967:127–133
Li JB, Church GM (2013) Deciphering the functions and regulation of brain-enriched A-to-I RNA editing. Nat Neurosci 16:1518–1522

Li X, Guan MX (2002) A human mitochondrial GTP binding protein related to tRNA modification may modulate phenotypic expression of the deafness-associated mitochondrial 12S rRNA mutation. Mol Cell Biol 22:7701–7711

Li X, Li R, Lin X et al (2002) Isolation and characterization of the putative nuclear modifier gene MTO1 involved in the pathogenesis of deafness-associated mitochondrial 12 S rRNA A1555G mutation. J Biol Chem 277:27256–27264

Li CH, Zuo ZC, Su JG et al (2013) The interactions and recognition of cyclic peptide mimetics of Tat with HIV-1 TAR RNA: a molecular dynamics simulation study. J Biomol Struct Dyn 31:276–287

Li X, Ma S, Yi C (2015) Pseudouridine chemical labeling and profiling. Methods Enzymol 560:247–272

Lilley DM (2012) The structure and folding of kink turns in RNA. Wiley Interdiscip Rev RNA 3:797–805

Lim VI (1995) Analysis of action of the wobble adenine on codon reading within the ribosome. J Mol Biol 252:277–282

Linder B, Grozhik AV, Olarerin-George AO et al (2015) Single-nucleotide-resolution mapping of m6A and m6Am throughout the transcriptome. Nat Methods 12:767–772

Lippow SM, Tidor B (2007) Progress in computational protein design. Curr Opin Biotechnol 18:305–311

Liu N, Pan T (2015) RNA epigenetics. Transl Res 165:28–35

Liu J, Yue Y, Han D et al (2014) A METTL3-METTL14 complex mediates mammalian nuclear RNA N6-adenosine methylation. Nat Chem Biol 10:93–95

Long KS, Crothers DM (1999) Characterization of the solution conformations of unbound and Tat peptide-bound forms of HIV-1 TAR RNA. Biochemistry 38:10059–10069

Luft R (1994) The development of mitochondrial medicine. Proc Natl Acad Sci U S A 91:8731–8738

Lunde BM, Moore C, Varani G (2007) RNA-binding proteins: modular design for efficient function. Nat Rev Mol Cell Biol 8:479–490

Luo GZ, MacQueen A, Zheng G et al (2014) Unique features of the m6A methylome in *Arabidopsis thaliana*. Nat Commun 5:5630

Machnicka MA, Milanowska K, Osman Oglou O et al (2013) MODOMICS: a database of RNA modification pathways—2013 update. Nucleic Acids Res 41(Database issue):D262–267

Mackereth CD, Sattler M (2012) Dynamics in multi-domain protein recognition of RNA. Curr Opin Struct Biol 22:287–296

Madore E, Florentz C, Giege R et al (1999) Effect of modified nucleotides on Escherichia coli tRNAGlu structure and on its aminoacylation by glutamyl-tRNA synthetase. Predominant and distinct roles of the mnm5 and s2 modifications of U34. Eur J Biochem 266:1128–1135

Maehigashi T, Dunkle JA, Miles SJ et al (2014) Structural insights into +1 frameshifting promoted by expanded or modification-deficient anticodon stem loops. Proc Natl Acad Sci U S A 111:12740–12745

Mandell DJ, Kortemme T (2009) Backbone flexibility in computational protein design. Curr Opin Biotechnol 20:420–428

Manna AK, Kumar A, Ray U et al (2013) A cyclic peptide mimic of an RNA recognition motif of human La protein is a potent inhibitor of hepatitis C virus. Antiviral Res 97:223–226

Martinez FJ, Lee JH, Lee JE et al (2012) Whole exome sequencing identifies a splicing mutation in NSUN2 as a cause of a Dubowitz-like syndrome. J Med Genet 49:380–385

Masliah G, Barraud P, Allain FH (2013) RNA recognition by double-stranded RNA binding domains: a matter of shape and sequence. Cell Mol Life Sci 70:1875–1895

Meng Z, Limbach PA (2006) Mass spectrometry of RNA: linking the genome to the proteome. Brief Funct Genomic Proteomic 5:87–95

Meyer KD, Jaffrey SR (2014) The dynamic epitranscriptome: N6-methyladenosine and gene expression control. Nat Rev Mol Cell Biol 15:313–326

Meyer KD, Saletore Y, Zumbo P et al (2012) Comprehensive analysis of mRNA methylation reveals enrichment in 3′ UTRs and near stop codons. Cell 149:1635–1646

Moore PB (1999) Structural motifs in RNA. Annu Rev Biochem 68:287–300

Moriya J, Yokogawa T, Wakita K et al (1994) A novel modified nucleoside found at the first position of the anticodon of methionine transfer-RNA from Bovine Liver-Mitochondria. Biochemistry 33:2234–2239

Moukadiri I, Prado S, Piera J et al (2009) Evolutionarily conserved proteins MnmE and GidA catalyze the formation of two methyluridine derivatives at tRNA wobble positions. Nucleic Acids Res 37:7177–7193

Mucha P, Szyk A, Rekowski P et al (2001) Anticodon domain methylated nucleosides of yeast tRNA(Phe) are significant recognition determinants in the binding of a phage display selected peptide. Biochemistry 40:14191–14199

Mucha P, Szyk A, Rekowski P et al (2002) Interaction of RNA with phage display selected peptides analyzed by capillary electrophoresis mobility shift assay. RNA 8:698–704

Mucha P, Szyk A, Rekowski P et al (2003) Using capillary electrophoresis to study methylation effect on RNA-peptide interaction. Acta Biochim Pol 50:857–864

Mucha P, Szyk A, Rekowski P et al (2004) Sequence-altered peptide adopts optimum conformation for modification-dependent binding of the yeast tRNAPhe anticodon domain. Protein J 23:33–38

Muramatsu T, Nishikawa K, Nemoto F et al (1988) Codon and amino-acid specificities of a transfer RNA are both converted by a single post-transcriptional modification. Nature 336:179–181

Murphy FV, Ramakrishnan V, Malkiewicz A et al (2004) The role of modifications in codon discrimination by tRNA(Lys)UUU. Nat Struct Mol Biol 11:1186–1191

Muto Y, Yokoyama S (2012) Structural insight into RNA recognition motifs: versatile molecular Lego building blocks for biological systems. Wiley Interdiscip Rev RNA 3:229–246

Nakanishi K, Bonnefond L, Kimura S et al (2009) Structural basis for translational fidelity ensured by transfer RNA lysidine synthetase. Nature 461:1144–1148

Ng MC, Saxena R, Li J et al (2013) Transferability and fine mapping of type 2 diabetes loci in African Americans: the Candidate Gene Association Resource Plus Study. Diabetes 62:965–976

Niu Y, Jones AJ, Wu H et al (2011) gamma-AApeptides bind to RNA by mimicking RNA-binding proteins. Org Biomol Chem 9:6604–6609

Numata T, Ikeuchi Y, Fukai S et al (2006) Snapshots of tRNA sulphuration via an adenylated intermediate. Nature 442:419–424

O'Connell M (2015) RNA modification and the epitranscriptome; the next frontier. RNA 21:703–704

Ofek G, Guenaga FJ, Schief WR et al (2010) Elicitation of structure-specific antibodies by epitope scaffolds. Proc Natl Acad Sci U S A 107:17880–17887

Ofengand J (2002) Ribosomal RNA pseudouridines and pseudouridine synthases. FEBS Lett 514:17–25

Ogle JM, Murphy FV, Tarry MJ et al (2002) Selection of tRNA by the ribosome requires a transition from an open to a closed form. Cell 111:721–732

Omori S, Tanaka Y, Takahashi A et al (2008) Association of CDKAL1, IGF2BP2, CDKN2A/B, HHEX, SLC30A8, and KCNJ11 with susceptibility to type 2 diabetes in a Japanese population. Diabetes 57:791–795

Pagliarini DJ, Calvo SE, Chang B et al (2008) A mitochondrial protein compendium elucidates complex I disease biology. Cell 134:112–123

Pascoe L, Tura A, Patel SK et al (2007) Common variants of the novel type 2 diabetes genes CDKAL1 and HHEX/IDE are associated with decreased pancreatic beta-cell function. Diabetes 56:3101–3104

Paska AV, Hudler P (2015) Aberrant methylation patterns in cancer: a clinical view. Biochem Med (Zagreb) 25:161–176

Patton JR, Bykhovskaya Y, Mengesha E et al (2005) Mitochondrial myopathy and sideroblastic anemia (MLASA): missense mutation in the pseudouridine synthase 1 (PUS1) gene is associated with the loss of tRNA pseudouridylation. J Biol Chem 280:19823–19828

Pavon-Eternod M, Gomes S, Geslain R et al (2009) tRNA over-expression in breast cancer and functional consequences. Nucleic Acids Res 37:7268–7280

Pavon-Eternod M, Wei M, Pan T et al (2010) Profiling non-lysyl tRNAs in HIV-1. RNA 16:267–273

Pavon-Eternod M, David A, Dittmar K et al (2013) Vaccinia and influenza A viruses select rather than adjust tRNAs to optimize translation. Nucleic Acids Res 41:1914–1921

Pierrel F, Bjork GR, Fontecave M et al (2002) Enzymatic modification of tRNAs: MiaB is an iron-sulfur protein. J Biol Chem 277:13367–13370

Pierrel F, Douki T, Fontecave M et al (2004) MiaB protein is a bifunctional radical-S-adenosylmethionine enzyme involved in thiolation and methylation of tRNA. J Biol Chem 279:47555–47563

Ping XL, Sun BF, Wang L et al (2014) Mammalian WTAP is a regulatory subunit of the RNA N6-methyladenosine methyltransferase. Cell Res 24:177–189

Pomerantz SC, McCloskey JA (1990) Analysis of RNA hydrolyzates by liquid chromatography-mass spectrometry. Methods Enzymol 193:796–824

Putz J, Florentz C, Benseler F et al (1994) A single methyl group prevents the mischarging of a tRNA. Nat Struct Biol 1:580–582

Ramser J, Winnepenninckx B, Lenski C et al (2004) A splice site mutation in the methyltransferase gene FTSJ1 in Xp11.23 is associated with non-syndromic mental retardation in a large Belgian family (MRX9). J Med Genet 41:679–683

Reichow SL, Hamma T, Ferre-D'Amare AR et al (2007) The structure and function of small nucleolar ribonucleoproteins. Nucleic Acids Res 35:1452–1464

Reiter NJ, Chan CW, Mondragon A (2011) Emerging structural themes in large RNA molecules. Curr Opin Struct Biol 21:319–326

Riley LG, Cooper S, Hickey P et al (2010) Mutation of the mitochondrial tyrosyl-tRNA synthetase gene, YARS2, causes myopathy, lactic acidosis, and sideroblastic anemia—MLASA syndrome. Am J Hum Genet 87:52–59

Rodnina MV, Wintermeyer W (2001) Fidelity of aminoacyl-tRNA selection on the ribosome: kinetic and structural mechanisms. Annu Rev Biochem 70:415–435

Rodriguez V, Chen Y, Elkahloun A et al (2007) Chromosome 8 BAC array comparative genomic hybridization and expression analysis identify amplification and overexpression of TRMT12 in breast cancer. Genes Chromosomes Cancer 46:694–707

Rose RE, Quinn R, Sayre JL et al (2015) Profiling ribonucleotide modifications at full-transcriptome level: a step toward MS-based epitranscriptomics. RNA 21:1361–1374

Russell SP, Limbach PA (2013) Evaluating the reproducibility of quantifying modified nucleosides from ribonucleic acids by LC-UV-MS. J Chromatogr B Analyt Technol Biomed Life Sci 923–924:74–82

Sakurai M, Suzuki T (2011) Biochemical identification of A-to-I RNA editing sites by the inosine chemical erasing (ICE) method. Methods Mol Biol 718:89–99

Samish I, MacDermaid CM, Perez-Aguilar JM et al (2011) Theoretical and computational protein design. Annu Rev Phys Chem 62:129–149

Satterlee JS, Basanta-Sanchez M, Blanco S et al (2014) Novel RNA modifications in the nervous system: form and function. J Neurosci 34:15170–15177

Scaglia F, Northrop JL (2006) The mitochondrial myopathy encephalopathy, lactic acidosis with stroke-like episodes (MELAS) syndrome: a review of treatment options. CNS Drugs 20:443–464

Schaefer M, Hagemann S, Hanna K et al (2009a) Azacytidine inhibits RNA methylation at DNMT2 target sites in human cancer cell lines. Cancer Res 69:8127–8132

Schaefer M, Pollex T, Hanna K et al (2009b) RNA cytosine methylation analysis by bisulfite sequencing. Nucleic Acids Res 37, e12

Schellenberg MJ, Dul EL, MacMillan AM (2011) Structural model of the p14/SF3b155. branch duplex complex. RNA 17(1):155–165

Schimmel P, Frugier M, Glasfeld E (1997) Peptides for RNA discrimination and for assembly of enzymes that act on RNA. Nucleic Acids Symp Ser 36:1

Schmidt PG, Sierzputowska-Gracz H, Agris PF (1987) Internal motions in yeast phenylalanine transfer RNA from 13C NMR relaxation rates of modified base methyl groups: a model-free approach. Biochemistry 26:8529–8534

Schwartz S, Agarwala SD, Mumbach MR et al (2013) High-resolution mapping reveals a conserved, widespread, dynamic mRNA methylation program in yeast meiosis. Cell 155:1409–1421

Schwartz S, Bernstein DA, Mumbach MR et al (2014) Transcriptome-wide mapping reveals widespread dynamic-regulated pseudouridylation of ncRNA and mRNA. Cell 159:148–162

Scott LJ, Mohlke KL, Bonnycastle LL et al (2007) A genome-wide association study of type 2 diabetes in Finns detects multiple susceptibility variants. Science 316:1341–1345

Seno T, Agris PF, Soll D (1974) Involvement of the anticodon region of *Escherichia coli* tRNAGln and tRNAGlu in the specific interaction with cognate aminoacyl-tRNA synthetase. Alteration of the 2-thiouridine derivatives located in the anticodon of the tRNAs by BrCN or sulfur deprivation. Biochim Biophys Acta 349:328–338

Sharma S, Lafontaine DL (2015) 'View From A Bridge': a new perspective on eukaryotic rRNA base modification. Trends Biochem Sci 40:560–575

Sierzputowska-Gracz H, Sochacka E, Malkiewicz A et al (1987) Chemistry and structure of modified uridines in the anticodon, wobble position of transfer RNA are determined by thiolation. J Am Chem Soc 109:7171–7177

Simos G, Tekotte H, Grosjean H et al (1996) Nuclear pore proteins are involved in the biogenesis of functional tRNA. EMBO J 15:2270–2284

Slotkin W, Nishikura K (2013) Adenosine-to-inosine RNA editing and human disease. Genome Med 5:105

Song CX, Yi C, He C (2012) Mapping recently identified nucleotide variants in the genome and transcriptome. Nat Biotechnol 30:1107–1116

Spears JL, Xiao X, Hall CK et al (2014) Amino acid signature enables proteins to recognize modified tRNA. Biochemistry 53:1125–1133

Spitale RC, Flynn RA, Zhang QC et al (2015) Structural imprints in vivo decode RNA regulatory mechanisms. Nature 519:486–490

Spriggs RV, Murakami Y, Nakamura H et al (2009) Protein function annotation from sequence: prediction of residues interacting with RNA. Bioinformatics 25:1492–1497

Squires JE, Patel HR, Nousch M et al (2012) Widespread occurrence of 5-methylcytosine in human coding and non-coding RNA. Nucleic Acids Res 40:5023–5033

Steinthorsdottir V, Thorleifsson G, Reynisdottir I et al (2007) A variant in CDKAL1 influences insulin response and risk of type 2 diabetes. Nat Genet 39:770–775

Stewart JB, Chinnery PF (2015) The dynamics of mitochondrial DNA heteroplasmy: implications for human health and disease. Nat Rev Genet 16:530–542

Stuart JW, Basti MM, Smith WS et al (1996) Structure of the trinucleotide D acp3U-A with Coordinated Mg2+ demonstrates that modified nucleosides contribute to regional conformations of RNA. Nucleosides Nucleotides 15:1009–1028

Stuart JW, Gdaniec Z, Guenther R et al (2000) Functional anticodon architecture of human tRNALys3 includes disruption of intraloop hydrogen bonding by the naturally occurring amino acid modification, t6A. Biochemistry 39:13396–13404

Stuart JW, Koshlap KM, Guenther R et al (2003) Naturally-occurring modification restricts the anticodon domain conformational space of tRNA(Phe). J Mol Biol 334:901–918

Su D, Chan CT, Gu C et al (2014) Quantitative analysis of ribonucleoside modifications in tRNA by HPLC-coupled mass spectrometry. Nat Protoc 9:828–841

Sudrik C, Arha M, Cao J et al (2013) Translational repression using BIV Tat peptide-TAR RNA interaction in mammalian cells. Chem Commun (Camb) 49:7457–7459

Suzuki T, Suzuki T (2014) A complete landscape of post-transcriptional modifications in mammalian mitochondrial tRNAs. Nucleic Acids Res 42:7346–7357

Suzuki T, Suzuki T, Wada T et al (2002) Taurine as a constituent of mitochondrial tRNAs: new insights into the functions of taurine and human mitochondrial diseases. EMBO J 21:6581–6589

Suzuki T, Nagao A, Suzuki T (2011) Human mitochondrial tRNAs: biogenesis, function, structural aspects, and diseases. Annu Rev Genet 45:299–329

Sylvers LA, Rogers KC, Shimizu M et al (1993) A 2-thiouridine derivative in tRNAGlu is a positive determinant for aminoacylation by *Escherichia coli* glutamyl-tRNA synthetase. Biochemistry 32:3836–3841

Takano K, Nakagawa E, Inoue K et al (2008) A loss-of-function mutation in the FTSJ1 gene causes nonsyndromic X-linked mental retardation in a Japanese family. Am J Med Genet B Neuropsychiatr Genet 147B:479–484

Takeoka S, Unoki M, Onouchi Y et al (2001) Amino-acid substitutions in the IKAP gene product significantly increase risk for bronchial asthma in children. J Hum Genet 46:57–63

Tao J, Frankel AD (1992) Specific binding of arginine to TAR RNA. Proc Natl Acad Sci U S A 89:2723–2726

Taylor RW, Pyle A, Griffin H et al (2014) Use of whole-exome sequencing to determine the genetic basis of multiple mitochondrial respiratory chain complex deficiencies. JAMA 312:68–77

Thapar R, Denmon AP, Nikonowicz EP (2014) Recognition modes of RNA tetraloops and tetraloop-like motifs by RNA-binding proteins. Wiley Interdiscip Rev RNA 5:49–67

Thomas SR, Keller CA, Szyk A et al (2011) Structural insight into the functional mechanism of Nep1/Emg1 N1-specific pseudouridine methyltransferase in ribosome biogenesis. Nucleic Acids Res 39:2445–2457

Torres AG, Batlle E, Ribas de Pouplana L (2014) Role of tRNA modifications in human diseases. Trends Mol Med 20:306–314

Towns WL, Begley TJ (2012) Transfer RNA methytransferases and their corresponding modifications in budding yeast and humans: activities, predications, and potential roles in human health. DNA Cell Biol 31:434–454

Tzima E, Schimmel P (2006) Inhibition of tumor angiogenesis by a natural fragment of a tRNA synthetase. Trends Biochem Sci 31:7–10

Ulyanov NB, James TL (2010) RNA structural motifs that entail hydrogen bonds involving sugar-phosphate backbone atoms of RNA. N J Chem 34:910–917

Umeda N, Suzuki T, Yukawa M et al (2005) Mitochondria-specific RNA-modifying enzymes responsible for the biosynthesis of the wobble base in mitochondrial tRNAs. Implications for the molecular pathogenesis of human mitochondrial diseases. J Biol Chem 280(2):1613–1624

UniProt C (2009) The Universal Protein Resource (UniProt) 2009. Nucleic Acids Res 37(Database issue):D169–174

Urbonavicius J, Qian O, Durand JMB et al (2001) Improvement of reading frame maintenance is a common function for several tRNA modifications. EMBO J 20(17):4863–4873

Vachon CM, Sellers TA, Carlson EE et al (2007) Strong evidence of a genetic determinant for mammographic density, a major risk factor for breast cancer. Cancer Res 67:8412–8418

van Buul CP, van Knippenberg PH (1985) Nucleotide sequence of the ksgA gene of Escherichia coli: comparison of methyltransferases effecting dimethylation of adenosine in ribosomal RNA. Gene 38:65–72

van Buul CP, Visser W, van Knippenberg PH (1984) Increased translational fidelity caused by the antibiotic kasugamycin and ribosomal ambiguity in mutants harbouring the ksgA gene. FEBS Lett 177:119–124

Vendeix FA, Murphy FV, Cantara WA et al (2012) Human tRNA(Lys3)(UUU) is pre-structured by natural modifications for cognate and wobble codon binding through keto-enol tautomerism. J Mol Biol 416:467–485

Villarroya M, Prado S, Esteve JM et al (2008) Characterization of human GTPBP3, a GTP-binding protein involved in mitochondrial tRNA modification. Mol Cell Biol 28:7514–7531

Voigt CA, Gordon DB, Mayo SL (2000) Trading accuracy for speed: a quantitative comparison of search algorithms in protein sequence design. J Mol Biol 299:789–803

Wallace DC (2005) A mitochondrial paradigm of metabolic and degenerative diseases, aging, and cancer: a dawn for evolutionary medicine. Annu Rev Genet 39:359–407

Wang C, Bradley P, Baker D (2007) Protein-protein docking with backbone flexibility. J Mol Biol 373(2):503–519

Wang X, Lu Z, Gomez A et al (2014) N6-methyladenosine-dependent regulation of messenger RNA stability. Nature 505:117–120

Wei FY, Tomizawa K (2011) Functional loss of Cdkal1, a novel tRNA modification enzyme, causes the development of type 2 diabetes. Endocr J 58:819–825

Wei FY, Tomizawa K (2012) Development of type 2 diabetes caused by a deficiency of a tRNA (lys) modification. Islets 4:71–73

Wei FY, Suzuki T, Watanabe S et al (2011) Deficit of tRNA(Lys) modification by Cdkal1 causes the development of type 2 diabetes in mice. J Clin Invest 121:3598–3608

Weiss MA, Narayana N (1998) RNA recognition by arginine-rich peptide motifs. Biopolymers 48:167–180

Xiao X, Hall CK, Agris PF (2014) The design of a peptide sequence to inhibit HIV replication: a search algorithm combining Monte Carlo and self-consistent mean field techniques. J Biomol Struct Dyn 32:1523–1536

Xiao X, Agris PF, Hall CK (2015) Molecular recognition mechanism of peptide chain bound to the tRNA(Lys3) anticodon loop in silico. J Biomol Struct Dyn 33:14–27

Xu C, Wang X, Liu K et al (2014) Structural basis for selective binding of m6A RNA by the YTHDC1 YTH domain. Nat Chem Biol 10:927–929

Yarian C, Marszalek M, Sochacka E et al (2000) Modified nucleoside dependent Watson-Crick and wobble codon binding by tRNALysUUU species. Biochemistry 39(44):13390–13395

Yarian C, Townsend H, Czestkowski W et al (2002) Accurate translation of the genetic code depends on tRNA modified nucleosides. J Biol Chem 277:16391–16395

Yasukawa T, Suzuki T, Ishii N et al (2000a) Defect in modification at the anticodon wobble nucleotide of mitochondrial tRNA(Lys) with the MERRF encephalomyopathy pathogenic mutation. FEBS Lett 467:175–178

Yasukawa T, Suzuki T, Ueda T et al (2000b) Modification defect at anticodon wobble nucleotide of mitochondrial tRNAs(Leu)(UUR) with pathogenic mutations of mitochondrial myopathy, encephalopathy, lactic acidosis, and stroke-like episodes. J Biol Chem 275:4251–4257

Yasukawa T, Suzuki T, Ishii N et al (2001) Wobble modification defect in tRNA disturbs codon-anticodon interaction in a mitochondrial disease. EMBO J 20:4794–4802

Yasukawa T, Kirino Y, Ishii N et al (2005) Wobble modification deficiency in mutant tRNAs in patients with mitochondrial diseases. FEBS Lett 579:2948–2952

Ye X, Gorin A, Frederick R et al (1999) RNA architecture dictates the conformations of a bound peptide. Chem Biol 6:657–669

Yu YT, Shu MD, Steitz JA (1998) Modifications of U2 snRNA are required for snRNP assembly and pre-mRNA splicing. EMBO J 17:5783–5795

Yu AT, Ge J, Yu YT (2011) Pseudouridines in spliceosomal snRNAs. Protein Cell 2:712–725

Yue Y, Liu J, He C (2015) RNA N6-methyladenosine methylation in post-transcriptional gene expression regulation. Genes Dev 29:1343–1355

Zeggini E, Weedon MN, Lindgren CM et al (2007) Replication of genome-wide association signals in UK samples reveals risk loci for type 2 diabetes. Science 316:1336–1341

Zeharia A, Shaag A, Pappo O et al (2009) Acute infantile liver failure due to mutations in the TRMU gene. Am J Hum Genet 85:401–407

Zhang T, Cooper S, Brockdorff N (2015) The interplay of histone modifications—writers that read. EMBO Rep 16:1467–1481

Zhao BS, He C (2015) Pseudouridine in a new era of RNA modifications. Cell Res 25:153–154

Zheng G, Dahl JA, Niu Y et al (2013) ALKBH5 is a mammalian RNA demethylase that impacts RNA metabolism and mouse fertility. Mol Cell 49:18–29

Zhou Y, Goodenbour JM, Godley LA et al (2009) High levels of tRNA abundance and alteration of tRNA charging by bortezomib in multiple myeloma. Biochem Biophys Res Commun 385:160–164

Zhou KI, Parisien M, Dai Q et al (2015) N-methyladenosine modification in a long noncoding RNA hairpin predisposes its conformation to protein binding. J Mol Biol pii: S0022-2836(15) 00486-6. doi: 10.1016/j.jmb.2015.08.021. [Epub ahead of print]

RNA Modification N^6-Methyladenosine in Post-transcriptional Regulation

Guifang Jia

Contents

Abstract N^6-methyladenosine (m^6A) is the most prevalent internal messenger RNA (mRNA) chemical modification in eukaryotes. This methylation has been shown to be reversible in mammals. It is installed by a methyltransferase complex (writers) of METTL3, METTL14, and Wilms' tumor 1-associating protein (WTAP) and can be removed by demethylases (erasers) FTO and ALKBH5, which are iron(II)- and α-ketoglutarate-dependent dioxygenases. The reversible and dynamic methylation exhibits significant functional roles in various biological processes. The m^6A modification as an RNA mark is recognized by reader proteins, such as YTH domain family proteins. YTHDF2 regulates the stability of the methylated transcripts in cytoplasm; YTHDF1 promotes protein synthesis of the methylated mRNA by

G. Jia (✉)
Synthetic and Functional Biomolecules Center, Beijing National Laboratory for Molecular Sciences, Key Laboratory of Bioorganic Chemistry and Molecular Engineering of Ministry of Education, College of Chemistry and Molecular Engineering, Peking University, Beijing 100871, China
e-mail: guifangjia@pku.edu.cn

S. Jurga et al. (eds.), *Modified Nucleic Acids in Biology and Medicine*, RNA Technologies, DOI 10.1007/978-3-319-34175-0_6

interacting with translation machinery. m^6A as a switch controls the RNA structure to affect RNA-protein interactions for biological regulation. Meanwhile, many m^6A detection techniques were developed and applied in biology and medicine. The total m^6A level in mRNA can be determined by ultra-performance liquid chromatography coupled with triple-quadrupole tandem mass spectrometry (UPLC-QQQ-MS/MS), two-dimensional thin-layer chromatography (2-D TLC), and dot blot. The m^6A antibody affinity sequencing (MeRIP-seq) was developed to map the m^6A site location in a transcriptome-wide manner. The single-base resolution methods for single gene or whole transcripts were also invented.

Keywords Epitranscriptome • N^6-methyladenosine (m^6A) • Methyltransferase • Demethylase • m^6A reader

1 Introduction

In the central dogma, genetic information is passed from DNA to RNA and then to protein. Reversible chemical modifications on DNA and histone proteins are termed epigenetic marks to regulate gene transcription (Fig. 1). More than 100 post-

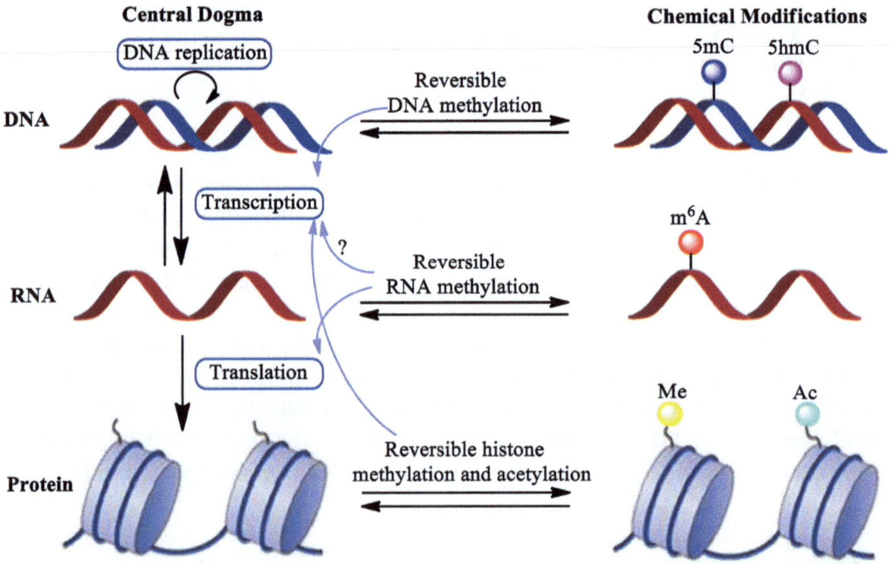

Fig. 1 The reversible chemical modifications that regulate the flow of genetic information in the central dogma. Genetic information flows from DNA to RNA and then to protein. Epigenetic DNA modifications (e.g., 5-methylcytosine (5mC) and 5-hydroxymethylcytosine (5hmC)) and histone modifications (e.g., methylation (Me) and acetylation (Ac)) play important roles in transcriptional gene regulation. Epitranscriptomic modification (N^6-methyladenosine (m^6A)) is a new mode of post-transcriptional gene expression regulation

transcriptional modifications occur on RNA (Cantara et al. 2011), although the functions of most of these RNA modifications remain unclear. Until 2011, m^6A was discovered as the first example of the reversible RNA modification (Jia et al. 2011), which prompted m^6A as epitranscriptomic mark to study its functions in post-transcriptional regulation. In this chapter, we summarize recent progress in the study of m^6A methylation in mRNA and discuss the newly discovered functional roles in regulation of mRNA fate.

2 m^6A RNA Methylation in Eukaryotic mRNA

The m^6A methylation is the most prevalent internal modification in eukaryotic mRNA including mammals, plants, flies, yeast in the meiotic state, and bacteria. It was first found in mammalian mRNA in the mid-1970s (Wei et al. 1975). m^6A methylation has been shown to preferentially occur at the consensus sequence [G/A/U][G>A]m^6AC[U>A>C] (Schibler et al. 1977; Wei et al. 1976), and the methylation in a specific site is nonstoichiometric, and only a portion of these consensus sequences are methylated in mRNA (Horowitz et al. 1984; Kane and Beemon 1985).

The transcriptome-wide sequencing of m^6A has been performed in mammalians (Dominissini et al. 2012; Meyer et al. 2012), yeasts (Schwartz et al. 2013), and plants (Luo et al. 2014). The m^6A RNA methylomes showed that m^6A is widely distributed on over 12,000 sites in the transcripts of more than 7000 human coding genes and 300 human noncoding genes and is preferentially enriched in near stop codons, in 3′ UTR (untranslated regions), and within long internal exons in mammals (Dominissini et al. 2012; Meyer et al. 2012). In the yeast *Saccharomyces cerevisiae*, m^6A occurs only during meiosis and plays a role in the initiation and nutritional control of meiosis (Bodi et al. 2010; Clancy et al. 2002; Shah and Clancy 1992). By comparing m^6A-seq results of wild-type stain with that of *IME4* (orthologue of human METTL3)-deficient stain, *ime4*Δ/Δ, the m^6A sequencing in yeast meiosis revealed that 1308 putatively methylated sites locate within 1183 transcripts and are strongly enriched in 3′ UTR (Schwartz et al. 2013). In *Arabidopsis thaliana*, m^6A is critical for normal plant development (Bodi et al. 2012; Zhong et al. 2008). m^6A is abundant and highly conserved across *A. thaliana* accessions with 4317 m^6A peaks detected within both Can-0 and Hen-16 (Luo et al. 2014). Although most m^6A peaks are shared between these two stains, a portion of strain-specific m^6A peaks are distributed on the genes with the functions on mRNA metabolic process, response to stimulus, and regulation of translational elongation (Luo et al. 2014). Distinct from the distribution patterns found in mammals and yeast, m^6A in these two stains of *A. thaliana* is enriched not only around the stop codon and within 3′-UTR but also around the start codon (Luo et al. 2014).

The modification m^6A is reversible and dynamically regulated and is essential for eukaryotic developments. The m^6A methylation determines stem cell fate by regulating pluripotency transition toward differentiation (Batista et al. 2014; Geula et al. 2015) and controls the speed of the circadian clock (Fustin et al. 2013). The m^{6}A methylation plays functions in mRNA metabolism to accelerate mRNA decay (Wang et al. 2014), promote translation efficiency (Wang et al. 2015b), affect nuclear RNA export (Fustin et al. 2013; Zheng et al. 2013), and regulate splicing (Dominissini et al. 2012; Ping et al. 2014) through the recognition of m^{6}A reader proteins.

3 m^{6}A Writers in Eukaryotes

m^{6}A mRNA methylation is installed during processing of the nascent pre-mRNA by a multicomponent N^6-Adenosine methyltransferase complex, which was originally isolated as ~200 kDa and ~800 kDa subcomplexes from HeLa nuclear extracts (Bokar et al. 1997; Narayan and Rottman 1988). The first well-characterized component is the SAM-binding subunit with 70 kDa molecular weight, MT-A70 (known as METTL3) (Fig. 1) (Bokar et al. 1997). METTL3 is highly conserved in most eukaryotes from yeast to humans. METTL3 is colocalized with nuclear speckles in human cells (Bokar 2005). Knockdown of METTL3 affects gene expression and alternative splicing patterns and modulates p53 signaling and induction of apoptosis (Bokar 2005; Dominissini et al. 2012). The deficiency in METTL3 blocked the differentiation of murine embryonic stem cells (mESCs) (Batista et al. 2014; Geula et al. 2015) and led to early embryonic lethality (Geula et al. 2015). Similar to mammals, complete deletion of *Dm ime4* (orthologue of human METTL3) in *Drosophila* (Hongay and Orr-Weaver 2011) and *MTA* (orthologue of human METTL3, encoded by At4g10760) in *A. thaliana* (Zhong et al. 2008) is lethal. Decrease of *Dm ime4* expression in *Drosophila* leads to semilethality and subfertility and reduces Notch signaling levels in follicle cells, which can be fully recovered by a wild-type transgenic copy of *Dm ime4*, but not the inactive one (Hongay and Orr-Weaver 2011). Expression of MTA driven by the embryo-specific *ABI3* promoter in *MTA* mutant during embryonic development rescues the embryonic lethality, reduces more than 90 % of m^{6}A level, and leads to abnormal growth with reduced apical dominance, abnormal organ definition, and an increased number of trichome branches (Bodi et al. 2012). These results suggest that m^{6}A is essential for eukaryotic viability.

A phylogenetic analysis identified METTL4 and METTL14 as close homologues of METTL3 in human (Bujnicki et al. 2002). Knockdown of METTL14, but not METTL4, decreases mRNA m^{6}A level in HeLa and 293FT cells (Liu et al. 2014), which revealed that METTL14 is another active component of the m^{6}A methyltransferase complex (Fig. 2). Biochemical characterization demonstrated that these two active components, METTL3 and METTL14, form a stable heterodimer and colocalize in nuclear speckles. The photoactivatableribonucleoside-enhanced crosslinking and immunoprecipitation (PAR-CLIP) data showed that the binding sites of METTL3 and METTL14 in substrate RNAs contain a similar consensus sequence to that known for m^6

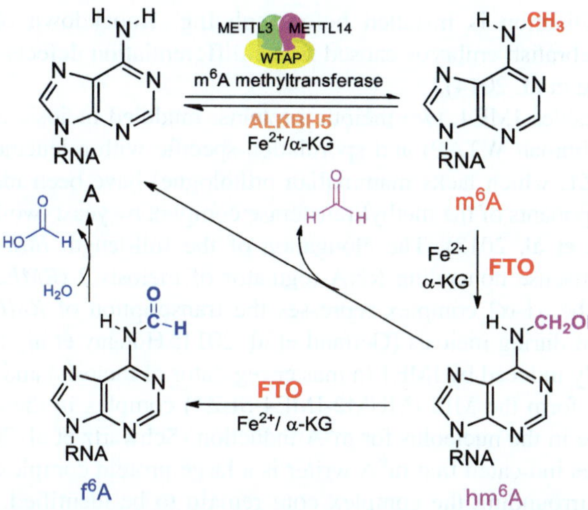

Fig. 2 The chemical scheme of methylation and demethylation of m^{6}A. The m^{6}A modification is installed by a methyltransferase complex of METTL3, METTL14, and WTAP and removed by demethylases FTO and ALKBH5. FTO can oxidize m^{6}A to N^6-hydroxymethyladenosine (hm^{6}A) and N^6-formyladenosine (f^{6}A) sequentially; hm^{6}A and f^{6}A can be hydrolyzed to adenine with half-lives of ~3 h under physiological conditions, while ALKBH5 oxidizes m^{6}A to hm^{6}A without further oxidation

A. Silencing of the m^{6}A methyltransferase complex led to an increase in the abundance and lifetime of their m^{6}A target transcripts, which is consistent with the m^{6}A-mediated mRNA decay function.

The third crucial component of the m^{6}A methyltransferase complex is the Wilms tumor 1-associating protein (WTAP) in human (Fig. 2) (Liu et al. 2014). Its orthologue protein in *Arabidopsis*, FKBP12-interacting protein of 37 kDa (FIP37, encoded by At3g54170) was firstly identified as the partner protein of MTA by yeast two-hybrid screening (Bodi et al. 2012). WTAP bound with METTL3-METTL14 heterodimer colocalizes in nuclear speckles to participate in m^{6}A RNA methylation (Liu et al. 2014). Compared with METTL3 and METTL14, silence of WTAP leads to the largest decrease in m^{6}A level in cell lines, indicating that WTAP plays important roles in cellular m^{6}A deposition. PAR-CLIP data also revealed that WTAP shares a similar binding sequence to that known for m^{6}A. The targets of METTL3, METTL14, and WTAP identified by PAR-CLIP have a ~50 % overlap with m^{6}A-containing transcripts, further indicating that these three components form the core of the major cellular writer complex of m^{6}A. As WTAP is a splicing factor, knockdown of WTAP and METTL3 yields different isoforms of m^{6}A-containing transcripts (Ping et al. 2014), suggesting that the methylation could affect splicing. In addition, the m^{6}A methylome shows that m^{6}A is present in introns (Dominissini et al. 2012; Meyer et al. 2012). Together, these data indicate that this modification is installed before splicing. Knockdown of WTAP and METTL3 in zebrafish embryos caused tissue differentiation defects and increased apoptosis (Ping et al. 2014).

In yeast, besides IME4, two meiotic proteins, muddled meiosis 2 (MUM2, the orthologue of human WTAP) and sporulation-specific with a leucine zipper motif protein 1 (SLZ1, which lacks mammalian orthologue) have been identified as the other two components of the methyltransferase complex by yeast two-hybrid screening (Agarwala et al. 2012). The elongation of the full-length *IME4* transcript is blocked by antisense noncoding RNA regulator of meiosis 2 (*RME2*), whereas in diploid cells, the a1-α2 complex represses the transcription of *RME2* and allows *IME4* induction during meiosis (Gelfand et al. 2011; Hongay et al. 2006). SLZ1 is transcriptionally induced by IME1 (a master regulator of meiosis) and recruits IME4 and MUM2 to form the MIS (MUM2-IME4-SLZ1) complex in the cytoplasm and then translocate in the nucleolus for m^{6}A induction (Schwartz et al. 2013).

Early studies indicated that m^{6}A writer is a large protein complex; hence, other components surrounding the complex core remain to be identified. Given only a portion of the consensus sequence is methylated, future research to identify other important components or protein-translational modifications for m^{6}A writers may elucidate the selectivity of m^{6}A deposition.

4 m^{6}A Erasers in Eukaryotes

In 2011, the discovery of the fat mass and obesity-associated (FTO) protein as an m^{6}A-demethylase was an important breakthrough in reigniting investigations of m^{6}A and envisioning RNA modification as epitranscriptomic mark parallel with DNA and histone epigenetic modifications (Jia et al. 2011). In 2007, *FTO* gene was verified to have a strong association with body mass index (BMI) and risk of obesity and type 2 diabetes by genome-wide association studies (GWAS) (Dina et al. 2007; Frayling et al. 2007; Hinney et al. 2007). Overexpression of *FTO* in mice caused increased food intake and increased adiposity in a dose-dependent manner (Church et al. 2010). Depletion of *FTO* in mice leads to the phenotypes with increased postnatal lethality, postnatal growth retardation, reduced fat mass, lower body weight, increased energy expenditure, and a relative increase in food intake (Church et al. 2009; Fischer et al. 2009; Gao et al. 2010). GWAS also shows that common variants in the *FTO* gene have been linked with other human diseases, such as cancers (Hernandez-Caballero and Sierra-Ramirez 2015; Kaklamani et al. 2011), cardiovascular diseases (Lappalainen et al. 2011), Alzheimer's disease (Keller et al. 2011), and so on.

Despite the strong genetic association between FTO and body weight through GWAS, less is known of the physiological mechanisms or pathways at the molecular level. FTO was identified as the ninth homologue of the nonheme Fe (II)/α-ketoglutarate (α-KG)-dependent AlkB family dioxygenases in human, and it oxidatively demethylates N^3-methylthymidine in single-stranded DNA (ssDNA) and N^3-methyluridine in single-stranded RNA (ssRNA) in vitro with a low activity (Gerken et al. 2007; Jia et al. 2008). GWAS study found that a loss-of-function homozygous mutation (R316Q) in the *FTO* gene typically leads to postnatal growth retardation, as well as multiple dysmorphisms and malformations (Boissel et al. 2009). R316 is one of the α-KG ligands, and R316Q mutant causes reduced FTO catalytic activity. The phenotypes associated with this mutation demonstrated that the enzymatic activity of FTO plays functional roles. However, the function of FTO in vivo remained unknown until we discovered FTO efficiently demethylates m^6A in DNA and RNA in vitro (Fig. 2). Knockdown or overexpression of FTO leads to the increase or decrease of m^6A in polyA-RNA in human cell lines, which confirmed in vivo m^6A demethylation activity of FTO. FTO partially colocalizes with nuclear speckles. These cell-based results indicated that nuclear RNA (including mRNA, lncRNA, and possibly other types of RNA) is the main substrate of FTO (Jia et al. 2011). A recent study found FTO can remove m^6A around the splice sites of adipogenic factor RUNX1T1 to prevent the binding of the splicing factor SRSF2 and to promote the production of a shorter isoform, which in turn acts to induce preadipocyte differentiation (Zhao et al. 2014).

The FTO-catalyzed oxidation of m^6A generates two intermediates, N^6-hydroxymethyladenosine (hm^6A) and N^6-formyladenosine (f^6A) (Fig. 2) (Fu et al. 2013). The hm^6A is a direct oxidation product of m^6A, and f^6A is the further oxidized product of hm^6A. Both hm^6A and f^6A can decompose in water to yield the unmethylated adenine and formaldehyde (from hm^6A) or formic acid (from f^6A). Surprisingly, these two modifications are metastable under physiological condition with half-lives of ~3 h and indeed exist in human cells and mouse tissues. Given median mammalian RNA half-lives are ~5 h, these new modifications may have functional roles and affect RNA-protein interactions.

The second demethylase identified in mammals is another AlkB family protein, ALKBH5 (Zheng et al. 2013). Distinct from FTO, ALKBH5 oxidizes m^6A to hm^6A without further oxidation (Fig. 2). ALKBH5 primarily colocalizes with nuclear speckles, and its demethylation activity affects nascent mRNA synthesis, the rate of splicing, and mRNA nuclear export. Unlike FTO, direct immunoprecipitation of ALKBH5 has identified bound RNA targets (Zheng et al. 2013), and ALKBH5 was shown to be part of the mRNA-bound proteome (Baltz et al. 2012), suggesting ALKBH5 binds mRNA and other RNAs tightly.

FTO is associated with obesity and human diseases, while the *alkbh5* knockout mice have impaired fertility. The distinct physiological functions of the two discovered m^6A demethylases demonstrate that m^6A plays important functional roles, and it is critical to balance the m^6A writer/eraser activities in mammals.

So far, only FTO and ALKBH5 were discovered as m^6A demethylase in mammals. As there are many Fe(II)/α-KG-dependent dioxygenases with unknown functions, we should not be surprised to see the discovery of more m^6A demethylases in mammals. In addition, FTO and ALKBH5 are only conserved in vertebrates (Kurowski et al. 2003; Robbens et al. 2008). To explore m^6A function widely, it is necessary and priority to discover m^6A demethylase in other organisms, such as yeast and plant.

5 m^{6}A Readers in Eukaryotes

Analogous to epigenetic DNA and histone modifications, the reversible and dynamic m^{6}A plays important functional roles through the recognition of m^{6}A-binding proteins (readers) (Fig. 3). The YTH family proteins have been identified as direct m^{6}A reader in RNA-affinity pull-down assay using methylated RNA probes (Dominissini et al. 2012). There are five homologues in mammals: YTH domain family (YTHDF) proteins 1, 2, and 3 and YTH domain-containing (YTHDC) proteins 1 and 2. YTHDF1, YTHDF2 and YTHDF3 in the cytoplasm and YTHDC1 in the nucleus have been validated to bind the methylated RNA more tightly than unmethylated RNA (Wang et al. 2014). In particular, YTHDF2 has been shown to bind m^{6}A through the C-terminal YTH domain and to transport the bound methylated transcripts to processing bodies (P-bodies) for accelerated degradation through its N-terminal Pro/Gln/Asn-rich domain (Wang et al. 2014). YTHDF1 selectively recognizes m^{6}A-modified mRNAs to promote translation efficiency through the interaction with translation-initiation factor (Wang et al. 2015b). YTHDC1 facilitates exon inclusion by recruiting RNA splicing factor SRSF3 and blocking SRSF10 for the access to the m6A binding regions of targeted mRNA (Xiao et al. 2016). The evidence of m^{6}A-mediated mRNA nuclear export has been observed, and it might also affect mRNA storage (Fig. 3). However, the mechanism and the related m^{6}A readers remain to be investigated. Another m^{6}A reader protein was found as a nuclear reader heterogeneous nuclear ribonucleoprotein A2B1 (HNRNPA2B1), which mediates m^{6}A-modified primary microRNA processing and alternative splicing (Alarcon et al. 2015).

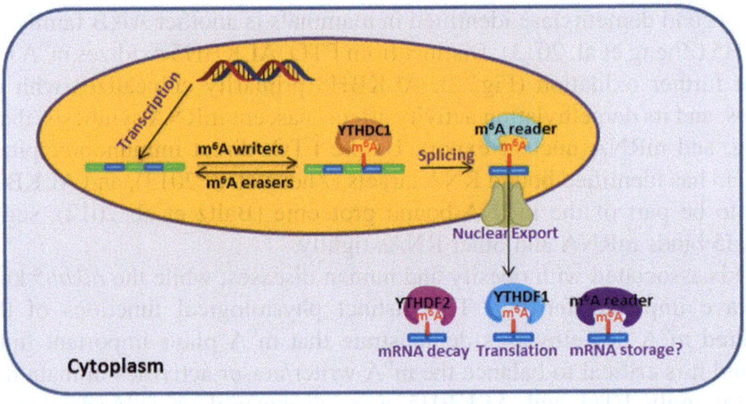

Fig. 3 The reversible m^{6}A functions in RNA metabolism and translation. The reversible and dynamic of m^{6}A is controlled by m^{6}A writers and erasers. m^{6}A plays functional roles in mRNA processing through the recognition of m^{6}A reader proteins. In the cell nucleus, m^{6}A affects mRNA splicing and nuclear export in aid of unknown m^{6}A readers. After mRNA is exported to the cytoplasm, YTHDF2 can bind the m^{6}A-containing mRNA to accelerate mRNA decay; YTHDF1 can recognize the m^{6}A-containing mRNA to promote their translation efficiency; YTHDC1 regulates pre-mRNA alternative splicing. Other m^{6}A reader proteins may bind the m^{6}A-containing mRNA to control their storage

Although the methyl group on N^6 position of adenosine cannot break Watson-Crick base pairing, m^6A can destabilize the base pairing or disturb the RNA triple base pairs or Hoogsteen pairs which rely on N^6 proton as a donor. Thus, m^6A could be a switch to control the RNA structure to affect RNA-protein interactions for biological regulation. A recent study indeed confirms that m^6A installed in a stem-loop of RNA destabilizes the base pairing to facilitate the binding of heterogeneous nuclear ribonucleoprotein C (HNRNPC) to the uridine track in the loop (Liu et al. 2015).

6 Biological Consequences of m^6A Methylation in Eukaryotic mRNA and the Underlying Mechanisms

6.1 *m^6A mRNA Methylation Steers Stem Cell Pluripotency*

Mammalian development commences with the zygote, which is "totipotent" capable of producing an entire embryo with all the specialized cells that make up a living being. During the subsequent rounds of cell division, cells rapidly lose this plasticity and become "pluripotent." In cell biology, the definition of pluripotency refers to a stem cell (Mitalipov and Wolf 2009). The pluripotent stage can be further divided into two states: a ground, naïve state (embryonic stem cells, ESCs) and a differentiation-prepared, primed state (epiblast stem cells, EpiSCs) (Hackett and Surani 2014). In order to investigate the molecular regulators that are critical for transitioning toward primed state, Geula and colleagues (Geula et al. 2015) started a siRNA screening from the pool for naïve pluripotency modulation and tested whether the primed EpiSCs might rely on some factors. Luckily, they found that METTL3 is one of the crucial components for regulation of the primed cells. The authors then separately investigated the impacts of m^6A on the naïve pluripotent state (ESCs) and the primed state (EpiSCs). To resolve the role of m6A in the naïve state, heterozygous *Mettl3*$^{+/-}$ mice were generated, and homozygous *Mettl3*$^{-/-}$ mESCs were obtained from embryos. Consistent with previous results (Batista et al. 2014), *Mettl3*$^{-/-}$ cells showed an almost complete loss of m^6A and kept at a hypernaïve state but failed to proceed into the primed EpiSC-like state, hence blocked the subsequent differentiation (Fig. 4). In contrast, the depletion of METTL3 and m^6A in the primed EpiSCs during a primed pluripotent state produced the opposite spectrum of effects—this depletion resulted in minimal self-renewal, attenuated stability of the primed state, and an enhanced tendency to lineage priming, which finally led to fast differentiation and/or cell death.

The balance between naïve pluripotency and lineage priming is fine-tuned by the relative expression of naïve pluripotent factors and linage commitment factors. It was found that 80 % of transcripts of these genes are m^6A methylated. m^6A contains the function to degrade mRNA assisted by m^6A reader protein YTHDF2 (Wang et al. 2014). For both types of regulators, loss of m^6A results in longer mRNA

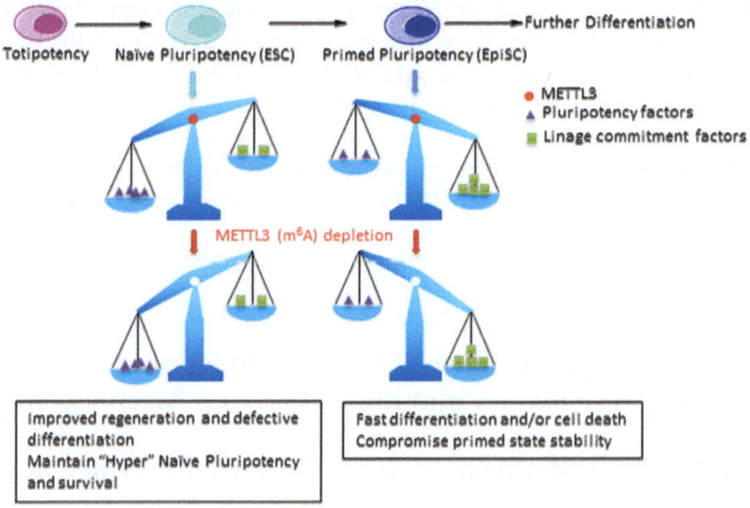

Fig. 4 m^{6}A steers the balance between expression levels of pluripotency genes and of linage commitment genes in naïve and primed states. In naïve cells the expression levels of pluripotency factors are dominant, while the expression levels of linage commitment factors increases with priming. As m^{6}A promotes mRNA decay, the m^{6}A methyltransferase (METTL3) depletion in naïve pluripotent cells further upregulates the already-high naïve pluripotency genes to create a "hyper" naïve pluripotent states, whereas m^{6}A depletion in the primed state further boosts the dominant differentiation priming genes to push the cells above the critical threshold toward differentiation and to lead to fast differentiation and/or cell death

lifetime and increased abundance of transcripts (Batista et al. 2014; Geula et al. 2015). For instance, in the ground naïve state, pluripotency genes predominate. Deletion of m^{6}A further amplifies the already highly expressed naïve pluripotency genes but leads to only a marginal increase in lineage commitment transcripts, which causes cells stuck at the so-called "hypernaïve" pluripotent state. By contrast, in the primed state, where lineage commitment factors dominate, loss of m^{6}A further tips the balance toward lineage priming and differentiation (Geula et al. 2015).

6.2 m^{6}A mRNA Methylation Is a New Circadian Pacesetter

The mechanism of circadian clock that drives rhythms is a delayed transcription-translation feedback loop, in which circadian "clock genes" are suppressed by their protein products. The circadian rhythm is thought to arise from the time lags involved in transcription, translation, and protein shutting into the nucleus (Koike et al. 2012). Around 10 % of the liver genes is rhythmic, but only one-fifth of them is driven by de novo transcription, which indicates that mRNA processing is a major circadian component. Recent work showed that m^{6}A mRNA methylation sets the pace of circadian RNA processing, thereby determines clock speed and stability (Fustin et al. 2013). In cultured cell lines and in mice, inhibition of methylation by 3-deazaadenosine (3-DAA) lengthens the circadian period by 3 h and 1 h, respectively. Through transcriptome-wide m^6A sequencing, they found many clock genes, including *Per1* and *Per2*, as well as clock output genes, carrying m^6A modification. The key result they found is that suppression of METTL3 delays the nuclear exit of mature *Per2* and *Bmal1* mRNA. Thus, reducing m^6A methylation by silencing METTL3 led to delay mRNA nuclear export and slowed the clock. These results reveal that m^6A-mediated mRNA processing is a new point control to circadian rhythms.

7 Detection Technology of m^6A

7.1 *Quantification of Total m^6A Level in mRNA*

The abundance of m^6A has been shown to be ~0.1 %–0.4 % of total adenosine residues in cellular mRNA. The total amount of m^6A in RNA can be probed by several methods, including UPLC-QQQ-MS/MS, dot blot, and 2-D TLC (Jia et al. 2011). The purity of mRNA is extremely important for the quantification because ribosomal RNA (rRNA), small nuclear RNA (snRNA), microRNA (miRNA), and transfer RNA (tRNA) also contain m^6A.

The femtomole sensitivity achieved by UPLC-QQQ-MS/MS makes it a quantitative tool for monitoring m^6A dynamics. The mRNA isolated from the total RNA using oligo (dT) followed by rRNA removal is digested to nucleosides by Nuclease P1 and alkaline phosphatase. The nucleosides are detected by UPLC-QQQ-MS/MS, and m^6A/A concentration ratio is calculated. In order to avoid rRNA contamination, $N^{6,6}$-dimethyladenosine is monitored because it only exists in rRNA but not mRNA.

Dot-blot assay is convenient without the requirement of expensive machine but is a semiquantitative method. Its sensitivity and detection limits depend on the quality of m^6A antibody. It is simple and easy to do by spotting mRNA on membrane followed by a routine western blotting with m^6A antibody.

2-D TLC requires isotope radioactive labeling and only detects partial m^6A. m^6 A is at the GAC (75 %) and AAC (25 %) sequence. mRNA is digested by ribonuclease T1, which cleaves ssRNA after guanine, to expose m^6A at the 5′ end. After [γ-^{32}P] ATP labeling, the radial-labeled m^6A is cleaved by Nuclease P1 and spotted on cellulose TLC plate to run two-dimensional chromatography.

7.2 *High-Throughput m^6A Sequencing*

Before 2012, the transcriptome-wide distribution of m6A was unknown, until two independent groups developed an m^6A RNA immunoprecipitation method followed by high-throughput sequencing (MeRIP-seq) to map the m^6A RNA methylomes with a ~200 bp resolution (Dominissini et al. 2012; Meyer et al. 2012). Briefly, the isolated mRNA was fragmented to ~100 nt and immunoprecipitated using an m^6A antibody. The eluted mRNA by m^6A nucleotide and the fragmented mRNA without immunoprecipitation as input were subjected to RNA-seq library generation and high-throughput sequencing. The sequencing data showed that m^6A is a widespread modification distributed in more than 7000 mRNAs and 300 non-coding RNAs (ncRNAs), prefers to locate around stop codons and in 3′ UTRs and the long internal exons, and is highly conserved between human and mouse. *Arabidopsis* m^6A methylome revealed that m^6A is additionally enriched around start codons (Luo et al. 2014).

The shortcoming of MeRIP-seq is not a single-base resolution. Recently two modified MeRIP-seqs, photo-crosslinking-assisted m^6A-sequencing (PA-m^6A-seq) (Chen et al. 2015) and m^6A individual-nucleotide-resolution cross-linking and immunoprecipitation (miCLIP) (Linder et al. 2015), improve the m^6A resolution by using two well-established methods for the identification of RNA-binding protein targets: photoactivatableribonucleoside-enhanced crosslinking and immunoprecipitation (PAR-CLIP) and iCLIP, respectively. Photo-crosslinking-assisted m^6A-sequencing (PA-m^6A-seq) efficiently improves the accuracy of the methylation site assignments and provides a ~23 nt high-resolution transcriptome-wide m^6A map; however, it is limited to be used in cells because it requires to incorporate photoactivatableribonucleoside4-thiouridine (4SU) into mRNA (Chen et al. 2015). miCLIP can induce mutational signatures at m^6A residues to achieve a single-base resolution using some specific m^6A antibody (Linder et al. 2015). The main drawback is miCLIP relies on the property of m^6A antibody and ultraviolet cross-linking efficiency.

7.3 Map m^6A Site with Single-Base Resolution

The antibody-based m^6A profiling could not provide precise m^6A position and the fraction of the modification at each specific site. Till now, two methods have been developed to determine the percentage of m^6A at a specific site with single-nucleotide resolution. Termed site-specific cleavage and radioactive-labeling followed by ligation-assisted extraction and thin-layer chromatography (SCARLET) combines two previously established techniques, site-specific cleavage and splint ligation, to identify m^6A position and modification fraction in mRNA and long noncoding RNA (lncRNA) (Liu et al. 2013). The second method identifies m^6A site in rRNA using *Bst* DNA polymerase-catalyzed primer extension (Wang et al. 2015a). m^6A is found to hinder RNA- and DNA-directed DNA synthesis of *Bst* DNA polymerase under specific conditions.

8 Conclusions and Future Prospects

The reversible mRNA m^6A modification is emerging as critical landmark of posttranscriptional gene regulation. mRNA is divided into two portions with and without m^6A methylation; consequently, mRNA processing is sorted into two tracks. m^6A modification is a bar code for fast track of mRNA processing, which is critical for fast response to signaling and stress stimuli, as well as circadian rhythms, cell differentiation, and development.

Although the key m^6A methyltransferase components, demethylases and readers have been identified, a thorough understanding of their functions remains. For example, how does the writer complex select m^6A deposition? What is the molecular mechanism and relationship between demethylase enzyme function and human disease? And how does m^6A affect nuclear export and other RNA metabolism steps? Beyond the known factors, identifying and characterizing more m^6A writers, erasers, and readers as well as their partner proteins and post-translational modifications will be essential for understanding and expanding the biological roles of m^6 A. The ability to perform transcriptome-wide profiling of m^6A promotes efficiently elucidating its biological function; however, quantitative and base resolution sequencing methods with the limited input for the whole transcriptome are still highly desirable for future research.

References

Agarwala SD, Blitzblau HG, Hochwagen A et al (2012) RNA methylation by the MIS complex regulates a cell fate decision in yeast. PLoS Genet 8, e1002732

Alarcon CR, Goodarzi H, Lee H et al (2015) HNRNPA2B1 is a mediator of m^6A-dependent nuclear RNA processing events. Cell 162:1299–1308

Baltz AG, Munschauer M, Schwanhausser B et al (2012) The mRNA-bound proteome and its global occupancy profile on protein-coding transcripts. Mol Cell 46:674–690

Batista PJ, Molinie B, Wang J et al (2014) m^6A RNA modification controls cell fate transition in mammalian embryonic stem cells. Cell Stem Cell 15:707–719

Bodi Z, Button JD, Grierson D et al (2010) Yeast targets for mRNA methylation. Nucleic Acids Res 38:5327–5335

Bodi Z, Zhong S, Mehra S et al (2012) Adenosine methylation in Arabidopsis mRNA is associated with the 3′ end and reduced levels cause developmental defects. Front Plant Sci 3:48

Boissel S, Reish O, Proulx K et al (2009) Loss-of-function mutation in the dioxygenase-encoding FTO gene causes severe growth retardation and multiple malformations. Am J Hum Genet 85:106–111

Bokar JA (2005) The biosynthesis and functional roles of methylated nucleosides in eukaryotic mRNA. In: Grosjean H (ed) Fine-tuning of RNA functions by modification and editing. Springer, Berlin, pp 141–147

Bokar JA, Shambaugh ME, Polayes D et al (1997) Purification and cDNA cloning of the AdoMet-binding subunit of the human mRNA (N^6-adenosine)-methyltransferase. RNA 3:1233–1247

Bujnicki JM, Feder M, Radlinska M et al (2002) Structure prediction and phylogenetic analysis of a functionally diverse family of proteins homologous to the MT-A70 subunit of the human mRNA:m^6A methyltransferase. J Mol Evol 55:431–444

Cantara WA, Crain PF, Rozenski J et al (2011) The RNA modification database, RNAMDB: 2011 update. Nucleic Acids Res 39:D195–201

Chen K, Lu Z, Wang X et al (2015) High-resolution N^6-methyladenosine (m^{6}A) map using photo-crosslinking-assisted m^{6}A sequencing. Angew Chem Int Ed Engl 54:1587–1590

Church C, Lee S, Bagg EA et al (2009) A mouse model for the metabolic effects of the human fat mass and obesity associated FTO gene. PLoS Genet 5, e1000599

Church C, Moir L, McMurray F et al (2010) Overexpression of Fto leads to increased food intake and results in obesity. Nat Genet 42:1086–1092

Clancy MJ, Shambaugh ME, Timpte CS et al (2002) Induction of sporulation in *Saccharomyces cerevisiae* leads to the formation of N^6-methyladenosine in mRNA: a potential mechanism for the activity of the IME4 gene. Nucleic Acids Res 30:4509–4518

Dina C, Meyre D, Gallina S et al (2007) Variation in FTO contributes to childhood obesity and severe adult obesity. Nat Genet 39:724–726

Dominissini D, Moshitch-Moshkovitz S, Schwartz S et al (2012) Topology of the human and mouse m^{6}A RNA methylomes revealed by m^{6}A-seq. Nature 485:201–206

Fischer J, Koch L, Emmerling C et al (2009) Inactivation of the Fto gene protects from obesity. Nature 458:894–898

Frayling TM, Timpson NJ, Weedon MN et al (2007) A common variant in the FTO gene is associated with body mass index and predisposes to childhood and adult obesity. Science 316:889–894

Fu Y, Jia G, Pang X et al (2013) FTO-mediated formation of N^6-hydroxymethyladenosine and N^6-formyladenosine in mammalian RNA. Nat Commun 4:1798

Fustin JM, Doi M, Yamaguchi Y et al (2013) RNA-methylation-dependent RNA processing controls the speed of the circadian clock. Cell 155:793–806

Gao X, Shin YH, Li M et al (2010) The fat mass and obesity associated gene FTO functions in the brain to regulate postnatal growth in mice. PLoS One 5, e14005

Gelfand B, Mead J, Bruning A et al (2011) Regulated antisense transcription controls expression of cell-type-specific genes in yeast. Mol Cell Biol 31:1701–1709

Gerken T, Girard CA, Tung YC et al (2007) The obesity-associated FTO gene encodes a 2-oxoglutarate-dependent nucleic acid demethylase. Science 318:1469–1472

Geula S, Moshitch-Moshkovitz S, Dominissini D et al (2015) Stem cells. m^{6}A mRNA methylation facilitates resolution of naïve pluripotency toward differentiation. Science 347:1002–1006

Hackett JA, Surani MA (2014) Regulatory principles of pluripotency: from the ground state up. Cell Stem Cell 15:416–430

Hernandez-Caballero ME, Sierra-Ramirez JA (2015) Single nucleotide polymorphisms of the FTO gene and cancer risk: an overview. Mol Biol Rep 42:699–704

Hinney A, Nguyen TT, Scherag A et al (2007) Genome wide association (GWA) study for early onset extreme obesity supports the role of fat mass and obesity associated gene (FTO) variants. PLoS One 2, e1361

Hongay CF, Orr-Weaver TL (2011) Drosophila Inducer of MEiosis 4 (IME4) is required for Notch signaling during oogenesis. Proc Natl Acad Sci U S A 108:14855–14860

Hongay CF, Grisafi PL, Galitski T et al (2006) Antisense transcription controls cell fate in *Saccharomyces cerevisiae*. Cell 127:735–745

Horowitz S, Horowitz A, Nilsen TW et al (1984) Mapping of N^6-methyladenosine residues in bovine prolactin mRNA. Proc Natl Acad Sci U S A 81:5667–5671

Jia G, Yang CG, Yang S et al (2008) Oxidative demethylation of 3-methylthymine and 3-methyluracil in single-stranded DNA and RNA by mouse and human FTO. FEBS Lett 582:3313–3319

Jia G, Fu Y, Zhao X et al (2011) N^6-methyladenosine in nuclear RNA is a major substrate of the obesity-associated FTO. Nat Chem Biol 7:885–887

Kaklamani V, Yi N, Sadim M et al (2011) The role of the fat mass and obesity associated gene (FTO) in breast cancer risk. BMC Med Genet 12:52

Kane SE, Beemon K (1985) Precise localization of m^{6}A in *Rous sarcoma* virus RNA reveals clustering of methylation sites: implications for RNA processing. Mol Cell Biol 5:2298–2306

Keller L, Xu W, Wang HX et al (2011) The obesity related gene, FTO, interacts with APOE, and is associated with Alzheimer's disease risk: a prospective cohort study. J Alzheimers Dis 23:461–469
Koike N, Yoo SH, Huang HC et al (2012) Transcriptional architecture and chromatin landscape of the core circadian clock in mammals. Science 338:349–354
Kurowski MA, Bhagwat AS, Papaj G et al (2003) Phylogenomic identification of five new human homologs of the DNA repair enzyme AlkB. BMC Genomics 4:48
Lappalainen T, Kolehmainen M, Schwab US et al (2011) Association of the FTO gene variant (rs9939609) with cardiovascular disease in men with abnormal glucose metabolism--the Finnish Diabetes Prevention Study. Nutr Metab Cardiovasc Dis 21:691–698
Linder B, Grozhik AV, Olarerin-George AO et al (2015) Single-nucleotide-resolution mapping of m^6A and m^6Am throughout the transcriptome. Nat Methods 12:767–772
Liu N, Parisien M, Dai Q et al (2013) Probing N^6-methyladenosine RNA modification status at single nucleotide resolution in mRNA and long noncoding RNA. RNA 19:1848–1856
Liu J, Yue Y, Han D et al (2014) A METTL3-METTL14 complex mediates mammalian nuclear RNA N^6-adenosine methylation. Nat Chem Biol 10:93–95
Liu N, Dai Q, Zheng G et al (2015) N^6-methyladenosine-dependent RNA structural switches regulate RNA-protein interactions. Nature 518:560–564
Luo GZ, MacQueen A, Zheng G et al (2014) Unique features of the m6A methylome in *Arabidopsis thaliana*. Nat Commun 5:5630
Meyer KD, Saletore Y, Zumbo P et al (2012) Comprehensive analysis of mRNA methylation reveals enrichment in 3′ UTRs and near stop codons. Cell 149:1635–1646
Mitalipov S, Wolf D (2009) Totipotency, pluripotency and nuclear reprogramming. Adv Biochem Eng Biotechnol 114:185–199
Narayan P, Rottman FM (1988) An in vitro system for accurate methylation of internal adenosine residues in messenger RNA. Science 242:1159–1162
Ping XL, Sun BF, Wang L et al (2014) Mammalian WTAP is a regulatory subunit of the RNA N^6-methyladenosine methyltransferase. Cell Res 24:177–189
Robbens S, Rouze P, Cock JM et al (2008) The FTO gene, implicated in human obesity, is found only in vertebrates and marine algae. J Mol Evol 66:80–84
Schibler U, Kelley DE, Perry RP (1977) Comparison of methylated sequences in messenger RNA and heterogeneous nuclear RNA from mouse L cells. J Mol Biol 115:695–714
Schwartz S, Agarwala SD, Mumbach MR et al (2013) High-resolution mapping reveals a conserved, widespread, dynamic mRNA methylation program in yeast meiosis. Cell 155:1409–1421
Shah JC, Clancy MJ (1992) IME4, a gene that mediates MAT and nutritional control of meiosis in *Saccharomyces cerevisiae*. Mol Cell Biol 12:1078–1086
Wang X, Lu Z, Gomez A et al (2014) N^6-methyladenosine-dependent regulation of messenger RNA stability. Nature 505:117–120
Wang S, Wang J, Zhang X et al (2015a) N^6-methyladenine hinders RNA- and DNA-directed DNA synthesis: application in human rRNA methylation analysis of clinical specimens. Chem Sci. doi:10.1039/c5sc02902c
Wang X, Zhao BS, Roundtree IA et al (2015b) N^6-methyladenosine modulates messenger RNA translation efficiency. Cell 161:1388–1399
Wei CM, Gershowitz A, Moss B (1975) Methylated nucleotides block 5′ terminus of HeLa cell messenger RNA. Cell 4:379–386
Wei CM, Gershowitz A, Moss B (1976) 5′-Terminal and internal methylated nucleotide sequences in HeLa cell mRNA. Biochemistry 15:397–401
Xiao W, Adhikari S, Dahal U et al (2016) Nuclear m6A reader YTHDC1 regulates mRNA splicing. Mol Cell 61:507–19
Zhao X, Yang Y, Sun BF et al (2014) FTO-dependent demethylation of N^6-methyladenosine regulates mRNA splicing and is required for adipogenesis. Cell Res 24:1403–1419
Zheng G, Dahl JA, Niu Y et al (2013) ALKBH5 is a mammalian RNA demethylase that impacts RNA metabolism and mouse fertility. Mol Cell 49:18–29
Zhong S, Li H, Bodi Z et al (2008) MTA is an Arabidopsis messenger RNA adenosine methylase and interacts with a homolog of a sex-specific splicing factor. Plant Cell 20:1278–1288

8-Hydroxyguanine, an Oxidative DNA and RNA Modification

Hiroshi Kasai and Kazuaki Kawai

Contents

Abstract Reactive oxygen species (ROS), produced by ionizing radiation and many other environmental agents, damage DNA and RNA. They are also endogenously generated in cells by oxygen metabolism. 8-Hydroxy-2′-deoxyguanine (8-OHdG) was first reported in 1983, as a major form of oxidative DNA damage produced by heated sugar, Fenton-type reagents, and ionizing radiation in vitro. 8-OHdG has been detected in cellular DNA by HPLC-ECD and LC/MS/MS methods in many laboratories. The increase in the 8-OHdG level in cellular DNA, detected by these chromatographic methods, is supported by its immunochemical detection and enhanced repair activity. Its analysis in human leukocyte DNA, and in urine and saliva, is a promising approach toward the assessment of an individual's oxidative stress level. The ribonucleoside 8-hydroxyguanosine (8-OHGuo), in tissue RNA and urine, is also a good marker of oxidative stress

H. Kasai (✉) • K. Kawai
Department of Environmental Oncology, Institute of Industrial Ecological Sciences, University of Occupational and Environmental Health, Japan, Fukuoka, Japan
e-mail: h-kasai@med.uoeh-u.ac.jp; kkawai@med.uoeh-u.ac.jp

S. Jurga et al. (eds.), *Modified Nucleic Acids in Biology and Medicine*, RNA Technologies, DOI 10.1007/978-3-319-34175-0_7

in vivo. The free 8-hydroxyguanine (8-OHGua) base is also detectable in biological samples, such as urine, serum, and saliva. In this chapter, the validity of the general use of 8-OHdG, 8-OHGuo, and 8-OHGua as markers of cellular oxidative stress is discussed.

Keywords Reactive oxygen species • 8-OHdG • 8-oxodG • 8-OHGuo • 8-OHGua • DNA damage • Oxidative stress • Biomarker

Abbreviations

8-OHdG	8-Hydroxy-2′-deoxyguanosine
8-OHGuo	8-Hydroxyguanosine
8-OHGua	8-Hydroxyguanine
ROS	Reactive oxygen species
HPLC-ECD	High performance liquid chromatography equipped with an electrochemical detector
ELISA	Enzyme-linked immunosorbent assay

1 Introduction

Many mutagens and carcinogens react with DNA and induce mutations in cancer-related genes. Reactive oxygen species (ROS) are implicated as a cause of cancer and lifestyle-related diseases. Ionizing radiation and many environmental chemicals generate ROS and damage DNA. ROS are also produced endogenously, as a by-product of oxygen metabolism. Therefore, ROS may also be involved in the aging process. A major form of oxidative DNA damage, 8-hydroxydeoxyguanosine (8-OHdG, 7,8-dihydro-8-oxodeoxyguanosine), was discovered in Japan in 1983, during a study of DNA modifications generated by heated glucose and ROS-forming agents (Kasai and Nishimura 1983; Kasai et al. 1984a) in vitro (Fig. 1). Since then, various aspects of this type of oxidative DNA damage, such as the mechanisms of its formation, its mutagenic effects, and its repair, have been studied worldwide, clarifying its biological significance. Floyd et al. first developed a sensitive method to analyze 8-OHdG, using an electrochemical detector with high performance liquid chromatography (HPLC-ECD) (Floyd et al. 1986). This method revealed that various ROS-forming carcinogens induce increased levels of 8-OHdG in cellular DNA (Kasai 1997). Ames and his collaborators were the first to detect 8-OHdG in animal and human urine samples by HPLC-ECD (Shigenaga et al. 1989). These discoveries triggered further studies on the analysis of 8-OHdG as a biomarker for risk assessment, the molecular epidemiology of ROS-related diseases, and aging. Patients with various diseases, such as cancer, diabetes, and Alzheimer's disease (urine), showed higher levels of 8-OHdG. In contrast, the consumption of antioxidants, vegetables, fruits, green tea, etc. was

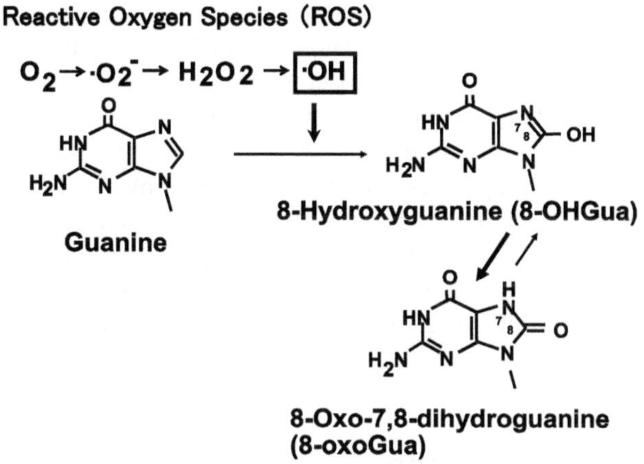

Fig. 1 Formation of 8-OHGua and its tautomeric change

correlated with a reduction in the amounts of 8-OHdG in urine, serum, and tissue DNA. Therefore, 8-OHdG seems to be a useful marker for monitoring the cellular oxidative stress involved in the induction of cancer and in lifestyle-related diseases and their prevention by antioxidants. In addition, the ribonucleoside 8-hydroxyguanosine (8-OHGuo), in tissue RNA and urine, is a good marker of oxidative stress in vivo. The free 8-hydroxyguanine base (8-OHGua) has also been detected in biological samples, such as urine, serum, and saliva. In this chapter, we summarize the studies on 8-OHdG and its related derivatives, reported over the past 32 years, with a particular focus on their usefulness as biomarkers.

2 Discovery of 8-OHdG and Mechanisms of Formation

The formation of 8-OHdG was first detected during a study on DNA modifications caused in vitro by mutagenic heated carbohydrates, which were being used as a model of cooked foods (Kasai et al. 1984a). Methylreductic acid and hydroxymethylreductic acid were later isolated and identified from heated carbohydrates as major ROS-forming mutagenic compounds (Kasai et al. 1989). Various ROS-forming agents, such as Fenton-type reagents (Kasai and Nishimura 1984b), ionizing radiation (Kasai et al. 1984b), metals (Kasai and Nishimura 1984c), cigarette smoke condensate (Kasai and Nishimura 1991), and asbestos (Kasai and Nishimura 1984a), also effectively promoted the formation of 8-OHdG in DNA in vitro. A hydroxyl radical (•OH) is involved in these reactions. The formation of 8-OHGua in vitro was most efficient with the monomer nucleoside, as compared to that in RNA and DNA polymers (described later in detail). A preliminary account of these results was reported in 1983 (Kasai and Nishimura 1983). Floyd and his collaborators found that methylene blue plus visible light specifically induces 8-OHdG in DNA without a strand break, suggesting the involvement of singlet oxygen in that reaction (Schneider et al. 1990). In collaboration with Cadet's group,

Kasai et al. found that riboflavin plus visible light induces 8-OHdG in DNA by a non-singlet oxygen mechanism; namely, via a guanine radical cation followed by a hydration reaction (Kasai et al. 1992). As an interesting example, Barton and his collaborators demonstrated that photoactivated metallointercalators induced 8-OHdG in DNA at sites 34–200 Å (10–60 base pairs) away from their binding sites, by long-range electron transfer along the DNA chain (Nunez et al. 1999). Kohda et al. reported that 8-OHdG is produced in cellular DNA by a treatment with the carcinogen 4-nitroquinoline 1-oxide, via N7-arylaminated dG followed by hydrolytic rearrangement (Kohda et al. 1986). Together, these results revealed that 8-OHdG is produced by a variety of mechanisms.

3 Nomenclature

8-OHdG is considered to exist mainly as the 8-oxo-form in aqueous solutions, because its UV spectrum resembles that of 7-methyl-8-oxoguanosine (Culp et al. 1989; Rizkalla et al. 1969), (Fig. 1). An X-ray crystallographic study of 8-OH-9-ethylguanine actually revealed the 8-oxo-structure (Kasai et al. 1987). In DNA, its 8-oxo-form mispairs with adenine and induces GC to TA transversion mutations (Shibutani et al. 1991) (Fig. 2). A repair enzyme that removes 8-OHGua in DNA was identified in mammalian cells and named oxoguanine glycosylase 1 (OGG1) (Lu et al. 1997). Therefore, many researchers, especially those studying the mutagenic effects and the repair enzymes, use the name 8-oxodG, rather than 8-OHdG. In fact, Cooke et al. recommended using the 8-oxodG nomenclature (Cooke et al. 2010). However, a drawback is that the correct name of 8-oxodG is rather complicated, as it is 7,8-dihydro-8-oxo-dG or 8-oxo-7,8-dihydro-dG, etc. The 7,8-double bond of the guanine skeleton must be saturated before the 8-oxo is added to the guanine name, in the systematic nomenclature rules used by Chemical Abstracts, IUPAC, etc. Surprisingly, the incorrect name, 8-oxodeoxyguanosine, is

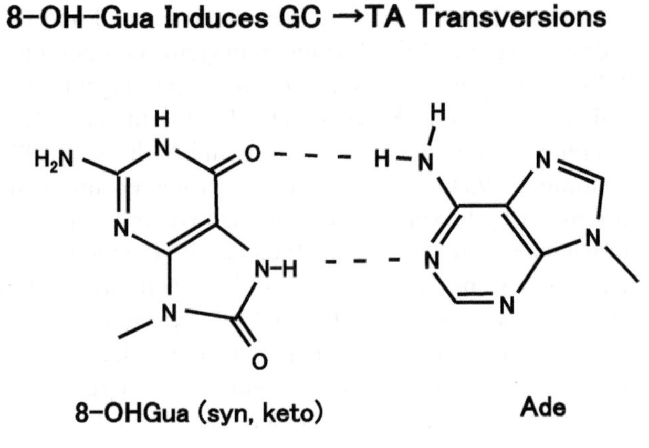

Fig. 2 Mismatched base pair caused by 8-OHGua

Table 1 The number of published reports with 8-OHdG in the title

Keywords	1983–2000[a]	2001–2010[b]	2011–2015[c]	Total[(a+b+c)]
8-Hydroxy-2′-deoxyguanosine (A)	120	215	107	442
8-Hydroxydeoxyguanosine (B)	156	140	44	340
8-OHdG (C)	16	87	58	161
8-OH-dG (D)	8	19	1	28
(A) or (B) or (C) or (D)	300	461	199*	960*
8-Oxo-7,8-dihydro-2-′-deoxyguanosine (E)	32	90	43	165
8-Oxo-7,8-dihydrodeoxyguanosine (F)	3	1	1	5
8-Oxo-2′-deoxyguanosine (G)	45	73	18	136
7,8-Dihydro-8-oxo-2-′-deoxyguanosine (H)	15	15	3	33
7,8-Dihydro-8-oxodeoxyguanosine (I)	0	7	1	8
8-Oxodeoxyguanosine (J)	17	23	1	41
8-oxodG (K)	7	22	15	44
8-oxo-dG (L)	4	16	11	31
(E) or (F) or (G) or (H) or (I) or (J) or (K) or (L)	123	247	93	463

*The numbers of these items are less than the sum of the above-described numbers, because some papers overlap

quite often used (Table 1). Therefore, the 8-oxo-type nomenclature is somewhat confusing. There are at least 6 different 8-oxo-type names, excluding abbreviations (Table 1). In contrast, 8-hydroxy-2′-deoxyguanosine is a simple, clear, and suitable name as a systematic nomenclature. In fact, 60–70 % of the published papers have consistently used 8-OHdG-type names in the titles throughout the past 32 years, as shown in Table 1. A major tautomeric structure in aqueous solution is not related to the systematic nomenclature of chemicals and is not always recommended as the nomenclature. For example, malondialdehyde (MDA, IUPAC name propanedial) is a widely used name, although it mainly exists in the 3-enol form in aqueous solution. However, the name 3-hydroxy-2-propenal (or ß-hydroxyacrolein) is not used for MDA (Marnett 2002).

4 Formation of 8-OHdG In Vivo

What kinds of carcinogenesis-related factors contribute to the generation of 8-OHdG? The relationship between well-known carcinogens and 8-OHdG generation in DNA has been investigated in animal experiments and human studies, to clarify the carcinogenic mechanism. The measured levels of 8-OHdG depend on the balance between its formation and repair, and thus the 8-hydroxyguanine

Table 2 The 8-OHdG levels are dependent on the balance between formation and repair

		8-OHdG	Repair activity
		Level in DNA	(OGG1 type)
Cultured cell	γ-Rays	↗	↘
	Arsenite	↗	↘
Animal organ	Chromium (VI)	↗	↘
	Cadmium (GSH depletion)	↗	↘
	Diesel exhaust particles	↗	↘
	Ethanol (Nutrition-deficient)	↗	↗
	3′-Me-4-DAB	↗	↗
	Fe-NTA	↗	↗
	Asbestos	↗	↗
	$KBrO_3$	↗	↗
Human leukocyte	Cigarette smoking	↗	↗
	Physical exercise	↘	↗

(8-OHGua) repair activity (OGG1 type) should also be assayed in evaluations of the cellular oxidative stress (Table 2). For example, when ethanol (under nutrition-deficient conditions) (Asami et al. 2000), 3′-methyl-4-dimethylaminoazobenzene (Hirano et al. 2000), ferric nitrotriacetate (Fe-NTA) (Yamaguchi et al. 1996), potassium bromate ($KBrO_3$) (Lee et al. 1996), and asbestos (Yamaguchi et al. 1999) were administered to rats, increases in both the 8-OHdG level and the repair activity were observed in the target organs, esophagus, liver, kidney, and lung. Cigarette smoking also increased the levels of both 8-OHdG and its repair activity in human leukocytes (Asami et al. 1996). In contrast, cancer preventive physical exercise induced an increase in the repair activity and a decrease in the 8-OHdG level (Asami et al. 1998a). The administration of cadmium (Cd) to rats, under conditions of glutathione depletion, impaired the 8-OHGua repair activity in the target organ, testis, while the 8-OHdG levels in the DNA were increased (Hirano et al. 1997). When rats were exposed to diesel exhaust particles (DEP) by intratracheal administration (Tsurudome et al. 1999), or to a hexavalent chromium (Cr) mist by inhalation (Maeng et al. 2003), again the repair activity was decreased in the lungs, while the 8-OHdG levels in the DNA were increased. These potent carcinogens, Cd, Cr, and DEP, may enhance the accumulation of 8-OHdG by impairing the repair activity.

One of the mechanisms of asbestos fiber genotoxicity appears to be the generation of ROS, either from its surface by reactions involving catalytic iron or from its phagocytosis by frustrated phagocytes (Kamp and Weitzman 1999). For example, increased levels of 8-OHdG were observed in rat and hamster lung DNA after the intra-tracheal instillation of crocidolite asbestos (Yamaguchi et al. 1999). These results agreed well with our prediction and suggested that one of the mechanisms of asbestos-induced lung cancer or mesothelioma is 8-OHdG generation in DNA. A positive correlation between the 8-OHdG levels in leukocyte DNA and the grades of asbestosis at a Chinese asbestos plant (Takahashi et al. 1997) was also observed.

A German group conducted a large-scale study of asbestos-exposed workers, to determine whether asbestos induces the formation of 8-OHdG in white blood cells (Marczynski et al. 2000). The data from that study revealed a 1.7–2-fold increase in 8-OHdG due to asbestos exposure ($p < 0.001$). These data support the hypothesis that asbestos fibers damage cells through an oxidative mechanism. Based on these results, preventive and therapeutic approaches using antioxidants may be possible. The various chemicals and environmental factors that induced increases in 8-OHdG levels are listed in Table 3.

5 Ionizing Radiation

Oxidative DNA damage is one of the major causes of radiation injury. 8-OHdG and 8-OHGua are increased in a linear fashion by 20–300 mGy of gamma irradiation to aqueous solutions of dG and Gua, respectively (Li et al. 2013b). These markers are considered to have sufficient sensitivity for detecting oxidative damage by ionizing radiation. The adverse health effects of radiation doses around 100 mSv have been vigorously discussed, especially in terms of cancer induction. Meanwhile, we reported that the threshold radiation level for increasing the 8-OHdG level in mouse urine was about 100–200 mGy (Li et al. 2013b), which supports the threshold theory of some reliable epidemiological studies on atomic bomb survivors (Land 1980; Shimizu et al. 1992). However, in most reports, an increase in 8-OHdG could be detected after irradiation with doses greater than a few Gy. Furthermore, most of the human data were collected from patients undergoing radiotherapy, who usually get quite high doses of radiation. It is essential to collect lower dose data to clarify the contribution of oxidative damage to the adverse health effects and to develop protective measures. In addition, the radiation health effects change with the radiation dose rate (Gy/min) (Tanooka 2011). At present, the evidence for the effects of low dose rate radiation is insufficient, especially for the human population. In cells, low molecular weight antioxidants and ROS-scavenging enzymes may process some of the ROS generated by radiation and prevent cellular DNA and nucleotide damage. In addition, the higher 8-OHdG levels induced in tissue DNA may decrease as time passes. This is due to the cellular DNA repair systems, such as nucleotide excision repair, base excision repair, and damaged nucleotide sanitization. As a result, oxidized nucleosides and bases accumulate in the urine. Therefore, urinary 8-OHdG is a sensitive marker for radiation-induced oxidative damage in vivo. The published data on the increased formation of 8-OHdG by ionizing radiation are summarized in Table 4.

Table 3 Occupational, environmental exposure and 8-OHdG

Chemicals or occupation	Ratio to control	Species	Sample	Method	Reference
Asbestos	1.31	Human	Urine	HPLC-ECD	Tagesson et al. (1993)
	1.45	Human	Urine	HPLC-ECD	Takahashi et al. (1997)
	1.37–1.64	Hamster, rat	Lung	HPLC-ECD	Yamaguchi et al. (1999)
	1.7	Human	White blood cell DNA		Marczynski et al. (2000)
PAHs	↗	Human	Urine	LC-MS/MS	Li et al. (2015)
	1.19	Human	Urine	HPLC-ECD	Harri et al. (2005)
	1.63	Human	White blood cell DNA	HPLC-ECD	Marczynski et al. (2009)
Refractory materials	1.54	Human	White blood cell DNA	HPLC-ECD	ibid.
Carbon electrodes	3.23	Human	White blood cell DNA	HPLC-ECD	ibid.
Converter workers	1.26	Human	White blood cell DNA	HPLC-ECD	ibid.
PAH, anthraquinone	↗	Human	Urine	ELISA	Wei et al. (2010)
DEP PM2.5	2.14	Human	Urine	ELISA	Lee et al. (2010)
PM 2.5	1.15	Human	Urine	ELISA	Kim et al. (2004)
	↗	Human	Urine	ELISA	Lee et al. (2012)
	>3	Human	Urine	ELISA	Wei et al. (2009)
	↗	Human	Urine	HPLC-ECD	Neophytou et al. (2013)
Coke oven worker	↗	Human	Urine		Guo et al. (2014)
	Top 1.20 Side 1.53	Human	Urine	HPLC-ECD	Nguyen et al. (2014)
	↗	Human	Urine	LC-MS/MS	Chao et al. (2008)
Coke production	1.4	Human	White blood cell DNA	HPLC-ECD	Marczynski et al. (2009)

(continued)

Table 3 (continued)

Chemicals or occupation	Ratio to control	Species	Sample	Method	Reference
Coke plant, policeman, taxi driver	↗	Human	Urine	Capillary electrophorasis-ECD	Zhang et al. (2013)
Bottom ash treatment plant, fly ash treatment plant	1.46	Human	Urine	ELISA	Liu et al. (2008)
Diesel exhaust emission inspector	>3	Human	Urine	ELISA	Wei et al. (2009)
Bus driver	1.3	Human	Urine	HPLC-ECD	Loft et al. (1999)
	1.27–1.45	Human	Urine	ELISA	Rossner et al. (2008)
Bus drivers, garagemen	2.59	Human	Urine	HPLC-ECD	Bagryantseva et al. (2010)
Long distance bus driver	Adjusted OR = 9.4	Human	Urine	ELISA	Han et al. (2010)
Traffic policeman	↗	Human	Leukocyte DNA	HPLC-ECD	Arayasiri et al. (2010)
Subway workers	1.07 ($p = 0.038$)	Human	Urine	ELISA	Mehrdad et al. (2015)
Foundry workers	2.72	Human	Urine	ELISA	Liu et al., (2009)
Toner-exposed	1.03	Human	Urine	HPLC-ECD	Kitamura et al. (2009)
Wildland firefighter	↗	Human	Urine	ELISA	Gaughan et al. (2014)
Temple workers, incense smoke	↗	Human	Leukocyte DNA	HPLC-ECD	Navasumrit et al. (2008)
Benzene	↗ 8-OHdG, 8-OHGuo, 8-OHGua	Human	Urine	LC-MS/MS	Manini et al. (2010)
	1.25–8.00	Human	Leukocyte DNA	HPLC-ECD	Liu et al. (1996)
	↗	Human	Urine	HPLC-ECD	Nilsson et al. (1996)
	↗ 8-OHdG, 8-OHGuo	Human	Urine	MS	Andreoli et al. (2012)
Styrene exposed workers	1.03–1.23	Human	Urine	LC-MS/MS	Manini et al. (2009)
Styrene	1.47	Human	Peripheral blood DNA	HPLC-ECD	Marczynski et al. (1997)
Gas station attendant, taxi driver	↗	Human	Urine	ELISA	Goeethel et al. (2014)

(continued)

Table 3 (continued)

Chemicals or occupation	Ratio to control	Species	Sample	Method	Reference
Ethylbenzene	4.21	Human	Urine	ELISA	Chang et al. (2011)
Di-(2-ethylhexyl) phthalate plastic recycling	1.27	Human	Urine	HPLC-ECD	Wang et al. (2011)
Smoker	2.15	Human	Urine	ELISA	Lu et al. (2014)
	1.5	Human	Urine	HPLC-ECD	Loft et al. (1992)
	1.43	Human	Lung	HPLC-ECD	Asami et al. (1997)
	2.13	Human	Leukocytes	HPLC-ECD	Lodovici et al. (2000)
Iron, smoking	↗	Human	Urine	LC-MS/MS	Hossain et al. (2014)
Environmental tobacco smoke	↗	Human	Plasma	LC-MS/MS	Chiang et al. (2012)
Hexavalent chromium electroplating worker	1.64	Human	Urine	ELISA	Zhang et al. (2011)
Chromium	1.57	Human	Urine	HPLC-ECD	Kuo et al. (2003)
Chromate	↗	Human	Urine	ELISA	Li et al. (2014)
Arsenic	4	Human	Saliva	LC-MS/MS	Hinhumpatch et al. (2013)
	1.12	Human	Urine	ELISA	Wong et al. (2005)
As, heavy metals	↗	Human	Serum	ELISA	Szymanska-Chabowska et al. (2009)
As, Cd	↗	Human	Urine	LC-MS/MS	Engstrom et al. (2010)
As, Cd, Ni, Se	↗	Human	Urine	ELISA	Lin et al. (2012)
Nickel-cadmium	1.05–2.55 8-OHGua	Human	Urine	HPLC-ECD	Yoshioka et al. (2008)
Manufacturing surgical instrument (nickel)	$r = 0.41$, $p < 0.0001$	Human	Urine	ELISA	Sughis et al. (2012)
Rubber	1.38	Human	Urine	HPLC-ECD	Tagesson et al. (1993)
Roofers	1.2	Human	Urine	HPLC-ECD	Toraason et al. (2001)

(continued)

Table 3 (continued)

Chemicals or occupation	Ratio to control	Species	Sample	Method	Reference
Azo-dye	1.79	Human	Urine	HPLC-ECD	Tagesson et al. (1993)
Cooking oil fume	1.46	Human	Urine	HPLC-ECD	Pan et al. (2008)
Cooking oil fume	↗	human	urine	HPLC-ECD	Ke et al. (2009)
Zinc oxide nanoparticle	↗	Rat	Blood		Chuang et al. (2014)
Nanoparticles from photocopiers	↗	Human	Urine		Khatri et al. (2013)
Metal nanoparticles	↗	Mouse	Urine, bone marrow, liver	HPLC-ECD	Song et al. (2012)
PCDD, dibenzofurans, etc.	3.84	Human	Urine	HPLC-ECD	Wen et al. (2008)
Aroclor 1254	>5	Mouse	Liver	HPLC-ECD	Faux et al. (1992)
TCDD	>20	Mouse	Urine	ELISA	Shertzer et al. (1998)
2,3,7,8-TCDD	↗	Human	Plasma	LC-MS/MS	Pelclova et al. (2011)
Trichloroethylene	↗	Human	Urine	LC-MS/MS	Abusoglu et al. (2014)
Agricultural worker	↗	Human	Urine	HPLC-ECD	Kisby et al. (2009)
Antineoplasic drugs	1.38	Human	Urine	ELISA	Huang et al. (2012)
VOCs*, hair salon	7.5	Human	Serum	ELISA	Ma et al. (2010)

*VOCs: volatile organic compounds

6 Diseases

Oxidative stress leads to many kinds of diseases. Examinations of the oxidative damage in connection with diseases are quite important for their treatment and prevention. In epidemiological studies, chronic oxidative stress is a cancer risk factor. For example, higher levels of 8-OHdG were observed in the stomach tissues of children (Baik et al. 1996) and cancer patients with *Helicobacter pylori* infection. Increased levels of 8-OHdG have been reported in various types of cancer. Oxidative stress engenders vascular complications and pancreatic beta cell damage, which induces insulin resistance and diabetes. In these patients, the 8-OHdG and 8-OHGua levels in urine or plasma were higher than those in the control group. In addition to patients with hypertension or cardiac infarcts, those with Alzheimer's or

Table 4 Ionizing radiation studies on 8-OHdG

Type of radiation	Dose rate	Total dose	8-OHdG induced ratio to control	Sample	Method	Reference
X-ray	0.5 Gy/min	1 Gy	1.62	Mouse liver	HPLC-ECD	Li et al. (2013b)
	0.5 Gy/min	0.5 Gy	1.61	Mouse urine	HPLC-ECD	Li et al. (2013b)
	0.59 Gy/min	3.9 Gy	2	Rat mammary gland	HPLC-ECD	Haegele et al. (1998)
	0.4 Gy/min	3 Gy	3.07	Rat bone marrow	HPLC-ECD	Umegaki et al. (2001)
	0.55 Gy/min	6 Gy	1.77	Mouse serum	ELISA	Manda et al. (2007)
		0.5 Gy	2.75	MT-I/II null mouse serum	ELISA	Shibuya et al. (2008)
		0.5 Gy	3.12	MT-I/II null mouse urine	ELISA	Shibuya et al. (2008)
γ-ray	2.4 mGy/min	0.3 Gy	2.62	F1 medaka muscle	HPLC-ECD	Grygoryev et al. (2013)
		3 Gy	1.3	Rat urine	ELISA	Inano and Onoda (2002)
	1 Gy/min	5 Gy	1.7–2.7	Rat liver	HPLC-ECD	Kaneko et al. (2003)
	0.7 Gy/min	3 Gy	2.5	Mouse bone marrow	ELISA	Rithidech et al. (2012)
	0.7 Gy/min	4 Gy	6.28	Mouse intestine	ELISA	Qian et al. (2010)

Radiotherapy, 6 MeV photons	2 Gy/d, 5 d/w	70 Gy	1.6 (8-OHGua)	Human urine	HPLC-ECD	Roszkowski et al. (2008)
		70 Gy	2.42	Human urine	HPLC-ECD	Roszkowski et al. (2008)
High LET ^{56}Fe beam, 500 MeV/nucleon	0.89 Gy/min	2 Gy	1.7	Mouse serum	ELISA	Manda et al. (2008)
Radiochemotherapy		12 Gy	>2.34	Human urine	HPLC-ECD	Bergman et al. (2004)

Parkinson's disease also have higher levels of 8-OHdG. Interestingly, patients with mental disorders, such as schizophrenia, bipolar disorder, and autism, also have higher levels of 8-OHdG and 8-OHGuo. Examples of recent publications describing increased 8-OHdG levels in various diseases are provided in Table 5.

7 Lifestyle

Lifestyle factors are closely related to the individual oxidative status. Epidemiological studies have suggested that lifestyle improvements can lead to the prevention of cancers and lifestyle-related diseases, such as diabetes. A well-balanced diet rich in vegetables and fruits reduced the 8-OHdG levels in the body, as an oxidative stress marker. In contrast, alcohol consumption and job stress increased the oxidative stress. Interestingly, the BMI of smokers showed an inverse correlation between the 8-OHdG level (Mizoue et al. 2006), which partly supports the U-shaped relation between BMI and cancer risk, concluded from epidemiological studies (Inoue et al. 2004). Namely, cancer risk increases with a very low BMI, especially in smokers (Kabat and Wynder 1992). A very thin state may induce oxidative stress, due to a high metabolic rate (Shah et al. 1988). Smoking seems to be one of the worst factors for inducing oxidative damage. Moderate exercise reduced the 8-OHdG levels in leukocyte DNA, by the induction of either ROS-scavenging enzymes (SOD, catalase, and glutathione peroxidase) (Mena et al. 1991) or repair enzymes [OGG1 and MTH1 (Sato et al. 2003)]. Representative references describing the effects of lifestyle factors on 8-OHdG levels are provided in Table 6.

8 Antioxidants

Antioxidants help to keep the body healthy. There are several methods for evaluating antioxidant activity. Among them, the measurement of 8-OHdG as an oxidative damage marker is the most widely used method for in vivo experiments, including human studies. The 8-OHdG reducing effects of typical antioxidants on induced oxidative stress are shown in Table 7. Vitamin C intake significantly decreased the 8-OHdG levels induced by periodontitis, ischemia, and chronic hemodialysis. Alpha-tocopherol reduced the increased 8-OHdG levels caused by heavy athletic training or iron therapy. The combined effects of alpha-tocopherol, ascorbic acid, beta-carotene, acetylsalicylic acid, and sesamin were reported. Many components in fruits or vegetables, such as astaxanthin (Aoi et al. 2003), lycopene (Devaraj et al. 2008), resveratrol (Sirerol et al. 2015), green tea polyphenols (Luo et al. 2006), quercetin (Ozyurt et al. 2014), and curcumin (Okada et al. 2001), were also reported to reduce 8-OHdG levels.

Table 5 Diseases and 8-OHdG

Disease	8-OHdG levels case to control	Note	Sample	Method	Reference
Cancer					
Stomach	5.3	H. pylori	Human stomach	Immunohistochemistry	Raza et al. (2014)
	2.1		Human stomach	HPLC-ECD	Borrego et al. (2013)
	7.5		Human urine	HPLC-ECD	Borrego et al. (2013)
	2.08		Human stomach	HPLC-ECD	Farinati et al. (2008)
	1.43		Human leukocyte	HPLC-ECD	Siomek et al. (2006)
	1.29		Human urine	HPLC-GC/MS	Siomek et al. (2006)
	1.32	8-OHGua	Human urine	HPLC-GC/MS	Siomek et al. (2006)
Intestine, colon	↗		Human urine	LC-MS/MS	Hsu et al. (2009)
	1.44		Human lymphocyte	HPLC-ECD	Gackowski et al. (2002)
Liver	2.7	8-OHGuo	Human urine	LC-MS/MS	Broedbaek et al. (2009)
	7.3		Human liver	HPLC-ECD	Kato et al. (2001)
Bladder	1.96		Human urine	ELISA	Chiou et al. (2003)
Prostate	1.63		Human urine	ELISA	Chiou et al. (2003)
Breast	↗ IRR* 1.08		Human urine	HPLC-ECD	Loft et al. (2013)
	↗		Human urine	LC-MS/MS	Cho et al. (2006)
Lung	↗ IRR* 9.94	Men, never-smoker	Human urine	HPLC-ECD	Loft et al. (2012)
	1.41		Human lung	HPLC-ECD	Inoue et al. (1998)

(continued)

Table 5 (continued)

Disease	8-OHdG levels case to control	Note	Sample	Method	Reference
Diabetes	2		Human plasma	LC-MS/MS	Waris et al. (2015)
	2.6	8-OHGua	Mouse urine	HPLC-ECD	Li et al. (2013a)
	2	8-OHGua	Mouse serum	HPLC-ECD	Li et al. (2013a)
	↗ Hazard ratio 1.72	8-OHGuo	Human urine	LC-MS/MS	Broedbaek et al. (2013)
	1.65		Human urine	ELISA	Dong et al. (2008)
	2.45		Human urine	CE-AD	Xu et al. (2004)
	2.19		Human urine	HPLC-ECD	Hinokio et al. (1999)
Hypertension	1.74		Human urine	HPLC-ECD	Espinosa et al. (2007)
Cardiac infarct	2.45		Human serum	ELISA	Suzuki et al. (2011)
	1.76		Human urine	ELISA	Nagayoshi et al. (2005)
Inflammation	1.79	6 % DSS**	Rat colonic mucosa	HPLC-ECD	Tardieu et al. (2000)
Inflammatory bowel disease	↗		Human blood	HPLC-ECD	D'Odorico et al. (2001)
Chronic hepatitis C	8		Human liver	HPLC-ECD	Kato et al. (2001)
Periodontitis	2		Human saliva	ELISA	Sezer et al. (2012)
Opisthorchis viverrini	↗		Hamster liver	Immunostaining	Pinlaor et al. (2004)
Cystic fibrosis	↗		Human urine	SPE-HPLC-ECD	Brown et al. (1995)
Alzheimer	5.15		Human cerebrospinal fluid	HPLC-ECD	Abe et al. (2002)
	1.76		Human lymphocyte	HPLC-ECD	Mecocci et al. (2002)
Parkinson	2.97		Human cerebrospinal fluid	HPLC-ECD	Abe et al. (2003)

Genetic diseases[***]	↗		Human leukocyte human urine	HPLC-ECD, ELISA	Lloret et al. (2008)
Mental					
Schizophrenia	1.21		Human urine	LC-MS/MS	Jorgensen et al. (2013)
	1.22	8-OHGuo	Human urine	LC-MS/MS	Jorgensen et al. (2013)
Bipolar disorder	1.4		Human urine		Munkholm et al. (2015)
	1.43	8-OHGuo	Human urine		Munkholm et al. (2015)
Autism	↗ $p < 0.01$		Human cerebellum	LC-MS	Rose et al. (2012)
Depression	1.13		Human serum	ELISA	Forlenza and Miller (2006)
Cortisol	$r^2 = 0.15$, $p < 0.001$	8-OHdG, 8-OHGuo	Human urine	LC-MS/MS	Joergensen et al. (2011)

[*]IRR: Incidence Rate Ratio
[**]DSS: dextran sodium sulfate
[***]Bloom syndrome, Fanconi anemia, Werner syndrome, Xeroderma pigmentosum

Table 6 Lifestyle and 8-OHdG

Habit	Ratio to control	Species	Sample	Method	Reference
Fruit and vegetable intake	Lower 8-OHdG (P for trend, 0.05)	Human	Urine	ELISA	Cocate et al. (2014)
Light-colored vegetable, soybean product, rice, BMI	↘	Human	Urine	HPLC-ECD	Irie et al. (2005)
Working hours, cigarette smoke	↗	Human	Urine	HPLC-ECD	Irie et al. (2005)
Fruit, daily physical activity, healthy meal	↘	Human	Urine	HPLC-ECD	Tamae et al. (2009)
Cigarette, alcohol	↗	Human	Urine	HPLC-ECD	Tamae et al. (2009)
Fish intake	↘	Human	Urine	ELISA	Muzembo et al. (2012)
Job stress	↗	Human	Urine	ELISA	Inoue et al. (2009)
Age	↗	Human	Urine	ELISA	Sakano et al. (2009)
BMI	↗ 2.7 % / unit BMI ↘	Human	Urine	HPLC-ECD	Mizoue et al. (2007)
Smoking	1.6	Human	Leukocyte DNA	HPLC-ECD	Asami et al. (1996)
	1.43	Human	Lung	HPLC-ECD	Asami et al. (1997)
	3.34	Human	Leukocyte DNA	HPLC-ECD	Lodovici et al. (2005)
Environmental smoke	1.55	Human	Leukocyte DNA	HPLC-ECD	Lodovici et al. (2005)
Physical activity (subject $n = 6{,}422$)	↘	Human	Urine	HPLC-ECD	Hara et al. (in press)
2 weeks, moderate intensity exercise after primary therapy	0.67	Human (colorectal cancer)	Urine	LC-MS	Allgayer et al. (2008)
Race (African American/caucasian)	1.3	Human	Urine	ELISA	Huang et al. (2000)
Regular exercise	1.16	Human	Urine	ELISA	Huang et al. (2000)
Wrestling exercise	↘	Human	Serum	ELISA	Hamurcu et al. (2010)
Exercise	0.53	Human	Leukocyte DNA	HPLC-ECD	Asami et al. (1998a)
Forced exercise	1.9–2.4	Rat	Heart, lung, liver	HPLC-ECD	Asami et al. (1998b)

(continued)

Table 6 (continued)

Habit	Ratio to control	Species	Sample	Method	Reference
Ultramarathon	↗ mid race ↘post race	Human	Urine	HPLC-ECD	Miyata et al. (2008)
Sunlight	OR: 4.35	Human	Urine	ELISA	Kato et al. (2011)
UVB, 280–350 nm	1.73	Rabbit	Eye	HPLC-ECD	Lodovici et al. (2009)
UVA1, 364 nm Ar laser	2.56	Mouse	Skin	HPLC-ECD	Ikehata et al. (2008)
UVA, 364 nm	2.7	Drosophila		HPLC-ECD	Negishi et al. (2007)

9 Formation of 8-OHGuo in RNA

In the previously mentioned in vitro experiments, the formation of 8-OHGuo in RNA was higher than that in DNA (Kasai and Nishimura 1984c). One reason for this may be the more open structure of single-stranded RNA than double-stranded DNA. In fact, Fiala et al. reported that the hepatocarcinogen 2-nitropropane induces 8-hydroxyguanosine (8-OHGuo) in rat liver RNA much more efficiently (11-fold as compared to control) than 8-OHdG in DNA (3.6-fold as compared to control) (Fiala et al. 1989). This may also be due to the rapid removal of 8-OHGua from DNA by repair enzymes, or to the higher reactivity of ROS, produced by the metabolism of 2-nitropropane, with cytoplasmic single-stranded RNA. When doxorubicin (adriamycin) was administered to rats, a significant increase of 8-OHGuo in the liver RNA, but not 8-OHdG in the DNA, was observed (Hofer et al. 2006). Malayappan et al. observed increased levels of 8-OHGuo and 8-OHdG in smoker's urine, as compared to control nonsmokers (Malayappan et al. 2007). As other examples, analyses of ribonucleoside 8-OHGuo levels in tissue RNA or biological fluids were reported in relation to aging, calorie restriction, exercise (rat liver RNA) (Seo et al. 2006), cisplatin treatment in cancer patients (urine) (Andreoli et al. 2012), Alzheimer's disease (cerebrospinal fluid) (Isobe et al. 2009), hereditary hemochromatosis (urine) (Broedbaek et al. 2009), exposure to benzene (human urine) (Manini et al. 2010), and the effect of antioxidants in cherry juice (human urine) (Traustadottir et al. 2009). In those studies, higher formation of 8-OHGuo than 8-OHdG was observed, which is compatible with the general tendency that the ultimate reactive forms of carcinogens, such as aflatoxin B1 (Garner and Wright 1975) or N-nitrosopyrrolidine (Wang and Hecht 1997), induced more modifications in RNA than in DNA. Therefore, the ribonucleoside 8-OHGuo is also a promising biomarker for oxidative stress (Poulsen et al. 2012).

High levels of the ribonucleoside triphosphate 8-OHGTP may also be produced in cells, in addition to 8-OHdGTP (see next paragraph). The MTH1 protein, a

Table 7 Antioxidants and 8-OHdG

Antioxidant	Ratio to control	oxidative stress	Sample	Method	Reference
Ascorbic acid	0.62	Atherosclerosis	Rat serum	ELISA	Ekuni et al. (2009)
	↘	Periodontitis	Rat gingival	ELISA	Tomofuji et al. (2009)
	↘	Ischemia	Human serum	ELISA	Cangemi et al. (2007)
	0.82	Hemodialysis	Human lymphocyte	HPLC-ECD	Tarng et al. (2004)
Alpha-tocopherol	↘	Down syndrome	Human urine	ELISA	Nachvak et al. (2014)
	0.48	Training	Human urine	ELISA	Tsakiris et al. (2006)
	↘		Human urine	ELISA	Lee and Wan (2000)
Alpha-tocopherol + ascorbic acid	0.75	Exercise	Human blood	ELISA	Bloomer et al. (2006)
	↘		Human urine	ELISA	Hong et al. (2013)
	↘		Human serum	ELISA	Goldfarb et al. (2007)
Alpha-tocopherol + ascorbic acid + beta-carotene	0.95	Nontreat	Rat liver	HPLC-ECD	Wawrzyniak et al. (2013)
Ascorbic acid + acetylsalicylic acid	↘	Neuroinflammation	Rat	ELISA	Kara et al. (2014)
alpha-tocopherol + sesamin	↘	Aging	Rat urine	ELISA	Noguchi et al. (2001)
Carotenoid	↘		Human urine	ELISA	Cocate et al. (2015)
Astaxanthin	↘	Exercise	Mouse heart	HPLC-ECD	Aoi et al. (2003)
Lycopene	↘		Human urine	ELISA	Devaraj et al. (2008)
Lycopene + beta-carotene	0.3	Iron	Rat prostate	HPLC-ECD	Matos et al. (2006)
Lycopene + genistein	0.55	DMBA	Rat serum	ELISA	Sahin et al. (2011)
Resveratrol	0.59	UV B	Mouse skin	LC-MS/MS	Sirerol et al. (2015)
	0.58	Exercise	Rat serum	ELISA	Xiao (2015)
	0.71	Periodontitis	Mouse urine	ELISA	Tamaki et al. (2014)
	0.62	Sodium fluoride	Rat plasma	ELISA	Atmaca et al. (2014)

Green tea	0.7		Rat urine	HPLC-ECD	Shen et al. (2008)
Green tea polyphenols	↘ Dose dependent		Human urine	HPLC-ECD	Luo et al. (2006)
Coffee	↘ Trend $p = 0.046$		Human urine	HPLC-ECD	Hori et al. (2014)
Polyphenol (vegetables, red wine)	↘		Human urine	ELISA	Pedret et al. (2012)
CoQ10	0.88	Placebo	Human urine	HPLC-ECD	Ito et al. (2015)
Pterostilbene	0.69	UV B	Mouse skin	LC-MS/MS	Sirerol et al. (2015)
Quercetin	↘	Ionizing radiation	Rat tissue	ELISA	Ozyurt et al. (2014)
Ursolic acid	0.64	Carbon tetrachloride	Mouse kidney	HPLC-ECD	Ma et al. (2014)
Vitex honey	0.8	Paracetamol	Mouse serum		Wang et al. (2015)
Curcumin	↘	Formaldehyde	Rat brain, urine		Ciftci et al. (2015)
	↘	Fe-NTA	Mouse kidney	ELISA	Okada et al. (2001)
Fermented papaya	0.6	Alzheimer	Human urine	ELISA	Barbagallo et al. (2015)
	↘	Hepatitis C virus	Human leukocyte	HPLC-ECD	Marotta et al (2007)
Almond	0.72	Smoker	Human urine	HPLC-ECD	Li et al. (2007)
Tomato sauce	0.79	Prostate cancer	Human leukocyte	HPLC-ECD	Chen et al. (2001)
Pigmented potato	↘		Human plasma	ELISA	Kaspar et al. (2011)

(continued)

Table 7 (continued)

Antioxidant	Ratio to control	oxidative stress	Sample	Method	Reference
Ginseng	0.68	Gentamicin	Rat urine	ELISA	Shin et al. (2014)
	↘		Human urine	HPLC-ECD	Lee et al. (1998)
Vegetable	p for trend < 0.001		Human urine	ELISA	Cocate et al. (2014)
Fruit and vegetable	p for trend = 0.028		Human urine	ELISA	Cocate et al. (2014)
	0.68		Human lymphocyte	HPLC-ECD	Thompson et al. (1999)
Ursolic acid, luteolin	↘		Caco-2 cell	comet assay	Ramos et al. (2010)
Olive oil	0.87		Human urine	LC-MS/MS	Machowetz et al. (2007)

mutT-related protein that catalyzes the hydrolysis of 8-OHdGTP to 8-OHdGMP, also hydrolyzed the ribonucleotide 8-OHGTP (Fujikawa et al. 2001). In oxidative stress-related diseases induced by aging, 8-OHGuo formation in RNA, by either the incorporation of 8-OHGTP or the direct oxidation of RNA, caused a reduction in translation or an increase in mistranslation, which induced the accumulation of nonfunctional proteins (Poulsen et al. 2012). For example, in Alzheimer's disease patients, increased 8-OHGuo in RNA- and reduced MTH1 activity were observed in their hippocampi (Song et al. 2011). These results suggested that increased oxidative stress and MTH1 deficiency during aging might be causative factors for this disease.

10 The Nucleotide Pool Is a Significant Target

In the initial in vitro study, the formation of 8-OHdG in the monomer nucleoside (dG) was 15 times higher than that in the DNA (Kasai and Nishimura 1984c). This suggests that the modification of dGTP to 8-OHdGTP in the nucleotide pool is more important than the formation of 8-OHdG in the DNA. In vivo, an *E. coli mutT*-deletion mutant, which lacks the 8-OHdGTP sanitization system, showed a 100–10,000 times higher spontaneous mutation rate than the wild type (Maki and Sekiguchi 1992), while the rate in a *mutM*-deletion mutant, which lacks the system to remove 8-OHGua from DNA, was only 6–14 times higher than that of the wild type (Cabrera et al. 1988; Michaels et al. 1992). In fact, Russo et al. reported that 8-OHdGTP is a significant contributor to genetic instability in mismatch repair-deficient cells (Russo et al. 2004). Harms-Ringdahl and his collaborators detected considerable amounts of 8-OHdG in the nucleotide pool fraction, which were much higher (35-fold) than those in the DNA fraction, and concluded that the nucleotide pool is a significant oxidative modification target (Haghdoost et al. 2006). They also reported that the reduction of 8-OHdGTP in the nucleotide pool by hMTH1 leads to fewer mutations in the human lymphoblastoid cell line TK6, exposed to UVA (Fotouhi et al. 2011). Kaczmarek et al. described the efficient formation of 8-OHdGTP from the Ni(II)-dGTP or Ni(II)-dGTP-His complex in the presence of H_2O_2, which may be an underlying mechanism of the potent carcinogenic effects of nickel compounds (Kaczmarek et al. 2005). Together, these results suggest that 8-OHdG formation in the nucleotide pool is more important than that in the DNA, in relation to mutagenesis and carcinogenesis (Fig. 3). It is worth mentioning that the nucleotide pool is also a significant target of alkylation in N-methyl-N-nitrosourea-induced mutagenesis and carcinogenesis (Topal and Baker 1982).

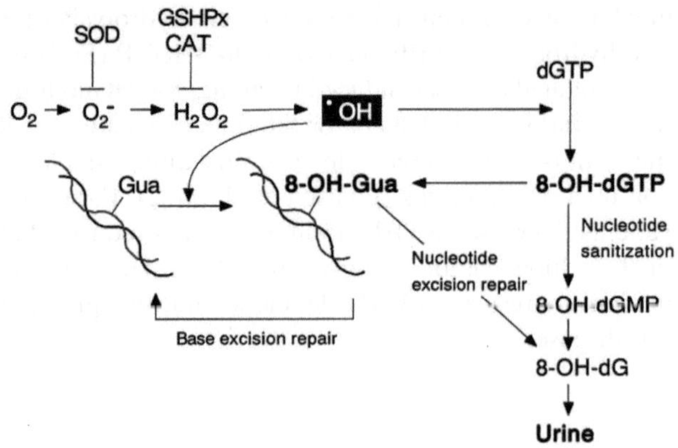

Fig. 3 8-OHdG formation and repair in vivo

11 Accurate Measurement of 8-OHdG as a Reliable Marker

The need to accurately measure 8-OHdG has long been discussed (Kasai 1997). In the case of 8-OHdG measurements in cellular DNA, special precautions must be taken to prevent sample auto-oxidation. An antioxidant (NaI) and a metal chelator (Desferal, EDTA) must be used during DNA isolation, especially in the lysis step. When a DNA digest was stored at 10 °C, the 8-OHdG levels significantly increased in a few hours. In contrast, when stored at −80 °C, no increase was observed (Kawai et al. 2007).

For the measurement of urinary 8-OHdG, an automated HPLC-ECD system to analyze urinary 8-OHdG with higher accuracy was developed (Kasai 2003). Disparity in the results has occurred frequently, depending upon the measurement methods (Shimoi et al. 2002). There are considerable discrepancies between the results obtained by the ELISA and HPLC methods. Usually, the 8-OHdG levels are 2–3 times higher in the ELISA methods, as compared to the HPLC methods, and the data observed with ELISA are quite variable. Recently, urea was recognized as a major cause of this problem (Song et al. 2009). A high concentration of urea in the sample (urine or blood) could cross-react with the anti-8-OHdG antibody in the ELISA. Although various approaches, including the performance of the ELISA at 4 °C, a pretreatment with urease, and a pretreatment by solid-phase extraction (SPE), have been taken to resolve this issue, satisfactory results have not been achieved (Rossner et al. 2013). Regarding this situation, Watanabe et al. reported a good correlation between the ELISA and HPLC methods for the 8-OHdG values by the ELISA method, following urease treatment and ethanol precipitation. Urea is considered to be a major interfering substance in ELISA, but there are still other cross-reacting components, such as 8-OHGuo (Song et al. 2012) and creatinine (Rossner et al. 2008). Most reports of the 8-OHdG levels in serum or saliva were obtained by the direct use of ELISA for these fluids. The levels of 8-OHdG in plasma and saliva measured by LC-MS/MS were several hundred times lower than those reported by scientists using the commercial ELISA kit (Hu et al. 2010b). Although serum and saliva are quite useful materials, pretreatments for concentration and cleanup, such as SPE treatment, are needed before ELISA and HPLC measurements because of the low concentration of 8-OHdG, and for the removal of cross-reacting materials, especially in ELISA. From an overall consideration, although ELISA measurements have revealed certain rough trends in large-scale analyses, the HPLC methods (HPLC-ECD or LC-MS/MS) are recommended for the accurate measurement of 8-OHdG. The urine analysis data obtained by our method (HPLC-ECD) are almost identical to those obtained by LC-MS/MS, as judged by ESCULA (European Standard Committee on Urinary Lesion Analysis) (Barregard et al. 2013). For urinary 8-OHGua analysis, diets containing 8-OHGua must be considered, because 90 % of the 8-OHGua administered to rats was excreted into the urine (Kawai et al. 2006). The CE-2 diet, which is generally used for animal experiments, contains a large amount of 8-OHGua. Therefore, in animal experiments, nucleic acid-free diets, such as those with egg white as the protein source, should be used. For human studies, the intake of various 8-OHGua-containing foods, especially fish products, must be minimized before urine collection.

It is also important to check the stability of 8-OHdG under various conditions, and to determine whether it is formed from dG in biological fluids, such as urine, for its general use as an oxidative stress marker. Urinary 8-OHdG is stable at −20 °C for 15 years (Loft et al. 2006) and at 25 °C for 24 h (nonsmokers) (Matsumoto et al. 2008). However, the levels of urinary 8-OHdG from smokers showed a tendency to increase over 24 h at 25 °C (Matsumoto et al.). This may occur because (1) smoker's urine contains lower levels of antioxidants than that of nonsmokers, and/or (2) smoking-related substances in urine generate ROS. Shigenaga et al. injected ^{3}H-8-OHdG into the tail veins of rats, and 24 h urine samples were analyzed by HPLC (Shigenaga et al. 1989). They found no degradation of 8-OHdG after administration and excretion. When ^{3}H-dG was stored in urine for 19 days at 4 °C, no 8-OHdG was produced, indicating that the chemical transformation of dG to 8-OHdG did not occur in rat urine (Shigenaga et al. 1989).

8-OHGua is rather unstable, as compared to 8-OHdG (Hu et al. 2010a). Its solution (pH 7) is stable at room temperature for 6 days, at 4 °C for 45 days, and at −20 °C for 87 days. After these periods, its degradation was observed.

12 Sources of 8-OHdG, 8-OHGuo, and 8-OHGua Generation and Validity of Their Analyses

Urinary 8-OHdG is generated by either nucleotide excision repair (NER) from oxidized DNA or hydrolysis of 8-OHdGTP by the sanitization enzyme MTH1. The free 8-OHGua base is produced by base excision repair (BER) from oxidized DNA or by the oxidation of guanine (formed by the hydrolytic degradation of DNA,

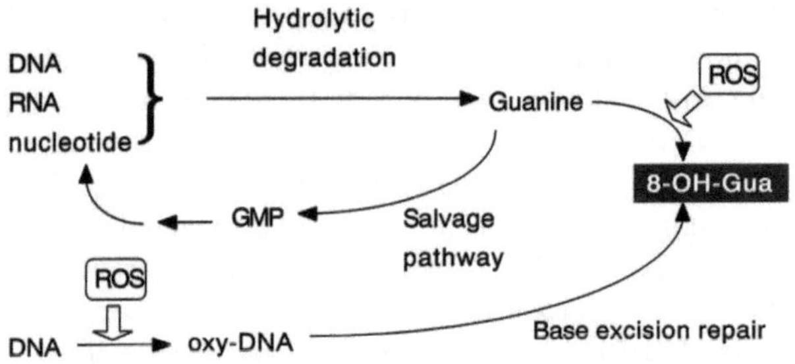

Fig. 4 8-OHGua formation and repair in vivo

RNA, and the nucleotide) before the salvage pathway (Kasai et al. 2008) (Fig. 4). 8-OHGuo may be produced by the hydrolysis of 8-OHGTP by MTH1 or by the degradation of oxidized RNA.

A human study revealed a correlation between the 8-OHdG levels in lymphocyte DNA and the urinary 8-OHdG levels (Gedik et al. 2002). A correlation between the concentrations of 8-OHdG in human urine, plasma, and saliva was also observed (Hu et al. 2010b), based on accurate LC-MS/MS measurements. In mouse experiments, correlations were observed between the urinary 8-OHdG and 8-OHGua levels, between the serum 8-OHGua and urinary 8-OHdG levels, and between the serum 8-OHGua and urinary 8-OHGua levels (Li et al. 2013a).

There are few direct studies of the relationship between 8-OHdG-related oxidative markers and cancer risks in humans, such as cohort studies, because of the length of time required to form conclusions (Loft et al. 2006). Considering the time needed to collect data for the large-scale analysis of 8-OHdG-related markers, the amounts of direct evidence for use as a predictor of cancer development are expected to increase in the future.

As described in this chapter, many chemical carcinogens, as well as UV- and ionizing radiation (UVA, gamma-ray, X-ray, etc.), induced 8-OHdG in animal experiments, while many antioxidants, which are known to suppress cancer, reduced the 8-OHdG levels, as indicated in Tables 4 and 7. In human studies, asbestos, azo-dyes, benzene, and chemicals used in the rubber industry, which were all concluded to be human carcinogens with sufficient evidence by the IARC (Lagorio et al. 1994; Tagesson et al. 1993), induced an increase in the urinary 8-OHdG level.

Furthermore, many lifestyle habits for cancer prevention, such as cessation of smoking, avoiding drinking and a high-fat diet, following the recommended levels of fish, fruit and vegetable consumption, and exercising moderately, are supported by the data showing increased or decreased 8-OHdG levels by these factors, in human studies. Urinary analyses of cancer high-risk groups (dermatomyositis, polymyositis, systemic sclerosis, cholangiocarcinogenesis) revealed higher levels of urinary 8-OHdG, as compared to those of the healthy control groups (Kasai et al. 2007; Thanan et al. 2008). In cancer- or aging-related genetic diseases, such as Fanconianemia, Bloom syndrome, and Xeroderma pigmentosum, the urinary 8-OHdG levels were also increased (Lloret et al. 2008).

Based on the direct and indirect evidence described herein, we consider 8-OHdG (or its related compounds) to be a useful marker for monitoring the oxidative stress involved in the induction of cancer and ROS-related diseases, if analyses are performed with the precautions mentioned above.

References

Abe T, Tohgi H, Isobe C et al (2002) Remarkable increase in the concentration of 8-hydroxyguanosine in cerebrospinal fluid from patients with Alzheimer's disease. J Neurosci Res 70:447–450

Abe T, Isobe C, Murata T et al (2003) Alteration of 8-hydroxyguanosine concentrations in the cerebrospinal fluid and serum from patients with Parkinson's disease. Neurosci Lett 336:105–108

Abusoglu S, Celik HT, Tutkun E et al (2014) 8-Hydroxydeoxyguanosine as a useful marker for determining the severity of trichloroethylene exposure. Arch Environ Occup Health 69:180–186

Allgayer H, Owen RW, Nair J et al (2008) Short-term moderate exercise programs reduce oxidative DNA damage as determined by high-performance liquid chromatography-electrospray ionization-mass spectrometry in patients with colorectal carcinoma following primary treatment. Scand J Gastroenterol 43:971–978

Andreoli R, Protano R, Manini P et al (2012) Association between environmental exposure to benzene and oxidative damage to nucleic acids in children. Med Lav 103:324–337

Aoi W, Naito Y, Sakuma K et al (2003) Astaxanthin limits exercise-induced skeletal and cardiac muscle damage in mice. Antioxid Redox Sign 5:139–144

Arayasiri M, Mahidol C, Navasumrit P et al (2010) Biomonitoring of benzene and 1,3-butadiene exposure and early biological effects in traffic policemen. Sci Total Environ 408:4855–4862

Asami S, Hirano T, Yamaguchi R et al (1996) Increase of a type of oxidative DNA damage, 8-hydroxyguanine, and its repair activity in human leukocytes by cigarette smoking. Cancer Res 56:2546–2549

Asami S, Manabe H, Miyake J et al (1997) Cigarette smoking induces an increase in oxidative DNA damage, 8-hydroxydeoxyguanosine, in a central site of the human lung. Carcinogenesis 18:1763–1766

Asami S, Hirano T, Yamaguchi R et al (1998a) Reduction of 8-hydroxyguanine in human leukocyte DNA by physical exercise. Free Radic Res 29:581–584

Asami S, Hirano T, Yamaguchi R et al (1998b) Effects of forced and spontaneous exercise on 8-hydroxydeoxyguanosine levels in rat organs. Biochem Biophys Res Commun 243:678–682

Asami S, Hirano T, Yamaguchi R et al (2000) Increase in 8-hydroxyguanine and its repair activity in the esophagi of rats given long-term ethanol and nutrition-deficient diet. Jpn J Cancer Res 91:973–978

Atmaca N, Atmaca HT, Kanici A et al (2014) Protective effect of resveratrol on sodium fluoride-induced oxidative stress, hepatotoxicity and neurotoxicity in rats. Food Chem Toxicol 70:191–197

Bagryantseva Y, Novotna B, Rossner P et al (2010) Oxidative damage to biological macromolecules in Prague bus drivers and garagemen: impact of air pollution and genetic polymorphisms. Toxicol Lett 199:60–68

Baik SC, Youn HS, Chung MH et al (1996) Increased oxidative DNA damage in Helicobacter pylori-infected human gastric mucosa. Cancer Res 56:1279–1282

Barbagallo M, Marotta F, Dominguez LJ (2015) Oxidative stress in patients with Alzheimer's disease: effect of extracts of fermented papaya powder. Mediators Inflamm

Barregard L, Moller P, Henriksen T et al (2013) Human and methodological sources of variability in the measurement of urinary 8-oxo-7,8-dihydro-2′-deoxyguanosine. Antioxid Redox Sign 18:2377–2391

Bergman V, Leanderson P, Starkhammar H et al (2004) Urinary excretion of 8-hydroxydeoxyguanosine and malondialdehyde after high dose radiochemotherapy preceding stem cell transplantation. Free Radic Biol Med 36:300–306

Bloomer RJ, Goldfarb AH, McKenzie MJ (2006) Oxidative stress response to aerobic exercise: comparison of antioxidant supplements. Med Sci Sports Exerc 38:1098–1105

Borrego S, Vazquez A, Dasi F et al (2013) Oxidative stress and DNA damage in human gastric carcinoma: 8-oxo-7′ 8-dihydro-2′-deoxyguanosine (8-oxo-dG) as a possible tumor marker. Int J Mol Sci 14:3467–3486

Broedbaek K, Poulsen HE, Weimann A et al (2009) Urinary excretion of biomarkers of oxidatively damaged DNA and RNA in hereditary hemochromatosis. Free Radic Biol Med 47:1230–1233

Broedbaek K, Siersma V, Henriksen T et al (2013) Association between urinary markers of nucleic acid oxidation and mortality in type 2 diabetes a population-based cohort study. Diabetes Care 36:669–676

Brown RK, McBurney A, Lunec J et al (1995) Oxidative damage to DNA in patients with cystic-fibrosis. Free Radic Biol Med 18:801–806

Cabrera M, Nghiem Y, Miller JH (1988) mutM, a second mutator locus in *Escherichia coli* that generates G.C—T.A transversions. J Bacteriol 170:5405–5407

Cangemi R, Angelico F, Loffredo L et al (2007) Oxidative stress-mediated arterial dysfunction in patients with metabolic syndrome: effect of ascorbic acid. Free Radic Biol Med 43:853–859

Chang FK, Mao IF, Chen ML et al (2011) Urinary 8-hydroxydeoxyguanosine as a biomarker of oxidative DNA damage in workers exposed to ethylbenzene. Ann Occup Hyg 55:519–525

Chao M-R, Wang C-J, Wu M-T et al (2008) Repeated measurements of urinary methylated/oxidative DNA lesions, acute toxicity, and mutagenicity in coke oven workers. Cancer Epidemiol Biomarkers Prev 17:3381–3389

Chen LW, Stacewicz-Sapuntzakis M, Duncan C et al (2001) Oxidative DNA damage in prostate cancer patients consuming tomato sauce-based entrees as a whole-food intervention. J Natl Cancer Inst 93:1872–1879

Chiang HC, Huang YK, Chen PF et al (2012) 4-(Methylnitrosamino)-1-(3-pyridyl)-1-butanone is correlated with 8-hydroxy-2′-deoxyguanosine in humans after exposure to environmental tobacco smoke. Sci Total Environ 414:134–139

Chiou CC, Chang PY, Chan EC et al (2003) Urinary 8-hydroxydeoxyguano sine and its analogs as DNA marker of oxidative stress: development of an ELISA and measurement in both bladder and prostate cancers. Clin Chim Acta 334:87–94

Cho S-H, Jung BH, Lee SH et al (2006) Direct determination of nucleosides in the urine of patients with breast cancer using column-switching liquid chromatography-tandem mass spectrometry. Biomed Chromatogr 20:1229–1236

Chuang H-C, Juan H-T, Chang C-N et al (2014) Cardiopulmonary toxicity of pulmonary exposure to occupationally relevant zinc oxide nanoparticles. Nanotoxicology 8:593–604

Ciftci G, Aksoy A, Cenesiz S et al (2015) Therapeutic role of curcumin in oxidative DNA damage caused by formaldehyde. Microsc Res Tech 78:391–395

Cocate PG, Natali AJ, de Oliveira A et al (2014) Fruit and vegetable intake and related nutrients are associated with oxidative stress markers in middle-aged men. Nutrition 30:660–665

Cocate PG, Natali AJ, Alfenas RCG et al (2015) Carotenoid consumption is related to lower lipid oxidation and DNA damage in middle-aged men. Br J Nutr 114:257–264

Cooke MS, Loft S, Olinski R et al (2010) Recommendations for standardized description of and nomenclature concerning oxidatively damaged nucleobases in DNA. Chem Res Toxicol 23:705–707

Culp SJ, Cho BP, Kadlubar FF et al (1989) Structural and conformational analyses of 8-hydroxy-2′-deoxyguanosine. Chem Res Toxicol 2:416–422

D'Odorico A, Bortolan S, Cardin R et al (2001) Reduced plasma antioxidant concentrations and increased oxidative DNA damage in inflammatory bowel disease. Scand J Gastroenterol 36:1289–1294

Devaraj S, Mathur S, Basu A et al (2008) A dose-response study on the effects of purified lycopene supplementation on biomarkers of oxidative stress. J Am Coll Nutr 27:267–273

Dong QY, Cui Y, Chen L et al (2008) Urinary 8-hydroxydeoxyguanosine levels in diabetic retinopathy patients. Eur J Ophthalmol 18:94–98

Ekuni D, Tomofuji T, Sanbe T et al (2009) Vitamin C intake attenuates the degree of experimental atherosclerosis induced by periodontitis in the rat by decreasing oxidative stress. Arch Oral Biol 54:495–502

Engstrom KS, Vahter M, Johansson G et al (2010) Chronic exposure to cadmium and arsenic strongly influences concentrations of 8-oxo-7,8-dihydro-2′-deoxyguanosine in urine. Free Radic Biol Med 48:1211–1217

Espinosa O, Jimenez-Almazan J, Chaves FJ et al (2007) Urinary 8-oxo-7,8-dihydro-2-′-deoxyguanosine (8-oxo-dG), a reliable oxidative stress marker in hypertension. Free Radic Res 41:546–554

Farinati F, Cardin R, Bortolami M et al (2008) Oxidative DNA damage in gastric cancer: cagA status and OGG1 gene polymorphism. Int J Cancer 123:51–55

Faux SP, Francis JE, Smith AG et al (1992) Induction of 8-hydroxydeoxyguanosine in Ah-responsive mouse-liver by iron and Aroclor 1254. Carcinogenesis 13:247–250

Fiala ES, Conaway CC, Mathis JE (1989) Oxidative DNA and RNA damage in the livers of Sprague-Dawley rats treated with the hepatocarcinogen 2-nitropropane. Cancer Res 49:5518–5522

Floyd RA, Watson JJ, Wong PK et al (1986) Hydroxyl free radical adduct of deoxyguanosine: sensitive detection and mechanisms of formation. Free Radic Res Commun 1:163–172

Forlenza MJ, Miller GE (2006) Increased serum levels of 8-hydroxy-2′-deoxyguanosine in clinical depression. Psychosom Med 68:1–7

Fotouhi A, Skiold S, Shakeri-Manesh S et al (2011) Reduction of 8-oxodGTP in the nucleotide pool by hMTH1 leads to reduction in mutations in the human lymphoblastoid cell line TK6 exposed to UVA. Mutat Res 715:13–18

Fujikawa K, Kamiya H, Yakushiji H et al (2001) Human MTH1 protein hydrolyzes the oxidized ribonucleotide, 2-hydroxy-ATP. Nucleic Acids Res 29:449–454

Gackowski D, Banaszkiewicz Z, Rozalski R et al (2002) Persistent oxidative stress in colorectal carcinoma patients. Int J Cancer 101:395–397

Garner RC, Wright CM (1975) Binding of -14C aflatoxin B1 to cellular macromolecules in the rat and hamster. Chem Biol Interact 11:121–131

Gaughan DM, Siegel PD, Hughes MD et al (2014) Arterial stiffness, oxidative stress, and smoke exposure in wildland firefighters. Am J Ind Med 57:748–756

Gedik CM, Boyle SP, Wood SG et al (2002) Oxidative stress in humans: validation of biomarkers of DNA damage. Carcinogenesis 23:1441–1446

Goethel G, Brucker N, Moro AM et al (2014) Evaluation of genotoxicity in workers exposed to benzene and atmospheric pollutants. Mutat Res 770:61–65

Goldfarb AH, McKenzie MJ, Bloomer RJ (2007) Gender comparisons of exercise-induced oxidative stress: influence of antioxidant supplementation. Appl Physiol Nutr Metab 32:1124–1131

Grygoryev D, Moskalenko O, Hinton TG et al (2013) DNA damage caused by chronic transgenerational exposure to low dose gamma radiation in medaka fish (*Oryzias latipes*). Radiat Res 180:235–246

Guo H, Huang K, Zhang X et al (2014) Women are more susceptible than men to oxidative stress and chromosome damage caused by polycyclic aromatic hydrocarbons exposure. Environ Mol Mutagen 55:472–481

Haegele AD, Wolfe P, Thompson HJ (1998) X-radiation induces 8-hydroxy-2′-deoxyguanosine formation in vivo in rat mammary gland DNA. Carcinogenesis 19:1319–1321

Haghdoost S, Sjolander L, Czene S et al (2006) The nucleotide pool is a significant target for oxidative stress. Free Radic Biol Med 41:620–626
Hamurcu Z, Saritas N, Baskol G et al (2010) Effect of wrestling exercise on oxidative DNA damage, nitric oxide level and paraoxonase activity in adolescent boys. Pediatr Exerc Sci 22:60–68
Han Y-Y, Donovan M, Sung F-C (2010) Increased urinary 8-hydroxy-2′-deoxyguanosine excretion in long-distance bus drivers in Taiwan. Chemosphere 79:942–948
Harri M, Svoboda P, Mori T et al (2005) Analysis of 8-hydroxydeoxyguanosine among workers exposed to diesel particulate exhaust: comparison with urinary metabolites and PAH air monitoring. Free Radic Res 39:963–972
Hinhumpatch P, Navasumrit P, Chaisatra K et al (2013) Oxidative DNA damage and repair in children exposed to low levels of arsenic in utero and during early childhood: application of salivary and urinary biomarkers. Toxicol Appl Pharmacol 273:569–579
Hinokio Y, Suzuki S, Hirai M et al (1999) Oxidative DNA damage in diabetes mellitus: its association with diabetic complications. Diabetologia 42:995–998
Hirano T, Yamaguchi Y, Kasai H (1997) Inhibition of 8-hydroxyguanine repair in testes after administration of cadmium chloride to GSH-depleted rats. Toxicol Appl Pharmacol 147:9–14
Hirano T, Higashi K, Sakai A et al (2000) Analyses of oxidative DNA damage and its repair activity in the livers of 3′-methyl-4-dimethylaminoazobenzene-treated rodents. Jpn J Cancer Res 91:681–685
Hofer T, Seo AY, Prudencio M et al (2006) A method to determine RNA and DNA oxidation simultaneously by HPLC-ECD: greater RNA than DNA oxidation in rat liver after doxorubicin administration. Biol Chem 387:103–111
Hong YC, Oh SY, Kwon SO et al (2013) Blood lead level modifies the association between dietary antioxidants and oxidative stress in an urban adult population. Br J Nutr 109:148–154
Hori A, Kasai H, Kawai K et al (2014) Coffee intake is associated with lower levels of oxidative DNA damage and decreasing body iron storage in healthy women. Nutr Cancer 66:964–969
Hossain MB, Barregard L, Sallsten G et al (2014) Cadmium, mercury, and lead in kidney cortex are not associated with urinary 8-oxo-7,8-dihydro-2′-deoxyguanosine (8-oxodG) in living kidney donors. Int Arch Occup Environ Health 87:315–322
Hsu W-Y, Chen WT-L, Lin W-D et al (2009) Analysis of urinary nucleosides as potential tumor markers in human colorectal cancer by high performance liquid chromatography/electrospray ionization tandem mass spectrometry. Clin Chim Acta 402:31–37
Hu CW, Chao MR, Sie CH (2010a) Urinary analysis of 8-oxo-7,8-dihydroguanine and 8-oxo-7,8-dihydro-2′-deoxyguanosine by isotope-dilution LC-MS/MS with automated solid-phase extraction: study of 8-oxo-7,8-dihydroguanine stability. Free Radic Biol Med 48:89–97
Hu CW, Huang YJ, Li YJ et al (2010b) Correlation between concentrations of 8-oxo-7,8-dihydro-2′-deoxyguanosine in urine, plasma and saliva measured by on-line solid-phase extraction LC-MS/MS. Clin Chim Acta 411:1218–1222
Huang HE, Helzlsouer KJ, Appel LJ (2000) The effects of vitamin C and vitamin E on oxidative DNA damage: results from a randomized controlled trial. Cancer Epidem Biomar 9:647–652
Huang YW, Jian L, Zhang MB et al (2012) An investigation of oxidative DNA damage in pharmacy technicians exposed to antineoplastic drugs in two Chinese hospitals using the urinary 8-OHdG assay. Biomed Environ Sci 25:109–116
Ikehata H, Kawai K, Komura J et al (2008) UVA1 genotoxicity is mediated not by oxidative damage but by cyclobutane pyrimidine dimers in normal mouse skin. J Invest Dermatol 128:2289–2296
Inano H, Onoda M (2002) Radioprotective action of curcumin extracted from Curcuma longa LINN: Inhibitory effect on formation of urinary 8-hydroxy-2′-deoxyguanosine, tumorigenesis, but not mortality, induced by gamma-ray irradiation. Int J Radiat Oncol 53:735–743
Inoue M, Osaki T, Noguchi M et al (1998) Lung cancer patients have increased 8-hydroxydeoxyguanosine levels in peripheral lung tissue DNA. Jpn J Cancer Res 89:691–695

Inoue M, Sobue T, Tsugane S et al (2004) Impact of body mass index on the risk of total cancer incidence and mortality among middle-aged Japanese: data from a large-scale population-based cohort study – The JPHC Study. Cancer Causes Control 15:671–680

Inoue A, Kawakami N, Ishizaki M et al (2009) Three job stress models/concepts and oxidative DNA damage in a sample of workers in Japan. J Psychosom Res 66:329–334

Irie M, Tamae K, Iwamoto-Tanaka N et al (2005) Occupational and lifestyle factors and urinary 8-hydroxydeoxyguanosine. Cancer Sci 96:600–606

Isobe C, Abe T, Terayama Y (2009) Homocysteine may contribute to pathogenesis of RNA damage in brains with Alzheimer's disease. Neurodegener Dis 6:252–257

Ito K, Watanabe C, Nakamura A et al (2015) Reduced coenzyme Q10 decreases urinary 8-oxo-7,8-dihydro-2′-deoxyguanosine concentrations in healthy young female subjects. J Med Food 18:835–840

Joergensen A, Broedbaek K, Weimann A et al (2011) Association between urinary excretion of cortisol and markers of oxidatively damaged DNA and RNA in humans. PLoS One 6, e20795

Jorgensen A, Broedbaek K, Fink-Jensen A et al (2013) Increased systemic oxidatively generated DNA and RNA damage in schizophrenia. Psychiatry Res 209:417–423

Kabat GC, Wynder EL (1992) Body-mass index and lung-cancer risk. Am J Epidemiol 135:769–774

Kaczmarek P, Jezowska-Bojczuk M, Bal W et al (2005) Determination of the stability constants and oxidation susceptibility of nickel(II) complexes with 2′-de oxyguano sine 5′-triphosphate and L-histidine. J Inorg Biochem 99:737–746

Kamp DW, Weitzman SA (1999) The molecular basis of asbestos induced lung injury. Thorax 54:638–652

Kaneko T, Tahara S, Tanno M et al (2003) Effect of age on the induction of 8-oxo-2′ deoxyguanosine-releasing enzyme in rat liver by gamma-ray irradiation. Arch Gerontol Geriatr 36:23–35

Kara Y, Doguc DK, Kulac E et al (2014) Acetylsalicylic acid and ascorbic acid combination improves cognition; Via antioxidant effect or increased expression of NMDARs and nAChRs? Environ Toxicol Pharmacol 37:916–927

Kasai H (1997) Analysis of a form of oxidative DNA damage, 8-hydroxy-2′-deoxyguanosine, as a marker of cellular oxidative stress during carcinogenesis. Mutat Res 387:147–163

Kasai H (2003) A new automated method to analyze urinary 8-hydroxydeoxyguanosine by a high-performance liquid chromatography-electrochemical detector system. J Radiat Res (Tokyo) 44:185–189

Kasai H, Nishimura S (1983) Hydroxylation of the C-8 position of deoxyguanosine by reducing agents in the presence of oxygen. Nucleic Acids Symp Ser 12:165–167

Kasai H, Nishimura S (1984a) DNA damage induced by asbestos in the presence of hydrogen peroxide. Gann 75:841–844

Kasai H, Nishimura S (1984b) Hydroxylation of deoxyguanosine at the C-8 position by polyphenols and aminophenols in the presence of hydrogen peroxide and ferric ion. Gann 75:565–566

Kasai H, Nishimura S (1984c) Hydroxylation of deoxyguanosine at the C-8 position by ascorbic acid and other reducing agents. Nucleic Acids Res 12:2137–2145

Kasai H, Nishimura S (1991) Formation of 8-hydroxydeoxyguanosine in DNA by oxygen radicals and its biological significance. In: Sies H (ed) Oxidative stress: oxidants and antioxidants. Academic, New York, pp 99–116

Kasai H, Hayami H, Yamaizumi Z et al (1984a) Detection and identification of mutagens and carcinogens as their adducts with guanosine derivatives. Nucleic Acids Res 12:2127–2136

Kasai H, Tanooka H, Nishimura S (1984b) Formation of 8-hydroxyguanine residues in DNA by X-irradiation. Gann 75:1037–1039

Kasai H, Nishimura S, Toriumi Y et al (1987) The crystal structure of 9-ethyl-8-hydroxyguanine. Bull Chem Soc Jpn 60:3799–3800

Kasai H, Nakayama M, Toda N et al (1989) Methylreductic acid and hydroxymethylreductic acid: oxygen radical-forming agents in heated starch. Mutat Res 214:159–164

Kasai H, Yamaizumi Z, Berger M et al (1992) Photosensitized formation of 7,8-dihydro-8-oxo-2′-deoxyguanosine (8-hydroxy-2′-deoxyguanosine) in DNA by riboflavin – a nonsinglet oxygen mediated reaction. J Am Chem Soc 114:9692–9694

Kasai H, Hirano T, Kawai K et al (2007) Analysis of 8-hydroxy-2′-deoxyguanosine as a marker of oxidatively damaged DNA in relation to carcinogenesis and aging. In: Cooke MS, Evans MD (eds) Oxidative damage to nucleic acids. Landes Bioscience Springer Science + Business Media, New York, pp 178–187

Kasai H, Kawai K, Li Y-S (2008) Analysis of 8-OH-dG and 8-OH-Gua as biomarkers of oxidative stress. Genes Environ 30:33–40

Kaspar KL, Park JS, Brown CR et al (2011) Pigmented potato consumption alters oxidative stress and inflammatory damage in men. J Nutr 141:108–111

Kato J, Kobune M, Nakamura T et al (2001) Normalization of elevated hepatic 8-hydroxy-2′-deoxyguanosine levels in chronic hepatitis C patients by phlebotomy and low iron diet. Cancer Res 61:8697–8702

Kato M, Iida M, Goto Y et al (2011) Sunlight exposure-mediated DNA damage in young adults. Cancer Epidem Biomar 20:1622–1628

Kawai K, Svoboda P, Kasai H (2006) Detection of genotoxic nucleosides, 8-hydroxydeoxyguanosine, 8-hydroxyguanosine and free base 8-hydroxyguanine, in fish food products. Genes Environ 28:120–122

Kawai K, Li Y-S, Kasai H (2007) Accurate measurement of 8-OH-dG and 8-OH-Gua in mouse DNA, urine and serum: effects of X-ray irradiation. Genes Environ 29:107–114

Ke Y, Cheng J, Zhang Z et al (2009) Increased levels of oxidative DNA damage attributable to cooking-oil fumes exposure among cooks. Inhal Toxicol 21:682–687

Khatri M, Bello D, Gaines P et al (2013) Nanoparticles from photocopiers induce oxidative stress and upper respiratory tract inflammation in healthy volunteers. Nanotoxicology 7:1014–1027

Kim JY, Mukherjee S, Ngo L et al (2004) Urinary 8-hydroxy-2′-deoxyguanosine as a biomarker of oxidative DNA damage in workers exposed to fine particulates. Environ Health Perspect 112:666–671

Kisby GE, Muniz JF, Scherer J et al (2009) Oxidative stress and DNA damage in agricultural workers. J Agromed 14:206–214

Kitamura H, Terunuma N, Kurosaki S et al (2009) Cross-sectional study on respiratory effect of toner-exposed work in manufacturing plants, Japan: pulmonary function, blood cells, and biochemical markers. Hum Exp Toxicol 28:331–338

Kohda K, Tada M, Kasai H et al (1986) Formation of 8-hydroxyguanine residues in cellular DNA exposed to the carcinogen 4-nitroquinoline 1-oxide. Biochem Biophys Res Commun 139:626–632

Kuo HW, Chang SF, Wu KY et al (2003) Chromium(VI) induced oxidative damage to DNA: increase of urinary 8-hydroxydeoxyguanosine concentrations (8-OHdG) among electroplating workers. Occup Environ Med 60:590–594

Lagorio S, Tagesson C, Forastiere F et al (1994) Exposure to benzene and urinary concentrations of 8-hydroxydeoxyguanosine, a biological marker of oxidative damage to DNA. Occup Environ Med 51:739–743

Land CE (1980) Estimating cancer risks from low doses of ionizing radiation. Science 209:1197–1203

Lee CYJ, Wan JMF (2000) Vitamin E supplementation improves cell-mediated immunity and oxidative stress of Asian men and women. J Nutr 130:2932–2937

Lee YS, Choi JY, Park MK et al (1996) Induction of oh(8)Gua glycosylase in rat kidneys by potassium bromate ($KBrO_3$), a renal oxidative carcinogen. Mutat Res/DNA Rep 364:227–233

Lee BM, Lee SK, Kim HS (1998) Inhibition of oxidative DNA damage, 8-OHdG, and carbonyl contents in smokers treated with antioxidants (vitamin E, vitamin C, beta-carotene and red ginseng). Cancer Lett 132:219–227

Lee M-W, Chen M-L, Lung S-CC et al (2010) Exposure assessment of PM2.5 and urinary 8-OHdG for diesel exhaust emission inspector. Sci Total Environ 408:505–510

Lee M-W, Chen M-L, Lung S-CC et al (2012) Increase of urinary concentrations of 8-hydroxy-2′-deoxyguanosine in diesel exhaust emission inspector exposed to polycyclic aromatic hydrocarbons. Int Arch Occup Environ Health 85:273–282

Li N, Jia X, Chen CYO et al (2007) Almond consumption reduces oxidative DNA damage and lipid peroxidation in male smokers. J Nutr 137:2717–2722

Li Y-S, Song M-F, Kasai H et al (2013a) 8-Hydroxyguanine in urine and serum as an oxidative stress marker: effects of diabetes and aging. J UOEH 35:119–127

Li Y-S, Song M-F, Kasai H et al (2013b) Generation and threshold level of 8-OHdG as oxidative DNA damage elicited by low dose ionizing radiation. Genes Environ 35:88–92

Li P, Gu Y, Yu S et al (2014) Assessing the suitability of 8-OHdG and micronuclei as genotoxic biomarkers in chromate-exposed workers: a cross-sectional study. BMJ Open 4, e005979

Li J, Lu S, Liu G et al (2015) Co-exposure to polycyclic aromatic hydrocarbons, benzene and toluene and their dose-effects on oxidative stress damage in kindergarten-aged children in Guangzhou, China. Sci Total Environ 524:74–80

Lin TS, Wu CC, Wu JD et al (2012) Oxidative DNA damage estimated by urinary 8-hydroxy-2′-deoxyguanosine and arsenic in glass production workers. Toxicol Ind Health 28:513–521

Liu L, Zhang Q, Feng J et al (1996) The study of DNA oxidative damage in benzene-exposed workers. Mutat Res 370:145–150

Liu H-H, Shih T-S, Chen IJ et al (2008) Lipid peroxidation and oxidative status compared in workers at a bottom ash recovery plant and fly ash treatment plants. J Occup Health 50:492–497

Liu HH, Lin MH, Liu PC et al (2009) Health risk assessment by measuring plasma malondialdehyde (MDA), urinary 8-hydroxydeoxyguanosine (8-OH-dG) and DNA strand breakage following metal exposure in foundry workers. J Hazard Mater 170:699–704

Lloret A, Calzone R, Dunster C et al (2008) Different patterns of in vivo pro-oxidant states in a set of cancer- or aging-related genetic diseases. Free Radic Biol Med 44:495–503

Lodovici M, Casalini C, Cariaggi R et al (2000) Levels of 8-hydroxydeoxyguanosine as a marker of DNA damage in human leukocytes. Free Radic Biol Med 28:13–17

Lodovici M, Caldini S, Luceri C et al (2005) Active and passive smoking and lifestyle determinants of 8-oxo-7,8-dihydro-2′-deoxyguanosine levels in human leukocyte DNA. Cancer Epidem Biomar 14:2975–2977

Lodovici M, Caldini S, Morbidelli L et al (2009) Protective effect of 4-coumaric acid from UVB ray damage in the rabbit eye. Toxicology 255:1–5

Loft S, Vistisen K, Ewertz M et al (1992) Oxidative DNA damage estimated by 8-hydroxydeoxyguanosine excretion in humans – influence of smoking, gender and body-mass index. Carcinogenesis 13:2241–2247

Loft S, Poulsen HE, Vistisen K et al (1999) Increased urinary excretion of 8-oxo-2-′-deoxyguanosine, a biomarker of oxidative DNA damage, in urban bus drivers. Mutat Res 441:11–19

Loft S, Svoboda P, Kasai H et al (2006) Prospective study of 8-oxo-7,8-dihydro-2-′-deoxyguanosine excretion and the risk of lung cancer. Carcinogenesis 27:1245–1250

Loft S, Svoboda P, Kawai K et al (2012) Association between 8-oxo-7,8-dihydroguanine excretion and risk of lung cancer in a prospective study. Free Radic Biol Med 52:167–172

Loft S, Olsen A, Moller P et al (2013) Association between 8-oxo-7,8-dihydro-2′-deoxyguanosine excretion and risk of postmenopausal breast cancer: nested case-control study. Cancer Epidem Biomar 22:1289–1296

Lu RZ, Nash HM, Verdine GL (1997) A mammalian DNA repair enzyme that excises oxidatively damaged guanines maps to a locus frequently lost in lung cancer. Curr Biol 7:397–407

Lu C-Y, Ma Y-C, Chen P-C et al (2014) Oxidative stress of office workers relevant to tobacco smoking and inner air quality. Int J Environ Res Public Health 11:5586–5597

Luo HT, Tang LL, Tang M et al (2006) Phase IIa chemoprevention trial of green tea polyphenols in high-risk individuals of liver cancer: modulation of urinary excretion of green tea polyphenols and 8-hydroxydeoxyguanosine. Carcinogenesis 27:262–268

Ma CM, Lin LY, Chen HW et al (2010) Volatile organic compounds exposure and cardiovascular effects in hair salons. Occup Med (Lond) 60:624–630

Ma JQ, Ding J, Xiao ZH et al (2014) Ursolic acid ameliorates carbon tetrachloride-induced oxidative DNA damage and inflammation in mouse kidney by inhibiting the STAT3 and NF-kappa B activities. Int Immunopharmacol 21:389–395

Machowetz A, Poulsen HE, Gruendel S et al (2007) Effect of olive oils on biomarkers of oxidative DNA stress in northern and southern Europeans. FASEB J 21:45–52

Maeng SH, Chung HW, Yu IJ et al (2003) Changes of 8-OH-dG levels in DNA and its base excision repair activity in rat lungs after inhalation exposure to hexavalent chromium. Mutat Res 539:109–116

Maki H, Sekiguchi M (1992) MutT protein specifically hydrolyzes a potent mutagenic substrate for DNA-synthesis. Nature 355:273–275

Malayappan B, Garrett TJ, Segal M et al (2007) Urinary analysis of 8-oxoguanine, 8-oxoguanosine, fapy-guanine and 8-oxo-2′-deoxyguano sine by high-performance liquid chromatography-electro spray tandem mass spectrometry as a measure of oxidative stress. J Chromatogr A 1167:54–62

Manda K, Ueno M, Anzai K (2007) AFMK, a melatonin metabolite, attenuates X-ray-induced oxidative damage to DNA, proteins and lipids in mice. J Pineal Res 42:386–393

Manda K, Ueno M, Anzai K (2008) Melatonin mitigates oxidative damage and apoptosis in mouse cerebellum induced by high-LET Fe-56 particle irradiation. J Pineal Res 44:189–196

Manini P, De Palma G, Andreoli R et al (2009) Biomarkers of nucleic acid oxidation, polymorphism in, and expression of, hOGG1 gene in styrene-exposed workers. Toxicol Lett 190:41–47

Manini P, De Palma G, Andreoli R et al (2010) Occupational exposure to low levels of benzene: Biomarkers of exposure and nucleic acid oxidation and their modulation by polymorphic xenobiotic metabolizing enzymes. Toxicol Lett 193:229–235

Marczynski B, Rozynek P, Elliehausen HJ et al (1997) Detection of 8-hydroxydeoxyguanosine, a marker of oxidative DNA damage, in white blood cells of workers occupationally exposed to styrene. Arch Toxicol 71:496–500

Marczynski B, Kraus T, Rozynek P et al (2000) Association between 8-hydroxy-2′-deoxyguanosine levels in DNA of workers highly exposed to asbestos and their clinical data, occupational and non-occupational confounding factors, and cancer. Mutat Res 468:203–212

Marczynski B, Pesch B, Wilhelm M et al (2009) Occupational exposure to polycyclic aromatic hydrocarbons and DNA damage by industry: a nationwide study in Germany. Arch Toxicol 83:947–957

Marnett LJ (2002) Oxy radicals, lipid peroxidation and DNA damage. Toxicology 181:219–222

Marotta F, Yoshida C, Barreto R et al (2007) Oxidative-inflammatory damage in cirrhosis: Effect of vitamin E and a fermented papaya preparation. J Gastroenterol Hepatol 22:697–703

Matos HR, Marques SA, Gomes OF et al (2006) Lycopene and beta-carotene protect in vivo iron-induced oxidative stress damage in rat prostate. Braz J Med Biol Res 39:203–210

Matsumoto Y, Ogawa Y, Yoshida R et al (2008) The stability of the oxidative stress marker, urinary 8-hydroxy-2′-deoxyguanosine (8-OHdG), when stored at room temperature. J Occup Health 50:366–372

Mecocci P, Polidori MC, Cherubini A et al (2002) Lymphocyte oxidative DNA damage and plasma antioxidants in Alzheimer disease. Arch Neurol 59:794–798

Mehrdad R, Aghdaei S, Pouryaghoub G (2015) Urinary 8-hydroxy-deoxyguanosine as a biomarker of oxidative DNA damage in employees of subway system. Acta Med Iran 53:287–292

Mena P, Maynar M, Gutierrez JM et al (1991) Erythrocyte free-radical scavenger enzymes in bicycle professional racers – adaptation to training. Int J Sports Med 12:563–566

Michaels ML, Cruz C, Grollman AP et al (1992) Evidence that mutY and mutM combine to prevent mutations by an oxidatively damaged form of guanine in DNA. Proc Natl Acad Sci U S A 89:7022–7025

Miyata M, Kasai H, Kawai K et al (2008) Changes of urinary 8-hydroxydeoxyguanosine levels during a two-day ultramarathon race period in Japanese non-professional runners. Int J Sports Med 29:27–33

Mizoue T, Kasai H, Kubo T et al (2006) Leanness, smoking, and enhanced oxidative DNA damage. Cancer Epidemiol Biomarkers Prev 15:582–585

Mizoue T, Tokunaga S, Kasai H et al (2007) Body mass index and oxidative DNA damage: a longitudinal study. Cancer Sci 98:1254–1258

Munkholm K, Poulsen HE, Kessing LV et al (2015) Elevated levels of urinary markers of oxidatively generated DNA and RNA damage in bipolar disorder. Bipolar Disord 17:257–268

Muzembo BA, Narongpon D, Ngatu NR et al (2012) Assessment of lifestyle effect on oxidative stress biomarkers in free-living elderly in rural Japan. Geriatr Gerontol Int 12:547–554

Nachvak SM, Neyestani TR, Mahboob SA et al (2014) alpha-Tocopherol supplementation reduces biomarkers of oxidative stress in children with Down syndrome: a randomized controlled trial. Eur J Clin Nutr 68:1119–1123

Nagayoshi Y, Kawano H, Hokamaki J et al (2005) Urinary 8-hydroxy-2′-deoxyguanosine levels increase after reperfusion in acute myocardial infarction and may predict subsequent cardiac events. Am J Cardiol 95:514–517

Navasumrit P, Arayasiri M, Hiang OMT et al (2008) Potential health effects of exposure to carcinogenic compounds in incense smoke in temple workers. Chem Biol Interact 173:19–31

Negishi T, Kawai K, Arakawal R et al (2007) Increased levels of 8-hydroxy-2′-deoxyguanosine in drosophila larval DNA after irradiation with 364-nm laser light but not with X-rays. Photochem Photobiol 83:658–663

Neophytou AM, Hart JE, Cavallari JM et al (2013) Traffic-related exposures and biomarkers of systemic inflammation, endothelial activation and oxidative stress: a panel study in the US trucking industry. Environ Health 12:105

Nguyen TT, Kawanami S, Kawai K et al (2014) Urinary 1-hydroxypyrene and 8-hydroxydeoxyguanosine levels among coke-oven workers for 2 consecutive days. J Occup Health 56:178–185

Nilsson RI, Nordlinder RG, Tagesson C et al (1996) Genotoxic effects in workers exposed to low levels of benzene from gasoline. Am J Ind Med 30:317–324

Noguchi T, Ikeda K, Sasaki Y et al (2001) Effects of vitamin E and sesamin on hypertension and cerebral thrombogenesis in stroke-prone spontaneously hypertensive rats. Hypertens Res 24:735–742

Nunez ME, Hall DB, Barton JK (1999) Long-range oxidative damage to DNA: effects of distance and sequence. Chem Biol 6:85–97

Okada K, Wangpoengtrakul C, Tanaka T et al (2001) Curcumin and especially tetrahydrocurcumin ameliorate oxidative stress-induced renal injury in mice. J Nutr 131:2090–2095

Ozyurt H, Cevik O, Ozgen Z et al (2014) Quercetin protects radiation-induced DNA damage and apoptosis in kidney and bladder tissues of rats. Free Radic Res 48:1247–1255

Pan C-H, Chan C-C, Wu K-Y (2008) Effects on Chinese restaurant workers of exposure to cooking oil fumes: a cautionary note on urinary 8-hydroxy-2′-deoxyguanosine. Cancer Epidemiol Biomarkers Prev 17:3351–3357

Pedret A, Valls RM, Fernandez-Castillejo S et al (2012) Polyphenol-rich foods exhibit DNA antioxidative properties and protect the glutathione system in healthy subjects. Mol Nutr Food Res 56:1025–1033

Pelclova D, Navratil T, Fenclova Z et al (2011) Increased oxidative/nitrosative stress markers measured non- invasively in patients with high 2,3,7,8-tetrachloro-dibenzo-p-dioxin plasma level. Neuroendocrinol Lett 32:71–76

Pinlaor S, Ma N, Hiraku Y et al (2004) Repeated infection with Opisthorchis viverrini induces accumulation of 8-nitroguanine and 8-oxo-7,8-dihydro-2′-deoxyguanine in the bile duct of hamsters via inducible nitric oxide synthase. Carcinogenesis 25:1535–1542

Poulsen HE, Specht E, Broedbaek K et al (2012) RNA modifications by oxidation: a novel disease mechanism? Free Radic Biol Med 52:1353–1361

Qian LR, Cao F, Cui JG et al (2010) Radioprotective effect of hydrogen in cultured cells and mice. Free Radic Res 44:275–282

Ramos AA, Pereira-Wilson C, Collins AR (2010) Protective effects of Ursolic acid and Luteolin against oxidative DNA damage include enhancement of DNA repair in Caco-2 cells. Mutat Res 692:6–11

Raza Y, Khan A, Farooqui A et al (2014) Oxidative DNA damage as a potential early biomarker of *Helicobacter pylori* associated carcinogenesis. Pathol Oncol Res 20:839–846

Rithidech KN, Tungjai M, Reungpatthanaphong P et al (2012) Attenuation of oxidative damage and inflammatory responses by apigenin given to mice after irradiation. Mutat Res 749:29–38

Rizkalla BH, Robins RK, Broom AD (1969) Purine nucleosides. XXVII. The synthesis of 1- and 7-methyl-8-oxoguanosine and related nucleosides. The use of the N-amino group as a selective blocking agent in nucleoside synthesis. Biochim Biophys Acta 195:285–293

Rose S, Melnyk S, Pavliv O et al (2012) Evidence of oxidative damage and inflammation associated with low glutathione redox status in the autism brain. Transl Psychiatry 2, e134

Rossner P Jr, Svecova V, Milcova A et al (2008) Seasonal variability of oxidative stress markers in city bus drivers – Part I. Oxidative damage to DNA. Mutat Res 642:14–20

Rossner P, Mistry V, Singh R, Sram RJ, Cooke MS (2013) Urinary 8-oxo-7,8-dihydro-2-′-deoxyguanosine values determined by a modified ELISA improves agreement with HPLC-MS/MS. Biochem Biophys Res Commun 440:725–730

Roszkowski K, Gackowski D, Rozalski R et al (2008) Small field radiotherapy of head and neck cancer patients is responsible for oxidatively damaged DNA/oxidative stress on the level of a whole organism. Int J Cancer 123:1964–1967

Russo MT, Blasi MF, Chiera F et al (2004) The oxidized deoxynucleoside triphosphate pool is a significant contributor to genetic instability in mismatch repair-deficient cells. Mol Cell Biol 24:465–474

Sahin K, Tuzcu M, Sahin N et al (2011) Inhibitory effects of combination of lycopene and genistein on 7,12-dimethyl benz(a)anthracene-induced breast cancer in rats. Nutr Cancer 63:1279–1286

Sakano N, Wang D-H, Takahashi N et al (2009) Oxidative stress biomarkers and lifestyles in Japanese healthy people. J Clin Biochem Nutr 44:185–195

Sato Y, Nanri H, Ohta M et al (2003) Increase of human MTH1 and decrease of 8-hydroxydeoxyguanosme in leukocyte DNA by acute and chronic exercise in healthy male subjects. Biochem Biophys Res Commun 305:333–338

Schneider JE, Price S, Maidt L et al (1990) Methylene-blue plus light mediates 8-hydroxy 2′-deoxyguanosine formation in DNA preferentially over strand breakage. Nucleic Acids Res 18:631–635

Seo AY, Hofer T, Sung B et al (2006) Hepatic oxidative stress during aging: effects of 8% long-term calorie restriction and lifelong exercise. Antioxid Redox Sign 8:529–538

Sezer U, Cicek Y, Canakci CF (2012) Increased salivary levels of 8-hydroxydeoxyguanosine may be a marker for disease activity for periodontitis. Dis Markers 32:165–172

Shah M, Miller DS, Geissler CA (1988) Lower metabolic rates of post-obese versus lean women: thermogenesis, basal metabolic rate and genetics. Eur J Clin Nutr 42:741–752

Shen CL, Wang P, Guerrieri J et al (2008) Protective effect of green tea polyphenols on bone loss in middle-aged female rats. Osteoporos Int 19:979–990

Shertzer HG, Nebert DW, Puga A et al (1998) Dioxin causes a sustained oxidative stress response in the mouse. Biochem Biophys Res Commun 253:44–48

Shibutani S, Takeshita M, Grollman AP (1991) Insertion of specific bases during DNA-synthesis past the oxidation-damaged base 8-oxodG. Nature 349:431–434

Shibuya K, Nishimura N, Suzuki JS et al (2008) Role of metallothionein as a protective factor against radiation carcinogenesis. J Toxicol Sci 33:651–655

Shigenaga MK, Gimeno CJ, Ames BN (1989) Urinary 8-hydroxy-2′-deoxyguanosine as a biological marker of in vivo oxidative DNA damage. Proc Natl Acad Sci U S A 86:9697–701

Shimizu Y, Kato H, Schull WJ et al (1992) Studies of the mortality of A-bomb survivors. 9. Mortality, 1950–1985: Part 3. Noncancer mortality based on the revised doses (DS86). Radiat Res 130:249–266

Shimoi K, Kasai H, Yokota N et al (2002) Comparison between high-performance liquid chromatography and enzyme-linked immunosorbent assay for the determination of 8-hydroxy-2′-deoxyguanosine in human urine. Cancer Epidem Biomar 11:767–770

Shin HS, Yu M, Kim M et al (2014) Renoprotective effect of red ginseng in gentamicin-induced acute kidney injury. Lab Invest 94:1147–1160

Siomek A, Rytarowska A, Szaflarska-Poplawska A et al (2006) Helicobacter pylori infection is associated with oxidatively damaged DNA in human leukocytes and decreased level of urinary 8-oxo-7,8-dihydroguanine. Carcinogenesis 27:405–408

Sirerol JA, Feddi F, Mena S et al (2015) Topical treatment with pterostilbene, a natural phytoalexin, effectively protects hairless mice against UVB radiation-induced skin damage and carcinogenesis. Free Radic Biol Med 85:1–11

Song M-F, Li Y-S, Ootsuyama Y et al (2009) Urea, the most abundant component in urine, cross-reacts with a commercial 8-OH-dG ELISA kit and contributes to overestimation of urinary 8-OH-dG. Free Radic Biol Med 47:41–46

Song X-N, Zhang L-Q, Liu D-G et al (2011) Oxidative damage to RNA and expression patterns of MTH1 in the Hippocampi of senescence-accelerated SAMP8 mice and Alzheimer's disease patients. Neurochem Res 36:1558–1565

Song MF, Li YS, Kasai H et al (2012) Metal nanoparticle-induced micronuclei and oxidative DNA damage in mice. J Clin Biochem Nutr 50:211–216

Sughis M, Nawrot TS, Haufroid V et al (2012) Adverse health effects of child labor: high exposure to chromium and oxidative DNA damage in children manufacturing surgical instruments. Environ Health Perspect 120:1469–1474

Suzuki S, Shishido T, Ishino M et al (2011) 8-Hydroxy-2′-deoxyguanosine is a prognostic mediator for cardiac event. Eur J Clin Invest 41:759–766

Szymanska-Chabowska A, Beck A, Poreba R et al (2009) Evaluation of DNA damage in people occupationally exposed to arsenic and some heavy metals. Pol J Environ Stud 18:1131–1139

Tagesson C, Chabiuk D, Axelson O et al (1993) Increased urinary excretion of the oxidative DNA adduct, 8-hydroxydeoxyguanosine, as a possible early indicator of occupational cancer hazards in the asbestos, rubber, and azo-dye industries. Pol J Occup Med Environ Health 6:357–368

Takahashi K, Pan G, Kasai H et al (1997) Relationship between asbestos exposures and 8-hydroxydeoxyguanosine levels in leukocytic DNA of workers at a Chinese asbestos-material plant. Int J Occup Environ Health 3:111–119

Tamae K, Kawai K, Yamasaki S et al (2009) Effect of age, smoking and other lifestyle factors on urinary 7-methylguanine and 8-hydroxydeoxyguanosine. Cancer Sci 100:715–721

Tamaki N, Orihuela-Campos RC, Inagaki Y et al (2014) Resveratrol improves oxidative stress and prevents the progression of periodontitis via the activation of the Sirtl/AMPK and the Nrf2/antioxidant defense pathways in a rat periodontitis model. Free Radic Biol Med 75:222–229

Tanooka H (2011) Meta-analysis of non-tumour doses for radiation-induced cancer on the basis of dose-rate. Int J Radiat Biol 87:645–652

Tardieu D, Jaeg JP, Deloly A et al (2000) The COX-2 inhibitor nimesulide suppresses superoxide and 8-hydroxy-deoxyguanosine formation, and stimulates apoptosis in mucosa during early colonic inflammation in rats. Carcinogenesis 21:973–976

Tarng DC, Liu TY, Huang TP (2004) Protective effect of vitamin C on 8-hydroxy-2-′-deoxyguanosine level in peripheral blood lymphocytes of chronic hemodialysis patients. Kidney Int 66:820–831

Thanan R, Murata M, Pinlaor S et al (2008) Urinary 8-oxo-7,8-dihydro-2′-deoxyguanosine in patients with parasite infection and effect of antiparasitic drug in relation to cholangiocarcinogenesis. Cancer Epidemiol Biomarkers Prev 17:518–524

Thompson HJ, Heimendinger J, Haegele A et al (1999) Effect of increased vegetable and fruit consumption on markers of oxidative cellular damage. Carcinogenesis 20:2261–2266

Tomofuji T, Ekuni D, Sanbe T et al (2009) Effects of vitamin C intake on gingival oxidative stress in rat periodontitis. Free Radic Biol Med 46:163–168

Topal MD, Baker MS (1982) DNA precursor pool: a significant target for N-methyl-N-nitrosourea in C3H/10T1/2 clone 8 cells. Proc Natl Acad Sci U S A 79:2211–2215

Toraason M, Hayden C, Marlow D et al (2001) DNA strand breaks, oxidative damage, and 1-OH pyrene in roofers with coal-tar pitch dust and/or asphalt fume exposure. Int Arch Occup Environ Health 74:396–404

Traustadottir T, Davies SS, Stock AA et al (2009) Tart cherry juice decreases oxidative stress in healthy older men and women. J Nutr 139:1896–1900

Tsakiris S, Parthimos T, Tsakiris T et al (2006) alpha-Tocopherol supplementation reduces the elevated 8-hydroxy-2-deoxyguanosine blood levels induced by training in basketball players. Clin Chem Lab Med 44:1004–1008

Tsurudome Y, Hirano T, Yamato H et al (1999) Changes in levels of 8-hydroxyguanine in DNA, its repair and OGG1 mRNA in rat lungs after intratracheal administration of diesel exhaust particles. Carcinogenesis 20:1573–1576

Umegaki K, Sugisawa A, Shin SJ et al (2001) Different onsets of oxidative damage to DNA and lipids in bone marrow and liver in rats given total body irradiation. Free Radic Biol Med 31:1066–1074

Wang MY, Hecht SS (1997) A cyclic N-7, C-8 guanine adduct of N-nitrosopyrrolidine (NPYR): formation in nucleic acids and excretion in the urine of NPYR-treated rats. Chem Res Toxicol 10:772–778

Wang Q, Wang L, Chen X et al (2011) Increased urinary 8-hydroxy-2′-deoxyguanosine levels in workers exposed to di-(2-ethylhexyl) phthalate in a waste plastic recycling site in China. Environ Sci Pollut Res Int 18:987–996

Wang Y, Li D, Cheng N et al (2015) Antioxidant and hepatoprotective activity of vitex honey against paracetamol induced liver damage in mice. Food Funct 6:2339–2349

Waris S, Winklhofer-Roob BM, Roob JM et al (2015) Increased DNA dicarbonyl glycation and oxidation markers in patients with type 2 diabetes and link to diabetic nephropathy. J Diabetes Res 2015:915486–915486

Wawrzyniak A, Gornicka M, Hamulka J et al (2013) alpha-Tocopherol, ascorbic acid, and beta-carotene protect against oxidative stress but reveal no direct influence on p53 expression in rats subjected to stress. Nutr Res 33:868–875

Wei Y, Han I-K, Shao M et al (2009) PM2.5 constituents and oxidative DNA damage in humans. Environ Sci Technol 43:4757–4762

Wei Y, Han IK, Hu M et al (2010) Personal exposure to particulate PAHs and anthraquinone and oxidative DNA damages in humans. Chemosphere 81:1280–1285

Wen S, Yang F-X, Gong Y et al (2008) Elevated levels of urinary 8-hydroxy-2′-deoxyguanosine in male electrical and electronic equipment dismantling workers exposed to high concentrations of polychlorinated dibenzo-p-dioxins and dibenzofurans, polybrominated diphenyl ethers, and polychlorinated biphenyls. Environ Sci Technol 42:4202–4207

Wong RH, Kuo CY, Hsu ML et al (2005) Increased levels of 8-hydroxy-2′-deoxyguanosine attributable to carcinogenic metal exposure among schoolchildren. Environ Health Perspect 113:1386–1390

Xiao N-N (2015) Effects of resveratrol supplementation on oxidative damage and lipid peroxidation induced by strenuous exercise in rats. Biomol Ther (Seoul) 23:374–378

Xu GW, Yao QH, Weng QF et al (2004) Study of urinary 8-hydroxydeoxyguanosine as a biomarker of oxidative DNA damage in diabetic nephropathy patients. J Pharm Biomed Anal 36:101–104

Yamaguchi R, Hirano T, Asami S et al (1996) Increase in the 8-hydroxyguanine repair activity in the rat kidney after the administration of a renal carcinogen, ferric nitrilotriacetate. Environ Health Perspect 104:651–653

Yamaguchi R, Hirano T, Ootsuyama Y et al (1999) Increased 8-hydroxyguanine in DNA and its repair activity in hamster and rat lung after intratracheal instillation of crocidolite asbestos. Jpn J Cancer Res 90:505–509

Yoshioka N, Nakashima H, Hosoda K et al (2008) Urinary excretion of an oxidative stress marker, 8-hydroxyguanine (8-OH-Gua), among nickel-cadmium battery workers. J Occup Health 50:229–235

Zhang X-H, Zhang X, Wang X-C et al (2011) Chronic occupational exposure to hexavalent chromium causes DNA damage in electroplating workers. BMC Public Health 11:224

Zhang S, Song X, Zhang W et al (2013) Determination of low urinary 8-hydroxy-2′-deoxyguanosine excretion with capillary electrophoresis and molecularly imprinted monolith solid phase microextraction. Sci Total Environ 450–451:266–270

Methods for Determination of 2′-O-Me in RNA

Ulf Birkedal, Nicolai Krogh, Kasper Langebjerg Andersen, and Henrik Nielsen

Contents

Abstract Ribose methylation is one of the most abundant RNA modifications and is found in all kingdoms of life and all major classes of RNA (rRNA, tRNA, and mRNA). Ribose methylations are introduced by stand-alone enzymes or by generic enzymes guided to the target by small RNA guides. The most abundant mechanism of ribose methylation is found in rRNA of Archaea and Eukarya where a methyltransferase (fibrillarin) use sRNA (Archaea) or box C/D snoRNA (Eukarya) as guide RNAs to specify the site of modification. The general function of these modifications is to promote ribosome biogenesis, in particular folding of the ribosomal RNA. Furthermore, some modifications affect the fidelity of translation.

U. Birkedal • N. Krogh • K.L. Andersen • H. Nielsen (✉)
Department of Cellular and Molecular Medicine, The Panum Institute, University of Copenhagen, 3 Blegdamsvej, 18.2.20, 2200N København, Denmark
e-mail: hamra@sund.ku.dk

S. Jurga et al. (eds.), *Modified Nucleic Acids in Biology and Medicine*, RNA Technologies, DOI 10.1007/978-3-319-34175-0_8

The function of individual modifications has in many cases remained elusive, because genetic deletion of the modification has a weak phenotype or no phenotype at all. Another problem is that methods for mapping modifications and quantitating the fraction of RNA molecules modified in a population until recently remained poorly developed. Here, we review the methods that have been used to study 2′-O-Me in RNA starting with the original approach employing in vivo isotope labeling followed by paper chromatography. The next generation of methods typically addressed one nucleotide at a time and was mostly based on primer extension. Finally, more recent mass spectrometry and high-throughput sequencing methods hold promise to reveal a new biology of this widespread type of nucleotide modification.

Keywords 2′-O-Me • Ribose methylation • Box C/D snoRNA • Fibrillarin • Ribosomal RNA

1 Introduction

More than a hundred different types of nucleotide modifications in RNA are currently known and catalogued in the RNA modification database at the State University of New York at Albany (http://mods.rna.albany.edu) and the Modomics database at University of Warsaw (http://modomics.genesilico.pl). The majority of the modifications are methylations and methylation of the ribose 2′-OH is the most abundant (Watkins and Bohnsack 2012). Ribose methylations are found in all kingdoms of life and have been documented in all major classes of RNAs. In prokaryotes and at some sites in archaea and eukaryotes, the methylation is introduced by a methyltransferase that recognizes a specific motif in the target RNA. However, for the majority of targets in Archaea and eukaryotes, a generic methyltransferase (fibrillarin) is guided to the target by sRNAs (archaea) or box C/D snoRNAs (Eukarya). The guide RNAs contains two sets of conserved sequences: boxes C and C′ (5′-RUGAUGA) and boxes D and D′ (CUGA). A stretch of nucleotides ending with the second nucleotide upstream of boxes D and D′, referred to as the guide sequence, typically form a continuous stretch of base pairs with the target sequence and the nucleotide opposite the fifth nucleotide upstream of boxes D or D′ becomes methylated (Fig. 1a).

Box C/D snoRNAs adopt a characteristic structure in which boxes C and D pair to form a terminal kink-turn motif, whereas the internal boxes C′ and D′ often form a kink-loop motif. The snoRNAs exist as snoRNP particles together with the 15.5K protein that binds to the kink motifs, NOP56, NOP58, and fibrillarin (human nomenclature for the proteins; see Watkins and Bohnsack 2012 for a review on the snoRNP). Ribose methylations contribute to chemical and structural stabilization of the target RNA, and the methylation blocks the possibility of activation of

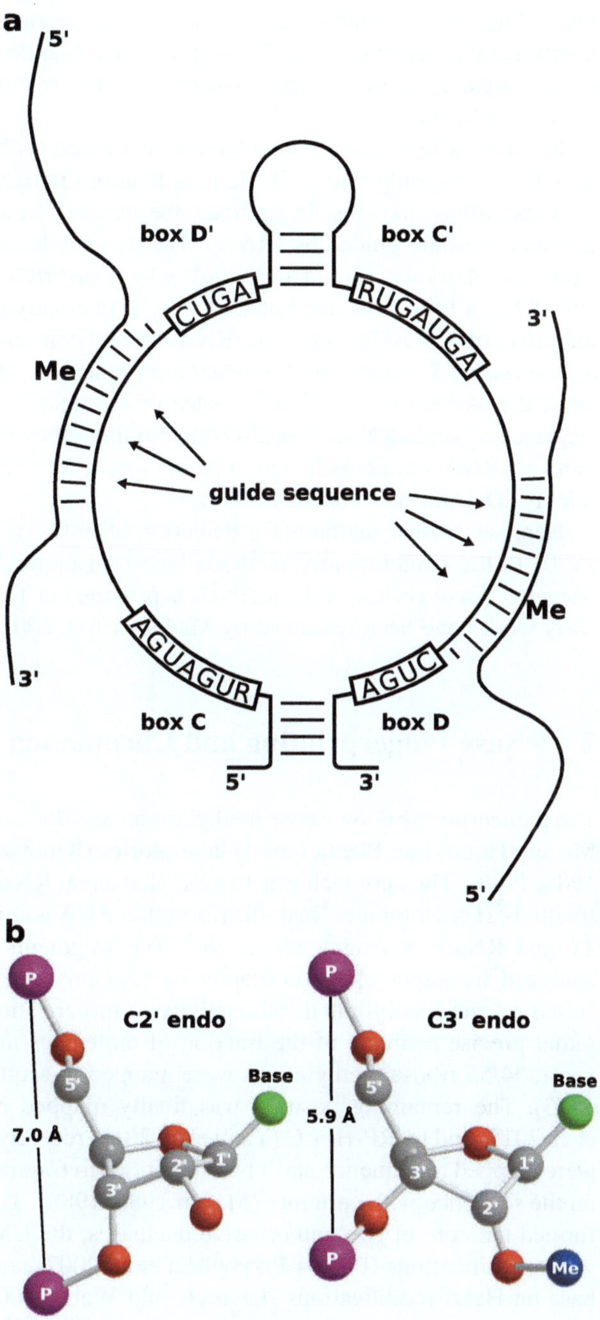

Fig. 1 (**a**) The configuration of the base pairing between a snoRNA (middle) and two target molecules. Ribose methylation catalyzed by the methyltransferase fibrillarin takes place in the target molecule at the nucleotide opposite the fifth nucleotide upstream of the D (or D′) box. The simultaneous pairing with the two targets in the drawing is speculative. (**b**) Methylation of the 2′-O enforces the 3′-endo conformation of the ribose and thereby affects the structure of the methylated RNA

the 2′-OH for nucleophilic attack on the neighboring phosphodiester bond. Furthermore, the presence of 2′-O-Me promotes 3′-endo conformation of the sugar, blocks sugar edge interactions, and changes the hydration sphere around the 2′O (Helm 2006) (Fig. 1b).

Ribosomal RNAs are particularly well studied with respect to ribose methylation. *E. coli* has only four 2′-O-Me modifications in rRNA, and these are introduced by stand-alone enzymes. In Archaea, the number varies considerably among species and most are guided by sRNAs. The overwhelming number of ribose methylations in eukaryotes is guided by snoRNAs. Yeast rRNA has 55 2′-O-Me and HeLa cell rRNA a little over one hundred. The only eukaryotic organisms for which the majority of 2′-O-Me sites in rRNA have been experimentally validated are *S. cerevisiae*, *S. pombe*, and humans represented by HeLa cells. For other organisms, the assignment of 2′-O-Me sites relies mostly on predictions based on the sequence of guide RNAs. The 2′-O-Me modifications are clustered in the conserved parts of rRNA and are believed to be important for rRNA folding and translational fidelity (Decatur and Fournier 2002).

Here we review methods for detection of 2′-O-Me from the early work in the 1970s to the contemporary methods based on parallel sequencing or mass spectrometry. An overview of the methods is provided in Table 1. Details of much of the early work have been reviewed by Maden (1990, 2001) and Maden et al. (1995).

2 RNase Fingerprinting and Chromatography

The pioneering work on ribose methylations was done in the 1970s and 1980s in the Maden (HeLa) and Planta (yeast) laboratories (Klootwijk and Planta 1973; Maden 1986, 1988). The approach was to label ribosomal RNA in vivo with $^{32}PO_4^-$ and L-[methyl-^{3}H]methionine. Next, the ribosomal RNA was isolated and a map of RNase T1 and RNase A fragments created. The fragments were further digested and analyzed by paper chromatography to map precisely the modified nucleotides. Based on the incorporated radioactivity, a molar ratio was calculated that was a rather precise estimate of the fraction of molecules modified at each position. In yeast, 54/55 ribose methylations were mapped to sequence (Klootwijk and Planta 1973). The remaining position was finally mapped by RiboMeth-seq (Birkedal et al. 2015) and by RP-HPLC (Yang et al. 2015) recently. In HeLa cells, 93 positions were mapped to sequence and 11–13 positions to fragments that could not be placed on the sequence with certainty (Maden 1986, 1988). The data from the two studies formed the core of two authoritative databases, the UMASS snoRNA database on yeast modifications (Piekna-Przybylska et al. 2007) and the snoRNA-LBME database on HeLa modifications (Lestrade and Weber 2006). These databases are no longer being updated but are still the main reference for studies on ribosomal RNA modifications in the two organisms.

The advantage of the fingerprinting approach is that it delivers both qualitative and quantitative information of high quality. However, the approach is very

Table 1 Comparison of the methods discussed in the text

Method	Section	No. of sites addressed	Sample throughput	Quantitative information	Labeling	Reference
RNase fingerprinting and chromatography	2	All	Single	Yes	In vivo $^{32}PO_4^-$ and ^3H-methionine	Maden (1986,1988), Klootwijk (1973)
Primer extension at low dNTP	3.1	Few	Many	No	In vitro ^{32}P or fluorescence	Maden (1995)
Do. followed by qRT-PCR	3.2	Few	Many	No	No	Belin (2009)
Primer extension after alkaline degradation	3.3	Few	Many	No	In vitro ^{32}P or fluorescence	Kiss-Laszlo (1996)
RNase H	3.4	Single	Many	Yes	In vitro ^{32}P or fluorescence	Yu (1997)
Deoxyribozymes	3.5	Single	Many	Yes	In vitro ^{32}P or fluorescence	Buchhaupt (2007)
Ligation	3.6	Single	Many	Yes	In vitro ^{32}P or fluorescence	Saikia (1993)
Mung Bean nuclease protection/RP-HPLC	3.7	Single	Many	(No)	No	Yang (2015)
Ab-IP-seq	4.1	All	Few	(No)	No	–
RiboMeth-seq	4.2	All	Few	Yes	No	Birkedal (2015)
LC-UV-MS/MS	5.1	Single	Few	Yes	No	Buchhaupt (2014)
SILAC-qMS	5.2	All	Single	Yes	Stable isotopes	Popova (2014)
SILNAS	5.2	All	Single	Yes	Stable isotopes	Taoka (2015)

laborious and is based on in vivo labeling with radioisotopes. This has gone out of fashion and it is unlikely that more organisms or cell types will be characterized in this way.

3 Methods Directed to Analysis of 2′-O-Me at Specific Locations

Once the groundwork had been done, a need for methods that could reveal changes in methylation at specific sites became apparent. The main driver of this was the discovery of the box C/D snoRNAs (Kiss-Laszlo et al. 1996). The fact that most sites are specified by particular guide RNAs and the modifications are executed by a single generic enzyme lends itself to genetic analysis and manipulation based on knockout and knock-in of snoRNAs, followed by determination of ribose methylation as a validation of the genetic manipulation. A range of methods was developed over the years, several of which are still in use. However, the number of similar methods also suggests that none of them were ideal.

3.1 Primer Extension at Limited dNTP Concentrations

The first method that came into use sprang from the observation of a spurious property of reverse transcriptase (RT). The principle of the method is based on pausing of reverse transcriptase at 2′-O-Me during reverse transcription (Fig. 2a; upper panel). Although RT is not impeded by 2′-O-Me at high (2 mM) dNTP concentrations (Lane et al. 1985), there is a pausing effect that is inversely correlated with the dNTP concentration at lower concentrations in the 0.004–1 mM range (Maden et al. 1995). The pausing signal can be visualized on gels immediately preceding the 2′-O-Me site (and sometimes even stuttering for one or two bands before the main band) (Fig. 2a; middle panel). Sequencing reactions using the same primer and a DNA template with identical sequence to the RNA used in the RT reaction are run in parallel to provide markers. The signal from the pausing effect differs from site to site, and some sites are not detected at all.

Since the ribose methylations in rRNA are clustered, several sites can be analyzed by extension of a single primer in many cases. The analysis requires primer labeling, and several samples can be analyzed in parallel. Unfortunately, anomalous bands that are not related to the presence of 2′-O-Me are frequently observed. This is a known phenomenon in RT reactions and can be caused, for instance, by strong secondary structure and modified nucleotides that interfere with Watson–Crick pairing of the incoming dNTP. Well-described examples of the latter are the $m^6{}_2A$–$m^6{}_2A$ doublet and the hypermodified $m^1acp_3\Psi$, both in small subunit (SSU) rRNA. Unfortunately, such effects are exacerbated at low dNTP concentration. Another problem for application of RT-based methods in detection of 2′-O-Me

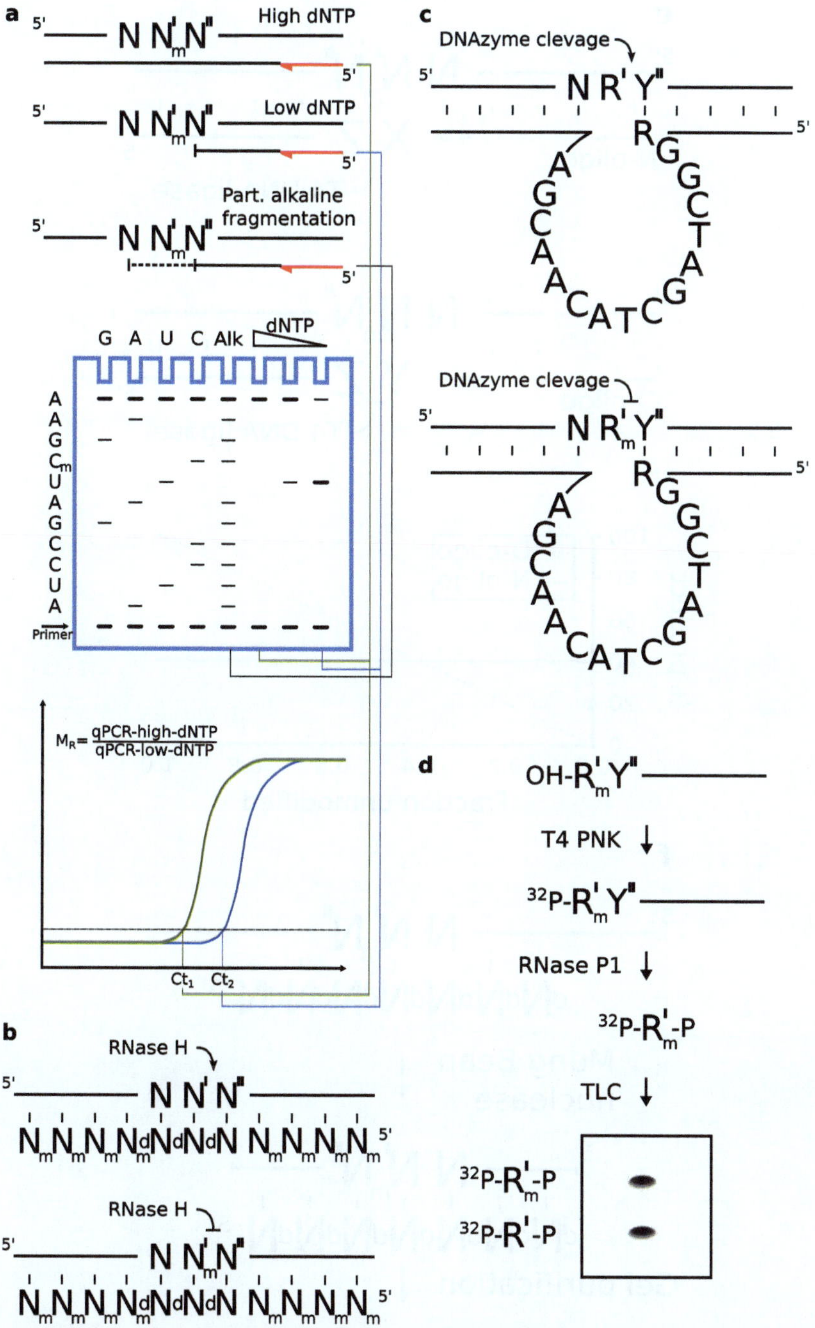

Fig. 2 Outline of methods for detection of 2′-O-me at specific sites. See text for explanation. (**a**) Primer extension at low dNTP (see 3.1) and after alkaline degradation (upper and middle panels; see 3.3), qRT-PCR after primer extension at low dNTP (lower panel; see 3.2), (**b**) RNase H method (see 3.4), (**c**) DNazyme method used as a cleavage assay and (**d**) in combination with labeling and TLC analysis for quantification (see 3.5), (**e**) Ligation method (see 3.6), (**f**) Mung Bean nuclease strategy for isolation of RNA fragments

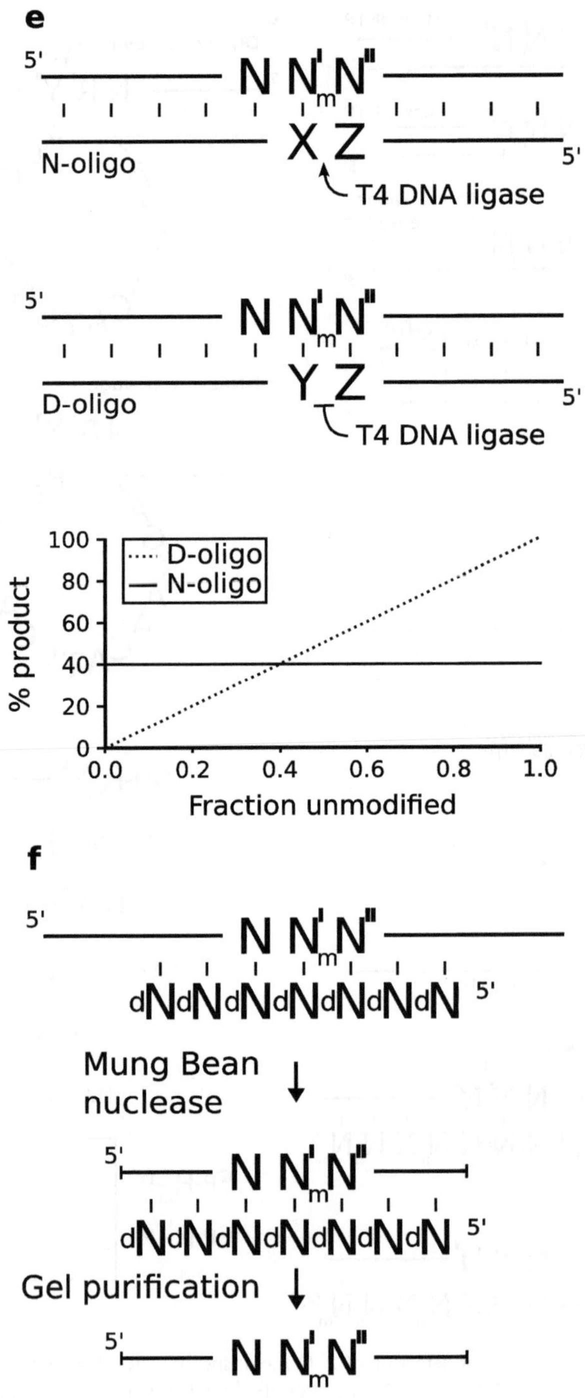

Fig. 2 (continued)

is difficulty with GC-rich RNAs that has hampered the use of this approach in some species of Archaea. Finally, the primer extension method is not quantitative although attempts have been made to improve the method by using fluorescent labeling and capillary electrophoresis on an automated genetic analyzer (Filippova et al. 2015).

3.2 *The Primer Extension/qRT-PCR Method*

The method based on primer extension at low dNTP concentration has been further developed by addition of a qRT-PCR step using primers flanking the queried position. The basic idea is that at low dNTP, unmodified molecules will allow cDNA synthesis to proceed past an upstream primer site in contrast to modified molecules that will impair cDNA synthesis. Thus, the amount of cDNA template available for PCR will depend on the degree of modification. This is compared to cDNA synthesis at high dNTP where RT is unimpeded and used to calculate a ratio (qPCR at high dNTP/qPCR at low dNTP) that expresses the "methylation ratio" (Fig. 2a; lower panel). This has been used in experimental settings to calculate a fold change in methylation ratio depending on, e.g., knockdown of factors (Belin et al. 2009). Although the methylation ratio has in some cases been taken as an indication of the methylation percentage, it is important to recall the huge variation that has been observed in signals from different sites in the underlying primer extension reaction. Furthermore, the fold changes in methylation reported in some papers are unrealistic given that some of the affected positions have been shown by other methods to be close to fully methylated in all cell types analyzed.

The advantage of the qRT-PCR method is that it has no requirement for labeling and that many samples can be processed in parallel in a qRT-PCR machine that is routinely used in most molecular biology laboratories. However, results based on this approach should be treated with caution until it has been rigorously tested in parallel with methods that are theoretically better founded.

3.3 *Primer Extension after Alkaline Hydrolysis*

Another RT-based method relies on the resistance to alkaline hydrolysis of the phosphodiester bond 3′ to the 2′-O-Me (Kiss-Laszlo et al. 1996). RNA is partially hydrolyzed into fragments with an average length of 200 nt and subjected to primer extension from a primer. The RT reaction will produce a ladder of bands when analyzed on polyacrylamide gels due to run-off of the RT at the ends of fragments. However, the ladder will have gaps at positions immediately preceding the 2′-O-Me sites because these bonds remain uncleaved (Fig. 2a; upper and middle panels). As with the limited dNTP method, sequencing reactions are run in parallel as markers. If the site is partially methylated, this will produce a weak band instead of a gap in the ladder.

3.4 The RNase H Method

RNase H cleaves RNA in DNA:RNA hybrids. The cleavage window is narrowed down to a single phosphodiester bond in the RNA by replacing deoxyribonucleotides in the oligo with 2′-O-Me ribonucleotides (Inoue et al. 1987). Thus, the oligos have a short stretch of deoxyribonucleotides flanked by 2′-O-Me ribonucleotides. Importantly, the presence of 2′-O-Me in RNA protects the 3′ neighboring phosphodiester bond from cleavage by RNase H. Thus, RNA that is unmethylated at the queried position will be cleaved whereas 2′-O methylated RNA will remain intact (Fig. 2b). This can be used to make a rough estimate of the fraction of molecules methylated at the queried site (Yu et al. 1997). It is not straightforward to predict which phosphodiester bonds will be cleaved by RNase H in the hybrids, and different sources of RNase H yield different results (Lapham et al. 1997). Thus, several oligos are tested in order to make a robust assay. An interesting example of application is the demonstration of a methylation defect at SSU-C463 upon depletion of the nucleolar phosphoprotein treacle (Gonzales et al. 2005).

3.5 The Deoxyribozyme (DNAzyme) Method

DNAzymes are catalysts made of DNA, and some have been designed to base pair to specific targets and cleave a predetermined site (Fig. 2c). The mechanism of cleavage requires a free 2′-OH at the target site and is blocked by 2′-O-methylation. DNAzymes consist of a catalytic core that has been selected by in vitro evolution and flanking sequences that are designed to base pair to the selected target. Their names derive from the number of selection rounds in the in vitro evolution experiment followed by the number of the clone isolated from the pool of molecules. 10–23 and 8–17 DNAzymes have been applied as site-specific cleavage reagents for detection of (lack of) ribose methylation in rRNA (Buchhaupt et al. 2007). Although each DNAzyme has constraints in terms of targets, a collection of DNAzymes can be applied to target all sites. The method requires selection of the appropriate DNAzyme and optimization of its cleavage. Full cleavage is rarely observed, probably due to structure in the RNA competing with the association of the DNAzyme. This problem was later addressed by inclusion of several rounds of denaturation, re-annealing, and cleavage in order to achieve full cleavage (Buchhaupt et al. 2014). Another improvement was to direct DNAzyme cleavage to the nucleotide immediately upstream of the modification. After cleavage, the 3′ cleavage fragment was ^{32}P-labeled at its 5′ end, digested with nuclease P1 to completion, and subjected to TLC analysis of the labeled nucleotides (Fig. 2d). In this way, the fraction of molecules modified at the queried site could be determined with precision. This approach was first developed with RNase H cleavage (see 3.4) by Zhao and Yu (2004) for pseudouridine and then adopted by Hengesbach et al. (2008) for DNAzymes in analysis of pseudouridines, but could easily be adapted for analysis of other modifications, including ribose methylations. Buchhaupt et al. used DNAzymes to demonstrate that a fraction of yeast rRNA is unmethylated at SSU-A100 and subsequently determined the fraction by LC-UV-MS/MS (see 5.1) (Buchhaupt et al. 2014).

3.6 The Ligation Method

The ligation method is a general method for detection of modified nucleotides in RNA. It relies on the ability of T4 DNA ligase to ligate two oligos annealed to a template consisting of the RNA carrying the modification (Saikia et al. 2006). Two sets of oligos are found by systematic screening of a pool of oligos with variable nucleotides (including modified nucleotides) flanking the site of ligation. One set is selected for ligation independent of the presence of the modification in the template, and one set is selected to be dependent on its presence. Once the oligos have been selected on a model template carrying the queried modification, specific sets are designed based on complementarity to the sites that are to be analyzed. The degree of modification is deduced by comparison of the accumulation of ligation product using the discriminating oligo set versus the non-discriminating set (Fig. 2e). The method has been validated by demonstrating the loss of three methylations in box C/D snoRNA deletion strains in yeast (Saikia et al. 2006).

3.7 Mung Bean Nuclease Protection and RP-HPLC

A convenient way to isolate RNA fragments carrying modifications is to anneal an oligo to a stretch of RNA comprising the modification and then digest the unprotected RNA with Mung Bean nuclease (Fig. 2f). This approach was used by Yang et al. to isolate overlapping fragments covering the entire SSU rRNA from *S. cerevisiae* (Yang et al. 2015). Then, the fragments were digested by nuclease P1 and alkaline phosphatase and the nucleosides analyzed by RP-HPLC. The analysis was consistent with previous mapping in the literature and resulted in mapping of a 2′-O-Me at SSU-G562 that was not mapped in the classical work from the Planta lab (Klootwijk and Planta 1973). The mapping was in accordance with results reported by a sequencing-based approach (Birkedal et al. 2015) (see 4.2). By comparison of methylated nucleosides in wt and snoRNA deletion strains, the level of methylation was estimated at a few sites. However, the approach can be combined with downstream methods that allow stringent quantification as demonstrated in related work (see 5.1).

4 Sequencing-Based Methods

With the decreasing costs for massive parallel sequencing, it has become attractive to develop methods based on sequencing that address nucleotide modifications in a transcriptome-wide fashion. The successful approaches have used either chemical modification of the RNA or an antibody directed towards the modified nucleotide as the central principle for detection. A classical method for detection of pseudouridines relies on selective modification of pseudouridines with CMC (N-cyclohexyl-N9-(2-morpholinoethyl)-carbodiimidemetho-p-toluenesulfonate) to generate a block to reverse transcriptase (Bakin and Ofengand 1993). This principle was used to develop a method (Pseudo-seq) for transcriptome-wide mapping based on comparative sequencing of libraries of mock (CMC-) and CMC treated RNA (Carlile et al. 2014). Transcriptome-wide mapping ofm^6A (Dominissini et al. 2012) has been accomplished by a strategy reminiscent of ChIP-seq used in characterization of chromatin. Specific antibodies directed towards the RNA modification in question are used to precipitate fragmented RNA. After sequencing and local alignment, the modified nucleotide is mapped by locating the nucleotide common to all sequences in each local alignment (Fig. 3a). Several methods have been developed for transcriptome-wide mapping of m^5C including methods based on chemical modification and immunoprecipitation (Hussain et al. 2013).

Due to the abundance of orphan box C/D snoRNAs encoded by the human genome, it is expected that 2′-O-Me will also be found in mRNA similarly to pseudouridines, m^6A, and m^5C. In principle, the immunoprecipitation approach should be applicable to 2′-O-Me, but so far there are no reports of antibodies directed towards this modification.

4.1 *Antibody-Based Detection of 2′-O-Me*

In the absence of antibodies directed towards 2′-O-Me, information can be obtained from PAR-CLIP (Photoactivatable-Ribonucleoside-enhanced Crosslinking and Immunoprecipitation) experiments (Hafner et al. 2010) with antibodies against snoRNP components. Kishore et al. (2013) used antibodies against fibrillarin (FBL) and NOP58 for analysis of HEK293 cells as well as a stable cell line expressing FLAG-tagged NOP56 analyzed with anti-FLAG antibodies. The aim of the study was to map RNAs that bind the snoRNP proteins; however the protocol resulted in co-purification of the targets of the box C/D snoRNAs as well. Importantly, the resolution of these experiments is insufficient to be used as a method for mapping of 2′-O-Me sites to single nucleotides.

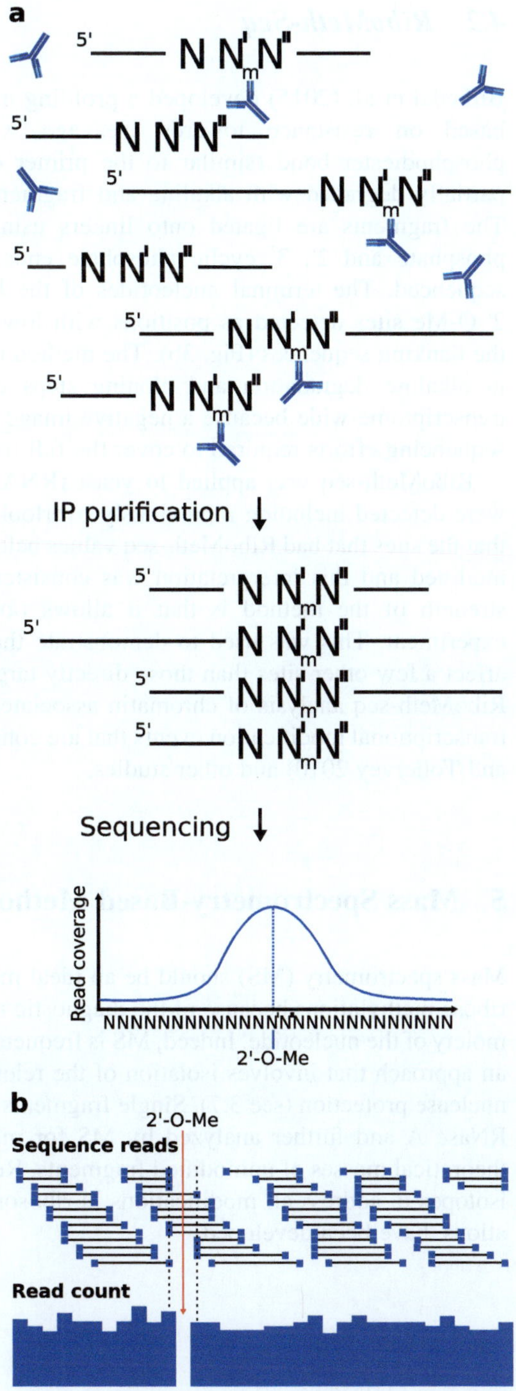

Fig. 3 Sequencing-based methods for 2′-O-Me profiling. (**a**) Outline of a hypothetical method based on the use of modification-specific antibodies similarly to what has been developed for other modifications (see 4.1), (**b**) RiboMeth-seq (see 4.2)

4.2 *RiboMeth-Seq*

Birkedal et al. (2015) developed a profiling method (RiboMeth-seq) for 2′-O-Me based on resistance towards cleavage with alkaline of the neighboring phosphodiester bond (similar to the primer extension method in 3.3). RNA is partially degraded with alkaline and fragments of 20–40 nt purified from gels. The fragments are ligated onto linkers using plant tRNA ligase that joins 5′ phosphate and 2′, 3′ cyclic phosphate ends. Then, cDNA is synthesized and sequenced. The terminal nucleotides of the library fragments are recorded and 2′-O-Me sites detected as positions with low number of read-ends compared to the flanking sequences (Fig. 3b). The method is quantitative to the extent that bias in alkaline degradation and cloning steps can be eliminated. It is not truly transcriptome-wide because a negative image of 2′-O-Me sites is created and the sequencing efforts required to cover the full transcriptome would be formidable.

RiboMeth-seq was applied to yeast rRNA. Almost all (53/55) 2′-O-Me sites were detected including a previously overlooked site (SSU-G562). It was argued that the sites that had RiboMeth-seq values below the cutoff value were fractionally modified and this interpretation was consistent with the existing literature. The strength of the method is that it allows observations of all sites in a single experiment. This was used to demonstrate that snoRNA knockout and knock-in affect a few other sites than those directly targeted by the snoRNA manipulation. RiboMeth-seq analysis of chromatin associated RNA demonstrated co- and post-transcriptional modification events that are consistent with metabolic labeling (Kos and Tollervey 2010) and other studies.

5 Mass Spectrometry-Based Methods

Mass spectrometry (MS) should be an ideal method for mapping and quantitating ribose methylations because of the diagnostic mass addition of 14 Da to the ribose moiety of the nucleotide. Indeed, MS is frequently used to validate 2′-O-Me sites by an approach that involves isolation of the relevant RNA fragment by Mung Bean nuclease protection (see 3.7). Single fragments are then digested with RNase T1 or RNase A and further analyzed by MS for mass additions in comparison to the theoretical masses of unmodified fragments. Recently, two methods that use stable isotopes to address all modifications in ribosomal RNA, including ribose methylations, have been developed.

5.1 LC-UV-MS/MS

Buchhaupt et al. (2014) used oligonucleotide-dependent Mung Bean nuclease protection (Fig. 2f) to isolate a fragment containing SSU-A100 from *S. cerevisiae.* The fragment was enzymatically digested into nucleosides and an internal standard containing ^{13}C-labeled 2′-O-methyladenosine and guanosine added. The RNA content was then determined by UV_{254nm} using the guanosine standard and the methylation level at SSU-A100 determined by MS/MS using the 2′-O-methyladenosine standard. The value obtained (68 %) confirms this position to be one of the 2′-O-Me sites in yeast that are partially methylated.

5.2 Labeling with Heavy Isotopes and qMS (SILAC-qMS and SILNAS)

Ribosomal RNA modifications in *E. coli* were studied by Popova and Williamson (Popova and Williamson 2014) using quantitative mass spectrometry (qMS) and a metabolic labeling strategy (SILAC-MS). Bacterial cells were grown in parallel with ^{14}N-(sample) or ^{15}N-(external standard)-labeled ammonium sulfate as the sole source of nitrogen. Mature ribosomes and pre-ribosomal particles were then isolated by sucrose-gradient centrifugation from the sample and spiked-in at a roughly 1:1 molar ratio with mature rRNA from the heavy-isotope-labeled culture. The RNA was then fragmented with various RNases and the fragments analyzed by LC-MS. Pairs of co-eluting ^{14}N- and ^{15}N-labeled fragments were identified and the isotope distributions fitted using a dedicated algorithm. In addition, the analysis was complemented by application of specific heavy-isotope labeling of methyl groups (using CD_3-methionine) and pseudouridines (using 5,6-D-uracil). The authors succeeded in precise quantification of 29/36 rRNA modifications in *E. coli*, including all four ribose methylations. Importantly, it was demonstrated that residues are modified at different time points during assembly and that this correlates with recruitment of ribosomal proteins.

A related approach was used by Taoka et al. (2015) to profile modifications in *S. pombe* (Stable Isotope-Labeled riboNucleic Acid as an internal Standard (SILNAS)). Here, in vitro transcripts corresponding to ribosomal RNA species were labeled with guanosine-$^{13}C_{10}$ 5′-triphosphate, cytidine-$^{13}C_9$ 5′-triphosphate, or uridine-$^{13}C_9$ 5′-triphosphate and used as external standards added at approximately 1:1 stoichiometry (Fig. 4). The RNA was fragmented by RNase T1 and fractionated by high-resolution liquid chromatography. In subsequent MS, sample and standard fragments are detected as pairs displaced by the isotopic mass differences. Non-modified fragments are used for calibration such that reduction in peak height resulting from a mass shift of (parts of) the sample peak can be used for quantification. The mass shift in the shifted peak can be used for identification of the type of modification and subsequent MS/MS used to verify this identity and

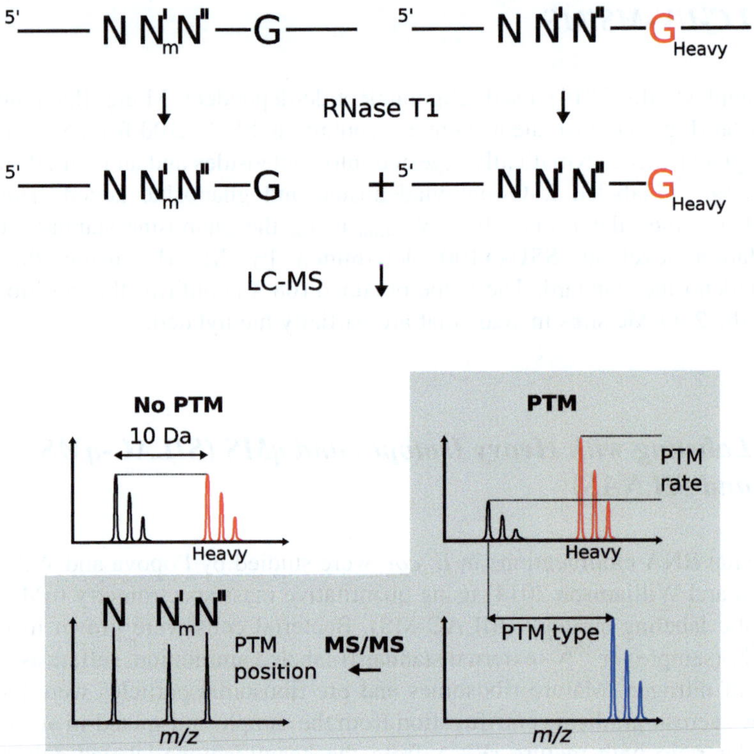

Fig. 4 Mass spectrometry by the SILNAS method (see 5.2). PTM: *P*ost-*T*ranscriptional *M*odification

precisely map the site of modification. Pseudouridines were addressed by chemical derivation of the sample RNA by cyanoethylation. Due to the large size of eukaryotic rRNA, an oligo-directed RNase H approach was used to digest the SSU and LSU rRNAs into smaller fragments for analysis. The method was used to study growth temperature-related changes in the stoichiometry of nucleotide modifications. Here, the levels at 2′-O-Me sites remained constant, in contrast to several pseudouridine sites.

The two methods differ in that SILAC-qMS use metabolic labeling in contrast to SILNAS that use an in vitro transcript as an external standard implying that it does not contain nucleotide modifications. Thus, the MS shifts that underlie modification detection are data dependent in the case of SILNAS and predicted from isotope differences in the SILAC-qMS. The quantification of modifications is absolute in comparison to an unmodified standard in SILNAS and relative to the reference in SILAC-qMS. Unfortunately, the two methods cannot be compared directly because they were applied to different organisms.

6 Discussion

Ideally, robust methods should be available for determination of 2′-O-Me both at a transcriptome-wide level and a more focused level aimed at precise quantification of ribose methylation at individual sites. Several methods for addressing individual sites have been developed, but unfortunately, there are very few papers that allow comparison of methods. The RNase H method (Inoue et al. 1987; Lapham et al. 1997; Yu et al. 1997) is currently applied by several laboratories. The drawbacks of this method are that the modified oligos are quite expensive and that careful design of oligos and optimization is required. The development of the DNAzyme method in which cleavage is directed to the phosphodiester bond immediately upstream of the modification followed by 5′ end labeling, hydrolysis, and TLC (Hengesbach et al. 2008) has been used successfully for quantification of pseudouridines in several laboratories and could be used equally for quantification of 2′-O-Me. Once again, the method is laborious and requires optimization, in this case of the DNAzymes.

The recent studies of ribosomal RNA modifications based on MS (Popova and Williamson 2014; Taoka et al. 2015) represent a major step forward. The main problem with these methods is that they require heavy instrumentation and specialized skills. However, the establishment of a 2′-O-Me profile for the main model systems by MS will be a valuable reference for results obtained with other methods. RiboMeth-seq (Birkedal et al. 2015) has been applied to rRNA and small RNAs from a number of organisms (unpublished data) and generally provides information that is consistent with the literature and has strong predictive power, e.g., for snoRNA finding. The prospects for going transcriptome-wide with the method are currently limited. First, the method generates a negative image (i.e., detect sites that are not methylated), and it would take a considerable sequencing effort to profile complex transcriptomes. Second, the background in the method is too high to allow detection of sites that are methylated in a small fraction of molecules only. The method could be combined with an enrichment step, e.g. if an antibody directed against 2′-O-Me could be applied to precipitate methylated RNA prior to profiling.

The incentive for development and optimization of methods for profiling of 2′-O-Me is high. After the discoveries of m^5C, m^6A, and pseudouridines in mRNAs, 2′-O-Me in mRNA is likely to be next. The modification is well known from cap-proximal nucleotides, and indications for its presence in the body of the mRNA are accumulating from biochemical studies. Furthermore, mRNA targets of snoRNAs have been predicted, and the physical association of snoRNAs with mRNAs has been demonstrated.

The biological implications of ribose methylation are not well understood. In rRNA, ribose methylation is believed to be important for folding and stability of the rRNA and for the fidelity of translation. However, it has generally been difficult to link individual methylation events to a biological function. Recent ribosome structures from X-ray crystallography have sufficiently high resolution to model the modifications and their interactions (Noeske et al. 2015; Polikanov et al. 2015) and holds promise for a deeper understanding of the molecular biology of this abundant RNA modification.

Acknowledgements The authors wish to thank Dr. Martin Jansson for critical reading of the manuscript. The work was supported by the Danish Research Council for Independent Research.

References

Bakin A, Ofengand J (1993) Four newly located pseudouridylate residues in *Escherichia coli* 23S ribosomal RNA are all at the peptidyltransferase center: analysis by the application of a new sequencing technique. Biochemistry 32:9754–9762

Belin S, Beghin A, Solano-Gonzalez E et al (2009) Dysregulation of ribosome biogenesis and translational capacity is associated with tumor progression of human breast cancer cells. PLoS One 4, e7147

Birkedal U, Christensen-Dalsgaard M, Krogh N et al (2015) Profiling of ribose methylations in RNA by high-throughput sequencing. Angew Chem Int Ed Engl 54:451–455

Buchhaupt M, Peifer C, Entian KD (2007) Analysis of 2′-O-methylated nucleosides and pseudouridines in ribosomal RNAs using DNAzymes. Anal Biochem 361:102–108

Buchhaupt M, Sharma S, Kellner S et al (2014) Partial methylation at Am100 in 18S rRNA of baker's yeast reveals ribosome heterogeneity on the level of eukaryotic rRNA modification. PLoS One 9, e89640

Carlile TM, Rojas-Duran MF, Zinshteyn B et al (2014) Pseudouridine profiling reveals regulated mRNA pseudouridylation in yeast and human cells. Nature 515:143–146

Decatur WA, Fournier MJ (2002) rRNA modifications and ribosome function. Trends Biochem Sci 27:344–351

Dominissini D, Moshitch-Moshkovitz S, Schwartz S et al (2012) Topology of the human and mouse m^6A RNA methylomes revealed by m^6A-seq. Nature 485:201–206

Filippova JA, Stepanov GA, Semenov DV et al (2015) Modified method of rRNA structure analysis reveals novel characteristics of box C/D RNA analogues. Acta Nat 7:64–73

Gonzales B, Henning D, So RB et al (2005) The Treacher Collins syndrome (TCOF1) gene product is involved in pre-rRNA methylation. Hum Mol Genet 14:2035–2043

Hafner M, Landthaler M, Burger L et al (2010) Transcriptome-wide identification of RNA-binding protein and microRNA target sites by PAR-CLIP. Cell 141:129–141

Helm M (2006) Post-transcriptional nucleotide modification and alternative folding of RNA. Nucleic Acids Res 34:721–733

Hengesbach M, Meusburger M, Lyko F et al (2008) Use of DNAzymes for site-specific analysis of ribonucleotide modifications. RNA 14:180–187

Hussain S, Aleksic J, Blanco S et al (2013) Characterizing 5-methylcytosine in the mammalian epitranscriptome. Genome Biol 14:215

Inoue H, Hayase Y, Iwai S et al (1987) Sequence-dependent hydrolysis of RNA using modified oligonucleotide splints and RNase H. FEBS Lett 215:327–330

Kishore S, Gruber AR, Jedlinski DJ et al (2013) Insights into snoRNA biogenesis and processing from PAR-CLIP of snoRNA core proteins and small RNA sequencing. Genome Biol 14:R45

Kiss-Laszlo Z, Henry Y, Bachellerie JP et al (1996) Site-specific ribose methylation of preribosomal RNA: a novel function for small nucleolar RNAs. Cell 85:1077–1088

Klootwijk JR, Planta J (1973) Analysis of the methylation sites in yeast ribosomal RNA. Eur J Biochem 39:325–333

Kos M, Tollervey D (2010) Yeast pre-rRNA processing and modification occur cotranscriptionally. Mol Cell 37:809–820

Lane DJ, Pace B, Olsen GJ et al (1985) Rapid determination of 16S ribosomal RNA sequences for phylogenetic analyses. Proc Natl Acad Sci U S A 82:6955–6959

Lapham J, Yu YT, Shu MD et al (1997) The position of site-directed cleavage of RNA using RNase H and 2′-O-methyl oligonucleotides is dependent on the enzyme source. RNA 3:950–951

Lestrade LM, Weber J (2006) snoRNA-LBME-db a comprehensive database of human H/ACA and C/D box snoRNAs. Nucleic Acids Res 34:D158–162

Maden BE (1986) Identification of the locations of the methyl groups in 18 S ribosomal RNA from *Xenopus laevis* and man. J Mol Biol 189:681–699

Maden BE (1988) Locations of methyl groups in 28 S rRNA of *Xenopus laevis* and man clustering in the conserved core of molecule. J Mol Biol 201:289–314

Maden BE (1990) The numerous modified nucleotides in eukaryotic ribosomal RNA. Prog Nucleic Acid Res Mol Biol 39:241–303

Maden BE (2001) Mapping 2′-O-methyl groups in ribosomal RNA. Methods 25:374–382

Maden BE, Corbett ME, Heeney PA et al (1995) Classical and novel approaches to the detection and localization of the numerous modified nucleotides in eukaryotic ribosomal RNA. Biochimie 77:22–29

Noeske JM, Wasserman R, Terry DS et al (2015) High-resolution structure of the *Escherichia coli* ribosome. Nat Struct Mol Biol 22:336–341

Piekna-Przybylska D, Decatur WA, Fournier MJ (2007) New bioinformatic tools for analysis of nucleotide modifications in eukaryotic rRNA. RNA 13:305–312

Polikanov YS, Melnikov SV, Soll D et al (2015) Structural insights into the role of rRNA modifications in protein synthesis and ribosome assembly. Nat Struct Mol Biol 22:342–344

Popova AM, Williamson JR (2014) Quantitative analysis of rRNA modifications using stable isotope labeling and mass spectrometry. J Am Chem Soc 136:2058–2069

Saikia M, Dai Q, Decatur WA et al (2006) A systematic ligation-based approach to study RNA modifications. RNA 12:2025–2033

Taoka M, Nobe Y, Hori M et al (2015) A mass spectrometry-based method for comprehensive quantitative determination of post-transcriptional RNA modifications: the complete chemical structure of *Schizosaccharomyces pombe* ribosomal RNAs. Nucleic Acids Res

Watkins NJ, Bohnsack MT (2012) The box C/D and H/ACA snoRNPs: key players in the modification processing and the dynamic folding of ribosomal RNA. Wiley Interdiscip Rev RNA 3:397–414

Yang J, Sharma S, Kotter P et al (2015) Identification of a new ribose methylation in the 18S rRNA of *S. cerevisiae*. Nucleic Acids Res 43:2342–2352

Yu YT, Shu MD, Steitz JA (1997) A new method for detecting sites of 2′-O-methylation in RNA molecules. RNA 3:324–331

Zhao XY, Yu T (2004) Detection and quantitation of RNA base modifications. RNA 10:996–1002

Diadenosine Tetraphosphate (Ap_4A) in Health and Disease

Suliman Boulos, Ehud Razin, Hovav Nechushtan, and Inbal Rachmin

Contents

Abstract Diadenosine oligophosphates (Ap*n*As) were initially discovered more than 50 years ago. This group of molecules form a class of compounds derived from ATP and consist of two adenosine molecules bridged by up to six phosphate groups. The first enzymatic production of these compounds was noted by Zamecnik and colleagues in their study with purified lysyl tRNA synthetase (KARS) in mammalian cells.

S. Boulos • H. Nechushtan
Sharett Institute of Oncology, Hadassah-Hebrew University Medical Center, EinKerem, Jerusalem 91120, Israel
e-mail: hovavnech@hadassah.org.il

E. Razin (✉)
Department of Biochemistry and Molecular Biology, Institute for Medical Research Israel-Canada, The Hebrew University of Jerusalem, Jerusalem 91120, Israel
e-mail: ehudr@ekmd.huji.ac.il

I. Rachmin
Cardiovascular Division, Department of Medicine, and Brigham Regenerative Medicine Center, Brigham and Women's Hospital, Boston, MA 02115, USA

Harvard Stem Cell Institute and Department of Stem Cell and Regenerative Biology, Harvard University, Cambridge, MA 02139, USA

S. Jurga et al. (eds.), *Modified Nucleic Acids in Biology and Medicine*, RNA Technologies, DOI 10.1007/978-3-319-34175-0_9

Multiple studies on the role of Ap*n*As have been published during the years following their initial discovery. However, technical difficulties hampered some of the studies, and the field has been abandoned for nearly 20 years, until the use of new molecular methods inspired new studies into the functional aspects of these nucleotides in bacterial and eukaryotic systems.

In this chapter, we will discuss the role of Ap*n*As in prokaryotic and eukaryotic cells and will focus on the most investigated member of the Ap*n*As family, namely diadenosine tetraphosphate (Ap_4A), and its role in a variety of tissues such as the heart and blood vessels, neurons, spermatocytes, neutrophils, and pancreatic cells.

We conclude our chapter with a description of a putative cell signaling pathway involving KARS, whose structure can be modulated so that it is no longer involved in translation but mainly in transcription, through its ability to produce the second messenger Ap_4A.

Keywords Ap*n*As: diadenosine oligophosphates • Ap_3A: diadenosine triphosphate • Ap_4A: diadenosine tetraphosphate • KARS: lysyl-tRNA-synthetase

1 Introduction

The first description of the Ap*n*A family was reported in the sixties by Paul Zamecnik and coworkers in mammalian cells. Zamecnik pioneered the in vitro synthesis of proteins and with Mahlon Hoagland discovered transfer RNA (tRNA) and later lysyl-tRNA-synthetase (KARS) and its product, Ap_4A (Zamecnik et al. 1966).

An area in which Ap*n*A research developed was in prokaryotes. Indeed, several interesting findings have emerged from these studies, some of which will be described in this chapter. In eukaryote, sAp*n*As may act not only intracellularly but may also have a role as extracellular mediators, most notably binding to adenosine receptors as will be explained.

Intracellular Ap*n*As can be produced by several mechanisms including DNA ligase (Marriott et al. 2015) and various tRNA synthetases. It is interesting in this regard that the same synthetase initially described as capable of producing Ap_4A, KARS, has been implicated as a source of Ap_4A both in bacteria and in mammalian cells. We also describe studies on the production of Ap_4A as a result of extracellular stimuli in mast cells, which was proposed for the first time by Ehud Razin and colleagues to be a part of a new signaling pathway in which Ap_4A acts as a second messenger (Carmi-Levy et al. 2008).

2 Ap*n*As in Bacteria

There are numerous examples of increased Ap*n*As under different conditions both in eukaryotes and prokaryotes. *Escherichia coli* (*E. coli*) is one of the most intensively studied microbes. Studies of this bacteria revealed that levels of Ap_4A increase rapidly when *E. coli* is exposed to heat shock or oxidative stress. When heat is applied to *E. coli* cells and temperatures are increased from 28 to 43 °C, the intracellular Ap_4A concentration increases from ~ 1 to 160 μM (Coste et al. 1987; Farr et al. 1989). Similar results have been observed following treatment with hydrogen peroxide, ethanol, and cadmium chloride. The source of the increase in *E. coli* Ap_4A has not been conclusively demonstrated, but it is most likely that it is a special kind of KARS. While in mammalian cells, phosphorylation creates a structural switch in the lysyl-tRNA molecule creating a form which is much more efficient at the production of Ap_4A and much less efficient at its translational function (Ofir-Birin et al. 2013), in *E. coli* it exists in two distinct isoforms, LysS and LysU (Charlier and Sanchez 1987). These two forms share a high degree of sequence identity (88 %) and have similar aminoacylation activities, but are regulated differently (Leveque et al. 1990). LysS is constitutively expressed under normal growth conditions and appears to be responsible for the tRNA charging activity, whereas LysU is the product of a normally silent gene that is induced to high expression during heat shock (Charlier and Sanchez 1987) as can be seen in Fig. 1.

In a series of studies, Miller and colleagues demonstrated that although LysU is also capable of tRNA charging, it is an unusually effective catalyst of Ap_4A, diadenosine, Ap_3A, and ATP (Wright et al. 2006, 2014; Chen et al. 2013). Interestingly, when studying bacteria isolated from Antarctic seawater, Kawamoto and colleagues identified LysU as one of the few proteins induced when growing the bacteria in very cold water (Kawamoto et al. 2007). The putative role of LysU under these conditions is not clear, but it may well be that Ap*n*As also have an important role in adaptation to conditions of extreme cold.

One environment which demands substantial adaptation from bacteria is a new intracellular environment. Intracellular pathogens are the leading cause of morbidity and mortality in underdeveloped countries. Outstanding is *Mycobacterium tuberculosis*, which according to the WHO 2014 global report has killed 1.5 million people around the globe. However, other intracellular pathogens such as *Salmonella enteritidi, Rickettsia* species, as well as others cause millions of severe infections worldwide. In order to hijack the host intracellular environment for its own benefit, many of these pathogens utilize the host cell machinery for this purpose (Coiffard et al. 2015). It has been observed that degradation of adenosine dinucleotides by their hydrolases has a critical role in the invasiveness of *E. coli* and *Rickettsiae* (Gaywee et al. 2002). Furthermore, Nudix hydrolases active on adenosine (5′)-pentaphospho-(5″)-adenosine were shown to be part of the virulence machinery in *Rickettsiae* species and Bartonella (Cartwright et al. 1999) and are considered essential for bacterial invasion. In addition, it has been suggested that

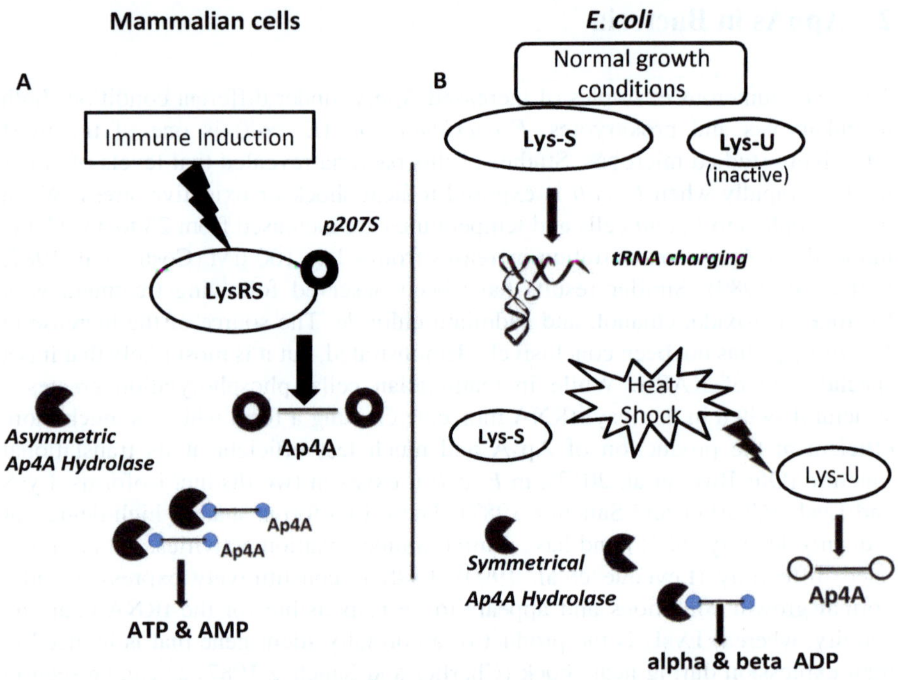

Fig. 1 (**a**) KARS-Ap$_4$A pathway in mammalian cells, in which immune induction results in KARS phosphorylation at Serine 207, enabling KARS to produce Ap$_4$A. The degradation of Ap$_4$A in these cells is via the asymmetric Ap$_4$A hydrolase to adenosine monophosphate (AMP) and adenosine triphosphate (ATP). (**b**) In bacteria, there are two forms of KARS, LysU and LysS. In normal growth conditions, LysS is active and is responsible for tRNA charging, whereas LysU is inactive. In the situation of heat shock, there is a switch and activation of LysU which produces Ap$_4$A. In bacteria, Ap$_4$A is degraded via the symmetric Ap$_4$A hydrolase to the alpha and beta forms of adenosine diphosphate (ADP)

these enzymes are involved in protection against damaging oxidative and other chemical stresses. Interestingly, almost all invasive bacterial pathogens have similar systems acting on either Ap$_5$A or Ap$_4$A. These include all the major intracellular pathogens such as *Mycobacteria* and *Salmonella* (Ismail et al. 2003).

While a comprehensive list of proteins modulated by Ap*n*As in bacteria has not been completed, a few prominent proteins have been already reported to be modulated in bacteria by Ap*n*As. For example, in *E. coli*, Ap$_4$A binds to several proteins, including the stress proteins *DnaK*, *GroEL*, *E89*, *C4*, and *C40* (Johnstone and Farr 1991). The oxidative stress protein *Dnak* plays a crucial role in the downregulation of the heat shock response. Moreover, a mutation in the *apaH* gene, which encodes the Ap$_4$A hydrolase in *E. coli*, results in a slower development of functional phage particles when exposed to 42 °C. Compared with wild type, *apaH* mutants, that lack the hydrolase responsible for Ap$_4$A degradation, show dramatically increased sensitivity to killing by heat and do not grow at 43 °C (Johnstone and Farr 1991). Preliminary data suggest that Ap$_4$A increases the phosphorylation of *Dnak* in bacteria. It has also been demonstrated that *Dnak* phosphorylation increases upon a shift from 30 to 43 °C (Sherman and Goldberg 1993), while in *apaH* mutants the level of phosphorylated *Dnak* is already high at 30 °C and does not increase with elevation of the temperature.

In conclusion, overproduction of Ap_4A and its binding to stress proteins such as *Dnak* and/or *GroEL* in prokaryotes may alter functions that are required for survival in thermal stress.

3 Ap_4A's Effect on Heart and Blood Vessels

Ap*n*As are well studied as endogenous regulators of cardiovascular function and are gaining more attention in this field of research. It has been shown that after β-adrenergic stimulation of the ventricles in both humans and animals, Ap_4A exerts negative isotropic effects via stimulation of A1-adenosine receptors. For example, in guinea pig and human cardiac preparations, Ap_4A concentrations between 0.1 and 100 μM reduced the force of contraction by ~ 36.5 %. On the other hand, Ap_4A alone can show positive inotropic effects in the same ventricular preparations by activating the P2-purinoceptors in ventricle myocardium (Vahlensieck et al. 1999).

It is worth mentioning that in rat mesenterial preparations lacking the endothelium layer, Ap_4A leads to vasoconstriction, whereas in endothelium-intact mesenteria Ap_4A causes vasodilatation (Busse et al. 1988). Also in rabbit, pig, and dog hearts, administration of Ap_4A leads to dilation of the coronary vessels and so reduces ischemic injury of the heart.

Ap_4A has also cardioprotective qualities, as it has been shown to improve cardiac function after hypothermic preservation in rat hearts (Vahlensieck et al. 1999). This protective effect was possible by the activation of the mitochondrial ATP-sensitive potassium channel (mK_{ATP} channel). When 80 μM Ap_4A was added to eurocollin solution, a well-known solution for heart preservation, the left ventricular systolic pressure, heart rate, and coronary flow rate were improved. Moreover, leakage of lactate dehydrate and creatine kinase during reperfusion was reduced. These effects of Ap_4A on the rate heart were reversed by adding 100 μM 5-hydroxydecanoic acid, a selective mK_{ATP} channel inhibitor (Ahmet et al. 2000).

The fact that Ap_4A reduces ischemic injury in threat heart led researchers to investigate the role of Ap_4A in stroke models of rat brain and in ischemic brain injuries. Treatment with Ap_4A (100 nM) prior to hypoxia induction provided protection to neuronal cells in vitro. In vivo pretreatment with Ap_4A also protected rat brain from ischemic damage due to stroke. This protection was not attributed to changes in blood pressure, blood gases, or blood flow in the infracted area (Wang et al. 2003). This protective effect was also observed with Ap_4A treatment after ischemic infarction. It is also well known that pro-apoptotic mechanisms are activated during ischemic injury, such as the translocation of cytochrome c from the mitochondria to the cytosol thus activating apoptosis via caspase-3 (Sasaki et al. 2000). Ap_4A has been shown to prevent this translocation of cytochrome c and therefore activation of caspase-3 (Wang et al. 2003).

In conclusion, these results imply that Ap_4A, a major representative of the Ap*n*As, plays an important role in cardiac function and ischemic damage and that further research in this direction has important therapeutic potential.

4 Ap*n*As in Neuronal Signaling and the Hippocampus

Ap_4A, Ap_5A, and Ap_6A have been shown to play a role in neuronal synaptic signaling (Ogilvie et al. 1996), and the release of Ap*n*As from synaptic vesicles via nerve terminal depolarization has been demonstrated in different brain areas of animals (Miras-Portugal et al. 2003).

Chromaffin cells were the first neural site found to contain Ap*n*As at various concentrations, and the presence of Ap_4A, Ap_5A, and Ap_6A in rat synaptic vesicles was demonstrated by HPLC (Pintor et al. 1992). When challenged, for example, with nicotinic agonists such as carbamylcholine, the chromaffin cells release Ap*n*As to the extracellular matrix. This release of nucleotides is calcium dependent as we will explain later.

The interaction of Ap*n*As with inotropic (P2X) and metabotropic (P2Y) ATP receptors plays an important role in their function in the CNS (Lazarowski et al. 1995; Communi et al. 1996; Schachter et al. 1996). This raised the question in the early nineties whether Ap*n*As also had their own receptors through which they influence neuronal transduction. Hilderman and coworkers managed to detect high-affinity binding sites for Ap*n*As in membranes of mouse brain tissue (Hilderman et al. 1991). Later, Pintor showed that when ATP and Ap*n*As were added, synaptic terminals from rat midbrain responded by increasing the inter-synaptosomal calcium concentration (Pintor et al. 1997). This effect was independent of the ATP receptors. The proof of the existence of these presynaptic receptors was based on the following findings: (1) cross-desensitization studies showing independent responses to Ap*n*As and ATP, (2) nerve terminal microfluorimetric studies revealing the existence of nerve terminals sensitive only to Ap_5A or ATP, (3) the dinucleotide receptor antagonist, Ip5I, abolished Ap_5A-induced calcium secretion without affecting ATP response (Pintor et al. 1997), whereas suramin, a P2 receptor antagonist, blocked the response to ATP but not to Ap_5A (Ralevic and Burnstock 1998).

The release of neurotransmitters such as gamma aminobutyric acid (GABA) is also influenced by the release of Ap_5A in rat spinal cord (Jo and Schlichter 1999). Released Ap_5A is able to induce a calcium influx that can lead to exocytotic release not only of GABA but also of other classical neurotransmitters such as acetylcholine and glutamate.

In hippocampal nerve terminals, it has been clearly demonstrated that activation of presynaptic adenosine receptors is able to modulate dinucleotide responses. When testing the effect of Ap_5A and ATP on hippocampal synaptic terminals,

Diaz-Hernandez et al. observed a concentration-dependent increase of calcium when hippocampal synaptosomes were incubated with Ap_5A. Moreover the activation of the dinucleotide receptors A1 and A2A potentiated Ap_5A response, whereas the A3 receptor depressed Ap_5A response (Diaz-Hernandez et al. 2002).

5 Ap*n*As and Fertility

Ap*n*As have also been studied in fertilization and sperm motility. Already in the early nineties, Ap_4A and Ap_3A were shown to be potential inhibitory signals of spermatozoa motility after four or more hours of exposure (Chan et al. 1991). On the other hand, Ap*n*A scan be degraded to AMP in rabbit seminal fluid, and the addition of diadenosine compounds to incapacitated rabbit spermatozoa increases fertility capacity (Minelli et al. 2003). Membrane vesicles were isolated from rabbit seminal plasma, enzyme activity was determined by HPLC, and Ap_3A, Ap_4A, and ATP degradation by the membrane vesicles in the seminal fluid was observed. Another interesting observation was that the spermatozoa capacitating effector bovine serum albumin was no longer needed when diadenosine compounds, that were previously incubated with membrane vesicles, were added to the medium (Minelli et al. 2003). In a review (Allegrucci et al. 2000), Minelli's group suggested that the capacitative effect of the degradation products of diadenosine compounds might be related to the activation of adenosine receptors by adenosine on the surface of rabbit sperm cells. In support of this hypothesis, Allegrucci's group demonstrated beneficial effects on fertilization when the A1 adenosine receptor was stimulated (Allegrucci et al. 2001). The activation of this receptor led to capacitation enhancement, enabling the acrosome reaction, a fundamental step in fertilization of the oocyte.

6 Ap*n*As and Neutrophils

The fact that Ap*n*As have various different physiological roles led researchers to investigate these nucleotides in immune-derived cells. In 1996, Gasmi et al. studied the interaction of Ap_3A and Ap_4A with granulocyte macrophage colony-stimulating factor (GM-CSF) in neutrophils (Gasmi et al. 1996a, b). The interest in these cells was because they play an important role in acute inflammation and are poorly responsive to patho-physiological agonists. Moreover, they possess a very short half-life and constitutively undergo apoptosis. Incubation of Ap_4A and Ap_3A with neutrophils resulted in delayed loss of function in acute inflammation, neutrophil apoptosis, and cellular morphology changes and was almost as effective as GM-CSF in delaying neutrophil apoptosis. The anti-apoptotic effect on neutrophils was greater when they were incubated with ATP, Ap_3A, and Ap_4A together with GM-CSF than when each nucleotide or cytokine alone was incubated with the neutrophils. Further investigation of other Ap*n*As showed that Ap_5A and Ap_6A also delayed neutrophil apoptosis, as assessed by morphology, function, CD16 expression, and chromatin integrity (Gasmi et al. 1996a, b).

7 The Role of Ap*n*As in Pancreatic Cells

It has been hypothesized that the Ap*n*As have a physiological role in the control of insulin and glucagon secretion. Ap_4A has been studied in pancreatic cells, especially in the insulin secreting β-cells. Stimulation of the glucose-sensitive pancreatic β-cells increased Ap_4A levels and inhibited the K_{ATP} channels, resulting in activation of calcium channels and release of insulin (Ripoll et al. 1996; Martin et al. 1998).

It has been also shown that Ap_4A directly targets pancreatic K_{ATP} channels and antagonizes their opening. Pancreatic K_{ATP} channel subunits, Kir6.2 and SUR1, were coexpressed in monkey kidney (COS-7) cells, and the effect of Ap_4A on the single channel behavior was examined (Jovanovic and Jovanovic 2001). Ap_4A inhibited channel opening in a concentration-dependent manner. Analysis of single channels demonstrated that Ap_4A did not change intraburst kinetic behavior of K_{ATP} channels, but rather decreased burst duration and increased between-burst duration. It was concluded that Ap_4A antagonizes K_{ATP} channel opening by targeting channel subunits themselves and by keeping channels longer in closed interburst states. Silverstre et al. studied the effects of Ap_3A and Ap_4A in rat pancreas and found that Ap_4A but not Ap_3A induced a 4-fold increase over the basal value of insulin release from β-cells (Silvestre et al. 1999). Glucagon release was also found to be stimulated by Ap_4A. Ap_4A decreases the probability of K_{ATP} channel opening without affecting the channel's amplitude nor the mean open or closed time. Therefore, the mechanism in which Ap_4A decreases the K_{ATP} channels opening is yet to be determined.

8 The Role of Ap_4A in Activated Mast Cells

Ap_4A also has an important role as a second messenger in mast cells after physiological stress. When mast cells are immunologically stimulated via IgE-antigen binding to the FcεRI receptor, KARS, which is situated in the cytoplasmic multi-synthetase complex (MSC) (Robinson et al. 2000; Han et al. 2003), is phosphorylated at serine 207, released from the MSC in a MAP kinase-dependent manner, and translocated to the nucleus (Yannay-Cohen et al. 2009). This phosphorylation at serine 207 switches the function of KARS from aminoacylation to Ap_4A production. In the nucleus, Ap_4A binds to the histidine triad nucleotide-binding protein 1 (Hint-1), causing its dissociation from the microphthalmia-associated transcription factor (MITF), allowing MITF to transcribe its target genes (Lee et al. 2004).

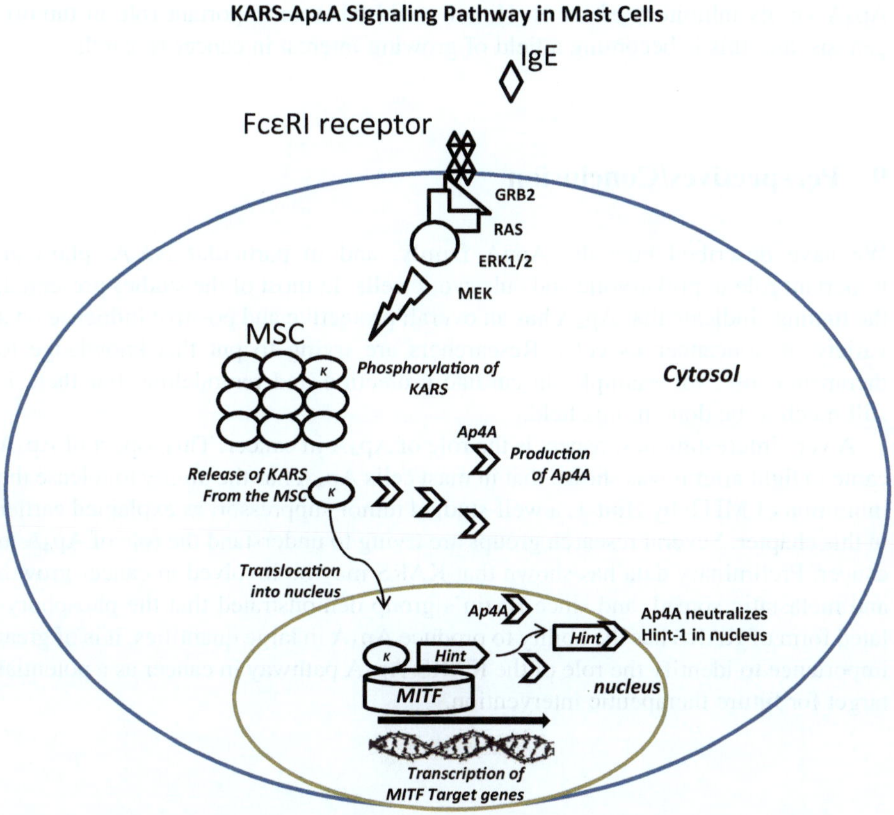

Fig. 2 Activation of the FcεRI receptor in mast cells via IgE results in the activation of the mitogene-activated protein kinase (MAPK) pathway, leading to phosphorylation of KARS in position 207S and its release from the multi-synthetase complex (MSC). On its way to the nucleus and after translocation to the nucleus, KARS produces Ap_4A. After KARS is translocated to the nucleus, it forms a complex with Hint-1 and MITF. Ap_4A on the other hand binds to the tumor suppressor Hint-1 and removes its inhibitory effect on microphthalmia transcription factor (MITF)

This finding implies that Ap_4A is a key player in gene regulation, as schematically shown in Fig. 2.

Other transcription factors that belong to the same family as MITF and that are also affected by the Ap_4A-Hint-1 binding are USF1 and USF2. These transcription factors are ubiquitously expressed in eukaryotic cells. They have a role in various cellular processes; for example, they are important transcription factors in mouse embryogenesis (Sirito et al. 1998) and are also involved in tumorigenesis and glucose metabolism. The binding of Ap_4A to Hint-1 removes the suppressor effect of Hint-1 on USF1 and USF2 allowing them to transcribe their target genes (Lee et al. 2004).

Since Hint-1 has been shown to be a tumor suppressor in melanoma (Genovese et al. 2012), in colon cancer (Wang et al. 2007) and in hepatoma (Wang et al. 2009),

Ap_4A via its inhibitory effect on Hint-1 could have an important role in tumorigenesis, and this is becoming a field of growing interest in cancer research.

9 Perspectives/Conclusion

We have described how the ApnA family, and in particular Ap_4A, plays an important role in prokaryotic and eukaryotic cells. In most of the studies presented, the findings indicate that Ap_4A has an overall protective and positive influence on a variety of noncancerous cells. Researchers are trying to put this knowledge to therapeutic use, for example, in cardiac protection and remodeling, but there is still much to be done in this field.

A very interesting new aspect is the role of Ap_4A in cancer. This aspect of Ap_4A came to light after it was shown that in mast cells Ap_4A has the ability to release the inhibition of MITF by Hint-1, a well-studied tumor suppressor, as explained earlier in this chapter. Several research groups are trying to understand the role of Ap_4A in cancer. Preliminary data has shown that KARS may be involved in cancer growth and metastatic spread, and since Razin's group demonstrated that the phosphorylated form of KARS has the ability to produce Ap_4A in large quantities, it is of great importance to identify the role of the KARS-Ap_4A pathway in cancer as a potential target for future therapeutic intervention.

References

Ahmet I, Sawa Y, Nishimura M et al (2000) Cardioprotective effect of diadenosine tetraphosphate (AP_4A) preservation in hypothermic storage and its relation with mitochondrial ATP-sensitive potassium channels. Transplantation 69:16–20

Allegrucci C, Liguori L, Mezzasoma I et al (2000) A1 adenosine receptor in human spermatozoa: its role in the fertilization process. Mol Genet Metab 71:381–386

Allegrucci C, Liguori L, Minelli A (2001) Stimulation by n6-cyclopentyladenosine of A1 adenosine receptors, coupled to galphai2 protein subunit, has a capacitative effect on human spermatozoa. Biol Reprod 64:1653–1659

Busse R, Ogilvie A, Pohl U (1988) Vasomotor activity of diadenosine triphosphate and diadenosine tetraphosphate in isolated arteries. Am J Physiol 254:H828–H832

Carmi-Levy I, Yannay-Cohen N, Kay G et al (2008) Diadenosine tetraphosphate hydrolase is part of the transcriptional regulation network in immunologically activated mast cells. Mol Cell Biol 28:5777–5784

Cartwright JL, Britton P, Minnick MF et al (1999) The IalA invasion gene of *Bartonella bacilliformis* encodes a (de)nucleoside polyphosphate hydrolase of the MutT motif family and has homologs in other invasive bacteria. Biochem Biophys Res Commun 256:474–479

Chan PJ, Su BC, Tredway DR (1991) Diadenosine tetraphosphate (Ap_4A) and triphosphate (Ap_3A) signaling of human sperm motility. Arch Androl 27:103–108

Charlier J, Sanchez R (1987) Lysyl-tRNA synthetase from *Escherichia coli* K12: chromatographic heterogeneity and the lysU-gene product. Biochem J 248:43–51

Chen X, Boonyalai N, Lau C et al (2013) Multiple catalytic activities of *Escherichia coli* lysyl-tRNA synthetase (LysU) are dissected by site-directed mutagenesis. FEBS J 280:102–114

Coiffard B, Soubeyran P, Ghigo E (2015) Editorial: Manipulation of the cellular microbicidal response and endocytic dynamic by pathogens membrane factors. Front Cell Infect Microbiol 5:42

Communi D, Motte S, Boeynaems JM et al (1996) Pharmacological characterization of the human P2Y4 receptor. Eur J Pharmacol 317:383–389

Coste H, Brevet A, Plateau P et al (1987) Non-adenylylated bis(5′-nucleosidyl) tetraphosphates occur in *Saccharomyces cerevisiae* and in *Escherichia coli* and accumulate upon temperature shift or exposure to cadmium. J Biol Chem 262:12096–12103

Diaz-Hernandez M, Pereira MF, Pintor J et al (2002) Modulation of the rat hippocampal dinucleotide receptor by adenosine receptor activation. J Pharmacol Exp Ther 301:441–450

Farr SB, Arnosti DN, Chamberlin MJ et al (1989) An apaH mutation causes AppppA to accumulate and affects motility and catabolite repression in *Escherichia coli*. Proc Natl Acad Sci USA 86:5010–5014

Gasmi L, McLennan AG, Edwards SW (1996a) The diadenosine polyphosphates Ap_3A and Ap_4A and adenosine triphosphate interact with granulocyte-macrophage colony-stimulating factor to delay neutrophil apoptosis: implications for neutrophil: platelet interactions during inflammation. Blood 87:3442–3449

Gasmi L, McLennan AG, Edwards SW (1996b) Neutrophil apoptosis is delayed by the diadenosine polyphosphates, Ap_5A and Ap_6A: synergism with granulocyte-macrophage colony-stimulating factor. Br J Haematol 95:637–639

Gaywee J, Radulovic S, Higgins JA et al (2002) Transcriptional analysis of *Rickettsia prowazekii* invasion gene homolog (invA) during host cell infection. Infect Immun 70:6346–6354

Genovese G, Ghosh P, Li H et al (2012) The tumor suppressor HINT1 regulates MITF and beta-catenin transcriptional activity in melanoma cells. Cell Cycle 11:2206–2215

Han JM, Kim JY, Kim S (2003) Molecular network and functional implications of macromolecular tRNA synthetase complex. Biochem Biophys Res Commun 303:985–993

Hilderman RH, Martin M, Zimmerman JK et al (1991) Identification of a unique membrane receptor for adenosine 5′, 5‴-P1, P4-tetraphosphate. J Biol Chem 266:6915–6918

Ismail TM, Hart CA, McLennan AG (2003) Regulation of dinucleoside polyphosphate pools by the YgdP and ApaH hydrolases is essential for the ability of *Salmonella enterica* serovar *typhimurium* to invade cultured mammalian cells. J Biol Chem 278:32602–32607

Jo YH, Schlichter R (1999) Synaptic corelease of ATP and GABA in cultured spinal neurons. Nat Neurosci 2:241–245

Johnstone DB, Farr SB (1991) AppppA binds to several proteins in *Escherichia coli*, including the heat shock and oxidative stress proteins DnaK, GroEL, E89, C45 and C40. EMBO J 10: 3897–3904

Jovanovic S, Jovanovic A (2001) Diadenosine tetraphosphate-gating of recombinant pancreatic ATP-sensitive K(+) channels. Biosci Rep 21:93–99

Kawamoto J, Kurihara T, Kitagawa M et al (2007) Proteomic studies of an Antarctic cold-adapted bacterium, *Shewanella livingstonensis* Ac10, for global identification of cold-inducible proteins. Extremophiles 11:819–826

Lazarowski ER, Watt WC, Stutts MJ et al (1995) Pharmacological selectivity of the cloned human P2U-purinoceptor: potent activation by diadenosine tetraphosphate. Br J Pharmacol 116: 1619–1627

Lee YN, Nechushtan H, Figov N et al (2004) The function of lysyl-tRNA synthetase and Ap_4A as signaling regulators of MITF activity in FcepsilonRI-activated mast cells. Immunity 20: 145–151

Leveque F, Plateau P, Dessen P et al (1990) Homology of lysS and lysU, the two *Escherichia coli* genes encoding distinct lysyl-tRNA synthetase species. Nucleic Acids Res 18:305–312

Marriott AS, Copeland NA, Cunningham R et al (2015) Diadenosine 5′, 5‴-P(1), P(4)-tetraphosphate (Ap_4A) is synthesized in response to DNA damage and inhibits the initiation of DNA replication. DNA Repair (Amst) 33:90–100

Martin F, Pintor J, Rovira JM et al (1998) Intracellular diadenosine polyphosphates: a novel second messenger in stimulus-secretion coupling. FASEB J 12:1499–1506

Minelli A, Liguori L, Bellezza I et al (2003) Effects of diadenosine polyphosphates and seminal fluid vesicles on rabbit sperm cells. Reproduction 125:827–835

Miras-Portugal MT, Pintor J, Gualix J (2003) Ca2+ signalling in brain synaptosomes activated by dinucleotides. J Membr Biol 194:1–10

Ofir-Birin Y, Fang P, Bennett SP et al (2013) Structural switch of lysyl-tRNA synthetase between translation and transcription. Mol Cell 49:30–42

Ogilvie A, Blasius R, Schulze-Lohoff E et al (1996) Adenine dinucleotides: a novel class of signalling molecules. J Auton Pharmacol 16:325–328

Pintor J, Diaz-Rey MA, Torres M et al (1992) Presence of diadenosine polyphosphates—Ap_4A and Ap_5A—in rat brain synaptic terminals. Ca2+ dependent release evoked by 4-aminopyridine and veratridine. Neurosci Lett 136:141–144

Pintor J, Puche JA, Gualix J et al (1997) Diadenosine polyphosphates evoke Ca2+ transients in guinea-pig brain via receptors distinct from those for ATP. J Physiol 504(2):327–335

Ralevic V, Burnstock G (1998) Receptors for purines and pyrimidines. Pharmacol Rev 50: 413–492

Ripoll C, Martin F, Manuel Rovira J et al (1996) Diadenosine polyphosphates: a novel class of glucose-induced intracellular messengers in the pancreatic beta-cell. Diabetes 45:1431–1434

Robinson JC, Kerjan P, Mirande M (2000) Macromolecular assemblage of aminoacyl-tRNA synthetases: quantitative analysis of protein-protein interactions and mechanism of complex assembly. J Mol Biol 304:983–994

Sasaki C, Kitagawa H, Zhang WR et al (2000) Temporal profile of cytochrome c and caspase-3 immunoreactivities and TUNEL staining after permanent middle cerebral artery occlusion in rats. Neurol Res 22:223–228

Schachter JB, Li Q, Boyer JL et al (1996) Second messenger cascade specificity and pharmacological selectivity of the human P2Y1-purinoceptor. Br J Pharmacol 118:167–173

Sherman MY, Goldberg AL (1993) Heat shock of *Escherichia coli* increases binding of dnaK (the hsp70 homolog) to polypeptides by promoting its phosphorylation. Proc Natl Acad Sci USA 90:8648–8652

Silvestre RA, Rodriguez-Gallardo J, Egido EM et al (1999) Stimulatory effect of exogenous diadenosine tetraphosphate on insulin and glucagon secretion in the perfused rat pancreas. Br J Pharmacol 128:795–801

Sirito M, Lin Q, Deng JM et al (1998) Overlapping roles and asymmetrical cross-regulation of the USF proteins in mice. Proc Natl Acad Sci USA 95:3758–3763

Vahlensieck U, Boknik P, Gombosova I et al (1999) Inotropic effects of diadenosine tetraphosphate (AP_4A) in human and animal cardiac preparations. J Pharmacol Exp Ther 288:805–813

Wang Y, Chang CF, Morales M et al (2003) Diadenosine tetraphosphate protects against injuries induced by ischemia and 6-hydroxydopamine in rat brain. J Neurosci 23:7958–7965

Wang L, Zhang Y, Li H et al (2007) Hint1 inhibits growth and activator protein-1 activity in human colon cancer cells. Cancer Res 67:4700–4708

Wang L, Li H, Zhang Y et al (2009) HINT1 inhibits beta-catenin/TCF4, USF2 and NFkappaB activity in human hepatoma cells. Int J Cancer 124:1526–1534

Wright M, Boonyalai N, Tanner JA et al (2006) The duality of LysU, a catalyst for both Ap_4A and Ap_3A formation. FEBS J 273:3534–3544

Wright M, Azhar MA, Kamal A et al (2014) Syntheses of stable, synthetic diadenosine polyphosphate analogues using recombinant histidine-tagged lysyl tRNA synthetase (LysU). Bioorg Med Chem Lett 24:2346–2352

Yannay-Cohen N, Carmi-Levy I, Kay G et al (2009) LysRS serves as a key signaling molecule in the immune response by regulating gene expression. Mol Cell 34:603–611
Zamecnik PC, Stephenson ML, Janeway CM et al (1966) Enzymatic synthesis of diadenosine tetraphosphate and diadenosine triphosphate with a purified lysyl-sRNA synthetase. Biochem Biophys Res Commun 24:91–97

Thinking Small: Circulating microRNAs as Novel Biomarkers for Diagnosis, Prognosis, and Treatment Monitoring in Breast Cancer

Yin-Long Yang

Contents

Abstract Cancer initiation and progression are governed by both genetic and epigenetic events. Epigenetic alterations which include changes in DNA methylation, histone modifications, and noncoding RNA-mediated gene silencing are reversible and heritable. Aberrant epigenetic modifications are believed to be essential players in cancer initiation and progression. Recent advances in epigenetics have offered not only a deeper understanding of the underlying mechanisms of carcinogenesis but also new avenues for identification of clinically relevant putative biomarkers for the early detection, prognosis, monitoring of treatment response, and risk stratification of cancer patients. Following identification of cell-free nucleic acids in systematic circulation, cumulating evidences have demonstrated the potential of cell-free epigenetic biomarkers in the body fluids for cancer. Recently, the emergence of microRNAs as biomarkers has added an extra dimension to the "molecular signatures" of breast cancer. In this chapter, we summarize the currently published state-of-the-art research on the role of the circulating microRNAs in clinical utility for breast cancer, the most common

Y.-L. Yang (✉)
Department of Breast Surgery, Fudan University Shanghai Cancer Center, Shanghai 200000, China

Department of Oncology, Shanghai Medical College, Fudan University, Shanghai 200000, China
e-mail: lancetyang@126.com

S. Jurga et al. (eds.), *Modified Nucleic Acids in Biology and Medicine*, RNA Technologies, DOI 10.1007/978-3-319-34175-0_10

cancer in women. In addition, we also discuss the current obstacles that have limited the routine use of epigenetic biomarkers and provide future perspectives, so that these novel cancer biomarkers can be readily developed for significant clinical improvement in the management of breast cancer patients.

Keywords Breast cancer • MicroRNAs • Biological markers • Early diagnosis • Prognosis

1 Introduction

Breast cancer (BC) is the most frequently diagnosed noncutaneous cancer and the second leading cause of cancer mortality in US women, accounting for 29 % of all cancer cases and 15 % of cancer-related deaths (Siegel et al. 2014). For 2014, an estimated 232,670 new cases of BC and 40,000 BC deaths were expected among US women (Siegel et al. 2014). Early diagnosis and treatment are of clinical significance to prevent and manage breast cancer. However, current diagnostic methods, mainly mammography and ultrasound, are not sensitive or sufficiently specific to diagnose BC at an early stage. MRI, despite of its potential excessive diagnosis for patients due to limited specificity, has thus become a new important diagnostic test in the evaluation of patients presenting with occult primary breast cancer. On the other hand, although two serum-based tumor biomarkers, CA15-3 and carcinoembryonic antigen (CEA), are used to assess the advanced breast cancer, neither is recommended for diagnostic use. Circulating tumor cells in blood have been used for prognostic assessment in patients with metastatic breast cancer (Cristofanilli et al. 2004), but they are of limited use for the detection of early diseases, because tumor burden is low and fewer cells are shed into the circulation (Taback et al. 2003). Consequently, finding a rapid and reliable serum assay for diagnostic as well as prognostic assessment of BC remains a major clinical challenge.

With the increasing understanding of the biological behavior of breast cancer, the treatment has entered an era of comprehensive mode including both local treatments (surgery and radiotherapy) and systemic therapy (chemotherapy, endocrine therapy, targeted therapy, and traditional Chinese medicine adjuvant therapy), resulting in a great reduction of tumor recurrence and mortality. Generally, chemotherapy is recommended for patients with a higher recurrence risk. However, though BC can respond to chemotherapy, its sensitivity to a given drug regimen varies with each patient. Certain patients, even with a high risk, are not sensitive to chemotherapy; therefore, it is necessary to find a predictive marker to select the sensitive BC patients to receive chemotherapy and avoid overtreatment.

Recently, the discovery of microRNAs (miRNAs) has brought new opportunities in cancer biology. miRNAs are regulatory, nonprotein-coding, 9- to

25-nucleotide-long RNA molecules that regulate the expression of a variety of genes by sequence-specific base pairing on the 3′ untranslated regions (3′UTR) of the target mRNA, resulting in mRNA degradation or inhibition of translation. Certain miRNAs are involved in the pathogenesis of tumors and function as oncogenes (oncomiRs) or tumor suppressors (Lee et al. 1993). These miRNAs can play a significant role in the regulation of cell development, metabolism, immunity, proliferation, differentiation, and apoptosis. Several reports have suggested that circulating miRNAs are stable and detectable in serum/plasma and the levels of some miRNAs specifically elevated in the patients with breast cancer (Heneghan et al. 2010). These findings suggest the possibility of blood-based miRNAs functioning as novel prognostic and predictive biomarkers and revolutionary sources of biomarker for BC diagnosis. Furthermore, expression profiles of serum miRNAs can distinguish patients with specific cancers. Thus, a comprehensive and systematic understanding of breast cancer-specific miRNAs profiling is of great benefit in the early diagnosis, prognosis, and personalized treatment of breast cancer. However, the clinical utility of miRNA has not been investigated in a well-defined breast cancer-related study. Here, we review our current knowledge about the involvement of circulating miRNAs in breast cancer and their potential as diagnostic, prognostic, and predictive biomarkers.

2 The Origin of Circulating miRNAs

Though the majority of miRNAs are observed intracellularly, a significant number of miRNAs have been found outside of cells, including various body fluids. Lawrie et al. (2008) were among the first to demonstrate the presence of circulating miRNAs in cell-free bodily fluids such as plasma and serum. Since then, circulating miRNAs have been constantly reported as being aberrantly expressed in blood plasma or serum in different types of cancer, e.g., prostate, colorectal, and esophageal carcinoma (Brase et al. 2011; Huang et al. 2010; Zhang et al. 2010). Their most important advantages include the possibility of repeated measurement in a noninvasive manner as well as remarkable stability in plasma/serum. Serum and other body fluids are known to contain ribonucleases (Weickmann and Glitz 1982), which suggests that secreted miRNAs are likely packaged in some manner to protect them against RNase digestion.

Despite the accumulating evidence for the presence of miRNAs in body fluids, the secretory mechanism of extracellular miRNAs remains poorly understood. miRNAs could be shielded from degradation by packaging in lipid vesicles, in complexes with RNA-binding proteins, or both (Gibbings et al. 2009). Recent studies have identified miRNAs in two types of cell-derived lipid vesicles: microvesicles and exosomes. Microvesicles are fragments of plasma membrane ranging from 100 to 1000 nm shed from almost all cell types. Exosomes, on the other hand, are cell-derived vesicles that are present in many and perhaps all biological fluids. Exosomes with its diameter ranging from 50 to 100 nm are released on exocytic fusion of multivesicular bodies (MVB) (Hasselmann et al. 2001) with plasma membranes (Cocucci et al. 2009). miRNAs have been identified in both exosomes and microvesicles derived from a variety of sources, including human and mouse mast cells, glioblastoma tumors, plasma, saliva, and urine. The underlying mechanism of the secretory process is that the release of miRNAs is controlled by neutral sphingomyelinase 2 (nSMase2) and through ceramide-dependent secretory machinery (Kosaka et al. 2010b). Apoptotic bodies are small membranous particles released during programmed cell death. Currently, accumulating evidence suggests that apoptotic bodies can function as intercellular transmitters to convey their contents, in particular, miRNA (Rechavi et al. 2009). Authors (Gidlof et al. 2013) also demonstrated that platelets activated during myocardial infarction could release functional miRNA into plasma, which can be taken up by endothelial cells and regulate intercellular adhesion molecule (ICAM-1) expression. Moreover, recent studies also suggest that circulating high-density lipoprotein (HDL) can bind and transport endogenous miRNAs and deliver them to recipient cells with functional targeting capabilities (Vickers et al. 2011). nSMase2, as also demonstrated, could regulate cellular export of miRNAs to HDL. According to another study, cells in culture predominantly exported miRNA in exosome-independent form (Wang et al. 2010). Turchinovich et al. (2011) study showed that most of the extracellular miRNAs in blood plasma and cell culture is independent of exosomes and is bound to Ago2 protein—a part of RNA-induced silencing complex. This result indicates that large parts of circulating miRNAs might be by-products of dead/dying cells which persist due to the stability of the miRNA/Ago2 complex. Other studies demonstrated that the circulating miRNAs also can be passively leaked from apoptotic or necrotic cells, which has been verified to occur in patients with HCV infection (Bala et al. 2012), acute myocardial infarction (Ai et al. 2010), and breast cancer (Roth et al. 2010) after tissue injury or cell death. The secretion of miRNAs by cells is associated with the microenvironment of cells. Figure 1 summarizes the secretary mechanism of circulating miRNAs.

Extracellular miRNAs, circulating miRNAs as a potential disease diagnostic biomarker in particular, has attracted much attention, and the clarification of the release and transportation mechanism of which is important for the understanding and discovery of the value of miRNAs as disease biomarkers. As mentioned above, a number of potential circulating miRNAs as diagnostic, prognostic, and predictive biomarkers in many cancers have been well recognized. However, the complexity of circulating miRNA export phenomenon determines the limited possibility of using one mechanism to completely explain the release of cellular miRNA into circulation. Much more studies are urged to determine how miRNAs are specifically targeted for release, recognized for uptake, and what information can be transmitted by this process.

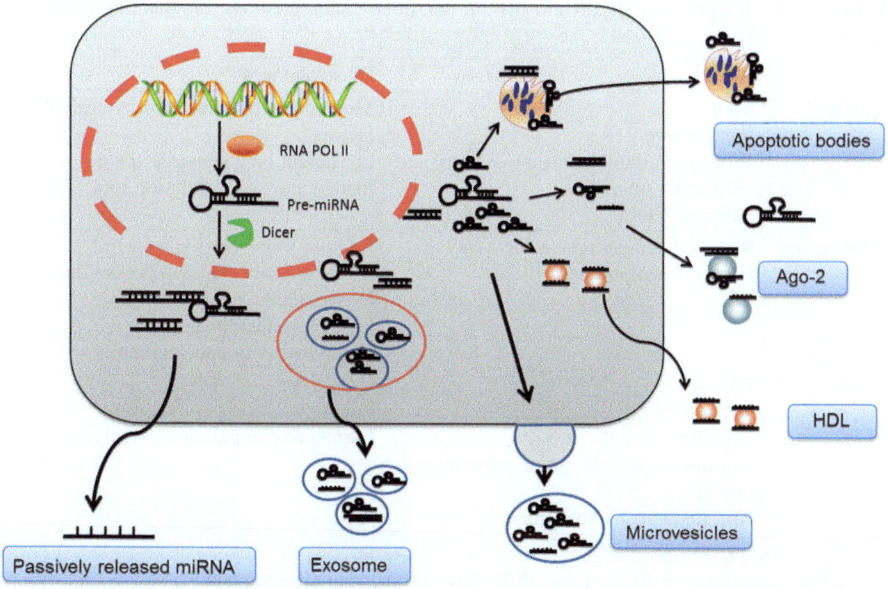

Fig. 1 Release of circulating miRNAs

3 Circulating miRNAs Measurement

Currently, real-time reverse transcription-PCR (qPCR), microarray, and massively parallel/next-generation sequencing (NGS) are three main methods to quantify circulating miRNAs. qPCR is often considered a "gold standard" in the detection and quantification of gene expression (Livak and Schmittgen 2001), for it not only provides us with a simple tool to efficiently determine the amount of a gene transcript in a given sample but is also relatively inexpensive, widely available, and allow measurements of very small quantities of miRNAs. However, the simplicity of this methodology can itself be problematic, as one tends to overlook critical factors that make this technique work. The rapid increase in number of miRNAs also renders qPCR inefficiency on a genomic scale, and it is probably better used as a validation rather than as a discovery tool.

Since the demonstration of microarray technology by DNA microarray paper, it has been utilized to analyze a comprehensive miRNA expression profiling. Generally speaking, microarray-based measurement methods require more starting material than qPCR, and it can be challenging to develop probes and hybridization conditions that work well to detect many different miRNAs at once. As with genomic RNA analysis, microarrays are still the best choice for a standardized genome-wide assay that is amenable to high-throughput applications. Although microarrays could improve the throughput of miRNA profiling, the method is relatively limited in terms of sensitivity and specificity (Krichevsky et al. 2003; Sato et al. 2009).

Table 1 Advantages and disadvantages of the three main methods to quantify circulating miRNAs

	Advantages	Disadvantages
qPCR	Simple Inexpensive Widely available measurements of very small quantities Discovery tool	Method easily influenced by critical factors Inefficient on a genomic scale Inadequate as a validation tool
Microarray	Standardized genome-wide assay High throughput	Require more starting material Develop probe develop probes and hybridization conditions Low reproducibility Low sensitivity and specificity
NGS	Precise Repeatable High throughput Not rely on the design of primers or probes Not hindered by variability Identify novel miRNAs	Expensive Laborious Immature computational analysis

qPCR Real-time reverse transcription-PCR; *NGS* Next-generation sequencing

Of these three methods, miRNA profiling by NGS may be the most promising, as it largely avoids many miRNA measurement pitfalls. NGS RNA sequencing is comparatively the most precise, repeatable, and of high throughput, representing the most advanced and expensive technology. This technology is neither hindered by variability in melting temperatures, coexpression of nearly identical miRNA family members, or post-transcriptional modifications, nor relies on the design of primers or probes specific to each miRNA. Instead, NGS sequencing provides the number of counts for each miRNA or transcript present in the sample. Its application in circulating miRNA includes new miRNAs exploration, detection of miRNAs, miRNAs editing, and isomiR and target mRNA detection. Despite these advantages, NGS remains expensive and labor intensive, both in the sample preparation and data analysis, and associated tools for computational analysis are in their infancy. Table 1 lists advantages and disadvantages of the three main methods. In addition to microarrays and qPCR, there are other less frequently used methods, such as traditional northern blotting and ligation-based measurement. As with mRNA measurement, each of these methods has benefits and drawbacks. To further develop miRNA-based biomarkers, it is urgently needed to set up a standardization process for sample preparation and develop a more accurate method to assess the quality and quantity of miRNAs.

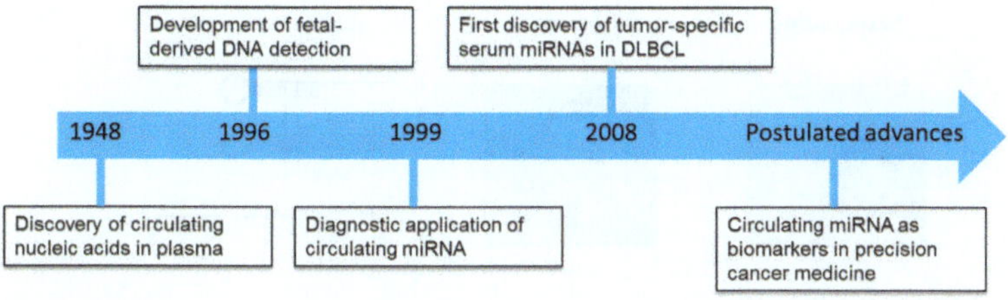

Fig. 2 Timeline of major progresses in the roles of circulating miRNAs in cancer

4 Progress on Circulating miRNAs in Cancer Application

One of the major challenges in cancer research is the identification of stable and routinely measurable biomarkers in easily accessible samples. Serum tumor markers, such as CEA and CA19-9, are being used in convenient diagnostic assays. However, these conventional serum markers lack sufficient sensitivity and specificity to facilitate early detection of cancer. Friel et al. (2010) reported the existence of circulating nucleic acids in plasma early in 1948. However, the diagnostic and prognostic utility of circulating DNA/RNA has not begun to be appreciated until the development of fetal-derived DNA detection in maternal plasma in 1996 which opened up the possibility of noninvasive prenatal diagnosis (Tsang and Lo 2007) and since then more discoveries of circulating tumor-derived RNA have proven to be equally important as their DNA counterparts. After 1999, the medical community started to pay attention to the diagnostic application of circulating miRNAs. In 2008, tumor-specific miRNAs were first discovered by Lawrie et al. (2008) in the serum of patients with diffuse large B-cell lymphoma (DLBCL): high level of miR-21 was associated with improved relapse-free survival. Following this initial finding, an increasing number of studies have been conducted during the past decade to analyze the clinical relevance of circulating miRNAs in peripheral blood and have successfully demonstrated, for one, the potential of circulating miRNAs to act as novel noninvasive biomarkers for early tumor detection, diagnosis, and prognosis of various cancers and other diseases, and for another the fact that miRNAs yields are higher in patients with malignant lesions than in healthy patients (Fig. 2).

miRNAs have emerged as critical players in cancer initiation and progression processes. The physiological events that lead to the increase of miRNAs during cancer development and progression are yet to be fully understood. However, analyses of circulating miRNAs have allowed the detection of tumor-related genetic and epigenetic alterations that are relevant to cancer development and progression. The circulating nucleic acids, present in serum and other body fluids, may represent potential biomarkers. The development of tumor-specific miRNA signatures as cancer biomarkers detectable in body fluids should help with early detection and more effective therapeutic intervention for individual patients. The detection and identification of potential miRNA biomarkers in BC are an emerging avenue of miRNA researches, which will be necessary prior to the application of miRNAs in BC diagnosis and prognosis (Figs. 3 and 4).

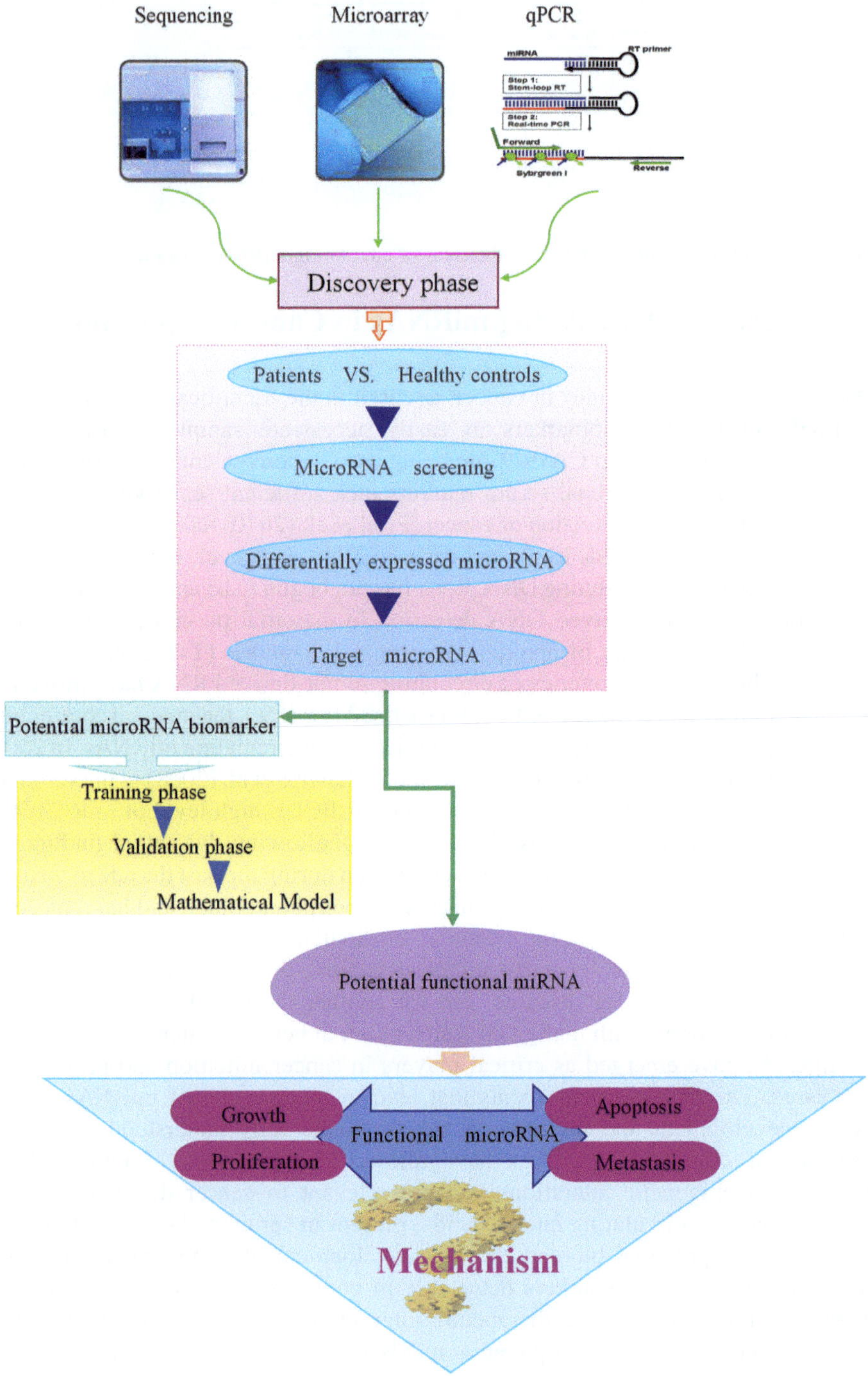

Fig. 3 Strategy for the detection and identification of circulating miRNA biomarkers

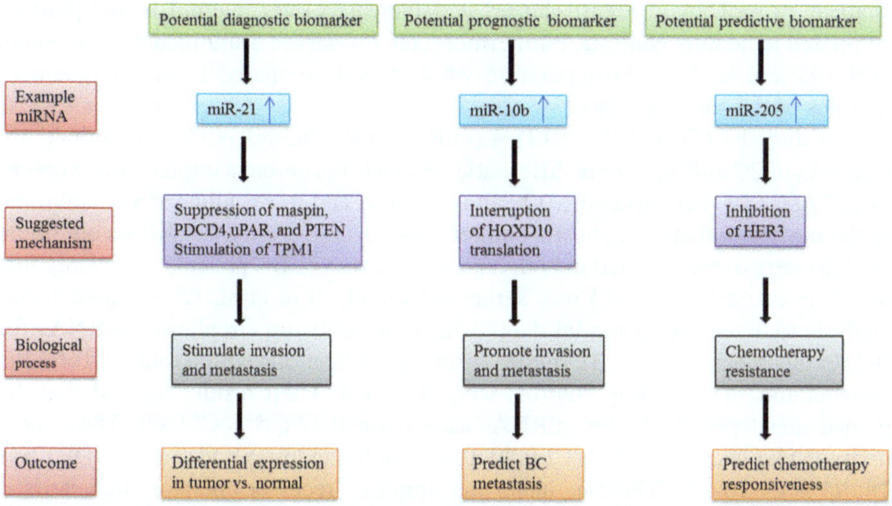

Fig. 4 Potential applications of circulating miRNAs in breast cancer. *PDCD4* programmed cell death 4; *uPAR* urokinase plasminogen activator surface receptor; *PTEN* phosphatase and tensin homolog; *TPM1* tropomyosin 1; *HOXD10* homeobox D10

5 Circulating miRNAs and Early Diagnosis of Breast Cancer

miRNAs are released from cells wrapped in proteins or microvesicles. Selectively, exported miRNAs are packaged in structures that are different from those that carry neutral released miRNAs (n-miRNAs). Malignant cells release greater quantities of selectively exported miRNAs, whereas the release of n-miRNAs is not affected by malignancy. The expression of circulating miRNAs in serum is stable, not easily degradable (Reid et al. 2011). Compared with exogenous synthetic miRNAs, endogenous cell-free miRNAs exist in serum more in form of RNase-resistant types and own self-renewable characteristics, resistance to DNase/RNase digestion and repeated frozen (Qu et al. 2011). The extreme stability of circulating miRNAs in the RNase-rich bloodstream environment is the basis of their value as biomarkers (Kosaka et al. 2010a).

Because of their remarkable stability in blood, circulating miRNAs are considered to be highly promising biomarkers for various tumors (Abba et al. 2012; Gilad et al. 2008), for it is a convenient noninvasive test method to diagnose tumors in early stages. The number of papers reporting that circulating miRNAs could serve as noninvasive biomarkers for BC detection is increasing. Heneghan et al. (2010) surveyed a panel of seven candidate miRNAs in whole blood RNAs from 148 BC patients and 44 age-matched and disease-free controls. They found the expression of miR-195 was significantly elevated in BC patients. In addition, they found the expression of let-7a, which is well regarded as a reliable endogenous control for analysis of miRNA in breast cancer, was increased over fivefold in BC patients compared to healthy controls. Furthermore, they observed a significant reduction in miR-195 and let-7a in postoperative whole blood compared to the preoperative samples of the same patients.

In a study by Chan et al. (2013), 4 (miR-1, miR-92a, miR-133a, and miR-133b) miRNAs of 20 miRNAs were differentiated in a set of serum samples from a cohort of 132 Asian BC patients and 101 healthy controls and were validated and identified as the most significant diagnostic markers. Interestingly, only 7 miRNAs out of the total 20 were overexpressed in both tumor and serum of BC patients, indicating that miRNAs could be released into serum selectively. Liu et al. (2014) used meta-analysis to assess the potential diagnostic value of using circulating miRNAs for BC. In this meta-analysis, 31 studies from 16 publications with a total of 1668 BC patients and 1111 healthy controls were included. Their results showed that the pooled sensitivity (SEN) for miRNAs assays was 0.77 (95 % CI 0.69–0.84), specificity (SPE) was 0.88 (95 % CI 0.79–0.93), and diagnostic odds ratio (DOR) was 18 (95 % CI 10–32). The area under the summary receiver operator characteristic (SROC) curve (AUC) was 0.89 (95 % CI 0.86–0.91). Subgroup analyses suggested that employing a combination of multiple miRNAs was better than using a single miRNA, suggesting that the expression profiles of circulating miRNAs, especially using a combination of them, would have potential to facilitate accurate breast tumor detection. Shen et al. (2014) performed a three-stage miRNA analysis with plasma samples from BC patients and healthy controls. In the discovery phase, they identified that plasma levels of three miRNAs, including miR-148b, miR-133a, and miR-409-3p, were significantly higher in BC cases than healthy controls. In the second validation phase, they substantiated the associations with BC detection for miR-148b and miR-133a. In the last in-vitro study phase, they found that both miR-148b and miR-133a were secreted from BC cell lines, showing their secretory potential and possible tumor origin. Their data thus suggest that both miR-148b and miR-133a have potential use as biomarkers for BC detection. Kodahl et al. (2014a) investigated the differential expression of miRNAs in the serum of patients with estrogen receptor-positive (ER+) early-stage BC and healthy controls. They identified a multivariable signature consisting of 9 miRNAs (miR-15a, miR-18a, miR-107, miR-133a, miR-139-5p, miR-143, miR-145, miR-365, miR-425) that provided considerable discrimination between BC patients and healthy controls. A 9 miRNA signature capable of discriminating between ER-positive BC and healthy controls was also presented in their research. This signature might be useful in the development of a blood-based multi-marker test to improve early detection of breast cancer. Recently, Park et al. (2014) used microarray-based expression profiling to compare the levels of circulating miRNAs in blood samples from 11 ER+/ human epidermal growth factor receptor 2 (HER2)-negative(HER2-) advanced BC patients plus 5 age-matched controls. Their results showed that miR-1280 levels increased significantly in BC patients and reflected tumor status (control $\ll$ early cancer $<$ metastatic cancer). Furthermore, they confirmed that miR-1280 was not a classic miRNA, but rather a tRNALeu-derived fragment. All these findings suggest that a circulating tRNA-derived miRNA, miR-1280, is differently expressed in BC patients and may serve as a biomarker for ER-positive breast cancer.

As is indicated in other profiling results from another study, the combination of circulating miR-145 and miR-451 seems capable of predicting BC patients from normal individuals (Ng et al. 2013). Seven serum miRNAs (miR-10b, miR-21, miR-125b, miR-145, miR-155, miR-191, and miR-382), identified by Fermín et al. (Mar-Aguilar et al. 2013), were significantly upregulated in BC patients compared to healthy controls. Asaga et al. (2011) also demonstrated an overexpression of miR-21 in plasma samples of BC patients. miR-181a, the key oncogenic regulator, is deregulated in many types of human cancer. It has been reported that serum miR-181a was downregulated in BC and may represent a novel biomarker for primary BC as well as for early stage BC diagnosis (Guo and Zhang 2012). Eichelser et al. (2013) found that the concentrations of circulating miR-34a, miR-93, and miR-373 were significantly higher in BC patients than in healthy women. Zeng et al. (2013) suggested that plasma miRNA-30a decreased in patients with BC. Hu et al. (2012) found four serum miRNAs (miR-16, miR-25, miR-222 and miR-324-3p) that were consistently differentially expressed between BC cases and controls. The area under the receiver operating characteristic curve is 0.954 for the four-miRNA signature in the discovery stage, suggesting the four-miRNA signature from serum may serve as a noninvasive prediction biomarker for breast cancer. Chen et al. (2014) reported that circulating levels of the muscle-enriched miR486 was lower in patients with BC compared with healthy individuals. The hypothesis was tested in this study that circulating miRNAs might serve as a surrogate of the effects of cancer on miRNA expression or release in distant organs and be emerging as important biomarkers of BC. Using microarray-based expression profiling followed by real time quantitative polymerase chain reaction (RT-qPCR), Zhao H and colleagues (Zhao et al. 2010) found deregulated expression levels of 49 miRNAs in plasma from 20 women with early stage BC compared to 20 matched controls. The authors also showed that both upregulated ($n = 26$) and downregulated ($n = 23$) miRNAs could discriminate patients from controls with acceptable specificity and sensitivity scores. To clearly present these important data, we summarized the differentially expressed miRNAs of each study in Table 2.

6 Circulating miRNAs and Prognosis of Breast Cancer

Metastatic breast cancer (MBC) is a leading cause of morbidity and mortality among females. There is an urgent need for prognostic biomarkers that can improve the quality of life for these patients. Circulating miRNAs have been identified in the plasma/serum and can be potentially used as biomarkers for tumor characterization and cancer prognosis (Heneghan et al. 2011; Yang et al. 2014; Zhao et al. 2010).

Roth et al. (2010) provided a pilot study in which the first evidence that tumor-associated circulating miRNAs were elevated in the blood of BC patients and associated with tumor progression was provided. The authors evaluated in

Table 2 Results of circulating miRNAs in breast cancer diagnosis

miRNA	Sample types	Expression	References
miR-195, let-7a	Blood-derived miRNA	Upregulated	Heneghan et al. (2010)
miR-10b, miR-125b, miR-145, miR-155, miR-191, miR-382	Serum miRNA	Upregulated	Mar-Aguilar et al. (2013)
miR-21	Serum miRNA	Upregulated	Asaga et al. (2011), Mar-Aguilar et al. (2013)
miR-181a	Serum miRNA	Downregulated	Guo and Zhang (2012)
miR-34a, miR-93, miR-373	Serum miRNA	Upregulated	Eichelser et al. (2013)
miR-30a	Plasma miRNA	Downregulated	Zeng et al. (2013)
miR-155, miR-19a, miR-181b, miR-24	Serum miRNA	Upregulated	Sochor et al. (2014)
miR-148b, miR-376c, miR-409-3p, miR-801	Plasma miRNA	Upregulated	Cuk et al. (2013)
miR-16, miR-25, miR-222, miR-324-3p	Serum miRNA	Upregulated	Hu et al. (2012)
miR-21, miR-146a	Plasma miRNA	Upregulated	Kumar et al. (2013)
miR-29a, miR-21	Serum miRNA	Upregulated	Wu et al. (2011)
miR-16, miR-21, miR-451, miR-145	Plasma miRNA	Upregulated Downregulated	Ng et al. (2013)
Let-7c, miR-589	Plasma miRNA	Downregulated Upregulated	Zhao et al. (2010)
miR-10b, miR-155	Serum miRNA	Upregulated	Wang et al. (2012)
miR-484	Serum miRNA	Upregulated	Zearo et al. (2014)
miR-1260	Blood-derived miRNA	Upregulated	Park et al. (2014)
miR-148b, miR-133a	Plasma miRNA	Upregulated	Shen et al. (2014)
miR486	Serum miRNA	Downregulated	Chen et al. (2014)

particular the relative concentrations of breast cancer-associated miR-10b, miR-34a, miR-141, and miR-155 in the blood serum of 89 patients with primary BC and metastatic disease and 29 healthy women, finding that miR-10b, miR-34a and miR-155 discriminated M1-patients from M0 controls. In BC patients, the changes in the levels of miR10b ($P = 0.01$), miR34a ($P = 0.003$), and miR155 ($P = 0.002$) were correlated with the presence of overt metastases. Within the

M0-cohort, patients at advanced tumor stages (pT3 to 4) had significantly more miR34a ($P = 0.01$) in their blood than that of patients at early tumor stages (pT1 to 2).

The use of circulating tumor cells (CTC) as a prognostic marker in metastatic BC has been well established. However, their efficacy and accuracy are still under scrutiny mainly because of the methods of their enrichment and identification. Madhavan et al. (2012) found that CTC-positive had significantly higher levels of circulating miR-141, miR-200a, miR-200b, miR-200c, miR-203, miR-210, miR-375, and miR-801 than CTC-negative MBC and controls, whereas miR-768-3p was present in lower amounts in MBC cases. miR-200b was singled out as the best marker for distinguishing CTC-positive from CTC-negative patients. They also identified combinations of miRNAs for differentiating MBC cases from controls, among which combinations of miRNAs and miR-200b alone were found to be promising prognostic marker for progression-free and overall survival. Furthermore, according to their research, the identified miRNAs seem to have a similar or even better prognostic value than CTCs, and combination of miRNAs and CTCs performs better than CTCs alone.

Increased concentrations of serum miR-373 were associated with negative HER2 status of the primary tumor (Eichelser et al. 2013). Deregulated concentrations of serum miR-17 and miR-34a were detected in patients with ER/progesterone receptor (PR)-positive and -negative status, respectively. These hormone receptor-negative tumors, which represent approximately 30 % of all breast cancers, have been reported to have a more aggressive clinical course (Sakr and Dizon 2011). Their findings indicate that serum concentrations of deregulated miRNAs may be linked to a particular biology of BC favoring progression and metastatic spread. Wang et al. (2012) demonstrated that expression levels of blood miR-125b and miR-155 were related to disease stage with higher miRNA levels in higher stage disease, whereas significant association with ER, PR, and HER2 status is wanted. Asaga et al. (2011) also showed that the concentration of circulating miR-21 was correlated with BC stage and was independent of ER status or age. Therefore, circulating miR-21 may be a potential biomarker for BC progression. Eichelser et al. (2013) found that serum miR-17 and miR-155 were differently expressed between M0 and M1 patients. Wu et al. (2012) indicated miR-122 prevalence in the circulation predicted BC metastasis in early-stage patients. Their finding is in accordance with several reports that miR-21 expression in breast tumors is correlated with the advanced clinical stage, lymph node metastasis, and poor prognosis (Qian et al. 2009; Yan et al. 2008). Chen et al. (2013) showed the plasma levels of circulating miR-10b and miR-373 were significantly higher in BC patients with lymph node metastasis compared to M0 patients and normal donors, suggesting that circulating miRNA-10b and miRNA-373 are potential biomarkers for detecting the lymph node status of breast cancer. Joosse et al. (2014) reported serum let-7b was significantly higher in patients with invasive carcinomas than in patients with benign breast diseases or healthy women ($P < 0.001$), whereas the level of serum miR-202 was elevated in both patient cohorts ($P < 0.001$). In uni- and multivariate

Table 3 Results of circulating miRNAs in breast cancer prognosis

miRNA	Sample types	Expression	Cohort	References
miR-10b miR-34a miR-155	Serum miRNA	Upregulated Downregulated	MI versus M0	Roth et al. (2010)
miR-155 miR-125b	Serum miRNA	Upregulated	Stage III versus Stage II	Wang et al. (2012)
miR-17 miR-155	Serum miRNA	Downregulated	MI versus M0	Eichelser et al. (2013)
miR-122	Serum miRNA	Upregulated	Stage II-III versus inflammatory BC	Wu et al. (2012)
miR-21	Serum miRNA	Upregulated	Stage IV versus Stage I, II, or III	Asaga et al. (2011)
miR-10b miR-373	Plasma miRNA	Upregulated	MI versus M0	Chen et al. (2013)
miR-202	Serum miRNA	Upregulated	MI versus M0	Joosse et al. (2014)

M0—patients with localized breast cancer. *M1*—patients with metastatic breast cancer

analyses, high levels of miR-202 significantly correlated with poor overall survival ($P = 0.0001$). Tables 3 summarized the results of circulating miRNAs in BC prognosis.

Circulating miRNAs closely relate to tumor progression or metastasis. Differential expression of circulating miRNAs can distinguish patients with BC from healthy females and further distinguish patients with distant metastases from those with locoregional disease, so they can act as valuable prognostic markers to predict clinical outcomes of patients suffering from cancer (Yu et al. 2011). The detected miRNAs hold promise as an early detection marker of metastasis in breast cancer.

7 Circulating miRNAs and Personalized Treatment of Breast Cancer

Therapeutic strategies based on modulation of miRNA activity hold great promise due to the ability of these small RNAs to potently influence cellular behaviors. Some studies evaluate the usefulness of miRNAs as both targets and tools in anticancer therapy Meng et al. (2006, 2007). The involvement of miRNAs in the biology of human cancer is supported by an increasing body of experimental evidence, the focus of which has gradually switched from profiling studies, as the first BC-specific signature reported in 2005 (Iorio et al. 2005) describing an aberrant miRNA expression in different tumor types, to biological demonstrations of the causal role of these small molecules in the tumorigenic process and the possible implications as biomarkers or therapeutic tools (Abba et al. 2012). These more recent studies have widely demonstrated that miRNAs can modulate oncogenic or tumor suppressor pathways, and that, at the same time, their expression can be regulated by oncogenes or tumor suppressor genes. Based on the theories above, using miRNAs as a target for molecular therapy will become a new direction for cancer treatment. Specific blocking or inducing miRNA-mediated biological processes could reverse the development of cancers (Budhu et al. 2010).

Treatment for BC is constantly evolving as new technologies, agents, and strategies are discovered. Systemic adjuvant chemotherapy improves outcomes and has advanced from first-generation regimens to modern dose-dense combinations. Advances in adjuvant treatment of BC have already led to a significant reduction in disease-related relapse and death (Davies et al. 2011). To remove the unresectable tumors, neoadjuvant chemotherapy (NAC), which is a promising platform for drug development, is recommended for inoperable locally advanced tumors. Chemotherapy remains a critical component of adjuvant therapy for BC patients. However, there is significant variation in drug response and survival outcomes in individuals treated with equivalent regimens. Traditional factors including tumor stage, tumor size, nodal status, and intra-tumoral characteristics such as grade, expression of ER/PR, and HER2 status have been used to guide choice of therapy. In recent years, advances in miRNAs research have aroused a significant interest in how differences in genetic makeup may be used to predict treatment safety and efficacy. However, the role of circulating miRNAs in therapy resistance of BC still remains poorly characterized. In recent years, some circulating miRNAs have been reported to be involved in chemotherapeutic resistance of breast cancers. Wang et al. (2012) demonstrated that higher circulating miR-125b level is correlated to more advanced ductal carcinoma of the breast and resistance for adjuvant chemotherapy. In addition, miR-125b expression in breast cancers was reversely correlated with apoptosis and proliferation inhibition induced by chemotherapy. Furthermore, BC cells expressing a higher level of miR-125b were more resistant; conversely, reducing miR-125b level sensitized BC cells to chemotherapy. E2F3, as a direct target of miR-125b in BC cells, may be involved in miR-125b-mediated chemotherapeutic response. These data suggest that circulating miR-125b, a potential marker predicting chemotherapeutic response and a target for overcoming chemotherapeutic resistance, contributes to chemotherapeutic resistance of breast cancer.

HER2 is amplified in 20–30 % of invasive BC cases, and its amplification is associated with poor patient prognosis. Trastuzumab is a humanized monoclonal antibody that binds specifically to the HER2 receptor and suppresses cell proliferation that is driven by overexpression of the HER2 protein. Since the advent of treatment with trastuzumab, survival of HER2-positive BC patients has significantly improved, and mortality in HER2 overexpressing or amplified tumors ("HER2-positive") has been significantly reduced (Slamon et al. 2011). Trastuzumab is now part of the standard treatment for HER2-positive breast cancer, but not all patients respond to trastuzumab, and metastatic patients treated with trastuzumab will eventually progress during therapy. Given variable response, significant risks associated with treatment, such as identification of genetic markers predictive of trastuzumab response is quite attractive. In patients receiving neoadjuvant chemotherapy combined with trastuzumab, Jung et al. (2012) reported that circulating miR-210 levels were significantly higher in those who had residual disease than in those who achieved a pathologic complete response. The mean expression ratio for miR-210 was significantly higher in trastuzumab-resistant BT474 cells, and miR-210 expression was significantly higher before surgery than after and in patients whose cancer metastasized to the lymph nodes. These results suggest that plasma miR-210 may be used to predict and perhaps monitor response to therapies that contain trastuzumab. Lapatinib is also an established treatment for patients with HER2-positive BC with different mechanisms of action. Muller et al. (2014) investigated whether altered expression levels of potentially relevant miRNAs in serum are associated with the response to trastuzumab or lapatinib. Circulating miR-21, miR-210, and miR-373 were quantified in serum of 127 HER2-postive BC patients receiving chemotherapy combined with either trastuzumab or lapatinib before and after NAC and in 19 healthy controls. Serum levels of miR-21, miR-210, and miR-373 were significantly higher in patients before and after chemotherapy than in healthy women. Concentrations of miR-21, miR-210, and miR-373 increased further after chemotherapy. A significant association of higher serum levels of miR-373 with advanced clinical tumor stage could be detected ($p < 0.002$). An association of miR-21 levels before ($p = 0.0091$) and after ($p = 0.037$) chemotherapy with overall survival of the patients could be detected, independent of type of anti-HER2 therapy. No association of circulating miRNAs with pathological response (pCR) was found. In all, their findings demonstrate a specific influence of neoadjuvant therapy on the serum levels of miR-21, miR-210, and miR-373 in BC patients together with a prognostic value of miR-21.

Nearly two-thirds of breast cancers are classified as ER+, which is prognostic for improved survival outcomes and predicts responsiveness to endocrine manipulations. By binding to either ER-alpha or -beta, estrogen regulates a wide variety of cellular effects and physiologic conditions including BC cell proliferation. Though not always considered in this manner, endocrine therapies are indeed BC-targeted therapies, as they treat cancer by blocking specific receptors, which prevents or inhibits tumor growth. Resistance to endocrine therapy is the major problem for ERα (+) BC patients. The luminal A subtype of BC is a type of BC that is ER- and/or PR-positive, HER2-negative, and Ki67 < 14 %. This type of BC has a relatively better prognosis but is less sensitive to chemotherapy compared to other subtypes of breast cancer (Coates et al. 2012). It has been controversial as to whether neoadjuvant chemotherapy should be given in luminal A subtype. Li et al. (2014) aimed to identify biomarkers to predict chemosensitivity for patients with luminal A subtype of BC who received neoadjuvant chemotherapy with epirubicin plus paclitaxel. The expression of serum miR-19a and miR-205 was highly expressed in BC patients with the luminal A subtype who are resistant to the epirubicin plus paclitaxel regimen. A predictive model of these two miRNAs was created by the logistic regression analysis. The probability of this model was 89.71 %. Based on the ROC curve, the specificity was 75.00 %, and the sensitivity was 81.25 %. The combination of serum miR-19a and miR-205 may predict the chemosensitivity of luminal A subtype of BC to epirubicin plus paclitaxel neoadjuvant chemotherapy. Zhao et al. (2011) explored the potential role of miR-221 as a biomarker for chemosensitivity in BC patients who previously received NAC. The expression level of plasma miR-221 was significantly associated with hormone receptor (HR) status. Patients with higher plasma miR-221 levels tended to be HR-negative. Patients with different miR-221 levels had significant differences in the overall response rate but not in the pathologic complete response rate. Their results indicate that plasma miR-221 may be a predictive biomarker for sensitivity to NAC in BC patients. Kodahl et al. (2014b) investigated changes in the levels of specific circulating miRNA following BC surgery in 24 postmenopausal women with ER-positive early-stage breast cancer. The results showed three circulating miRNAs (miR-338-3p, miR-223, and miR-148a) exhibited significantly lower and 1 miRNA (miR-107) higher levels in postoperative versus preoperative samples ($p < 0.05$). No miRNAs were consistently undetectable in the postoperative samples compared to the preoperative samples. Their findings demonstrates that these specific miRNAs may be involved in tumorigenesis and could potentially be used to monitor whether all cancer cells have been removed at surgery and/or, subsequently, whether the patients develop recurrence. According to Wu et al. (2012), two circulating miRNAs, miR-375 and miR-122, exhibited strong correlations with clinical outcomes, including NAC response and relapse with metastatic disease. These results may allow optimized chemotherapy treatments and preventive anti-metastasis interventions in future clinical applications. Freres et al. (2015) described the modifications of circulating miRNAs profile after NAC for breast cancer. Their research results showed serum miR-34a and miR-122 were highly upregulated after NAC. Studying the kinetics of circulating miR-34a and miR-122 expression during NAC revealed that their levels were especially increased after anthracycline-based chemotherapy. This study demonstrates that NAC specifically induces miRNA expression in plasma and tumor tissue, which might be involved in the antitumor effects of chemotherapy in BC patients. Gezer et al. (2014) assessed the effect of NAC on the levels of a panel of BC-associated miRNAs, which were at relatively low (let-7, miR-10b, miR-34, miR-155, miR-200c, and miR-205) or abundant (miR-21, miR-195, and miR-221) levels in the circulation. Their study demonstrated that highly expressed miRNAs were affected most frequently by chemotherapy, particularly in patients with early stage tumors. This information may be valuable in assessing the response of the patients to therapy. Table 4 presents the predictive function of circulating miRNAs in BC treatment.

8 Conclusions and Perspectives

From all the studies illustrated above, we know that circulating miRNA plays a key role in carcinogenesis and in the development of tumor metastasis. Circulating miRNA can be used as noninvasive biomarker for diagnosis of breast cancer. The

Table 4 Effects of circulating miRNAs on chemotherapeutic applications

miRNA	NAC	Effects	Expression	Target	References
miR-125b	Yes	5-fluorouracil resistance	Upregulated	E2F3	Wang et al. (2012)
miR-375 miR-122	Yes	Trastuzumab resistance	Upregulated Downregulated		Wu et al. (2012)
miR-210	Yes	Trastuzumab resistance	Upregulated	Ephrin-A3 E2F3 RAD52 FGFR1	Jung et al. (2012)
miR-221	Yes	Taxane and anthracycline resistance	Upregulated		Zhao et al. (2011)
miR-19a miR-205	Yes	Epirubicin and paclitaxel resistance	Upregulated		Li et al. (2014)
miR-21 miR-210 miR-373	Yes	Trastuzumab or lapatinib resistance	Upregulated		Muller et al. (2014)
miR-34a miR-122	Yes		Upregulated		Mar-Aguilar ct al. (2013)

NAC Neoadjuvant chemotherapy

stability of circulating miRNAs and the relatively cheap methods of their isolation and detection increase their usefulness as a biomarker. However, there are still challenges to be addressed to establish these new biomarkers before their application to routine clinical procedures. Several limitations existing in these results are as follows: (1) Given the limited sample size in these researches, further validations in large cohorts or in different ethnic groups are recommended; (2) Whether these plasma miRNAs elevation are specific for certain subtypes of BC remains uncertain. Thus, additional studies are necessary to compare their plasma levels with different subtypes. Nevertheless, it is worthy in the future to identify novel plasma miRNAs which could be used as a screening tool for BC detection. Differential expression signatures in plasma could discriminate BC from normal subject, which raises the possibility of using such marker to develop a blood-based screening test for BC in the future. Screening for BC allows early stage diagnosis of the malignancy and hence has potential to reduce mortality. As a screening tool acceptable for general population, it would be desirable to detect cancer accurately without resorting to an invasive procedure.

The circulating miRNAs in BC diagnosis and prognosis will become a promising method to deal with malignancies in a more specific and an efficient manner. However, it is still very challenging to deliver tumor cell-specific miRNA application via an affordable and convenient approach. Crucial issues need to be resolved before establishing circulating miRNAs as diagnostic biomarkers and predictive tools for BC. To bring miRNAs into clinical application, limitations are all urged to be overcome such as the lack of larger prospective clinical trials with standardized analyzing methods, the necessity for clarifying the real origin of circulating

Table 5 Challenges of circulating miRNAs research

※ Larger prospective clinical trials
※ Standardized analyzing methods
※ Clarifying the real origin of circulating miRNAs
※ Validation of a well-characterized BC-specific signature
※ Specific miRNAs for certain subtypes of BC
※ Target genes of circulating miRNA
※ Biological functions
※ Role in BC initiation and progression
※ Role in the regulation of CSCs self-renewal
※ Regulation of circulating miRNAs expression

BC Breast cancer; *CSC* Cancer stem cells

miRNAs, and the validation of a well-characterized BC-specific signature of circulating miRNAs. Furthermore, although the locations of multiple target genes of circulating miRNA and their biological functions are fully recognized, more pathogenesis, carcinogenesis, and biological mechanisms of miRNA on BC still call for further study. Meanwhile, lots of problems are emerging based on new role of miRNAs, including the understanding of the role of miRNAs in both the regulation of cancer stem cells (CSCs) self-renewal and the development of tumor invasion and metastasis, as well as their way of consisting a complex network to regulate miRNAs expression. Additionally, more advanced and perfected methods for diagnosis and treatment of cancer based on circulating miRNA need to be explored. We can draw a figure to describe the challenges regarding circulating miRNAs research in the future (Table 5).

In conclusion, being one of the major scientific breakthroughs, the discovery of circulating miRNA has revolutionized current medical science. With the in-depth study of circulating miRNA, it will definitely lead us to a totally new world of life sciences. We are sure that more tumor patients will get benefit from this kind of novel biomarker.

References

Abba M, Mudduluru G, Allgayer H (2012) MicroRNAs in cancer: small molecules, big chances. Anti-Cancer Agent Med Chem 12:733–743

Ai J, Zhang R, Li Y et al (2010) Circulating microRNA-1 as a potential novel biomarker for acute myocardial infarction. Biochem Biophys Res Commun 391:73–77

Asaga S, Kuo C, Nguyen T et al (2011) Direct serum assay for microRNA-21 concentrations in early and advanced breast cancer. Clin Chem 57:84–91

Bala S, Tilahun Y, Taha O et al (2012) Increased microRNA-155 expression in the serum and peripheral monocytes in chronic HCV infection. J Transl Med 10:151

Brase JC, Johannes M, Schlomm T et al (2011) Circulating miRNAs are correlated with tumor progression in prostate cancer. Int J Cancer 128:608–616

Budhu A, Ji J, Wang XW (2010) The clinical potential of microRNAs. J Hematol Oncol 3:37

Chan M, Liaw CS, Ji SM et al (2013) Identification of circulating microRNA signatures for breast cancer detection. Clin Cancer Res 19:4477–4487
Chen W, Cai F, Zhang B et al (2013) The level of circulating miRNA-10b and miRNA-373 in detecting lymph node metastasis of breast cancer: potential biomarkers. Tumour Biol 34:455–462
Chen D, Goswami CP, Burnett RM et al (2014) Cancer affects microRNA expression, release, and function in cardiac and skeletal muscle. Cancer Res 74:4270–4281
Coates AS, Colleoni M, Goldhirsch A (2012) Is adjuvant chemotherapy useful for women with luminal a breast cancer? J Clin Oncol 30:1260–1263
Cocucci E, Racchetti G, Meldolesi J (2009) Shedding microvesicles: artefacts no more. Trends Cell Biol 19:43–51
Cristofanilli M, Budd GT, Ellis MJ et al (2004) Circulating tumor cells, disease progression, and survival in metastatic breast cancer. N Engl J Med 351:781–791
Cuk K, Zucknick M, Heil J et al (2013) Circulating microRNAs in plasma as early detection markers for breast cancer. Int J Cancer 132:1602–1612
Davies C, Godwin J, Gray R et al (2011) Relevance of breast cancer hormone receptors and other factors to the efficacy of adjuvant tamoxifen: patient-level meta-analysis of randomised trials. Lancet 378:771–784
Eichelser C, Flesch-Janys D, Chang-Claude J et al (2013) Deregulated serum concentrations of circulating cell-free microRNAs miR-17, miR-34a, miR-155, and miR-373 in human breast cancer development and progression. Clin Chem 59:1489–1496
Freres P, Josse C, Bovy N et al (2015) Neoadjuvant chemotherapy in breast cancer patients induces miR-34a and miR-122 expression. J Cell Physiol 230:473–481
Friel AM, Corcoran C, Crown J et al (2010) Relevance of circulating tumor cells, extracellular nucleic acids, and exosomes in breast cancer. Breast Cancer Res Treat 123:613–625
Gezer U, Keskin S, Igci A et al (2014) Abundant circulating microRNAs in breast cancer patients fluctuate considerably during neoadjuvant chemotherapy. Oncol Lett 8:845–848
Gibbings DJ, Ciaudo C, Erhardt M et al (2009) Multivesicular bodies associate with components of miRNA effector complexes and modulate miRNA activity. Nat Cell Biol 11:1143–1149
Gidlof O, van der Brug M, Ohman J et al (2013) Platelets activated during myocardial infarction release functional miRNA, which can be taken up by endothelial cells and regulate ICAM1 expression. Blood 121(3908–3917):S3901–S3926
Gilad S, Meiri E, Yogev Y et al (2008) Serum microRNAs are promising novel biomarkers. PLoS One 3, e3148
Guo L-J, Zhang Q-Y (2012) Decreased serum miR-181a is a potential new tool for breast cancer screening. Int J Mol Med 30:680–686
Hasselmann DO, Rappl G, Tilgen W et al (2001) Extracellular tyrosinase mRNA within apoptotic bodies is protected from degradation in human serum. Clin Chem 47:1488–1489
Heneghan HM, Miller N, Lowery AJ et al (2010) Circulating microRNAs as novel minimally invasive biomarkers for breast cancer. Ann Surg 251:499–505
Heneghan HM, Miller N, Kerin MJ (2011) Circulating microRNAs: promising breast cancer biomarkers. Breast Cancer Res 13:402
Hu Z, Dong J, Wang LE et al (2012) Serum microRNA profiling and breast cancer risk: the use of miR-484/191 as endogenous controls. Carcinogenesis 33:828–834
Huang Z, Huang D, Ni S et al (2010) Plasma microRNAs are promising novel biomarkers for early detection of colorectal cancer. Int J Cancer 127:118–126
Iorio MV, Ferracin M, Liu CG et al (2005) MicroRNA gene expression deregulation in human breast cancer. Cancer Res 65:7065–7070
Joosse SA, Muller V, Steinbach B et al (2014) Circulating cell-free cancer-testis MAGE-A RNA, BORIS RNA, let-7b and miR-202 in the blood of patients with breast cancer and benign breast diseases. Br J Cancer 111:909–917
Jung EJ, Santarpia L, Kim J et al (2012) Plasma microRNA 210 levels correlate with sensitivity to trastuzumab and tumor presence in breast cancer patients. Cancer 118:2603–2614

Kodahl AR, Lyng MB, Binder H et al (2014a) Novel circulating microRNA signature as a potential non-invasive multi-marker test in ER-positive early-stage breast cancer: a case control study. Mol Oncol 8:874–883

Kodahl AR, Zeuthen P, Binder H et al (2014b) Alterations in circulating miRNA levels following early-stage estrogen receptor-positive breast cancer resection in post-menopausal women. PLoS One 9:e101950

Kosaka N, Iguchi H, Ochiya T (2010a) Circulating microRNA in body fluid: a new potential biomarker for cancer diagnosis and prognosis. Cancer Sci 101:2087–2092

Kosaka N, Iguchi H, Yoshioka Y et al (2010b) Secretory mechanisms and intercellular transfer of microRNAs in living cells. J Biol Chem 285:17442–17452

Krichevsky AM, King KS, Donahue CP et al (2003) A microRNA array reveals extensive regulation of microRNAs during brain development. RNA 9:1274–1281

Kumar S, Keerthana R, Pazhanimuthu A et al (2013) Overexpression of circulating miRNA-21 and miRNA-146a in plasma samples of breast cancer patients. Indian J Biochem Biophys 50:210–214

Lawrie CH, Gal S, Dunlop HM et al (2008) Detection of elevated levels of tumour-associated microRNAs in serum of patients with diffuse large B-cell lymphoma. Br J Haematol 141:672–675

Lee RC, Feinbaum RL, Ambros V (1993) The C. elegans heterochronic gene lin-4 encodes small RNAs with antisense complementarity to lin-14. Cell 75:843–854

Li Q, Liu M, Ma F et al (2014) Circulating miR-19a and miR-205 in serum may predict the sensitivity of luminal: a subtype of breast cancer patients to neoadjuvant chemotherapy with epirubicin plus paclitaxel. PLoS One 9:e104870

Liu L, Wang S, Cao X et al (2014) Analysis of circulating microRNA biomarkers for breast cancer detection: a meta-analysis. Tumour Biol 35:12245–12253

Livak KJ, Schmittgen TD (2001) Analysis of relative gene expression data using real-time quantitative PCR and the 2(-Delta Delta C(T)) Method. Methods 25:402–408

Madhavan D, Zucknick M, Wallwiener M et al (2012) Circulating miRNAs as surrogate markers for circulating tumor cells and prognostic markers in metastatic breast cancer. Clin Cancer Res 18:5972–5982

Mar-Aguilar F, Mendoza-Ramirez JA, Malagon-Santiago I et al (2013) Serum circulating microRNA profiling for identification of potential breast cancer biomarkers. Dis Markers 34:163–169

Meng F, Henson R, Lang M et al (2006) Involvement of human micro-RNA in growth and response to chemotherapy in human cholangiocarcinoma cell lines. Gastroenterology 130:2113–2129

Meng F, Henson R, Wehbe-Janek H et al (2007) MicroRNA-21 regulates expression of the PTEN tumor suppressor gene in human hepatocellular cancer. Gastroenterology 133:647–658

Muller V, Gade S, Steinbach B et al (2014) Changes in serum levels of miR-21, miR-210, and miR-373 in HER2-positive breast cancer patients undergoing neoadjuvant therapy: a translational research project within the Geparquinto trial. Breast Cancer Res Treat 147:61–68

Ng EK, Li R, Shin VY et al (2013) Circulating microRNAs as specific biomarkers for breast cancer detection. PLoS One 8:e53141

Park IH, Kang JH, Lee KS et al (2014) Identification and clinical implications of circulating microRNAs for estrogen receptor-positive breast cancer. Tumour Biol 35:12173–12180

Qian B, Katsaros D, Lu L et al (2009) High miR-21 expression in breast cancer associated with poor disease-free survival in early stage disease and high TGF-beta1. Breast Cancer Res Treat 117:131–140

Qu KZ, Zhang K, Li H et al (2011) Circulating microRNAs as biomarkers for hepatocellular carcinoma. J Clin Gastroenterol 45:355–360

Rechavi O, Erlich Y, Amram H et al (2009) Cell contact-dependent acquisition of cellular and viral nonautonomously encoded small RNAs. Genes Dev 23:1971–1979

Reid G, Kirschner MB, van Zandwijk N (2011) Circulating microRNAs: association with disease and potential use as biomarkers. Crit Rev Oncol Hematol 80:193–208
Roth C, Rack B, Muller V et al (2010) Circulating microRNAs as blood-based markers for patients with primary and metastatic breast cancer. Breast Cancer Res 12:R90
Sakr BJ, Dizon DS (2011) Breast cancer: adjuvant modalities. Clin Obstet Gynecol 54:150–156
Sato F, Tsuchiya S, Terasawa K et al (2009) Intra-platform repeatability and inter-platform comparability of microRNA microarray technology. PLoS One 4, e5540
Shen J, Hu Q, Schrauder M et al (2014) Circulating miR-148b and miR-133a as biomarkers for breast cancer detection. Oncotarget 5:5284–5294
Siegel R, Ma J, Zou Z et al (2014) Cancer statistics, 2014. CA Cancer J Clin 64:9–29
Slamon D, Eiermann W, Robert N et al (2011) Adjuvant trastuzumab in HER2-positive breast cancer. N Engl J Med 365:1273–1283
Sochor M, Basova P, Pesta M et al (2014) Oncogenic microRNAs: miR-155, miR-19a, miR-181b, and miR-24 enable monitoring of early breast cancer in serum. BMC Cancer 14:448
Taback B, Giuliano AE, Hansen NM et al (2003) Detection of tumor-specific genetic alterations in bone marrow from early-stage breast cancer patients. Cancer Res 63:1884–1887
Tsang JC, Lo YM (2007) Circulating nucleic acids in plasma/serum. Pathology 39:197–207
Turchinovich A, Weiz L, Langheinz A et al (2011) Characterization of extracellular circulating microRNA. Nucleic Acids Res 39:7223–7233
Vickers KC, Palmisano BT, Shoucri BM et al (2011) MicroRNAs are transported in plasma and delivered to recipient cells by high-density lipoproteins. Nat Cell Biol 13:423–433
Wang K, Zhang S, Weber J et al (2010) Export of microRNAs and microRNA-protective protein by mammalian cells. Nucleic Acids Res 38:7248–7259
Wang H, Tan G, Dong L et al (2012) Circulating MiR-125b as a marker predicting chemoresistance in breast cancer. PLoS One 7:e34210
Weickmann JL, Glitz DG (1982) Human ribonucleases: quantitation of pancreatic-like enzymes in serum, urine, and organ preparations. J Biol Chem 257:8705–8710
Wu Q, Lu Z, Li H et al (2011) Next-generation sequencing of microRNAs for breast cancer detection. J Biomed Biotechnol 2011:597145
Wu X, Somlo G, Yu Y et al (2012) De novo sequencing of circulating miRNAs identifies novel markers predicting clinical outcome of locally advanced breast cancer. J Transl Med 10:42
Yan LX, Huang XF, Shao Q et al (2008) MicroRNA miR-21 overexpression in human breast cancer is associated with advanced clinical stage, lymph node metastasis and patient poor prognosis. RNA 14:2348–2360
Yang Y, Qian J, Chen Y et al (2014) Prognostic role of circulating microRNA-21 in cancers: evidence from a meta-analysis. Tumour Biol 35:6365–6371
Yu DC, Li QG, Ding XW et al (2011) Circulating microRNAs: potential biomarkers for cancer. Int J Mol Sci 12:2055–2063
Zearo S, Kim E, Zhu Y et al (2014) MicroRNA-484 is more highly expressed in serum of early breast cancer patients compared to healthy volunteers. BMC Cancer 14:200
Zeng RC, Zhang W, Yan XQ et al (2013) Down-regulation of miRNA-30a in human plasma is a novel marker for breast cancer. Med Oncol 30:477
Zhang C, Wang C, Chen X et al (2010) Expression profile of microRNAs in serum: a fingerprint for esophageal squamous cell carcinoma. Clin Chem 56:1871–1879
Zhao H, Shen J, Medico L et al (2010) A pilot study of circulating miRNAs as potential biomarkers of early stage breast cancer. PLoS One 5:e13735
Zhao R, Wu J, Jia W et al (2011) Plasma miR-221 as a predictive biomarker for chemoresistance in breast cancer patients who previously received neoadjuvant chemotherapy. Onkologie 34:675–680

Modified Antisense Oligonucleotides and Their Analogs in Therapy of Neuromuscular Diseases

Patryk Konieczny, Ewa Stepniak-Konieczna, and Krzysztof Sobczak

Contents

Abstract Neuromuscular diseases (NMDs) affect the musculature due to pathologies in muscles or nerves that manage their function. Potential therapeutics based on antisense oligonucleotides (AONs), i.e., short, single-stranded nucleic acids or their analogs complementary to specific sequences in RNA targets, are among the most promising. Depending on the chemistry, modified AONs can be used to either sterically block regions in RNA that would otherwise serve as binding sites for other cellular components or degrade RNA transcripts through activation of RNase-H. Both strategies can be employed for therapeutic treatment of NMDs and we

P. Konieczny • E. Stepniak-Konieczna • K. Sobczak (✉)
Department of Gene Expression, Institute of Molecular Biology and Biotechnology, Adam Mickiewicz University, Umultowska 89, 61-614, Poznan, Poland
e-mail: ksobczak@amu.edu.pl

S. Jurga et al. (eds.), *Modified Nucleic Acids in Biology and Medicine*, RNA Technologies, DOI 10.1007/978-3-319-34175-0_11

describe them in the context of loss-of-function diseases, such as Duchenne muscular dystrophy and spinal muscular atrophy, and gain-of-function disorders, including myotonic dystrophy and Huntington's disease. Modified AONs vary depending on the modification site. Owing to the chemical alteration, which can be either in the phosphate linkage, ribose sugar, or the whole nucleoside, AONs gain certain properties that decide on their stability, specificity, mode of action, in vivo distribution, and non-antisense effects. Delineation of these properties in preclinical studies has led to numerous clinical trials that currently hold great promise in the future therapeutic treatment of NMDs.

Keywords Antisense oligonucleotides • Neuromuscular diseases • Duchenne muscular dystrophy • Myotonic dystrophy • Spinal muscular atrophy • Huntington's disease • Pre-mRNA splicing • RNA degradation

1 Introduction

In this review, we described therapeutic strategies employing chemically modified antisense oligonucleotides (AONs) in the context of neuromuscular diseases (NMDs). NMDs constitute a heterogeneous group of genetic disorders that affect the musculature due to pathologies in either the nervous system or directly in striated muscle. Among them, Duchenne muscular dystrophy (DMD) and spinal muscular atrophy (SMA) are classified as loss-of-function diseases, in which mutations in specific genes lead to loss-of-functional proteins. In contrast, myotonic dystrophy (DM) and Huntington's disease (HD) are classified as gain-of-function disorders, in which mutated genes gain a deleterious function upon transcription. In all of these diseases, therapies with modified AONs target mutant RNA transcripts, omitting the direct cause of the disease, i.e., mutations in DNA. One caveat of such an approach could be that a single treatment would presumably never be permanent. However, the ease of usage, relatively low toxicity, and a broad range of action speak in favor of AON analogs. None of the abovementioned diseases is curable at the moment, neither with a modified AON nor any other treatment. Nevertheless, therapies grounded on AONs are currently among the most advanced and also most promising for NMDs. One of the biggest advantages of AON-based therapy is its flexibility. Modified AONs can be used to include or excise an exon, block binding sites for other molecules on RNA particles, or induce degradation of targeted transcript. As described below, all these AON features have been used in designing therapeutic approaches for genetic disorders. Also, in several studies, DNA-binding AON analogs have been used to affect transcription. Such broad properties of AONs are a result of chemical modifications that not only lead to their various functions but also affect their half-life time in extracellular space, mode of cell entry, resistance to nucleases, as well as their safety profile.

AONs are single-stranded nucleic acid molecules or their derivatives that bind to complementary regions in RNA by forming classical Watson-Crick base pairs

(Barciszewski et al. 2015; Bennett and Swayze 2010; Havens et al. 2013). AON length depends on the chemistry, specificity, and probability of interaction with a given sequence in one transcriptome region. Intrinsic short oligonucleotides are rapidly degraded in cellular environment by nucleases, which are usually divided into two classes—exonucleases and endonucleases that cleave the phosphodiester bond either in the flanking nucleotide residues or inside the AON chain, respectively. Chemical modifications of the backbone or sugar moieties can increase AON stability against nucleases, which significantly prolongs their life span and facilitates therapeutic usage. Heavy modifications applied to AONs can significantly reduce or even completely abrogate recognition and cleavage by cellular nucleases. In contrast, deoxyribonucleotide-based AONs containing phosphorothioate residues (DNA-PS) and their derivatives, including gapmers, allow for recruitment of a specific endonuclease (RNase-H) that recognizes the AON/RNA hybrid complex and cleaves the RNA molecule without affecting the structure of the modified oligonucleotide. Noteworthy, in mammalian cells, RNase-H recognizes DNA/RNA heteroduplexes and cleaves the RNA strand, releasing the DNA molecule (Cerritelli and Crouch 2009). Following target strand degradation, released AON analog can find another target molecule containing complementary sequence and altogether one particle can induce degradation of many RNA targets. Importantly, a perfect match between a modified AON and its target is necessary for RNase-H-induced efficient cleavage of the target RNA (Lima et al. 2007). Both types of AONs, the ones that do not lead to RNA cleavage by nucleases and the ones that allow for RNase-H incorporation, can be used in therapy. The former can be employed to sterically block other molecules from recognizing a specific sequence in the bound RNA chain, a strategy that has been successfully used, e.g., in alteration of splicing patterns in DMD and SMA (see below). On the other hand, modifications that allow for recruitment of RNase-H enabled, e.g., degradation of toxic RNA targets in DM and HD. We will start our description with modified AONs that do not induce significant target degradation and then depict oligo analogs that allow for RNase-H-based cleavage of RNA.

2 Modified AONs that Sterically Block Binding of Proteins and Other Molecules to RNA

At least 90 % of human pre-mRNA transcripts are spliced (Wang et al. 2008), which markedly adds to the proteome variety that could be calculated based on the number of 20,000 genes in the genome. Modified AONs enable control over splicing of not only alternative but also constitutive exons. In both DMD and SMA, therapeutic AON-based inclusion or exclusion of constitutive exons has been achieved (see below). Generally, AONs can affect splicing by masking a specific 5′ or 3′ splice site (5′SS/3′SS), exonic/intronic splicing enhancer (ESE/ISE), or exonic/intronic splicing silencer (ESS/ISS), i.e., sequences that are recognized by either

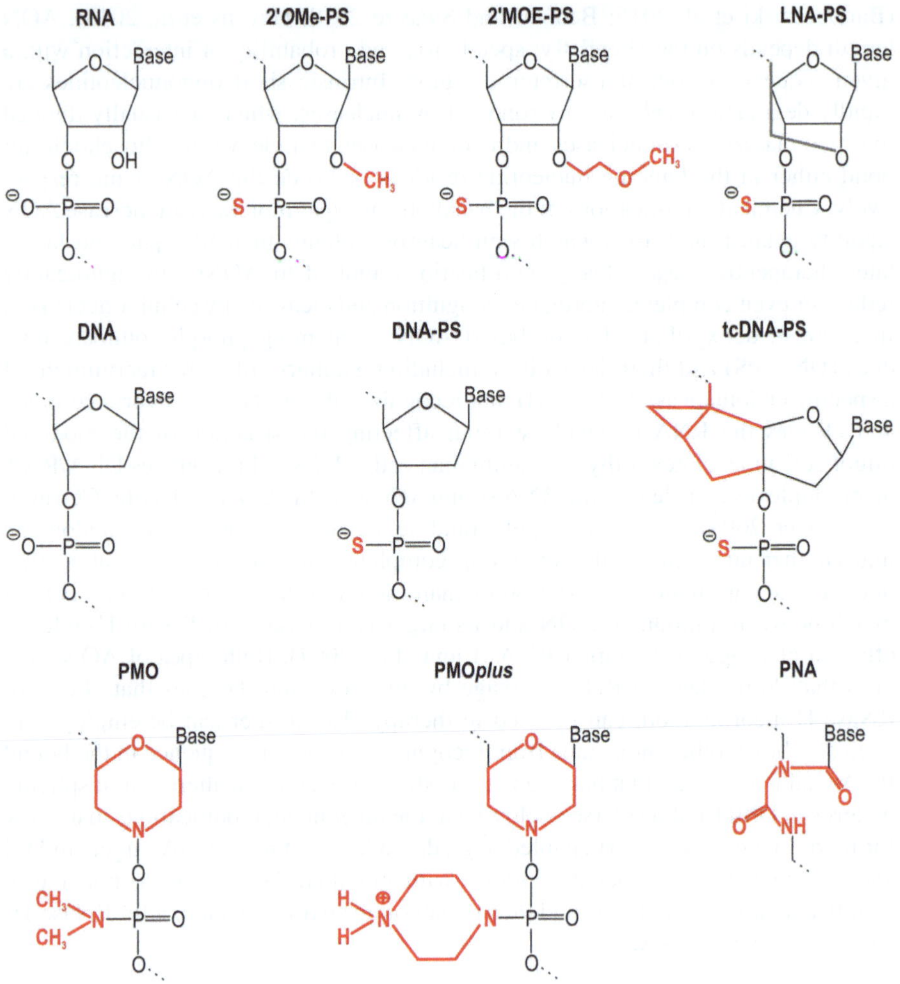

Fig. 1 Nucleotide chemistries of RNA, DNA, and their modified analogs utilized in antisense therapy of NMDs. Differences between RNA or DNA and their analogs are marked with *red*. Nucleotide charges are depicted with – or +

spliceosome or other trans-acting splicing regulators, respectively (Disterer et al. 2014). Furthermore, AONs could be used to select a specific polyadenylation site in pre-mRNA, block translation, strengthen or modify RNA structure, target miRNAs, or even alter transcription at the DNA level as has been reported for AONs with PNA modification (see below) (DeVos and Miller 2013). The most commonly used AON chemistries that do not induce target degradation include fully modified 2′-substituted phosphorothioate (PS) oligonucleotides, morpholino-based oligomers (PMOs), alternating locked nucleic acids (LNAs), peptide nucleic acids (PNAs), and tricyclo-DNAs (tcDNAs) (Falzarano et al. 2015) (Fig. 1).

In principle, more extensive chemical modifications of AONs often improve their drug-like qualities (Kole et al. 2012).

2.1 AONs with Alkyl Modifications at the 2′ Position of the Ribose

Modifications of the ribose ring at 2′ position with an alkyl group enhance target binding, increase AON resistance against nucleases, and generally prevent RNase-H incorporation to the AON/RNA duplex (Prakash and Bhat 2007). The most currently used 2′ sugar modifications are 2′-*O*-methyl (2′*O*Me) and 2′-*O*-methoxyethyl (2′MOE), with the methoxyethyl group, compared to the simple methyl group, showing increased nuclease resistance and increased binding affinity. Resistance to nucleases could be further enhanced by substituting the non-bridging oxygen atom of the phosphate group with a sulfur atom (PS modification) (Burnett and Rossi 2012). 2′*O*Me, 2′MOE, and their PS analogs (2′*O*Me-PS and 2′MOE-PS) are negatively charged. This aids with their solubility and dissemination in the blood due to binding to plasma proteins (Bennett and Swayze 2010). They can also be easily complexed with cationic lipids and proteins although they cannot be readily conjugated covalently to proteins as PMOs (see below). Their safety has been assessed in a number of preclinical and clinical trials (Goemans et al. 2011; van Deutekom et al. 2007; Yokota et al. 2009a).

2.2 Morpholino-Based Oligomers (PMOs)

In PMO-based AONs, the ribose sugar is replaced with a six-membered morpholino ring moiety, and the phosphodiester bond is substituted by a phosphorodiamidate linkage (DeVos and Miller 2013). As they show strong affinity to RNA and high resistance against nucleases, PMOs, along with 2′*O*Me-PS-modified oligoribonucleotides, are currently among the most commonly used AON analogs (Disterer et al. 2014). PMOs alone are neutral, which limits their in vivo usage as they do not bind plasma proteins and are rapidly disseminated from the body. To address this drawback, PMOs have been covalently bound to charged cell-penetrating peptides (PPMOs), including muscle-targeting peptides, and delivery octa-guanidine dendrimers (vivo-PMO), which have markedly improved intracellular uptake in animal studies. However, both of these modifications are potentially toxic (Ferguson et al. 2014; Moulton and Moulton 2010). More recently, a limited number of positively charged linkages have been added to PMOs resulting in generation of PMO *plus*. A direct comparison of PPMOs and PMO *plus* showed that the latter possess more potent antiviral capability while being better tolerated in nonhuman primates (Iversen et al. 2012).

2.3 Locked Nucleic Acids (LNAs)

LNAs contain a methylene bridge that connects the 2′-oxygen of ribose with the 4′-carbon (Jepsen et al. 2004; Kaur et al. 2007). They are resistant to nuclease digestion and demonstrate high thermodynamic stability and affinity to target RNAs based on the relatively high melting temperature of the AON/RNA complex. Several LNA analogs have been developed differing in activities and toxicity profiles in animals, including tiny LNAs that contain a PS modification. In contrast to unmodified LNAs, LNA-PS can enter cells and tissues without carriers, in the process known as gymnosis, and without induction of nonspecific immune response (Obad et al. 2011; Stein et al. 2010).

2.4 Peptide Nucleic Acids (PNAs)

PNAs are synthetic DNA/RNA analogs composed of repeating N-(2-aminoethyl)-glycine units to which nucleobases are attached via carbonyl linkages (Karkare and Bhatnagar 2006). PNAs show exceptional binding affinity to RNA targets and extreme stability against nucleases. However, PNAs have neutral charge, which limits their solubility in water and cellular uptake (McMahon et al. 2002). This can be largely omitted by combining PNAs with liposomes and microspheres or conjugation with polylysine or polyarginine peptides (Fabani et al. 2010; Zhou et al. 2006), or other positively charged groups (Dragulescu-Andrasi et al. 2006; Sforza et al. 2010). PNAs stand out among other AONs as they can bind to DNA by forming a triple helix, which is followed by strand invasion. Among other applications, this property of PNAs has been used to stop transcription by targeting a promoter sequence.

2.5 Tricyclo-DNAs (tcDNAs)

TcDNAs are modified DNA oligonucleotides with three additional carbon atoms between C5′ and C3′ (Renneberg et al. 2002). Their sulfur-containing analogs, tcDNA-PS, are characterized by high-affinity binding to RNA, hydrophobicity, and stability. Importantly, they spontaneously form 40- to 100-nm-sized nanoparticles, a feature which presumably improves their cellular uptake compared to 2′*O*Me-PS and PMO and enables their crossing through the blood–brain barrier of adult mice (Goyenvalle et al. 2015), making tcDNA-PS of particular interest to the treatment of NMDs.

3 AON Analogs that Induce Degradation of RNA targets

As indicated previously, AONs can be customized in a manner that induces recruitment of RNase-H into the AON/RNA complex and degradation of the target RNA strand, but still prevents AON degradation by exonucleases. This can be achieved by either modifying the entire DNA oligo backbone with PS residues (Fig. 1) or by using gapmers, i.e., AON analogs that contain fully modified nucleotides only at their sides. AON-triggered degradation of RNA has clear advantages over RNAi-based strategy, especially in animal or human studies, in which synthetic microRNAs (miRNAs) and short-hairpin RNAs (shRNAs) have to be expressed from viral vectors and processed into cytoplasmic small interfering RNAs (siRNAs), inducing RISC-mediated degradation or translational repression of target RNA molecules (Sibley et al. 2010). Synthetic siRNAs can also be delivered to animals; however, they are rapidly degraded in the circulatory system and cannot easily pass the cell membrane because of their physicochemical properties and thus require carriers or specific modifications in order to be potentially therapeutic (Burnett and Rossi 2012; Kole et al. 2012). In contrast, AONs can be usually administered to cells without delivery vehicles and induce RNA degradation without the risk of saturating the miRNA biogenesis pathway.

3.1 Phosphorothioate-Modified DNA AONs (DNA-PS)

In DNA-PS, the non-bridging phosphate oxygen atom is replaced with sulfur, which enhances the oligonucleotide stability against nucleases without losing RNase-H-induced cleavage of target RNAs (Cerritelli and Crouch 2009). Additionally, PS-modified AONs show higher affinity to plasma proteins, which is reflected by their increased half-life time in the circulatory system, and, similarly to 2′*O*Me and 2′MOE-modified AONs, carry a negative charge, which facilitates their binding to cells and uptake into specific tissues. On the downside, interaction with cell-surface and intracellular proteins can induce immune stimulation and complement activation (Muntoni and Wood 2011). Furthermore, similarly to tcDNAs, DNA-PS have been shown to cross the blood–brain barrier (Kazantsev and Thompson 2008).

3.2 Gapmers

Compared to DNA-PS AON analogs, gapmers display higher affinity for RNA, enhanced tissue uptake, increased resistance to nucleases, longer in vivo half-life, and lesser toxicity (Kole et al. 2012). A typical gapmer is composed of a central core, made of about ten-nucleotide-long DNA-PS that allows for the cleavage of targeted mRNA by RNase-H, and flanked by usually three to five more extensively modified nucleotides at each side that protect the gapmer from exonucleases and improve binding to RNA at the same time reducing the possibility of toxic off-target effects associated with phosphorothioates. 2′-Modified residues including 2′MOE have been used as gap surrounding nucleotides (Keiser et al. 2015, Wheeler et al. 2012). Potent antisense effects have also been obtained with administration of tcDNA and LNA gapmers (Renneberg et al. 2002; Stein et al. 2010).

4 AON-Based Splicing Regulation in Therapeutic Treatment of DMD and SMA

Splice-switching oligonucleotides were first described in the context of the human beta-globin pre-mRNA (Dominski and Kole 1993). Since then, numerous studies have followed, including those in which modified AONs were used to treat DMD and SMA (Fig. 2).

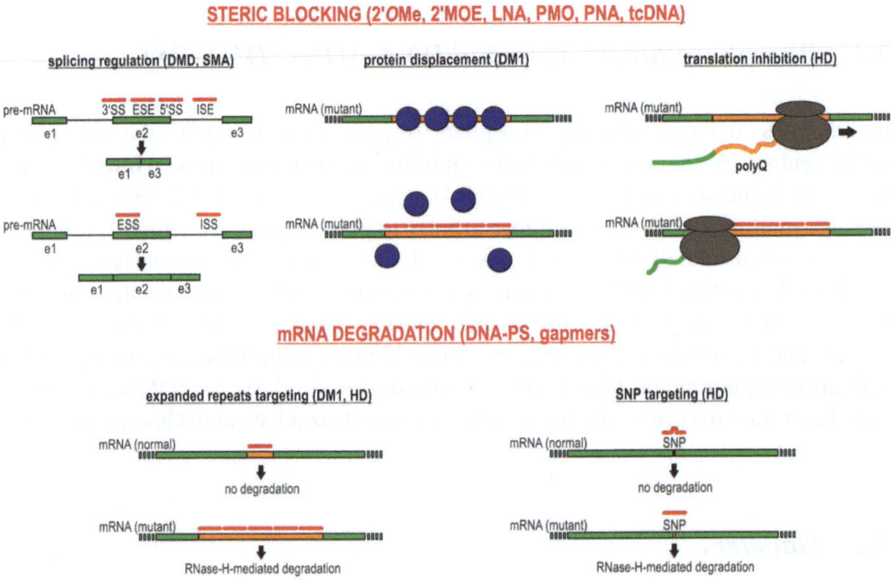

Fig. 2 Therapeutic strategies based on steric blocking or RNA target degradation with modified AONs. The figure color codes and abbreviations are as follows: *red*, AON analogs; *green*, exons; *orange*, expanded repeats; *brown*, ribosome; 3′/5′SS, 3′/5′ splice site; *ESE*/*ISE*, exonic/intronic splicing enhancer; *ESS*/*ISS*, exonic/intronic splicing silencer; *SNP*, single nucleotide polymorphism

4.1 *Duchenne Muscular Dystrophy (DMD)*

4.1.1 Clinical, Genetic, and Molecular Features of DMD

DMD is an X-linked disease that affects 1/5000 boys (Mendell and Lloyd-Puryear 2013). The most evident feature of the disease is the progressive musculature disability underlined by loss of myofibers and their replacement with adipocytes, as well as extensive fibrosis. DMD is usually diagnosed in 3–5-year-old boys. In their teens, the extent of muscle fiber loss is so vast that the patients are forced to use a wheelchair, and in their twenties, a ventilator system to aid with the breathing. Up until now, the disease is incurable, and although treatment with anti-inflammatory-acting corticosteroids aids to alleviate the symptoms, the disease irreversibly leads to premature death due to respiratory or cardiac failure. Years of research showed that dystrophin at the sarcolemma assembles a multimeric dystrophin glycoprotein complex that transmits forces between the extracellular matrix and the cytoskeleton (Le Rumeur et al. 2010). Absence of dystrophin causes contraction-induced damage of myofibers leading to protease activation, calcium inflows, and altered calcium homeostasis. Typically, myofiber damage induces activation of stem cells of muscle, particularly satellite cells, which leads to regeneration of myofibers. In DMD, this process is highly inefficient, causing replacement of myofibers with fibro-adipose tissue. Shadowing progressive muscular dystrophy, DMD patients also suffer from cognitive deficits and psychiatric symptoms that are interlinked to specific dystrophin functions in the brain (Perronnet and Vaillend 2010).

The advance in understanding the molecular alterations that trigger the disease has been possible through the use of animal models. *Mdx* mouse, the most frequently used animal model of DMD, has a spontaneous nonsense mutation in exon 23 of the *DMD* gene that leads to lack of dystrophin expression. Since the discovery of the original *mdx* mouse, multiple other variants have been generated (McGreevy et al. 2015). However, after a period of extensive fiber degeneration and regeneration, the progress of the muscle disease is somehow halted and the mice do not show extensive pathology. Because of their relatively mild phenotype, scientists generated another mouse model, which, in addition to dystrophin, lacks also utrophin, a protein that to some extent can compensate for the lack of dystrophin. The double knockout mouse shows very severe and progressive muscular dystrophy and dies prematurely, usually within 20 weeks (McGreevy et al. 2015). Canine models of the disease have also been extensively used in the experimental therapy of DMD, including the golden retriever with muscular dystrophy (GRMD) and the beagle with canine X-linked muscular dystrophy (CXMD) that harbor a mutation in the splice acceptor site of exon 7 leading to its skipping and generation of a premature stop codon in exon 8 (McGreevy et al. 2015). Importantly, muscular dystrophy observed in canine models closely resembles that of DMD patients.

Mutations in the *DMD* gene generally lead to changes in the open reading frame or substitutions that generate a stop codon. As a result, patients lack dystrophin protein, with the exception of a small number of so-called revertant fibers, in which an open reading frame is restored due to spontaneous exon skipping. A fraction of deletions do not change the open reading frame, and shorter but still partially functional dystrophins, usually containing unaltered N- and C-terminal portions, are expressed. As a consequence, the patients suffer from a much milder form of DMD, called Becker muscular dystrophy (BMD). Based on this principle, multiple viral vectors carrying micro- and mini-dystrophins have been generated (Konieczny et al. 2013). Similarly, modified AONs can be used to exclude exons containing nonsense mutations and to restore the dystrophin reading frame. Sometimes the latter can only be achieved with exclusion of more than one exon. AON-based therapy that force single exon skipping could be theoretically applied to 35 % of all DMD patients and this number can be further increased to approximately 90 % when taking into the account the possibility of multiple exon skipping events (Aartsma-Rus et al. 2009; Yokota et al. 2009b). As approximately 45 % mutations in the *DMD* gene accumulate in a hot spot region encoding the middle part of dystrophin that covers exons 45–55, multi-exon strategy targeting this specific region would be particularly beneficial (Béroud et al. 2007).

4.1.2 Modified AONs in Therapy of DMD

Initial studies employed AONs carrying a 2′*O*Me-PS modification to induce *DMD* exon exclusion in cultured cells. Exon skipping was achieved first in cultured *mdx* muscle cells following masking of exon 23 splice sites (Dunckley et al. 1998; Wilton et al. 1999). AON analogs directed to a polypurine-rich sequence within exon 46 also efficiently skipped the exon and restored dystrophin expression in up to 82 % myotubes differentiated from DMD patients' primary myoblasts carrying deletions in exon 45 (van Deutekom et al. 2001). Similar strategy was applied to muscle cells isolated from other patients, which resulted in efficient exclusions of exons 19, 44, 49, 50, 53, and 51 (Aartsma-Rus et al. 2003; Takeshima et al. 2001). Cell culture studies were followed by delivery of AON analogs to animals, first by local, i.e., intramuscular, and then by systemic injections through the vasculature. Becker-like dystrophin was expressed in muscles from *mdx* mice following intramuscular delivery of a 25-mer 2′*O*Me complementary to the 5′ splice site in intron 23 (Mann et al. 2001). Importantly, oligo-treated muscles showed the presence of dystrophin at the sarcolemma of myofibers for several weeks, were able to properly assemble the dystrophin glycoprotein complex, and displayed improved muscle force generation without immune response following AON re-administration (Lu et al. 2003). Subsequent direct comparison of 2′*O*Me-PS with 2′MOE-PS and either unmodified or peptide-conjugated PNAs injected intramuscularly into *mdx* mice revealed that the latter molecules might have a higher therapeutic potential (Yang et al. 2013; Yin et al. 2008a).

Animal studies demonstrated that systemic delivery of AON analogs to skeletal muscles can also be successful as shown upon repeated injections of *mdx* mice with a 20-mer 2′*O*Me-PS along with a delivery vehicle, a block copolymer, or a 25-mer

PMO (Alter et al. 2006; Lu et al. 2005). Particularly, seven weekly intravenous injections of PMO restored dystrophin localization at the sarcolemma of more than 70 % fibers in hind leg muscles (Lu et al. 2005). In comparison, systemic delivery of a 25-mer PNA by three weekly injections at 15 mg/kg induced exclusion at the level of 10 % without any phenotypic improvement (Yin et al. 2010a). The PMO treatment was also well tolerated in *mdx* mice injected biweekly with doses up to 1.5 g/kg in a long-term study (Wu et al. 2011), and following delivery of three PMOs targeting exons 6–9 in CXMD dogs that induced multiple exon skipping (Yokota et al. 2009a). Unfortunately, neither of the PMO treatments resulted in satisfactory dystrophin expression in the heart (Lu et al. 2003; 2005; Wu et al. 2011), limiting their therapeutic potential.

In contrast to PMOs, their derivatives conjugated with arginine-rich peptides (PPMOs) upon systemic delivery to *mdx* mice at a single dose of 25–30 mg/kg resulted in almost normal dystrophin levels in skeletal muscles and approximately 50 % in the heart, improving not only muscle strength but also cardiac function (Jearawiriyapaisarn et al. 2008; Wu et al. 2008; Yin et al. 2009; Yin et al. 2008b). Subsequent studies showed, however, high levels of acute toxicity of PPMOs. Mice deaths were observed at doses above 30 mg/kg, with the dose of 120 mg/kg killing 80 % of injected animals (Wu et al. 2012). PPMOs were toxic also for nonhuman primates, in which AVI-5038 targeting human exon 50 caused tubular degeneration after four weekly injections at a dose of 9 mg/kg (Moulton and Moulton 2010). Subsequently established protocol for repeated nontoxic PPMO injection required lower doses of approximately 6 mg/kg (Wu et al. 2012; Yin et al. 2010b). Applied biweekly for a year, these amounts induced protein expression up to 50 % in skeletal muscles and below 5 % in the heart (Wu et al. 2012). A new series of PPMOs, called Pips, characterized by a central hydrophobic motif flanked by arginine-rich domains, has been extensively tested in the Wood lab (Betts et al. 2012; 2015; Yin et al. 2011). In one of the most recent study, several 10 mg/kg systemic injections restored dystrophin protein levels up to 100 % in skeletal muscles and 28 % in the heart and prevented cardiomyopathy in a forced exercised *mdx* mice with no apparent toxicity (Betts et al. 2015). Five biweekly intravenous injections at a dose of 6 mg/kg of another PMO conjugate, so-called vivo-PMO, in which modified nucleotides are bound to dendrimeric octa-guanidines, resulted in expression of dystrophin in nearly all *mdx* myofibers and more than 40 % of cardiomyocytes, which corresponded to 50 and 10 % of normal dystrophin levels on Western blot, respectively (Wu et al. 2009). Vivo-PMOs were also tested in excision of exons 45–55 in exon 52-deficient *mdx* mice (Aoki et al. 2012; Echigoya et al. 2015) and showed a therapeutic potential in CXMD dogs in which three intramuscularly delivered oligo targeting exons 6–9 led to the presence of dystrophin in 75 % of myofibers. Importantly, the vivo-PMOs resulted in a better therapeutic outcome compared to unmodified PMOs (Yokota et al. 2012).

A recent study tested systemic potency of a 15-nt-long tcDNA-PS in exon 23 skipping in *mdx* mice compared to a 20-mer 2′*O*Me and a 25-mer unmodified PMO (Goyenvalle et al. 2015). Following 12 weekly injections, 200 mg/kg each, tcDNA-PS treatment revealed superior dystrophin expression compared to other chemistries, 20–30 % in the majority of tested skeletal muscles, 50 % in the diaphragm, and 40 % in the heart. The mice also showed dystrophin expression in the central nervous system, ranging from 2 to 4 % depending on the brain region. Surprisingly, the treatment not only improved striated muscle function but also corrected behavioral phenotype typical of *mdx* mice. Mild toxicity, comparable to that of 2′*O*Me-PS, was associated with tcDNA administration (Goyenvalle et al. 2015).

Two AON-based drugs against DMD are currently competing in clinical trials, drisapersen (PRO051, kyndrisa; Prosensa/BioMarin), a 2′*O*Me-PS-modified oligo, and eteplirsen (AVI-4658; Sarepta Therapeutics), a PMO-based analog. Both AON analogs induce skipping of exon 51 that could theoretically restore dystrophin expression in approximately 13 % of DMD patients by binding to the same sequence, with eteplirsen covering 10 more, 30 in total, nucleotides. Although their single intramuscular injections to patients' muscles induced extensive dystrophin expression (Kinali et al. 2009; van Deutekom et al. 2007), systemic delivery has met some challenges (Cirak et al. 2011; Goemans et al. 2011; Mendell et al. 2013; Voit et al. 2014). Nevertheless, both drugs are currently under review by the FDA's advisory panel.

4.2 Spinal Muscular Atrophy (SMA)

4.2.1 Clinical, Genetic, and Molecular Features of SMA

SMA is an autosomal recessive neurodegenerative disorder with an incidence ratio of up to 1/6000 live births (Hamilton and Gillingwater 2013; Viollet and Melki 2013). The disease results from the absence of survival motor neuron (SMN) protein, which leads to progressive degeneration of spinal cord alpha-motor neurons, responsible for innervation of muscle fibers. The symptoms of the disease usually manifest as muscle weakness, muscular atrophy, and paralysis. In the most severe form, the SMA type I, the disease leads to early infant death, with children rarely living beyond 2 years of age. The role of SMN protein is still elusive. SMN has been shown to have important roles in motor neurons and neuromuscular junctions (Fallini et al. 2012). However, restoration of SMN levels in neuronal cells only partially ameliorates the phenotype of SMA mice, with additional abnormalities appearing in other tissues, including skeletal and cardiac muscles (Hamilton and Gillingwater 2013). Strikingly, the most recent report showed that increasing SMN expression exclusively in peripheral tissues was sufficient to rescue motor neurons and extend survival rates of the severely affected SMA mouse model (Hua et al. 2015).

Typically, individuals have two almost indistinguishable genes, *SMN1* and *SMN2*, which differ in 11 nucleotide substitutions but have the same coding sequence. More than 90 % of *SMN1* mature transcripts contain exon 7, which is required for production of a functional SMN protein. In *SMN2*, on the other hand, a translationally silent C to T transition in exon 7 leads to preferential exclusion of exon 7. The mutation inactivates an ESE that serves as a binding site for an essential sequence-specific splicing factor, SF2/ASF (Cartegni and Krainer 2002), and/or creates a de novo ESS that induces binding of a splicing inhibitor hnRNP-A1 (Kashima et al. 2007). Only about 10 % of *SMN2* mature transcripts contain exon 7. The remaining mRNAs, depleted of exon 7, translate into a very unstable protein isoform showing low level of activity (Le et al. 2005).

Mutations in *SMN1* leading to loss of SMN1 protein expression result in SMA. Variability in the disease phenotype is a consequence of (i) small amounts of exon 7-containing SMN2 product and (ii) the fact that *SMN2* is subject to copy number variation (usually 1–4). In the most severe SMA form, only 1 or 2 *SMN2* copies are present (Viollet and Melki 2013). Several mouse models of SMA have been generated, including *SMN* knockout mice carrying different copy numbers of the human *SMN2* gene and correspondingly mild or severe disease phenotype (Bogdanik et al. 2015; Hamilton and Gillingwater 2013). These SMA mouse models show a range of pathologies that consist of motor neuron loss as well as abnormalities in other organs (Hamilton and Gillingwater 2013). No large animal models of SMA exist as they lack the *SMN2* gene.

4.2.2 Modified AONs in Therapy of SMA

Therapeutic strategies for SMA are based on the protective role of SMN2 full-length protein, whose expression can be increased by forcing exon 7 inclusion to the mature mRNA. Initiated by the early study showing that C/T transition in *SMN2* exon 7 might affect SF2/ASF interaction with pre-mRNA (Cartegni and Krainer 2002), Cartegni and Krainer targeted a PNA-modified AON complementary to the disrupted region, coupled to a peptide mimicking the natural domain of SF2/ASF, which resulted in efficient exon 7 inclusion in in vitro tests (Cartegni and Krainer 2003). In a subsequent study, an alternative bifunctional molecule, a 2′*O*Me AON carrying an oligo tail with binding sites for SF2/ASF, rescued incorporation of *SMN2* exon 7 in patients' fibroblasts (Skordis et al. 2003). In both cases, administered molecules also blocked the putative binding site for the negative regulator of splicing, hnRNP-A1.

Administration of AONs carrying alkyl modifications at the 2′ position of the ribose sugar allowed for identification of a couple of ISSs important for splicing of *SMN2* exon 7 (Hua et al. 2008; Miyajima et al. 2002; Singh et al. 2006). Particularly, screening of proximal parts of introns flanking exon 7 using an AON-tiling method resulted in identification of cis-acting inhibitory elements, including two weak tandem hnRNP-A1/A2 motifs (Hua et al. 2008). Both elements can be simultaneously and efficiently masked with a single AON, as shown, e.g., with a systemic treatment of an SMA mouse model with a 2′MOE-PS. Four weekly injections at 25 mg/kg resulted in approximately 90, 75, and 45 % inclusion rates of *SMN2* exon 7 in the liver, kidney, and skeletal muscles, respectively. The site consisting of hnRNP-A1/A2 binding motifs, named ISS-N1, was also targeted in the following studies (Hua et al. 2010; 2011; 2015; Passini et al. 2011; Porensky et al. 2012; Zhou et al. 2013). Intracerebroventricular infusion of an 18-nt-long 10–27 2′MOE-PS into adult mice for 7 days at 50 μg/day increased *SMN2* exon 7 inclusion in spinal cord motor neurons to approximately 85 %, with the treatment lasting for several months (Hua et al. 2010). Interestingly, a direct comparison of 10–27 2′MOE-PS with a 20-mer 2′*O*Me-PS showed that the latter is less potent and more toxic (Hua et al. 2010; Williams et al. 2009). Injection of neonatal mice with 10–27 2′MOE-PS led to various results depending on the route of administration. 20 μg intracerebroventricular delivery increased SMA mouse viability from median 10 to 16 days, while systemic administration (two 50 μg subcutaneous injections) or combined treatment extended survival rates to 108 and 173 days, respectively (Hua et al. 2011). The authors attributed these results to nonessential role of SMN in the central nervous system (CNS), which was also confirmed by subsequent studies. Particularly, systemic treatment of newborn mice with 10–27 2′MOE-PS, paralleled by administration of a decoy oligo that neutralized the 10–27 2′MOE-PS action in the CNS, showed that exclusive restoration of *SMN2* exon 7 in peripheral tissues improves the motor function and extends survival rates of SMA mice. The most recent study also showed that systemic delivery of a 2′MOE-PS targeted to ISS-N1 at the adolescence or adulthood can be therapeutic in a mildly affected SMA mouse model (Bogdanik et al. 2015).

PMO chemistry has also been utilized in the treatment of SMA. Intracerebroventricular delivery of a 20-mer PMO against ISS-N1 into newborn mice increased survival rates of mice from 15 to 100 days (Porensky et al. 2012). In a following study, researchers compared efficiency of *SMN* exon 7 inclusion upon delivery of PMOs with various lengths (Zhou et al. 2013). The longest PMO tested, a 25-mer, was found the most efficient in patients' fibroblasts and in injected mice. Intracerebroventricular single injection of 40 mg/kg resulted in an average survival of 85.5 days, while mice administered systemically once with 40 mg/kg PMO were viable for more than 93 days. Interestingly, systemic injection increased markedly *SMN2* exon 7 inclusion in the brain, spinal cord, and liver but not in skeletal muscles or the heart, showing efficient PMO crossing through the blood–brain barrier in neonatal mice. Furthermore, 25-mer vivo-PMOs at 10 mg/kg were tested systemically; however, they failed to improve the exon 7 inclusion rates in mice compared to unmodified PMO and induced toxicity following intracerebroventricular injection. At 10 mg/kg vivo-PMO injection to the CNS, all pups died overnight. Decreasing the dose to 2 mg/kg prevented the lethality but no significantly extended survival rates were observed.

The feasibility of the 10–27 2′MOE-PS delivery was tested by intrathecal and intracerebroventricular delivery to nonhuman primates and showed robust immunostaining of 10–27 2′MOE-PS in all regions of the spinal cord, somewhat more pronounced following intrathecal administration (Passini et al. 2011). Following encouraging data, 10–27 2′MOE-PS, also known as IONIS-SMNRx, has entered clinical trials that are being carried out by Ionis Pharmaceuticals (formerly Isis Pharmaceuticals), in which infants and children are treated by intrathecal administration.

5 Steric Blocking and Degradation of Toxic Repeats with Modified AONs

Therapeutic approach with modified AONs for DMD and SMA is solely based on splicing alterations. For DM and HD, two different strategies can be utilized, based either on steric blocking of toxic repeats or targeted degradation of the mutated allele (Fig. 2).

5.1 Myotonic Dystrophy (DM)

5.1.1 Clinical, Genetic, and Molecular Features of DM

Myotonic dystrophies (DM) comprise a group of NMDs inherited in an autosomal dominant manner, in which the underlying mutation constitutes an expansion of unstable noncoding microsatellite repeats which are transcribed into RNA, but not translated into protein. The major pathogenic effect of expanded RNA is sequestration of distinct binding proteins and disruption of several alternative pre-mRNA splicing processes. Such RNA gain of function, or RNA dominance, results in a severe cellular toxicity leading to a wide range of phenotypic alterations. The clinical features of the disease were first identified over a 100 years ago, and DM is today recognized as the most prevalent form of adult-onset muscular dystrophy, affecting roughly 1/8000 live births (Ashizawa and Harper 2006; Cho and Tapscott 2007). The majority of DM symptoms are connected primarily to, but not restricted to, the progressive muscle pathology. Although the two types of DM, type 1 (DM1) and type 2 (DM2), are caused by mutations in two different genes, both have distinct yet largely overlapping clinical features (Cho and Tapscott 2007). The most prominent features of DM include myotonia and muscle wasting and atrophy; however, DM is in principle a complex and slowly progressing multisystem disorder that apart from muscle may affect distinct tissues body-wide and demonstrate a striking clinical heterogeneity (Thornton 2014).

In DM1, the mutation is attributed to the expansion of a trinucleotide CTG repeat within the 3′UTR of the *DMPK* gene (Cho and Tapscott 2007). In normal individuals, CTG repeat alleles range from 5 to 37, while DM1-affected individuals have from fifty to hundreds or even several thousand repeats. Age-dependent and tissue-specific somatic mosaicism of the repeats is often observed in DM patients due to instability of the repeats and their strong bias towards expansion over time (Cho and Tapscott 2007). Notably, the length of the expansion is strongly correlated to patients' age at onset as well as the severity of the disease. Infants with severe congenital DM1, as well as their mothers, have on average a greater amplification of the CTG repeat than is seen in the non-congenital (childhood-onset or adult-onset) DM1 population. In addition, there is evidence for anticipation in DM1, in which successive generations inherit increasing disease severity with decreasing age of onset (Ashizawa and Harper 2006; Cho and Tapscott 2007).

RNA dominance, or gain of function by the mutant noncoding RNA, is considered a major player in the molecular pathogenesis of DM. Once transcribed into RNA, the expanded repeats (CUG^{exp}) tend to fold into thermodynamically stable hairpin structures, so-called toxic RNAs, which are not exported to the cytoplasm, but instead are retained in the nuclei of diseased cells, where they accumulate as discrete clumps or "foci" (Cho and Tapscott 2007; Magaña and Cisneros 2011; Wheeler and Thornton 2007). Importantly, the causal role of toxic RNA was supported by experiments with several distinct transgenic mouse models expressing CUG^{exp} either in the natural context of DMPK, or when inserted in an unrelated transcript (Gomes-Pereira et al. 2011). For example, the most widely used *HSA*^{LR} mice reproduce cardinal biochemical and clinical features of DM1 by expression of ~250 CTG repeats in the 3′UTR of the human skeletal actin transgene (Mankodi et al. 2000). The CUG^{exp}-containing toxic RNAs bind with high affinity and sequester much of the available cellular pool of several CUG-binding proteins including, among others, crucial alternative splicing regulators from the muscle blind-like protein family (MBNL), MBNL1-3 (Fardaei et al. 2002; Jiang et al. 2004; Mankodi et al. 2001; Miller et al. 2000). Experimental data also indicates inappropriate activation of the PKC pathway as a possible mechanism leading to abnormal hyperphosphorylation and stabilization of another splicing regulator *CUG-binding protein 1*/Elav-like family member 1 (CUGBP1) in DM1 tissues (Kuyumcu-Martinez et al. 2007). MBNLs and CUGBP1 proteins antagonistically regulate hundreds of splicing events (Wang et al. 2015), and when their normal function is compromised by the toxic RNA, the splicing patterns become consistent with a loss of MBNL1 and a gain of CUGBP1 function (Ho et al. 2004; Lin et al. 2006; Philips et al. 1998). Misregulation of a subset of developmentally regulated alternative splicing events that are controlled by these proteins shifts the splicing in DM1 from adult to the embryonic pattern and results in expression of splice products that are developmentally inappropriate for a particular tissue (Osborne and Thornton 2006). Summarizing, altered localization, regulation, and function of splicing regulatory proteins in DM result in impaired alternative splicing of hundreds of target pre-mRNAs, which ultimately lead to a wide range of clinical manifestations of the disease.

5.1.2 Modified AONs in Therapy of DM

As of today, there is still no cure for DM and patients' therapy is limited to symptomatic treatment and supportive care. Several distinct intermediates of the disease may constitute possible points of therapeutic intervention (Magaña and Cisneros 2011); however, the vast majority of current experimental therapeutics targets DM "downstream" of the mutant DMPK allele, i.e., at the level of the toxic RNA or RNA–protein complex, which constitute the core of the disease. Such is the case with rationally designed small chemical compounds "neutralizing" the toxic

CUG^{exp} hairpins by virtue of high-affinity binding and spatial hindrance, hence preventing MBNL proteins' entrapment in the nuclear foci and spliceopathy (Konieczny et al. 2015). An alternative approach is based on an antisense strategy, for example, using chemically modified antisense oligonucleotides (AON) or expressed antisense RNAs, designed to either neutralize toxic RNA by steric blocking of RNA–MBNL interaction or reduce the expression of mutant *DMPK* transcripts by distinct molecular pathways. Both approaches presume that the final effect would be the release of MBNLs from pathogenic sequestration by CUG^{exp} that would boost the pool of proteins available for natural pre-mRNA targets and hence correct the DM spliceopathy (Gao and Cooper 2013; Gomes-Pereira et al. 2011; Konieczny et al. 2015; Magaña and Cisneros 2011; Mulders et al. 2010; Wheeler and Thornton 2007).

In overview, two major AON-based strategies have been developed to target DM: (i) splice-switching AONs targeting DM-related aberrant transcripts in an effort to restore their normal alternative splicing and (ii) AONs designed to either eliminate/degrade the toxic CUG^{exp} RNA or inhibit the toxic RNA–protein interaction in an effort to block the sequestration of splicing regulatory proteins. The latter strategy is much more prevalent owing to the fact that it targets the disease more upstream, i.e., at the level of the RNA triggering the disease, while wide-ranging elimination of DM-related splicing defects using splice-switching AONs would require targeting of at least several transcripts and many repeated rounds of injections. Nevertheless, splice-switching AONs may be very effective in reversing cardinal features of DM, as demonstrated in a study with a PMO-based AON which was designed to target *Clc-1* pre-mRNA whose abnormal alternative splicing in DM triggers myotonia (Wheeler et al. 2007). By targeting the AON to 3′ splice site of exon 7a and suppressing its inclusion into mature transcript, the authors were able to reverse the *Clc-1* spliceopathy and eliminate myotonic discharges in two distinct mouse models of DM, the *HSA*^{LR} mice and the Mbnl1 knockout mice (Kanadia et al. 2003; Wheeler et al. 2007).

In contrast, the feasibility of the second AON-based strategy to bind CUG^{exp} RNA in an effort to either block protein sequestration or degrade the toxic RNA, or otherwise modify its metabolism, was shown in cellular and animal DM models by several independent studies using oligomers with distinct length and chemical modifications, but with a common trinucleotide CAG repeat motif. To this end, much data is available on therapeutic AONs able to degrade toxic RNA, while only two therapeutic AONs were described so far in the literature that efficiently bind CUG repeat RNA in vivo and preclude MBNL sequestration without inducing significant degradation of the toxic RNA, namely, a 25-mer morpholino CAG25 (Wheeler et al. 2009) and short, 8- or 10-unit all-LNAs (Wojtkowiak-Szlachcic et al. 2015). Local intramuscular injection/electroporation of CAG25 was used to target the CUG^{exp}RNA in an *HSA*^{LR} mouse model of DM1, leading to reversal of *Clcn1* splicing defects and elimination of myotonia (Wheeler et al. 2009). PMOs are not recognized by RNase-H, yet CAG25 caused a 50 % reduction of CUG repeat RNA, presumably by releasing RNA from foci and enhancing its degradation indirectly via other mechanisms. More recently, CAG25 was modified by attaching it to a cell-penetrating peptide in order to improve its uptake by muscle fibers, increase its effectiveness, and allow for systemic delivery (Leger et al. 2013). Indeed, intravenous injection of such peptide-linked morpholino (PPMO) into HSA^{LR} mice was able to neutralize or eliminate several cardinal features of DM like myotonia, MBNL1 sequestration, and aberrant RNA splicing (Leger et al. 2013). Likewise, alternative splicing abnormalities of MBNL-sensitive exons and myotonic discharges in skeletal muscles of HSA^{LR} mice were mitigated upon local intramuscular injection of short, ten-nucleotide-long LNA targeting CUG repeat motif which inhibited RNA–protein interaction without strong induction of targeted transcript degradation (Wojtkowiak-Szlachcic et al. 2015).

More data is available on experimental AON-based therapeutics that exert their effect via degradation of toxic RNA in patients' DM1 cells as well as in mouse models. In overview, efficient degradation of CUG^{exp} transcripts can be achieved by short, chemically modified gapmer-based AONs compatible with RNase-H-dependent cleavage (Lee et al. 2012; Nakamori et al. 2011; Pandey et al. 2015; Wheeler et al. 2012) or inducing CUG^{exp} degradation via other unknown mechanisms (Gonzalez-Barriga et al. 2013; Mulders et al. 2009). In vivo administration of PS58, a modified 2′*O*Me-PS-CAG7 AON (a 21-mer oligonucleotide with a 2′-*O*-methyl-phosphorothioate modification along the entire backbone, composed of CAG repeated sequence) into muscle of two distinct DM1 mouse models caused, for example, a significant reduction in toxic RNA levels in conjunction with a partial shift in aberrant splicing pattern of several transcripts (Mulders et al. 2009). Importantly, in DM1 mouse models as well as in DM1 patient cells, PS58 preferentially silenced the *DMPK* transcripts with expanded repeats, leaving the normal-sized allele intact. Chemistry of 2′*O*Me-PS-CAG7 AON is incompatible with RNase-H, and the mechanism underlying the silencing phenomenon as well as selectivity towards expanded transcripts is not entirely clear. A follow-up study confirmed that degradation of long *DMPK* transcripts induced by PS58-type AONs is indeed an RNase-H-independent process and does not involve oligo-intrinsic RNase-H activity nor interfere with splicing of *DMPK* transcripts (Gonzalez-Barriga et al. 2013). The same study further analyzed a comprehensive collection of $(CUG)_n$-directed DNA and RNA-based triplet repeat $(CAG)_n$ AONs differing in length and chemistry, including modifications like 2′*O*Me-PS, PS alone, ENA (2′-*O*,4′-*C*-ethylene-bridged nucleic acid), and PMO (coupled to octa-guanidine dendrimer, so-called VivoPorter). Results clearly demonstrated that for a significant reduction of expanded *DMPK* mRNAs, a minimal length of five CAG triplets was required and, importantly, the nature of chemical modification strongly influenced the silencing capacity of tested AONs. Briefly, among DNA-based (CAG)n AONs, only PS-modified and PS gapmer with 2′*O*Me wings were able to silence expanded *hDMPK* mRNA by ~50 and 80 %, respectively, leaving the endogenous mouse *Dmpk* mRNA intact, while pure DNA-based AON did not show any silencing activity as it was quickly degraded inside cells. ENA-modified AONs appeared not effective, probably due to lack of nuclear internalization and concentration in vesicle-like structures and, as a consequence, inability to reach muscle cell nuclei, where most expanded $(CUG)_n$ transcripts reside. AONs with 2′*O*Me-PS modifications were the most efficient, and similarly, PMO-type AONs significantly reduced *hDMPK* without affecting endogenous murine *Dmpk* (Gonzalez-Barriga et al. 2013).

Degradation of toxic RNA via RNase-H-compatible AONs is another strategy that exploits the nuclear retention phenomenon to gain a therapeutic advantage, while posing less risk of off-target effects by avoiding other CUG repetitive sequences. Recently, chimeric LNA/2′MOE-PS-$(CAG)_n$ gapmers containing an RNase-H-compatible PS-DNA fragment with CAG repeat sequences and additional LNA or 2′MOE modifications that convey increased affinity to RNA and render them resistant to nucleases were shown to selectively knockdown expanded CUG transcripts in cell culture and in a DM1 mouse model (Lee et al. 2012). Interestingly, lowering the dose of intramuscularly injected/electroporated 14-nt-long LNA/2′MOE-PS-$(CAG)_n$ gapmer from 2 to 0.5 μg resulted in a reversal of splicing defects even without reduction of toxic RNA, suggesting that this particular AON may also work through other RNase-H-independent mechanisms to affect downstream splicing events or, as the authors speculated, through displacement of MBNL1 from the hairpins. In addition, this work also indicated that combining AONs that help release CUG transcripts from nuclear foci, such as PMOs, with 2′MOE-gapmers that target the CUGs for degradation, can have a synergistic effect at certain concentrations and enhance the knockdown effect. Though local administration used in these studies restricts the systemic distribution of therapeutic AONs and thus is less clinically applicable, there are reports of AONs capable of exerting a systemic effect. For example, a chemically modified antisense oligomer termed "5-10-5 2′MOE gapmer" has been developed recently, containing a PS backbone and 2′MOE modification of 5 residues at the 5′ and 3′ ends, separated by a central gap of ten DNA nucleotides to support RNase-H activity (Wheeler et al. 2012). Upon systemic delivery via subcutaneous injection into interscapular region of an HSA^{LR} mouse model (8 injections—25 mg/kg 2 × weekly), this AON targeted specific sequences within the human skeletal actin transgene and resulted in a very efficient (by as much as ~80 %) and long-term silencing of nuclear-retained transcripts with expanded repeats. Knockdown of toxic RNA persisted up to 31 weeks post last injection, and in the case of one of the tested oligomers, even ~ 50 % reduction was still observed 1 year after last injection. Furthermore, this AON corrected the alternative splicing of four MBNL1-dependent exons and reduced myotonic discharges in hind limb muscles of HSA^{LR} mice. Recently, a novel gapmer-based AON referred to as "ISIS 486178" has been developed, which targets mouse, monkey, and human *DMPK* mRNA and shows potent activity when administered systemically by subcutaneous injection (Pandey et al. 2015). The central ten nucleotides are deoxynucleotide sugars with a phosphorothioate backbone, flanked by three nucleotides at the 5′ and 3′ ends containing a 2′-4′constrained ethyl (cEt) modifications that allow for significantly increased RNA binding affinity and in vivo potency relative to those modified with other 2′ chemistries ("3–10–3 gapmer" design). This cEt-modified AON inhibited *DMPK* transcript expression by about 80 % in several cell lines and substantially eliminated RNA foci in patient-derived DM1 myoblast cells. Moreover, reduction of *DMPK* mRNA levels in a

DM1 mouse model as well as in monkeys after subcutaneous injection of ISIS 486178 was sustained for several weeks after cessation of dosing (Pandey et al. 2015). Another study tested PS-LNA-based (CAG)n oligonucleotides, namely, RNase-H compatible 4-10-4 DNA/LNA gapmer (with a central stretch of ten DNA monomers flanked by four LNA nucleotides on either end) and RNase-H incompatible DNA/LNA mixmer of the same length (with eight LNA nucleotides interspersed throughout the AON) (Nakamori et al. 2011). Interestingly, both types of anti-CUG AONs, the gapmer and the mixmer, had a similar capacity to reduce CUG^{exp} transcripts and a comparable ability to suppress CTG repeat instability in DM1 cell and murine model.

As a final point, previous clinical trials have only tested drugs aimed at relieving particular symptoms of DM such as muscle stiffness and myotonia (mexiletine—currently in Phase 2 clinical trials) or insulin resistance (SomatoKine/iPlex—recombinant insulin-like growth factor 1). However, owing to recent developments in antisense-based technology, the first clinical trial tackling the expanded RNA using AON is currently underway. IONIS-DMPK-2.5$_{Rx}$, developed by Ionis Pharmaceuticals in cooperation with Biogen, is presently evaluated in a Phase 1/2 clinical study in approximately 36 adult DM1 patients in a randomized, placebo-controlled, dose-escalation study. IONIS-DMPK-2.5$_{Rx}$ is a 2′MOE gapmer, comprising 2′MOE modifications at both ends to maximize biostability and a central gap of ten unmodified nucleotides to support RNase-H activity (Wheeler et al. 2012). This drug is designed to reduce the expression of mutant DMPK RNA by targeting sequences outside the CUG expansion region, and currently, it is the only AON-based drug for DM in clinical trials.

5.2 Huntington's Disease (HD)

5.2.1 Clinical, Genetic, and Molecular Features of HD

Huntington's disease (HD) is an adult-onset neurodegenerative disorder with prevalence of about 1 in 20,000 worldwide (Kumar et al. 2015). The symptoms start on average at the age of 38, although the onset can vary from the age of 2 to 85 (Karadima et al. 2012; Zielonka et al. 2014). Clinically, HD manifests itself with involuntary movements, chorea, dystonia, and cognitive deficits, which relate to degeneration of small projecting neurons in the striatum and, at later time points, cortical atrophy (Kay et al. 2014; Vonsattel et al. 2011). HD is caused by an expansion of CAG trinucleotides in exon 1 of *HTT* gene (also known as *IT15*). Typically, patients have more than 35 repeats on one *HTT* allele; however, individuals with 27–35 repeats may transmit larger CAG expansions to their children. The size of CAG repeats correlates with the onset of the disease, with the juvenile form starting with more than 60 repeats, and the largest observed number of trinucleotides reaching above 120. Translation of expanded repeats that encode glutamines is essential for HD pathology and, unlike in DM, the protein not the RNA constitutes the main origin of cellular toxicity.

Huntingtin protein (HTT) is mainly expressed in the brain and testes localizing to various cellular structures, including the nucleus, ER, Golgi apparatus, and endosomes (Kumar et al. 2015; Zielonka et al. 2014). Although the exact role of HTT is still elusive, multiple roles have been indicated, based on hundreds of its binding partners. Some of the HTT's functions include regulation of gene transcription, intracellular signaling, trafficking, and endocytosis. It is also unclear how expansion of glutamines in mutant HTT (mtHTT) might lead to toxicity in HD (Yu et al. 2014). On one hand, complete loss of HTT in full knockout mice leads to early lethality and heterozygous mice show abnormalities in the brain as well as behavioral alterations, and on the other, the amounts of mtHTT protein correspond with the severity of HD phenotype, which indicates that the larger isoform, containing an increased number of glutamines, gains new detrimental functions. The gain-of-function hypothesis is also supported by experiments showing that loss of HTT in adult mice and rhesus monkeys is well tolerated (Boudreau et al. 2009; Grondin et al. 2012).

As HTT has numerous functions that might be important for long-term neuronal function, successful therapy against HD should employ allele-specific elimination of the mtHTT transcript or translation inhibition of the mutated protein. Both approaches can be tackled with modified AONs. Elimination of the *mtHTT* mRNA can be achieved by targeting DNA-PS AONs or gapmers to either CAG repeats or sequences containing single nucleotide polymorphisms (SNPs) linked to expanded trinucleotide repeats. Conversely, translation inhibition is utilized with fully modified AONs binding to CAG repeats that sterically block formation of the ribosomal complex. Therapeutic efficacy of AONs is generally tested in two transgenic mouse models, YAC128 and BACHD that in addition to the intrinsic HTT express human full-length mtHTT (Ehrnhoefer et al. 2009). Both HD mouse models show progressive motor abnormalities followed by striatal and cortical atrophy.

5.2.2 Modified AONs in Therapy of HD

Potential application of AON analogs towards HD was first shown by targeting a fragment of the first exon of *HTT* with PS-modified AONs in NT2N neurons (Nellemann et al. 2000). Selective knockdown of the mutant protein was achieved next in HD patient's fibroblast cells and cultured striatal neurons derived from YAC128 mice with the use of PNA and LNA oligomers targeting directly CAG repeat sequence, but not other regions of *HTT* mRNA (Hu et al. 2009). To allow for AON uptake, PNAs were conjugated to a peptide, and LNAs were delivered with lipids. Interestingly, as short as 13-nt-long PNAs with CTG repeat sequence were effective. No significant decrease in the amount of mRNA upon delivery of PNAs and LNAs was observed, suggesting that the major therapeutic effect is due to inhibition of translation. Importantly, both tested AON analogs not only discriminated between mutant and normal *HTT* variants but also did not affect expression of other CAG-containing mRNAs.

A spectrum of other AON analogs directed against CAG repeats were shown to selectively inhibit expression of mtHTT protein in a subsequent study (Gagnon et al. 2010). Those included LNAs, 2′MOEs, and RNase-H-activating gapmers. The comparison showed that the most effective AON analogs reduced *mtHTT* through steric interaction rather than transcript degradation, with, e.g., LNA gapmers showing relatively high toxicity compared to the fully modified LNAs, presumably due to inhibition of other CAG repeat-containing genes. Importantly, allele-specific inhibition of toxic mRNA was compatible with PS-modified AONs favorably used in in vivo studies due to their increased stability against nucleases and better cellular uptake. In a subsequent study comparing various therapeutic strategies directed at *mtHTT* transcript, PMOs and PNAs binding to CAG repeats showed relatively high allele selectivity in fibroblasts from affected individuals, decreasing the mtHTT and normal HTT protein isoforms to approximately 25 and 70 % of control levels, respectively (Fiszer et al. 2012).

Researchers identified a number of SNPs associated with expanded CAG repeats that allowed for selective downregulation of the *mtHTT* transcript (Warby et al. 2009). 2′MOE-PS gapmers binding to an SNP efficiently silenced mtHTT expression in patient's fibroblasts and in primary neurons from mouse models of HD (Carroll et al. 2011). Furthermore, the authors showed that intraparenchymal delivery of 2′MOE-PS gapmer analogs with S-constrained ethyl motifs to the BACHD mouse strain can selectively lower mtHTT protein to 47 %. In the following study, a marked decrease of the mutant protein in brains of YAC128 and BACHD mice following intracerebroventricular 2-week infusion with 50 μg/day gapmers carrying 2′MOE-PS chemistry was achieved (Kordasiewicz et al. 2012). Importantly, lowered mtHTT content was accompanied by sustained reversal of motor deficits that persisted for at least 12 months, despite the mutant protein restoration to initial levels 9 months after the treatment. Surprisingly, a similar phenotype was obtained upon reduction of the total HTT (the human mtHTT and the murine wild-type form) in the BACHD mouse model. Based on these data, as well as the fact that infusion of AONs at a dose of 4 mg/day for 21 days into rhesus monkeys' brains also effectively lowered huntingtin expression in many brain regions, the first clinical trial for treatment of HD was initiated, with the use of a nonselective 2′MOE-PS gapmer (IONIS-HTT$_{RX}$) targeting total HTT (Keiser et al. 2015).

Although the experiments showed that reduction of both mutant and wild-type HTT is therapeutic in rodents, the outcome of a long-term decrease of the wild-type form in humans is unpredictable. Based on this principle, enhanced single nucleotide discrimination of RNase-H was achieved by positional incorporation of chemical modifications within AONs (Østergaard et al. 2013). This strategy limited RNase-H-mediated cleavage of the nontargeted transcript and resulted in a near-complete silencing of mtHTT with no significant suppression of the normal form in patient's cells and a humanized mouse model of HD carrying only human *HTT* alleles (Southwell et al. 2013).

6 Future Perspectives

Over the recent years, studies exploring the therapeutic application of chemically modified AONs have moved from in vitro and cell culture-based experiments to mouse and large animal studies. Currently, several of the selected molecules are being tested in clinical trials. For DMD and HD, however, only a limited number of patients can benefit from a treatment with a single AON, and thus, clinical trials combining several AONs in one therapeutic treatment will presumably follow. In DMD, this would require, e.g., several AONs targeting exons 45–55 (Béroud et al. 2007). In HD, CAG repeat targeting AON analogs could presumably be only used in juvenile forms of HD characterized by longer expansions. Other HD patients might be more amenable for therapy with AONs targeting SNPs (Kay et al. 2014), in which at least three different SNPs will have to be targeted to achieve selective treatment in more than 80 % of HD patients (Warby et al. 2009). The pursuit of clinically relevant AON treatments targeting NMDs is still ongoing.

Funding This work was supported by the Foundation for Polish Science-TEAM program cofinanced by the European Union within the European Regional Development Fund (to K.S.), by the National Centre for Research and Development grant ERA-NET-E-Rare-2/III/DRUG_FXSPREMUT/01/2016 (to K.S.), and by the Polish National Science Centre grant 2014/15/B/NZ5/00142 (to E.S.K.). This project was also supported by the Ministry of Science and Higher Education of the Republic of Poland, from the quality-promoting subsidy, under the Leading National Research Centre (KNOW) program for the years 2012–2017 [KNOW RNA Research Centre in Poznan (No. 01/KNOW2/2014)].

The authors declare no conflict of interest.

References

Aartsma-Rus A, Fokkema I, Verschuuren J et al (2009) Theoretic applicability of antisense-mediated exon skipping for Duchenne muscular dystrophy mutations. Hum Mutat 30:293–299

Aartsma-Rus A, Janson AA, Kaman WE et al (2003) Therapeutic antisense-induced exon skipping in cultured muscle cells from six different DMD patients. Hum Mol Genet 12:907–914

Alter J, Lou F, Rabinowitz A et al (2006) Systemic delivery of morpholino oligonucleotide restores dystrophin expression bodywide and improves dystrophic pathology. Nat Med 12:175

Aoki Y, Yokota T, Nagata T et al (2012) Bodywide skipping of exons 45–55 in dystrophic mdx52 mice by systemic antisense delivery. Proc Natl Acad Sci USA 109:13763–13768

Ashizawa T, Harper PS (2006) Myotonic dystrophies: an overview. In: Ashizawa T, Wells RD (eds) Genetic instabilities and neurological disorders, 2nd edn. Elsevier, Burlington, pp 21–36

Barciszewski J, Marquez VE, Vasseur JJ et al (2015) Chemical biology of nucleic acids. ACS Chem Biol 10:1358–1361

Bennett CF, Swayze EE (2010) RNA targeting therapeutics: molecular mechanisms of antisense oligonucleotides as a therapeutic platform. Annu Rev Pharmacol Toxicol 50:259–293

Béroud C, Tuffery-Giraud S, Matsuo M et al (2007) Multiexon skipping leading to an artificial DMD protein lacking amino acids from exons 45 through 55 could rescue up to 63% of patients with Duchenne muscular dystrophy. Hum Mutat 28:196–202

Betts C, Saleh AF, Arzumanov AA et al (2012) Pip6-PMO, a new generation of peptide-oligonucleotide conjugates with improved cardiac exon skipping activity for DMD treatment. Mol Ther Nucleic Acids 1, e38

Betts CA, Saleh AF, Carr CA et al (2015) Prevention of exercised induced cardiomyopathy following Pip-PMO treatment in dystrophic mdx mice. Sci Rep 5:8986

Bogdanik LP, Osborne MA, Davis C et al (2015) Systemic, postsymptomatic antisense oligonucleotide rescues motor unit maturation delay in a new mouse model for type II/III spinal muscular atrophy. Proc Natl Acad Sci USA 112:72

Boudreau RL, McBride JL, Martins I et al (2009) Nonallele-specific silencing of mutant and wild-type huntingtin demonstrates therapeutic efficacy in Huntington's disease mice. Mol Ther 17:1053–1063

Burnett JC, Rossi JJ (2012) RNA-based therapeutics: current progress and future prospects. Chem Biol 19:60–71

Carroll JB, Warby SC, Southwell AL et al (2011) Potent and selective antisense oligonucleotides targeting single-nucleotide polymorphisms in the Huntington disease gene/allele-specific silencing of mutant huntingtin. Mol Ther 19:2178–2185

Cartegni L, Krainer AR (2002) Disruption of an SF2/ASF-dependent exonic splicing enhancer in SMN2 causes spinal muscular atrophy in the absence of SMN1. Nat Genet 30:377–384

Cartegni L, Krainer AR (2003) Correction of disease-associated exon skipping by synthetic exon-specific activators. Nat Struct Mol Biol 10:120–125

Cerritelli SM, Crouch RJ (2009) Ribonuclease H: the enzymes in eukaryotes. FEBS J 276:1494–1505. doi:10.1111/j.1742-4658.2009.06908.x

Cho DH, Tapscott SJ (2007) Myotonic dystrophy: emerging mechanisms for DM1 and DM2. Biochim Biophys Acta 1772:195–204. doi:10.1016/j.bbadis.2006.05.013

Cirak S, Arechavala-Gomeza V, Guglieri M et al (2011) Exon skipping and dystrophin restoration in patients with Duchenne muscular dystrophy after systemic phosphorodiamidate morpholino oligomer treatment: an open-label, phase 2, dose-escalation study. Lancet 378:595–605

DeVos SL, Miller TM (2013) Antisense oligonucleotides: treating neurodegeneration at the level of RNA. Neurotherapeutics 10:486–497

Disterer P, Kryczka A, Liu Y et al (2014) Development of therapeutic splice-switching oligonucleotides. Hum Gene Ther 25:587–598

Dominski Z, Kole R (1993) Restoration of correct splicing in thalassemic pre-mRNA by antisense oligonucleotides. Proc Natl Acad Sci USA 90:8673–8677

Dragulescu-Andrasi A, Rapireddy S, Frezza BM et al (2006) A simple gamma-backbone modification preorganizes peptide nucleic acid into a helical structure. J Am Chem Soc 128:10258–10267

Dunckley MG, Manoharan M, Villiet P et al (1998) Modification of splicing in the dystrophin gene in cultured Mdx muscle cells by antisense oligoribonucleotides. Hum Mol Genet 7:1083–1090

Echigoya Y, Aoki Y, Miskew B et al (2015) Long-term efficacy of systemic multiexon skipping targeting dystrophin exons 45–55 with a cocktail of vivo-morpholinos in mdx52 mice. Mol Ther Nucleic Acids 4, e225

Ehrnhoefer DE, Butland SL, Pouladi MA et al (2009) Mouse models of Huntington disease: variations on a theme. Dis Model Mech 2:123–129

Fabani MM, Abreu-Goodger C, Williams D (2010) Efficient inhibition of miR-155 function in vivo by peptide nucleic acids. Nucleic Acids Res 38:4466–4475

Fallini C, Bassell GJ, Rossoll W (2012) Spinal muscular atrophy: the role of SMN in axonal mRNA regulation. Brain Res 1462:81–92

Falzarano MS, Scotton C, Passarelli C et al (2015) Duchenne muscular dystrophy: from diagnosis to therapy. Molecules 20:18168–18184

Fardaei M, Rogers MT, Thorpe HM et al (2002) Three proteins, MBNL, MBLL and MBXL, co-localize in vivo with nuclear foci of expanded-repeat transcripts in DM1 and DM2 cells. Hum Mol Genet 11:805–814

Ferguson DP, Dangott LJ, Lightfoot JT (2014) Lessons learned from vivo-morpholinos: how to avoid vivo-morpholino toxicity. BioTechniques 56:251–256

Fiszer A, Olejniczak M, Switonski PM et al (2012) An evaluation of oligonucleotide-based therapeutic strategies for polyQ diseases. BMC Mol Biol 13:6

Gagnon KT, Pendergraff HM, Deleavey GF et al (2010) Allele-selective inhibition of mutant huntingtin expression with antisense oligonucleotides targeting the expanded CAG repeat. Biochemistry 49:10166–10178

Gao Z, Cooper TA (2013) Antisense oligonucleotides: rising stars in eliminating RNA toxicity in myotonic dystrophy. Hum Gene Ther 24:499–507

Goemans NM, Tulinius M, van den Akker JT et al (2011) Systemic administration of PRO051 in Duchenne's muscular dystrophy. N Engl J Med 364:1513–1522

Gomes-Pereira M, Cooper TA, Gourdon G (2011) Myotonic dystrophy mouse models: towards rational therapy development. Trends Mol Med 17:506–517

Gonzalez-Barriga A, Mulders SA, van de Giessen J et al (2013) Design and analysis of effects of triplet repeat oligonucleotides in cell models for myotonic dystrophy. Mol Ther Nucleic Acids 2, e81

Goyenvalle A, Griffith G, Babbs A et al (2015) Functional correction in mouse models of muscular dystrophy using exon-skipping tricyclo-DNA oligomers. Nat Med 21:270–275

Grondin R, Kaytor MD, Ai Y et al (2012) Six-month partial suppression of Huntingtin is well tolerated in the adult rhesus striatum. Brain 135:1197–1209

Hamilton G, Gillingwater TH (2013) Spinal muscular atrophy: going beyond the motor neuron. Trends Mol Med 19:40–50

Havens MA, Duelli DM, Hastings ML (2013) Targeting RNA splicing for disease therapy. Wiley Interdiscp Rev RNA 4:247–266

Ho TH, Charlet-B N, Poulos MG et al (2004) Muscleblind proteins regulate alternative splicing. EMBO J 23:3103–3112

Hu J, Matsui M, Gagnon KT et al (2009) Allele-specific silencing of mutant huntingtin and ataxin-3 genes by targeting expanded CAG repeats in mRNAs. Nat Biotech 27:478–484

Hua Y, Liu YH, Sahashi K et al (2015) Motor neuron cell-nonautonomous rescue of spinal muscular atrophy phenotypes in mild and severe transgenic mouse models. Genes Dev 29:288–297

Hua Y, Sahashi K, Hung G et al (2010) Antisense correction of SMN2 splicing in the CNS rescues necrosis in a type III SMA mouse model. Genes Dev 24:1634–1644

Hua Y, Sahashi K, Rigo F et al (2011) Peripheral SMN restoration is essential for long-term rescue of a severe spinal muscular atrophy mouse model. Nature 478:123–126

Hua Y, Vickers TA, Okunola HL et al (2008) Antisense masking of an hnRNP A1/A2 intronic splicing silencer corrects SMN2 splicing in transgenic mice. Am J Hum Genet 82:834–848

Iversen PL, Warren TK, Wells JB et al (2012) Discovery and early development of AVI-7537 and AVI-7288 for the treatment of Ebola virus and Marburg virus infections. Viruses 4:2806–2830

Jearawiriyapaisarn N, Moulton HM, Buckley B (2008) Sustained dystrophin expression induced by peptide-conjugated morpholino oligomers in the muscles of mdx mice. Mol Ther 16:1624–1629

Jepsen JS, Sørensen MD, Wengel J (2004) Locked nucleic acid: a potent nucleic acid analog in therapeutics and biotechnology. Oligonucleotides 14:130–146

Jiang H, Mankodi A, Swanson MS et al (2004) Myotonic dystrophy type 1 is associated with nuclear foci of mutant RNA, sequestration of muscleblind proteins and deregulated alternative splicing in neurons. Hum Mol Genet 13:3079–3088

Kanadia RN, Johnstone KA, Mankodi A et al (2003) A muscleblind knockout model for myotonic dystrophy. Science 302:1978–1980

Karadima G, Dimovasili C, Koutsis G (2012) Age at onset in Huntington's disease: replication study on the association of HAP1. Parkinsonism Relat Disord 18:1027–1028

Karkare S, Bhatnagar D (2006) Promising nucleic acid analogs and mimics: characteristic features and applications of PNA, LNA, and morpholino. Appl Microbiol Biotechnol 71:575–586

Kashima T, Rao N, David CJ et al (2007) hnRNP A1 functions with specificity in repression of SMN2 exon 7 splicing. Hum Mol Genet 16:3149–3159

Kaur H, Babu BR, Maiti S (2007) Perspectives on chemistry and therapeutic applications of Locked Nucleic Acid (LNA). Chem Rev 107:4672–4697

Kay C, Skotte NH, Southwell AL et al (2014) Personalized gene silencing therapeutics for Huntington disease. Clin Genet 86:29–36

Kazantsev AG, Thompson LM (2008) Therapeutic application of histone deacetylase inhibitors for central nervous system disorders. Nat Rev Drug Discov 7:854–868

Keiser MS, Kordasiewicz H, McBride J (2015) Gene suppression strategies for dominantly inherited neurodegenerative diseases: lessons from huntington's disease and spinocerebellar ataxia. Hum Mol Genet 25(R1):R53–R64

Kinali M, Arechavala-Gomeza V, Feng L et al (2009) Local restoration of dystrophin expression with the morpholino oligomer AVI-4658 in Duchenne muscular dystrophy: a single-blind, placebo-controlled, dose-escalation, proof-of-concept study. Lancet Neurol 8:918–928

Kole R, Krainer AR, Altman S (2012) RNA therapeutics: beyond RNA interference and antisense oligonucleotides. Nat Rev Drug Discov 11:125–140

Konieczny P, Stepniak-Konieczna E, Sobczak K (2015) MBNL proteins and their target RNAs, interaction and splicing regulation. Nucleic Acids Res 42:10873–87

Konieczny P, Swiderski K, Chamberlain JS (2013) Gene and cell-mediated therapies for muscular dystrophy. Muscle Nerve 47:649–663

Kordasiewicz HB, Stanek LM, Wancewicz EV et al (2012) Sustained therapeutic reversal of Huntington's disease by transient repression of huntingtin synthesis. Neuron 74:1031–1044

Kumar A, Kumar Singh S, Kumar V et al (2015) Huntington's disease: an update of therapeutic strategies. Gene 556:91–97

Kuyumcu-Martinez NM, Wang G-SS, Cooper TA (2007) Increased steady-state levels of CUGBP1 in myotonic dystrophy 1 are due to PKC-mediated hyperphosphorylation. Mol Cell 28:68–78

Le Rumeur E, Winder SJ, Hubert J-FF (2010) Dystrophin: more than just the sum of its parts. Biochim Biophys Acta 1804:1713–1722

Le TT, Pham LT, Butchbach MER (2005) SMNDelta7, the major product of the centromeric survival motor neuron (SMN2) gene, extends survival in mice with spinal muscular atrophy and associates with full-length SMN. Hum Mol Genet 14:845–857

Lee J, Bennett C, Cooper T (2012) RNase H-mediated degradation of toxic RNA in myotonic dystrophy type 1. Proc Natl Acad Sci USA A109:4221–4226

Leger AJ, Mosquea LM, Clayton NP et al (2013) Systemic delivery of a peptide-linked morpholino oligonucleotide neutralizes mutant RNA toxicity in a mouse model of myotonic dystrophy. Nucleic Acid Ther 23:109–117

Lima WF, Rose JB, Nichols JG et al (2007) Human RNase H1 discriminates between subtle variations in the structure of the heteroduplex substrate. Mol Pharmacol 71:83–91

Lin X, Miller JW, Mankodi A et al (2006) Failure of MBNL1-dependent post-natal splicing transitions in myotonic dystrophy. Hum Mol Genet 15:2087–2097

Lu QL, Mann CJ, Lou F et al (2003) Functional amounts of dystrophin produced by skipping the mutated exon in the mdx dystrophic mouse. Nat Med 9:1009–1014

Lu QL, Rabinowitz A, Chen YC et al (2005) Systemic delivery of antisense oligoribonucleotide restores dystrophin expression in body-wide skeletal muscles. Proc Natl Acad Sci USA 102:198–203

Magaña JJ, Cisneros B (2011) Perspectives on gene therapy in myotonic dystrophy type 1. J Neurosci Res 89:275–285

Mankodi A, Logigian E, Callahan L et al (2000) Myotonic dystrophy in transgenic mice expressing an expanded CUG repeat. Science 289:1769–1773

Mankodi A, Urbinati CR, Yuan QP et al (2001) Muscleblind localizes to nuclear foci of aberrant RNA in myotonic dystrophy types 1 and 2. Hum Mol Genet 10:2165–2170

Mann CJ, Honeyman K, Cheng AJ et al (2001) Antisense-induced exon skipping and synthesis of dystrophin in the mdx mouse. Proc Natl Acad Sci USA 98:42–47

McGreevy JW, Hakim CH, McIntosh MA et al (2015) Animal models of Duchenne muscular dystrophy: from basic mechanisms to gene therapy. Dis Model Mech 8:195–213

McMahon BM, Mays D, Lipsky J et al (2002) Pharmacokinetics and tissue distribution of a peptide nucleic acid after intravenous administration. Antisense Nucleic Acid Drug Dev 12:65–70

Mendell JR, Lloyd-Puryear M (2013) Report of MDA muscle disease symposium on newborn screening for Duchenne muscular dystrophy. Muscle Nerve 48:21–26

Mendell JR, Rodino-Klapac LR, Sahenk Z et al (2013) Eteplirsen for the treatment of Duchenne muscular dystrophy. Ann Neurol 74:637–647. doi:10.1002/ana.23982

Miller JW, Urbinati CR, Teng-Umnuay P et al (2000) Recruitment of human muscleblind proteins to (CUG)(n) expansions associated with myotonic dystrophy. EMBO J 19:4439–48. doi:10.1093/emboj/19.17.4439

Miyajima H, Miyaso H, Okumura M et al (2002) Identification of a cis-acting element for the regulation of SMN exon 7 splicing. J Biol Chem 277:23271–23277

Moulton HM, Moulton JD (2010) Morpholinos and their peptide conjugates: therapeutic promise and challenge for Duchenne muscular dystrophy. Biochim Biophys Acta 1798:2296–2303

Mulders SA, van den Broek WJ, Wheeler TM et al (2009) Triplet-repeat oligonucleotide-mediated reversal of RNA toxicity in myotonic dystrophy. Proc Natl Acad Sci USA 106:13915–13920

Mulders SA, van Engelen BG, Wieringa B et al (2010) Molecular therapy in myotonic dystrophy: focus on RNA gain-of-function. Hum Mol Genet 19(R1):R90–R97

Muntoni F, Wood MJ (2011) Targeting RNA to treat neuromuscular disease. Nat Rev Drug Discov 10:621–637

Nakamori M, Gourdon G, Thornton CA (2011) Stabilization of expanded (CTG)•(CAG) repeats by antisense oligonucleotides. Mol Ther 19:2222–2227

Nellemann C, Abell K, Nørremølle A et al (2000) Inhibition of huntingtin synthesis by antisense oligodeoxynucleotides. Mol Cell Neurosci 16:313–323

Obad S, dos Santos CO, Petri A et al (2011) Silencing of microRNA families by seed-targeting tiny LNAs. Nat Genet 43:371–378

Osborne RJ, Thornton CA (2006) RNA-dominant diseases. Hum Mol Genet 2:9

Østergaard ME, Southwell AL, Kordasiewicz H et al (2013) Rational design of antisense oligonucleotides targeting single nucleotide polymorphisms for potent and allele selective suppression of mutant Huntingtin in the CNS. Nucleic Acids Res 41:9634–9650. doi:10.1093/nar/gkt725

Pandey SK, Wheeler TM, Justice SL et al (2015) Identification and characterization of modified antisense oligonucleotides targeting DMPK in mice and nonhuman primates for the treatment of myotonic dystrophy type 1. J Pharmacol Exp Ther 355:310–321

Passini MA, Bu J, Richards AM et al (2011) Antisense oligonucleotides delivered to the mouse CNS ameliorate symptoms of severe spinal muscular atrophy. Sci Transl Med 3:72ra18

Perronnet C, Vaillend C (2010) Dystrophins, utrophins, and associated scaffolding complexes: role in mammalian brain and implications for therapeutic strategies. J Biomed Biotechnol 2010:849426

Philips AV, Timchenko LT, Cooper TA (1998) Disruption of splicing regulated by a CUG-binding protein in myotonic dystrophy. Science 280:737–741

Porensky PN, Mitrpant C, McGovern VL et al (2012) A single administration of morpholino antisense oligomer rescues spinal muscular atrophy in mouse. Hum Mol Genet 21:1625–1638

Prakash TP, Bhat B (2007) 2′-Modified oligonucleotides for antisense therapeutics. Curr Top Med Chem 7:641–649

Renneberg D, Bouliong E, Reber U et al (2002) Antisense properties of tricyclo-DNA. Nucleic Acids Res 30:2751–2757

Sforza S, Tedeschi T, Calabretta A (2010) A peptide nucleic acid embedding a pseudopeptide nuclear localization sequence in the backbone behaves as a peptide mimic. Eur J Org Chem 2010(13):2441–2444

Sibley CR, Seow Y, Wood MJ (2010) Novel RNA-based strategies for therapeutic gene silencing. Mol Ther 18:466–476

Singh NK, Singh NN, Androphy EJ (2006) Splicing of a critical exon of human Survival Motor Neuron is regulated by a unique silencer element located in the last intron. Mol Cell Biol 26:1333–1346

Skordis LA, Dunckley MG, Yue B (2003) Bifunctional antisense oligonucleotides provide a trans-acting splicing enhancer that stimulates SMN2 gene expression in patient fibroblasts. Proc Natl Acad Sci USA 100:4114–4119

Southwell AL, Warby SC, Carroll JB et al (2013) A fully humanized transgenic mouse model of Huntington disease. Hum Mol Genet 22:18–34. doi:10.1093/hmg/dds397

Stein CA, Hansen JB, Lai J et al (2010) Efficient gene silencing by delivery of locked nucleic acid antisense oligonucleotides, unassisted by transfection reagents. Nucleic Acids Res 38, e3

Takeshima Y, Wada H, Yagi M et al (2001) Oligonucleotides against a splicing enhancer sequence led to dystrophin production in muscle cells from a Duchenne muscular dystrophy patient. Brain Dev 23:788–790

Thornton CA (2014) Myotonic dystrophy. Neurol Clin 32:705–719

van Deutekom JC, Bremmer-Bout M, Janson AA et al (2001) Antisense-induced exon skipping restores dystrophin expression in DMD patient derived muscle cells. Hum Mol Genet 10:1547–1554

van Deutekom JC, Janson AA, Ginjaar IB et al (2007) Local dystrophin restoration with antisense oligonucleotide PRO051. N Engl J Med 357:2677–2686. doi:10.1056/NEJMoa073108

Viollet L, Melki J (2013) Spinal muscular atrophies. Handb Clin Neurol 113:1395–1411

Voit T, Topaloglu H, Straub V et al (2014) Safety and efficacy of drisapersen for the treatment of Duchenne muscular dystrophy (DEMAND II): an exploratory, randomised, placebo-controlled phase 2 study. Lancet Neurol 13:987–996

Vonsattel JP, Keller C, Cortes Ramirez EP (2011) Huntington's disease—neuropathology. Handb Clin Neurol 100:83–100

Wang ET, Sandberg R, Luo S et al (2008) Alternative isoform regulation in human tissue transcriptomes. Nature 456:470–476

Wang ET, Ward AJ, Cherone JM et al (2015) Antagonistic regulation of mRNA expression and splicing by CELF and MBNL proteins. Genome Res 25:858–871

Warby SC, Montpetit A, Hayden AR et al (2009) CAG expansion in the Huntington disease gene is associated with a specific and targetable predisposing haplogroup. Am J Hum Genet 84:351–366

Wheeler TM, Leger AJ, Pandey SK et al (2012) Targeting nuclear RNA for in vivo correction of myotonic dystrophy. Nature 488:111–115

Wheeler TM, Lueck JD, Swanson MS et al (2007) Correction of ClC-1 splicing eliminates chloride channelopathy and myotonia in mouse models of myotonic dystrophy. J Clin Invest 117:3952–3957

Wheeler TM, Sobczak K, Lueck JD et al (2009) Reversal of RNA dominance by displacement of protein sequestered on triplet repeat RNA. Science 325:336–339

Wheeler TM, Thornton CA (2007) Myotonic dystrophy: RNA-mediated muscle disease. Curr Opin Neurol 20:572–576

Williams JH, Schray RC, Patterson CA et al (2009) Oligonucleotide-mediated survival of motor neuron protein expression in CNS improves phenotype in a mouse model of spinal muscular atrophy. J Neurosci 29:7633–7638

Wilton SD, Lloyd F, Carville K et al (1999) Specific removal of the nonsense mutation from the mdx dystrophin mRNA using antisense oligonucleotides. Neuromuscul Disord 9:330–338

Wojtkowiak-Szlachcic A, Taylor K, Stepniak-Konieczna E et al (2015) Short antisense-locked nucleic acids (all-LNAs) correct alternative splicing abnormalities in myotonic dystrophy. Nucleic Acids Res 43:3318–3331

Wu B, Li Y, Morcos PA et al (2009) Octa-guanidine morpholino restores dystrophin expression in cardiac and skeletal muscles and ameliorates pathology in dystrophic mdx mice. Mol Ther 17:864–871

Wu B, Lu P, Cloer C et al (2012) Long-term rescue of dystrophin expression and improvement in muscle pathology and function in dystrophic mdx mice by peptide-conjugated morpholino. Am J Pathol 181:392–400

Wu B, Moulton HM, Iversen PL (2008) Effective rescue of dystrophin improves cardiac function in dystrophin-deficient mice by a modified morpholino oligomer. Proc Natl Acad Sci USA 105:14814–14819

Wu B, Xiao B, Cloer C et al (2011) One-year treatment of morpholino antisense oligomer improves skeletal and cardiac muscle functions in dystrophic mdx mice. Mol Ther 19:576–583

Yang L, Niu H, Gao X et al (2013) Effective exon skipping and dystrophin restoration by 2′-o-methoxyethyl antisense oligonucleotide in dystrophin-deficient mice. PLoS One 8, e61584

Yin H, Betts C, Saleh AF et al (2010a) Optimization of peptide nucleic acid antisense oligonucleotides for local and systemic dystrophin splice correction in the mdx mouse. Mol Ther 18:819–827

Yin H, Lu Q, Wood M (2008a) Effective exon skipping and restoration of dystrophin expression by peptide nucleic acid antisense oligonucleotides in mdx mice. Mol Ther 16:38–45

Yin H, Moulton HM, Betts C et al (2010b) Functional rescue of dystrophin-deficient mdx mice by a chimeric peptide-PMO. Mol Ther 18:1822–1829

Yin H, Moulton HM, Betts C et al (2009) A fusion peptide directs enhanced systemic dystrophin exon skipping and functional restoration in dystrophin-deficient mdx mice. Hum Mol Genet 18:4405–4414

Yin H, Moulton HM, Seow Y et al (2008b) Cell-penetrating peptide-conjugated antisense oligonucleotides restore systemic muscle and cardiac dystrophin expression and function. Hum Mol Genet 17:3909–3918

Yin H, Saleh AF, Betts C et al (2011) Pip5 transduction peptides direct high efficiency oligonucleotide-mediated dystrophin exon skipping in heart and phenotypic correction in mdx mice. Mol Ther 19:1295–1303

Yokota T, Lu Q-LL, Partridge T et al (2009a) Efficacy of systemic morpholino exon-skipping in Duchenne dystrophy dogs. Ann Neurol 65:667–676

Yokota T, Nakamura A, Nagata T et al (2012) Extensive and prolonged restoration of dystrophin expression with vivo-morpholino-mediated multiple exon skipping in dystrophic dogs. Nucleic Acid Ther 22:306–315. doi:10.1089/nat.2012.0368

Yokota T, Takeda Si LQ-LL et al (2009b) A renaissance for antisense oligonucleotide drugs in neurology: exon skipping breaks new ground. Arch Neurol 66:32–38

Yu S, Liang Y, Palacino J et al (2014) Drugging unconventional targets: insights from Huntington's disease. Trends Pharmacol Sci 35:53–62

Zhou H, Janghra N, Mitrpant C et al (2013) A novel morpholino oligomer targeting ISS-N1 improves rescue of severe spinal muscular atrophy transgenic mice. Hum Gene Ther 24:331–342

Zhou P, Dragulescu-Andrasi A, Bhattacharya B et al (2006) Synthesis of cell-permeable peptide nucleic acids and characterization of their hybridization and uptake properties. Bioorg Med Chem Lett 16:4931–4935

Zielonka D, Mielcarek M, Landwehrmeyer BG (2014) Update on Huntington's disease: advances in care and emerging therapeutic options. Parkinsonism Relat Disord 21:169–178. doi:10.1016/j.parkreldis.2014.12.013

Effect of Depurination on Cellular and Viral RNA

Kass A. Jobst, Alexander Klenov, Kira C.M. Neller, and Katalin A. Hudak

Contents

Abstract Abasic sites in DNA arise from spontaneous damage and, if left unrepaired, contribute to mutations. Dedicated pathways have evolved to repair abasic DNA; however, the biological implications and cellular fate of RNA possessing these lesions are not well characterized. In this chapter, we review advances in the area of RNA depurination, which produces the dominant form of abasic RNA in vivo. We discuss the biochemistry of RNA depurination, the varied impacts of cellular and viral RNA depurination, and the numerous methods available for detecting abasic RNA. Given the abundance of RNA in cells and the relative stability of abasic RNA compared with DNA, depurinated RNA likely

K.A. Jobst • A. Klenov • K.C.M. Neller • K.A. Hudak (✉)
Department of Biology, York University, 4700 Keele Street, Toronto, ON M3J 1P3, Canada
e-mail: hudak@yorku.ca

S. Jurga et al. (eds.), *Modified Nucleic Acids in Biology and Medicine*, RNA Technologies, DOI 10.1007/978-3-319-34175-0_12

has physiological significance. Recent work suggests a link between RNA depurination and some age-related diseases as well as downstream signalling pathways that induce apoptosis. In addition, pathogens with RNA genomes are influenced by depurination. For example, abasic sites of HIV-1 RNA may contribute to adaptation by mutation and recombination. Therefore, effects of RNA depurination have medical relevance that warrants continued study.

Keywords Abasic RNA • Glycosidase • Ribotoxic stress response • Ribosome-inactivating protein • RNA depurination • Translation inhibition • Virus resistance

1 Introduction

An abasic site, which is the lack of a purine or pyrimidine base in a nucleic acid, arises from spontaneous damage and is also produced as an intermediate during the process of DNA repair. It is estimated that over 10,000 apurinic sites in DNA are generated in a cell daily (Lindahl and Nyberg 1972). It has long been known that abasic lesions in DNA are readily repaired through the concerted action of several enzymes. However, little is known about the biology of abasic RNA, the most intriguing question being whether such RNA is degraded or repaired. Compared to DNA, RNA is thought to be rather expendable—it has a short half-life, exists as many copies, and any errors in sequence are not permanently detrimental to the cell. Still, there are important biological consequences of abasic lesions in RNA molecules. This review summarizes what is known about RNA depurination, including active areas of research and remaining questions in the field. We begin with the biochemistry of RNA depurination, followed by a discussion of the effects of cellular and viral RNA depurination, and finally, we summarize available methods of detecting abasic RNA.

2 Biochemistry of RNA Depurination

2.1 *The Mechanism of Depurination*

Depurination is the chemical process of removing an adenine or guanine base from DNA or RNA. Depicted in Fig. 1, the reaction involves hydrolysis of the β-N-glycosidic bond between the ribose (or deoxyribose) sugar and nitrogenous base. Depurination begins with protonation of the purine base at any of the N3 or N7 positions (**1**), which forms a positive charge on the ring structure and destabilizes the N-glycosidic bond (Zoltewicz et al. 1970; Shapiro and Danzig 1972; An

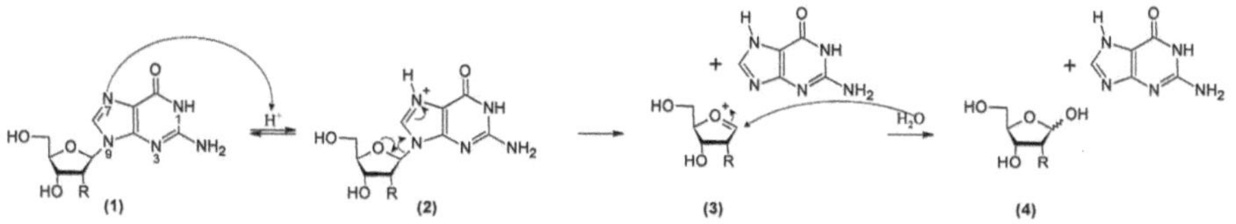

Fig. 1 A typical depurination reaction: acid hydrolysis of adenine from adenosine

Table 1 Kinetic parameters of acid hydrolysis of nucleosides at pH 1 and 37 °C. k_1 is the velocity of hydrolysis (of the first order). Adapted from Venner 1964

Nucleoside	k_1 [Sec^{-1}]
2′Deoxyadenosine	4.3×10^{-4}
2′Deoxyguanosine	8.3×10^{-4}
2′Deoxycytidine	1.1×10^{-7}
2′Deoxyuridine	$<10^{-7}$
2′Deoxythymidine	2×10^{-8}
Adenosine	3.6×10^{-7}
Guanosine	9.36×10^{-7}
Cytidine	$<10^{-9}$
Uridine	$<10^{-9}$

et al. 2014) (**2**). This leads to β-elimination of the base and generation of a (deoxy) ribofuranosyl oxocarbenium ion intermediate (Zoltewicz et al. 1970; Chen et al. 2000) (**3**). Finally, hydrolysis is completed with the formation of a (deoxy) ribofuranose hemiacetal through nucleophilic attack of water at the C1′ position (**4**).

Whether the nucleoside substrate is RNA or DNA has a significant effect on the rate of depurination. The hydroxyl group of RNA generates a negative dipole that stabilizes the C-N glycosidic bond, making it more difficult to eliminate the nitrogenous base (Venner 1964; Kochetkov and Budovskii 1972). Additionally, purine nucleosides tend to hydrolyze faster than pyrimidine derivatives (Venner 1964; An et al. 2014) because purines have the capacity to become diprotonated, which further destabilizes the N-glycosidic bond (Lindahl and Nyberg 1972). In addition, dual ring purines delocalize positive charge more effectively than single-ring pyrimidines due to the number of contributing resonance structures. The rates of base hydrolysis under acidic conditions are shown in Table 1.

The reaction rates indicate that 2-deoxyribosyl derivatives are hydrolyzed 100–1000 times faster than ribosyl derivatives, and purine derivatives are hydrolyzed 100–1000 times faster than pyrimidine derivatives. Depurination is enhanced by decreasing the pH to favor protonated forms of the nitrogenous base (Zoltewicz et al. 1970; Hevesi et al. 1972). The resultant positive charge or positive dipole on the base withdraws electron density from the N-glycosidic bond, thus weakening it. In a similar fashion, depurination is enhanced by modifying the nucleoside with alkylating agents that permanently induce a positive charge on the nitrogenous base (Brookes and Lawley 1961; Zoltewicz et al. 1970; Gates 2010). As shown in Fig. 2, simple alkylating agents such as dimethyl sulfate and methylnitrosourea,

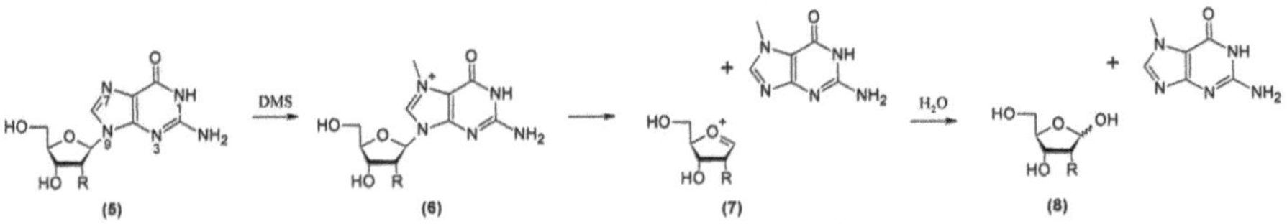

Fig. 2 Methylation of guanosine by dimethyl sulfate followed by hydrolysis of the N-glycosidic bond

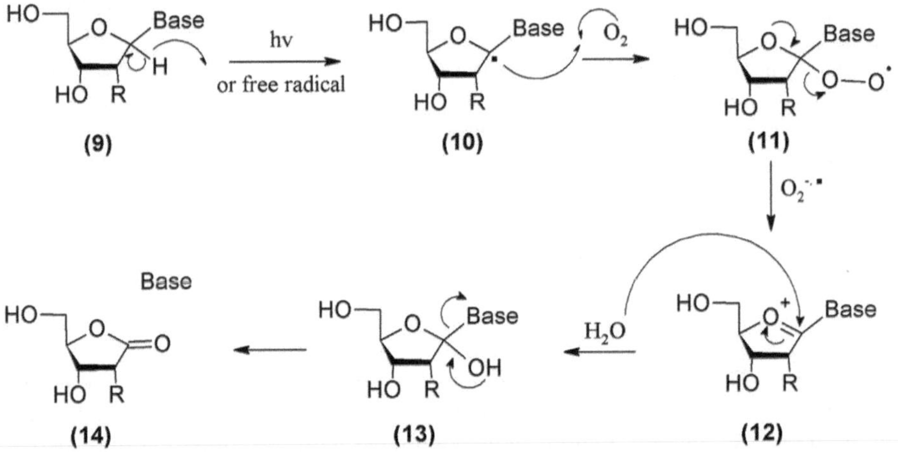

Fig. 3 Free radical oxidation of the nucleoside C1′ and subsequent elimination of the nitrogenous base

or nucleotide-modifying enzymes like methyltransferases (Drabløs et al. 2004), form adducts at nucleophilic N- and O-atoms of the base. Alkylation of endocyclic nitrogens, such as the N7 position of purines, can form a positive charge (**6**) that weakens the N-glycosidic bond and enhances depurination (Zoltewicz et al. 1970; Drabløs et al. 2004; Gates 2010).

Nucleosides can also be depurinated through free radical oxidation at the C1′ position (Fig. 3). This can occur by direct hydrogen atom abstraction (**9**) followed by peroxide radical formation of the C1′ carbon (**10**). The peroxyl radical is then ejected as a superoxide radical anion (**11**), leaving behind the ribofuranosyl oxocarbenium ion (**12**). Water attacks the C1′ position to form an acetal intermediate (**13**) which subsequently eliminates the base to generate a ribonolactone (**14**) (Gates 2010). There are many initiators of free radical reactions, but those of biological importance include hydroxyl radicals, superoxides, and reactive oxygen or reactive nitrogen species (ROS or NOS) (Kong and Lin 2010). These radicals stem from a number of sources, including ionizing radiation and chemical initiators, as well as biological processes like metabolism and pathogen defense.

2.2 Enzymes of RNA Depurination

Although there are several examples of DNA glycosylase enzymes, few RNA specific glycosidases have been identified. The most well-known class of RNA glycosidases is the ribosome-inactivating protein or RIP (EC 3.2.2.22) that exists in a wide range of plants and bacteria (Stirpe and Battelli 2006). For example, ricin is a well-studied RIP synthesized by the castor bean plant (Endo and Tsurugi 1987). RIPs catalyze the depurination of adenine from the invariant GAGA sequence within the sarcin/ricin loop (SRL) of the large ribosomal RNA (Endo et al. 1987; Endo and Tsurugi 1988; Stirpe 2004). The SRL adopts a conserved hairpin structure that recruits elongation factors eEF1 and eEF2 to the ribosome (Obrig et al. 1973). Depurination of the SRL disrupts the binding of these factors, resulting in protein translation inhibition at the translocation step (Obrig et al. 1973). Depurinated ribosomes trigger a ribotoxic stress response that activates apoptosis and can result in cell death (Jandhyala et al. 2012), as described in Sect. 2.1. In this manner, endogenous RIPs are thought to act as defense proteins that inhibit microbial pathogen replication. Interestingly, some RIPs also depurinate non-ribosomal RNA substrates, most notably viral RNA, with potential applications in medicine. The antiviral roles of RIPs are discussed in depth in Sect. 3.

With the exception of RIPs, little is known about other types of RNA glycosidases. However, similar DNA glycosylases have been discovered that participate in base excision repair (BER) pathways (Kim and Wilson 2011). This group of enzymes excises damaged bases from DNA; such damage may occur from mechanisms including deamination, alkylation, or oxidation. If left unrepaired, these lesions can alter base-pairing interactions and lead to mutations during DNA replication.

Surprisingly, some DNA glycosylases also demonstrate activity on RNA, suggesting that some RNA lesions may be repaired in a similar manner. One such example is single-strand-selective monofunctional uracil-DNA glycosylase 1, or SMUG1, which removes uracil and oxidized pyrimidines from DNA (Masaoka et al. 2003; Matsubara et al. 2004; Jobert and Nilsen 2014). SMUG1 directly interacts with an essential rRNA biogenesis enzyme, DKC1 (dyskeratosis congenita 1), which catalyzes the isomerization of uridine to pseudouridine in rRNA (Nakhoul et al. 2014) and is thought to be involved in rRNA surveillance and degradation by the nuclear exosome (Hoskins and Scott Butler 2008; Jobert et al. 2013). SMUG1 associates with 47S precursor rRNA in the nucleolus and depletion of SMUG1 reduces levels of 28S, 18S, and 5.8S rRNAs and increases the presence of 5-hydroxymethyl uracil (5-hmU) in rRNA (Jobert et al. 2013). The 5-hmU glycosylase activity of SMUG1 is thought to contribute to a quality control mechanism during rRNA biogenesis and degradation but how this may occur is unclear (Pettersen et al. 2011). Furthermore, the origin of 5-hmU nucleosides remains unknown; however it has been proposed to arise from hydrolytic deamination products of 5-hydroxymethylcytosine, a natural modification of 18S and 28S rRNA (Jobert et al. 2013). Although SMUG1 depyrimidates RNA, further investigation into the crossover of DNA and RNA metabolism may reveal the existence of novel RNA purine glycosidases and additional roles of abasic RNA.

2.3 Reactivity of an Abasic Site

After depurination, the resulting hemiacetal (**15**) is the predominant form of an abasic site. However, as illustrated in Fig. 4, a small percentage (~1 %) isomerizes into a straight chain aldehyde (**16**) (Manoharan et al. 1989).

As shown in Fig. 5, the aldehyde isoform is sensitive to β-elimination of the C3′ phosphate under mildly basic conditions (**17**), resulting in strand cleavage (**18**) (Suzuki et al. 1994; Lhomme et al. 1999). The aldehyde group is also reactive with hydrazine or other weak amine bases like spermine, aniline, or N,N-dimethylethylenediamine, which form a Schiff base (**21**) and promote β-elimination of the C3′ phosphate (**22**) (Kochetkov and Budovskii 1972). Subsequent δ-elimination can occur under strong alkaline conditions, resulting in the formation of an enol (**19**) at the C5′ and C4′ position that eventually tautomerizes to the ketone (**20**) (Kochetkov and Budovskii 1972). Under physiological conditions, the δ-elimination reaction is slow in comparison to β-elimination (Bailly and Verly 1988a, b), but δ-elimination is thought to occur following β-elimination of abasic

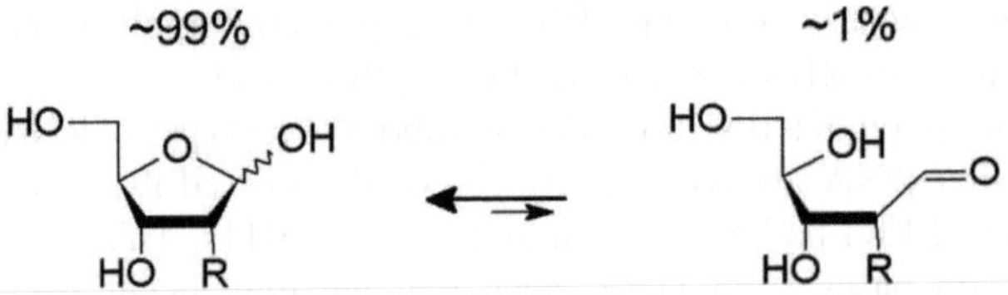

(15) **(16)**

Fig. 4 Isomers of abasic sites: *left* is the hemiacetal form and *right* is the aldehydic form. R is OH or H for RNA or DNA, respectively

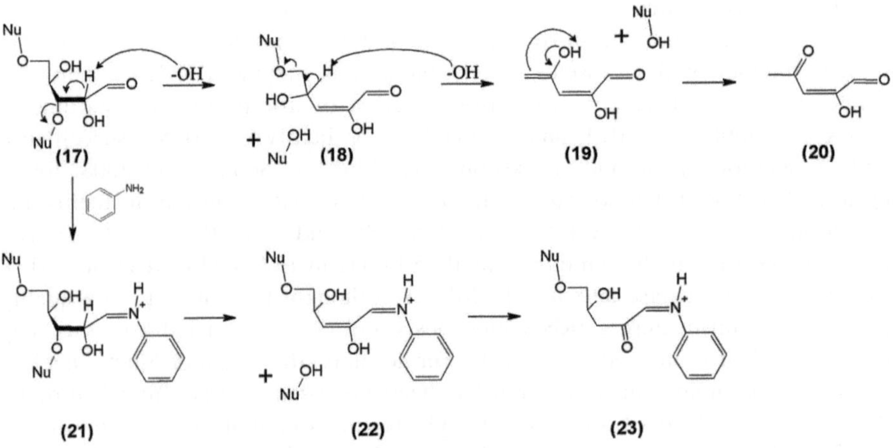

Fig. 5 β-elimination and subsequent δ-elimination of an abasic nucleotide. Schiff base formation enhances the β-elimination reaction

sites during DNA repair (Bailly et al. 1989; Morland et al. 2002). Interestingly, abasic RNA was found to be 153 times more resistant to β-elimination than abasic DNA (Küpfer and Leumann 2007), indicating that depurinated RNA may persist in the cell longer than depurinated DNA. The stability of abasic RNA may have interesting biological consequences when investigating long lived RNAs like rRNA, tRNA, or even viral RNA (Küpfer and Leumann 2007). For example, an abasic site may alter the structure of an RNA molecule and its cellular interactions.

Abasic RNA can also be a substrate for enzymatic RNA cleavage. RNA abasic lyase, or RAlyase, is an enzyme isolated from monocot plants such as wheat and rice and is thought to be involved in apoptosis (Sawasaki et al. 2008). This enzyme cleaves the phosphodiester bond at the 3′ end of an abasic site and acts in conjunction with RIPs that depurinate the SRL (Ozawa et al. 2003). It has been suggested that elongation factors bind weakly to depurinated ribosomes, which would allow viruses to maintain their replication in infected cells (Ozawa et al. 2003). RAlyase cleavage of RIP-induced depurination would abolish the binding of elongation factors and strengthen virus resistance in the plant.

2.4 Section Summary

Abasic sites in DNA are considered mutagenic because they lack coding information and are susceptible to 3′ strand cleavage. As a result, dedicated repair pathways exist to maintain the integrity of the genetic code. An increasing body of evidence reveals that some DNA repair enzymes also function on RNA, suggesting the existence of new RNA surveillance and/or repair mechanisms. Unlike DNA, single-stranded RNA lacks a complementary strand to utilize as a template, so repair of an mRNA transcript may result in mutations. Given that both long-lived (rRNA and tRNA) and short-lived (mRNA) species are more abundant than DNA, and abasic RNA is more stable than abasic DNA, depurinated RNA must be prevalent in a cell. Nevertheless, how the cell detects and manages abasic RNA is not well understood. It is conceivable that RNA depurination could represent an additional layer of gene regulation. For example, depurination may allow dynamic structural changes to mediate environmental response and facilitate new interactions. Further studies on the biological relevance of abasic RNA will improve our understanding of this common but poorly characterized nucleic acid modification.

3 Effect of Depurination on Cellular and Viral RNA

Removal of purine bases occurs to several types of cellular RNAs; those investigated to date include rRNA, mRNA, and tRNA. A summary of these effects is presented in Fig. 6. As discussed above, the chemical stability of abasic RNA is significantly greater than that of DNA. Importantly, the effect of RNA depurination

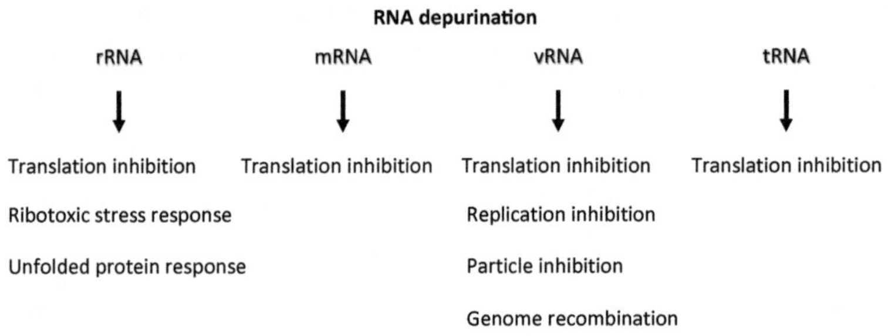

Fig. 6 Effects of RNA depurination within cells. Depurination inhibits translation by decreasing the activity of ribosomes, which may occur by base removal from rRNA or tRNA. Depurination of mRNA/vRNA also stalls elongating ribosomes. Depurination of rRNA triggers a ribotoxic stress response to signal this damage to the cell. Depurination of rRNA may contribute to the unfolded protein response caused by aberrant protein synthesis. Depurination of vRNA also inhibits the replication of viral RNA because the replicase stalls at the abasic nucleotide. Depurinated vRNA is not packaged into particles. Finally, depurination of vRNA may enhance genome recombination

on a cell is measurable and biologically relevant even though RNA is generally a shorter-lived molecule than DNA.

3.1 Ribosomal RNA Depurination

3.1.1 Effect on Translation

As outlined above in Sect. 2.2, depurination of rRNA occurs within the 28S at the SRL, a conserved stem-loop region. A4324 of mammalian ribosomes is removed by glycosidases and this activity is the defining feature of this group of enzymes (Endo et al. 1987). Early studies characterizing the effects of glycosidases showed that they slowed the rate of translation of cell-free ribosome mixtures. The translocation step during elongation was primarily inhibited, due to decreased ribosome-dependent hydrolysis of GTP by elongation factor (eEF2) (Gessner and Irvin 1980). Subsequent structural analyses of intact ribosomal subunits confirmed eEF2 binding in close proximity to the SRL and overlap in the binding locations of eEF1 and eEF2 (Moazed et al. 1988; Spahn et al. 2004). Though the binding of eEF1 and aminoacyl-tRNA to the acceptor site of the ribosome is also inhibited by depurination (Irvin et al. 1980), the large molar excess of eEF1 compared to ribosomes means that eEF1 is not likely the rate-limiting factor during protein synthesis (Browning et al. 1990).

The resulting inhibition of translation due to rRNA depurination is considered the main reason for toxicity of many glycosidases and the primary way in which they function in defense against pathogens. That is, expression of these enzymes would either limit pathogen invasion by targeting ribosomes of the invader or cause host cell death which would limit pathogen spread. Though several examples exist to illustrate the correlation between rRNA depurination and defense against pathogens (Taylor et al. 1994; Watanabe et al. 1997), not all RIPs depurinate ribosomes endogenous to the plant. For example, the ribosomes of *Beta vulgaris*, sugar beet, are resistant to depurination by beetin, a RIP synthesized by this plant (Iglesias et al. 2008). It is suggested that the enzyme may function to directly attack the RNA of an invading plant virus, as beetin has been reported to depurinate tobacco mosaic virus RNA (Iglesias 2005). The effects of viral RNA depurination by glycosidases are described below in Sect. 3.

Glycosidases have also been studied for their anticancer effects, as incubation of these enzymes with mammalian cells often causes apoptosis. Apoptosis has been linked to the unfolded protein response, a cellular response activated by the accumulation of misfolded proteins in the endoplasmic reticulum (ER) lumen (Tesh 2012). Depending on the glycosidase and cell type, the unfolded protein response may or may not be correlated with the depurination of rRNA. For example, the addition of low concentrations of the glycosidase riproximin or volkensin to HCT116 and MDA-MB231 cells induces the unfolded protein response and is accompanied by the transcription and translation of related proteins required for the response, despite rRNA depurination (Horrix et al. 2011). Other studies have measured indicators of the unfolded protein response hours before complete translation inhibition (Brigotti et al. 2002; Li et al. 2007). However, this does not preclude the possibility that the signal to induce this response originates from the depurinated ribosome before detectable protein translation inhibition or that a gradual decline in the rate of translation is sensed by the endoplasmic reticulum.

3.1.2 Ribotoxic Stress Response

The mechanism by which depurination of rRNA is communicated to the cell is called ribotoxic stress response. This term was originally coined by Iordanov et al. (1997) in describing the activation of jun-N-terminal kinase (JNKs) in response to inhibition of actively translating ribosomes. Not all forms of translation inhibition trigger the kinase cascade; for example, ricin, anisomycin, and α-sarcin do, whereas emetine, cycloheximide, and T-2 toxin do not. Also, treatment of cells with emetine, which stalls ribosomes in a pretranslocation state, prevented subsequent induction of the ribotoxic stress response. These results indicate that specific interactions of the toxin with ribosomes are required to initiate this response, rather than any form of translation inhibition. In addition, actively translating ribosomes were required. Since this work, activation of p38 and ERKs has also been shown to mediate the signal for damage at the 28S rRNA (Shifrin and Anderson 1999; Zhou 2003; Colpoys et al. 2005).

One candidate for the upstream mediator of the ribotoxic stress response is double-stranded RNA-activated protein kinase (PKR). Damage of 28S rRNA enhances eIF2α phosphorylation, a target of PKR (Zhou et al. 2003), and PKR inhibitors or antisense knockdown of PKR reduces MAPKinase activation associated with 28S rRNA damage (Gray et al. 2008). In addition, PKR is constitutively expressed and is known to associate with the ribosome (Garcia et al. 2006; Kang and Tang 2012). PKR has two dsRNA-binding domains and binds rRNA in a sequence-independent manner (Garcia et al. 2006), suggesting the model that damage to 28S rRNA may prompt dimerization and autophosphorylation of already bound PKR, which would initiate the kinase cascade (Zhou et al. 2014).

Despite the obvious inhibition of cellular translation due to rRNA depurination, the ribotoxic stress response, through its activation of MAP kinases, upregulates the transcription and expression of many proteins. Recent phosphoproteome analysis showed altered levels of phosphorylation for 188 proteins, 20 % of which were involved in transcriptional regulation. Other biological processes impacted included the cell cycle, RNA processing, and cytoskeletal organization (Pan et al. 2013). Therefore, the cellular effects of damage to 28S rRNA are extensive. Complementary proteome analysis of proteins associated with ribosomes following 28S rRNA damage shows a decrease in proteins involved in translation and a concurrent compensatory increase in chaperones and proteins involved in biosynthesis and cellular organization (Pan et al. 2014). Therefore, the ribosome plays a central role in coordinating the cellular response to 28S rRNA damage through its recruitment of proteins that mediate survival to stress.

3.2 Messenger RNA Depurination

Though rRNA is the predominant RNA of cells, mRNAs are also subject to depurination. The demonstrated consequences of this type of damage to mRNA are decreased translation of the message and enhanced signaling to downstream effectors of the damage.

The effect of abasic RNA on translation has been investigated. An in vitro translation assay comprised of a short synthetic mRNA containing an abasic nucleotide in a predefined position of the open reading frame was ^{32}P-labeled at the 5′ end and linked to puromycin at the 3′ end. Upon in vitro translation in a cell-free ribosome mixture, the encoded peptide chain was transferred to the puromycin, thereby releasing it from the ribosome, and the products analyzed by gel electrophoresis. Results indicated that abasic nucleotides completely halted protein synthesis at the site of modification (Calabretta et al. 2015). In addition, oxidation of nucleotides also affected translation. For example, ε-rC and ε-rA completely terminated peptide synthesis at the site of modification, whereas 8-oxo-rG, 8-oxo-rA, 5-HO-rC, and 5-HO-rU significantly inhibited full-length peptide synthesis, leading to some abortive peptides at the site of modification (Calabretta et al. 2015). Therefore, abasic and oxidized nucleotides inhibit mRNA translation.

Reactive oxygen species oxidize nucleic acids, leading to a loss or change in their biological functions. DNA is largely protected from oxidative damage by binding proteins and repair enzymes. Less is known about the nature of RNA oxidation and it is generally accepted that RNA is less protected against reactive oxygen species. Growing evidence suggests that age-related diseases, such as atherosclerosis and Parkinson and Alzheimer diseases, are correlated with elevated levels of oxidized RNA and that oxidized RNA is a contributing factor for disease pathogenesis (Zhang et al. 1999; Martinet et al. 2004; Lovell and Markesbery 2007). In vitro incubation of an RNA transcript with cytochrome C and hydrogen peroxide resulted in the preferential oxidation of guanosine bases (Tanaka et al. 2012). Moreover, these oxidized residues became depurinated, forming abasic sites in the RNA. When cytochrome C was associated with liposomes enriched in mitochondrial membrane lipids and incubated with RNA and hydrogen peroxide, it cross-linked with the oxidized RNA and dissociated from the liposomes. These experiments illustrate the possibility that oxidized RNA may facilitate the release of cytochrome C from mitochondrial membranes. This release would induce apoptosis in response to oxidative damage. In this way, oxidized and depurinated RNAs may contribute to protective signaling to promote cell death during stress.

3.3 Transfer RNA Depurination

Depurination of tRNA also occurs in cells. Yeast Phe-tRNA is subject to cleavage in its anticodon loop by several aminoglycoside antibiotics (Kirk and Tor 1999; Szczepanik et al. 2003). This process was thought to rely on the presence of a hypermodified guanine residue, called wybutine, at position 37 of the tRNA chain (Krzyzosiak et al. 1988). However, subsequent in vitro analysis has shown that it is the removal of the base (wybutine) that renders the tRNA sensitive to cleavage (Wrzesiński et al. 2005). The cleavage of depurinated Phe-tRNA by physiological levels (10 μM) of the antibiotic neomycin B was 160-fold faster than non-depurinated tRNA. It remains unknown how antibiotics cleave tRNA at the abasic site; however, given that even very low levels (0.1 μM) of antibiotic resulted in measurable tRNA cleavage, the process may have biological relevance. Cleaved tRNA would not function in translation. Moreover, depurinated RNA molecules, beyond tRNA, may be susceptible to cleavage by antibiotics as a mechanism of antibiotic-producing organisms to damage RNA of competing cells.

3.4 Viral RNA Depurination

The genomes of RNA viruses are also subject to depurination, with consequences for different stages of the virus life cycle. For example, the brome mosaic virus (BMV) is a plant virus with a (+)-sense, tripartite RNA genome comprised of RNAs 1, 2, and 3 (Ahlquist 1992; Kao and Sivakumaran 2000). We have shown that PAP, an RNA glycosidase, depurinates BMV RNA3 and inhibits replication of the viral genome (Karran and Hudak 2008). Depending on the location of the abasic nucleotide, the effect of this RNA damage is either to stall the viral replicase during synthesis of (−)-sense RNA or to prevent initiation altogether. That is, an abasic nucleotide will prevent the advance of an elongating replicase, and if depurination occurs at a replicase binding site, it will also prevent initiation of the enzyme.

The behavior of retroviral reverse transcriptases (RTs) at an abasic nucleotide in an RNA template has also been investigated. Though several groups have studied the activity of reverse transcriptases, the work we cite here was performed using a true abasic RNA template. Three different enzymes were tested, the human immunodeficiency (HIV) RT, avian myeloblastosis virus (AMV) RT, and Moloney murine leukemia virus (MMLV), for their ability to extend a cDNA primer complementary to a single-stranded RNA containing one abasic nucleotide. Trans-lesion synthesis occurred with HIV RT, to a lesser extent with AMV RT, whereas MMLV RT aborted DNA synthesis at the site of the missing base (Küpfer et al. 2007). dATP was preferentially inserted by HIV RT across from the missing base; however, there was less discrimination among the four dNTPs for incorporation across from abasic RNA compared with abasic DNA. In addition, A-incorporation was only twofold slower than extension along an undamaged template. The AMV RT slowed significantly when at or within one nucleotide of the abasic site, which increased the odds that the RNase H activity of the enzyme functioned to abort DNA synthesis by degrading the RNA template. In the case of HIV RT, trans-lesion synthesis was faster than RNA degradation, implying that for HIV, an abasic site in its genomic RNA may increase the possibility for mutation and perhaps selective advantage for the virus. Therefore, the fidelity among viral reverse transcriptases varies, and hence, the probability of genome mutation, rather than inhibition of cDNA synthesis at an abasic site, correlates accordingly.

In addition to mutation of the HIV genome by insertion of an adenine at the site of an abasic template nucleotide during cDNA synthesis, an abasic nucleotide in the template RNA genome may cause viral recombination. Viral recombination, resulting from internal strand transfer, occurs during reverse transcription (Basu et al. 2008). The enzymatic activities of the DNA polymerase and RNase H subunits of the viral reverse transcriptase function at an equilibrium, with cDNA synthesis and degradation of the RNA template at rates that maintain the stability of the polymerase on the RNA template (Svarovskaia et al. 2000). Modifications to either subunit that alter their respective rates of reaction have been shown to change the probability of internal strand transfer. For example, a slower polymerase will enhance strand transfers, whereas an inefficient RNase H will reduce their frequency (Operario et al. 2006). The biological relevance of this dynamic relationship is apparent when considering situations that cause the reverse transcriptase to stall or slow.

Internal strand transfer was initially described as a mechanism by which HIV could still transcribe its full genome even if one of two packaged genomic RNAs was damaged (Smyth et al. 2012). Internal strand transfer would occur when the reverse transcriptase is stalled at the damaged site. This stalling may be due to strand breakage and has also been enhanced by high amounts of secondary structure, which causes the polymerase to pause (Roda et al. 2002; Gao et al. 2006). We suggest that in addition to these possible scenarios, depurination of the RNA template may also enhance internal strand transfer. Even if the polymerase responds to an abasic site by inserting an adenine, this activity decreases its rate of migration along the template by twofold (Küpfer et al. 2007), which may be sufficient to increase the probability of template switching.

Therefore, depurination of HIV genomic RNA will contribute to mutation, either by the insertion of an adenine at the site of the abasic nucleotide or by enhancing genome recombination. Both of these events increase the diversity of viral RNA, which may have positive or negative effects. High genetic diversity will allow the virus to evolve quickly in response to selective pressures.

We have shown previously that co-expression of a glycosidase, along with BMV RNA3 in yeast, which is a surrogate host for the virus, results in depurination of the viral RNA. The depurinated RNA was associated with polysomes and caused elongating ribosomes to stall at the missing base. The mRNA was subsequently targeted for accelerated degradation by components of the no-go decay pathway (Gandhi et al. 2008). Several mRNA surveillance pathways have been described and outline how cells detect and degrade aberrant transcripts. The no-go decay pathway occurs in *cis* when elongating ribosomes are confronted by substantial RNA secondary structure, rare codons, or premature stop codons (Doma and Parker 2006). The Dom34:Hbs1 complex recognizes the stalled ribosome and recruits an uncharacterized endonuclease to cleave the mRNA close to the stalled ribosome (Tsuboi et al. 2012). The complex also facilitates release of the ribosome subunits from the mRNA at the expense of ATP hydrolysis (Shoemaker and Green 2011). The mRNA fragments are degraded at their cleaved 5′ ends by Xrn1 and at their 3′ ends by the exosome (Inada 2013).

In vitro incubation of BMV RNA with the glycosidase and subsequent translation in a cell-free ribosome mixture also resulted in synthesis of truncated protein, indicative of ribosome stalling (Gandhi et al. 2008). We suggest that the depurinated nucleotide at the wobble base of a codon in the ribosomal A-site would interfere with tRNA base pairing and subsequent peptidyl transfer. Therefore, depurination contributed to decreased protein product, due to both accelerated RNA decay and translation elongation arrest.

Apart from the effect on viral enzymes, depurination of viral RNA also influences the behavior of viral structural proteins. For example, we have shown that abasic sites on BMV RNA3 inhibit the binding of coat protein (Karran and Hudak 2011). Coat protein associates with strands of genomic viral RNA, in an incompletely understood fashion, to build a viral particle. This packaging is an essential step in the virus life cycle that prevents degradation of the genomic RNA by nucleases and allows for transfer of the virus from one host to the next. Depurination decreased the affinity of coat protein for binding to viral RNA, and we attribute this decrease to change in local RNA structure and enhanced flexibility of the viral RNA. Detection of depurinated RNA by coat protein suggests a quality control mechanism whereby coat protein can distinguish between intact and damaged RNA and only package intact RNA.

3.5 Section Summary

The SRL of rRNA is the most conserved sequence of RNA in nature. The removal of a base from this loop has detrimental effects on the binding and activity of protein elongation factors. Given recent progress on the structure of ribosomal subunits and their association with these factors, the opportunity exists for a detailed study to examine how the depurination of a single nucleotide changes the shape of the RNA loop and its specific contact points with eEF1 and eEF2. The biological impact of rRNA depurination extends beyond inhibition of protein translation, as many glycosidases also induce an ER stress response. Though the link between rRNA depurination and the UPR leading to apoptosis has not been mapped, we suggest that improper folding of proteins resulting from aberrant synthesis by damaged ribosomes is what triggers the stress response. Depurination of tRNAs can also have substantial impact on protein translation. Prokaryotes that synthesize antibiotics capable of cleaving abasic sites in relatively long-lived molecules such as tRNAs equip themselves with a tool to outcompete other species. mRNAs are generally considered as ephemeral molecules, not subject to repair like DNA, but rather degraded. However, given that the rate of cleavage of abasic RNA is slower than its DNA counterpart, depurinated mRNAs may persist in cells long enough to have physiological impact. For viruses whose genome is RNA, the effect of depurination is significant and ranges from inhibition of replication and protein translation to lack of viral particle formation. For retroviruses, depurination of genomic RNA may also increase recombination, with the potential for enhanced survival through favorable mutation.

4 Methods of Detecting Depurination

Several methods have been developed to determine the location as well as extent of depurination of an RNA substrate. A diagram summarizing some of these methods is illustrated in Fig. 7. Although these methods have been used to characterize RIP activity, most are applicable to the detection of other sources of depurination. Generally, methods to measure depurination fall into two categories, identifying the presence of an abasic site or the enrichment of free adenine in solution.

4.1 Detection of an Abasic Site

4.1.1 Translation Inhibition Assay

Endo and Tsurugi (1986) were the first to discover that ricin A-chain depurinates mammalian 28S ribosomal RNA at a specific adenine residue within the SRL. As

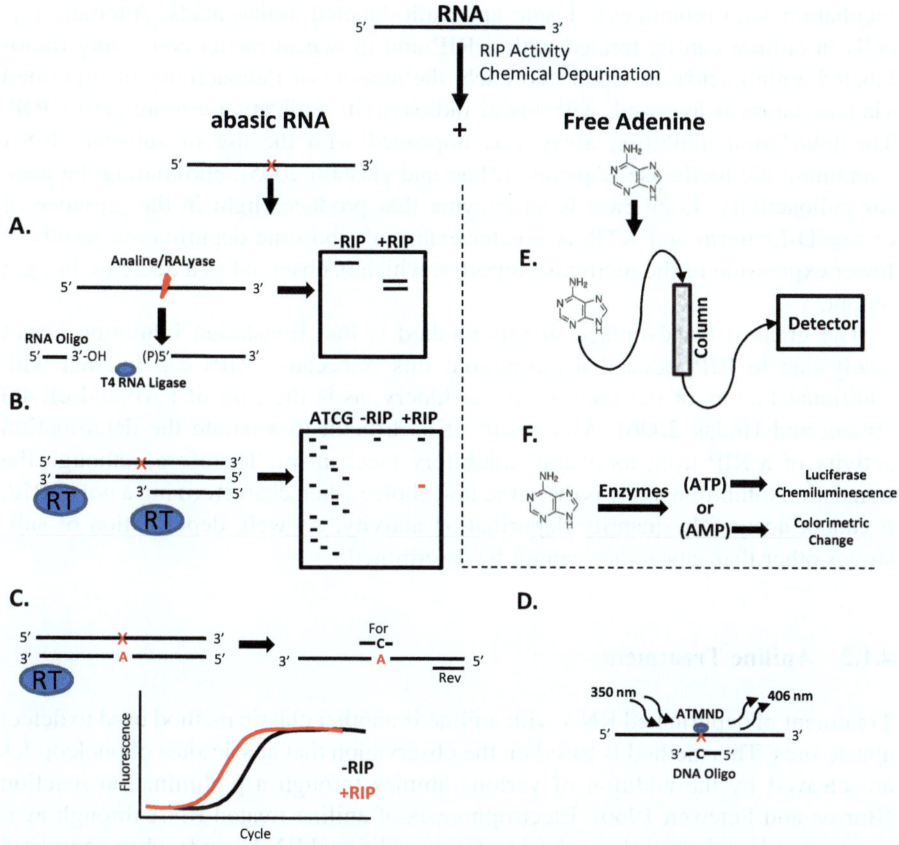

Fig. 7 Methods of detecting RNA depurination. Detection can be delineated into two categories: detecting the presence of an abasic site or the levels of free adenine. (**a**) Both aniline and RALyase cleave RNA at abasic sites, allowing electrophoretic separation of products and quantitation based on band intensity. Cleavage leaves RNA with a 5′ phosphate for ligation and cloning. (**b**) RTs can stall at abasic sites, generating truncated cDNA. When separated beside a sequencing ladder, the depurinated nucleotide can be identified. Primer extension can be quantitative with the addition of a primer that binds upstream of the depurination site. (**c**) Some RTs preferentially insert an adenine opposite an abasic site. By designing qRT-PCR primers with a mismatch to the adenine, the level of depurination can be determined by amplification of the template. (**d**) When a DNA oligo bearing a cytosine opposite the abasic site binds to the RNA strand, ATMND binds to the abasic site and fluoresces at 406 nm. (**e**) After passing the enzymatic mixture through a column, both fluorescently derivatized and unmodified adenine can be passed through a variety of detectors. (**f**) Free adenine can be converted to AMP or ATP by enzymes involved in adenine scavenging pathways. These substrates are then used to drive luciferase-based chemiluminescence or a colorimetric change

described previously, depurination within this region inhibits translation at the translocation step. Thus, the translation inhibition assay is the simplest method of detecting depurination by a RIP. The assay can be performed two ways, the most common of which involves treating purified ribosomes with a RIP, followed by incubation with reticulocyte lysate and radio-labeled amino acids. Alternatively, cells in culture can be treated with a RIP and grown in media containing radio-labeled amino acids. In either approach, the amount of radioactivity incorporated via translation is assessed, with lower radioactivity indicating a more active RIP. The translation inhibition assay was improved with the use of substrate RNA containing the luciferase sequence (Zhao and Haslam 2005), eliminating the need for radioactivity. Luciferase is an enzyme that produces light in the presence of excess D-luciferin and ATP. A greater extent of ribosome depurination results in lower expression of the luciferase reporter, which is observed as a decrease in light output.

The greatest disadvantage of this method is that translation inhibition is not solely due to RIP-induced depurination; this is because RIPs can interact with additional factors of the translation machinery, as is the case of PAP and eIF4G (Wang and Hudak 2006). As a result, it is difficult to separate the depurination activity of a RIP from its overall inhibitory mechanism. Therefore, although the translation inhibition assay is often the first choice when characterizing a novel RIP, it cannot accurately quantify depurination activity. As well, depurination of substrates other than ribosomes cannot be determined.

4.1.2 Aniline Treatment

Treatment of depurinated RNA with aniline is another classic method used to detect abasic sites. This method is based on the observation that abasic sites of nucleotides are cleaved by the addition of various amines through a β-elimination reaction (Burton and Petersen 1960). Electrophoresis of aniline-treated RNA through agarose or acrylamide will show abasic sites as additional RNA bands when compared with untreated RNA. Although this method suffers from poor sensitivity due to the detection limits of intercalating dyes, aniline treatment of RNA is a straightforward procedure that can be made semiquantitative with the addition of loading controls and densitometry. A substitute for aniline treatment has been developed which utilizes the enzyme RALyase (Sawasaki et al. 1999), discussed above in Sect. 2.3. Much like aniline, RALyase cleaves the phosphodiester bond at abasic sites through a β-elimination reaction.

4.1.3 Primer Extension

The primer extension assay was originally used to study RNA secondary structure by detecting bases that had been chemically modified (Holmberg et al. 1994). The assay was later applied to detect depurination of the SRL with ricin (Iordanov et al. 1997). Due to the low detection limit of phosphor screens, primer extension is more sensitive than the intercalating dyes used to observe aniline-treated RNA. More importantly, this assay can identify the precise location of an abasic site. As discussed above, low fidelity RTs such as HIV RT read through an abasic site and preferentially insert an adenine into the cDNA strand. However, higher fidelity RTs like MMLV RT stall upon encountering abasic sites on an RNA template and generate truncated cDNA products (Küpfer et al. 2007). The primer extension method uses radio-labeled primers to generate cDNAs of an RNA template after treatment with a RIP or other depurinating agent. With electrophoretic separation of the cDNA products alongside a corresponding DNA sequencing reaction of the template, the depurinated nucleotide can be identified.

A drawback of primer extension is that reverse transcriptases can also stall due to secondary structures, generating a false signal. This can be overcome through the addition of aniline or RALyase, which would cleave the depurinated site and provide a definitive stopping point for the RT. While classic primer extension is not quantitative, an improvement to the method uses a second primer upstream of the depurination site to provide an internal reference for quantification (Parikh et al. 2002). However, quantification using this method is based on the assumption that the two primers bind with equal affinity to the template. While still commonly used, primer extension is reliant on prior knowledge of the approximate location of depurination, as this information is necessary for primer design.

A difficulty in studying non-ribosomal RNA substrates is that it is tedious to determine the location of an abasic site. Previously in our lab, this was accomplished by performing a series of primer extensions along the length of a viral genome, 200 base pairs at a time (Mansouri et al. 2009; Zhabokritsky et al. 2014). We suggest an improvement to this method that will aid in the discovery of novel depurination sites. Specifically, both aniline and RALyase treatment cleave RNA at depurinated sites, leaving the upstream portion with a reactive ribose ring on its 3′ end. We suggest that an RNA adapter primer be ligated to the cleaved RNA by T4 RNA ligase, followed by reverse transcription with either a gene specific primer or a combination of poly-T/random hexamers. PCR with adapter-specific and gene-specific primers would allow all depurinated sites to be cloned and sequenced, alleviating the need to methodologically examine long templates with primer extension.

4.2 Detection of Free Adenine Enrichment

4.2.1 HPLC and Mass Spectrometry

The second approach for studying depurination is to detect the release of adenine into solution, which can be accomplished with high performance liquid chromatography (HPLC). An HPLC machine operates by passing a solution through a chromatography column, which is coupled to a detector that assays the flow-through. In the context of measuring adenine release, the column material is typically octadecyl carbon chain (C18)-bonded silica and the detector, a UV–vis spectrophotometer. An RNA substrate is incubated in the presence or absence of a RIP, which releases adenine into solution. The reaction can be injected onto the column directly (Chen et al. 1998), or the adenine is extracted and reacted with chloroacetaldehyde (McCann et al. 1983) to generate a fluorescent ethano derivative that is detected fluorometrically at 254 nm (Barrio et al. 1972; Zamboni et al. 1989). Both variations utilize a UV spectrophotometer as the detector. Mass spectrometers are also used as detectors for adenine release and have the same 0.1–10 pmol sensitivity as fluorescence HPLC (Fabris 2000; Hines et al. 2004).

Recently, a new technology has emerged known as direct analysis in real-time (DART) mass spectrometry. The advantage of DART-MS over traditional MS is that samples can be analyzed under ambient conditions without radioactive or other high-energy ionization sources (Cody et al. 2005). This technique has already been used to measure adenine release by ricin (Bevilacqua et al. 2010). The advantage of DART-MS over HPLC-MS is that DART does not require isotopic labeling of an internal standard and the reaction need not be quenched. The main drawback of HPLC and leading edge techniques like DART-MS is that both are expensive technologies that may be inaccessible to many basic research laboratories. As well, HPLC and MS are not yet amenable to high-throughput applications. Such limitations need to be addressed before these techniques will be widely adopted to study depurination.

4.2.2 Enzyme-Based Methods

Several enzymatic alternatives of measuring adenine release have been developed and are based on the activity of adenine phosphoribosyl transferase (APRT). This enzyme was originally isolated from the intracellular parasite *Leishmania donovani*, which is auxotrophic for purines (Allen et al. 1989). In the context of quantifying adenine release, adenine is converted to AMP by APRT, with 5-phosphorylribose 1-pyrophosphate (PRPP) as a substrate (Heisler et al. 2002). A blend of 5′ nucleotidase and pyrophosphatase liberates free phosphate ions from AMP, driving the enzymatic activity of purine nucleoside phosphorylase (PNP) to phosphorylase 2-amino-6- mercapto-7-methylpurine ribonucleoside (MESG), producing a colorimetric change that is observed at 355 nm. While amenable to high-throughput and sensitive to nanomolar concentrations of adenine, this assay relies on enzymes that must be assembled piecemeal from commercial kits or isolated, as is the case of APRT. Another group reduced the number of required enzymes by coupling APRT with pyruvate orthophosphate dikinase (PPDK), which converts AMP to ATP (Sturm and Schramm 2009). The ATP is then used as a substrate for luciferase, which produces luminescence and AMP. Finally, AMP is recycled back into the reaction, generating a steady signal of fluorescence for several minutes and allowing sub-picomole detection of adenine.

4.3 Recent Approaches

Novel approaches to detect the presence of abasic sites have also been proposed. An interesting example is the use of a fluorescent ligand, 2-amino-5,6,7-trimethyl-1,8-naphthyridine (ATMND) (Sato et al. 2007), which pseudo-base pairs with a cytosine that has been introduced opposite of an abasic site (Tanpure et al. 2012). This is accomplished by designing a DNA primer that is complementary to the depurination region and has a cytosine across from the depurinated nucleotide. Upon binding, the fluorescence of ATMND is quenched and the drop measured spectrophotometrically. The simplest method of detecting an abasic site utilizes an inherent property of reverse transcriptases, which can insert an AMP residue into the nascent cDNA opposite an abasic site (Takeshita et al. 1987). By designing qRT-PCR primers that bear a deliberate mismatch at the newly introduced adenine residue, levels of depurination can be quantified based on amplification of a template with an A to T mutation (Pierce et al. 2011). Both techniques have the same disadvantage wherein the site of depurination must already be known.

4.4 Section Summary

Significant developments have been made when considering the tools available for studying depurination. While indirect techniques like translational inhibition have largely fallen out of use, classic methods like aniline treatment and primer extension are still employed due to their simplicity, as well as the availability and economy of reagents. While the gold standard of quantitation and enzyme kinetics remains HPLC and MS, enzymatic-coupled reactions may be a viable alternative if the enzymes are made widely available. Novel methods of detecting abasic sites chemically or through PCR have been developed; however both reveal an existing limitation, wherein the site of depurination must already be known. Furthermore, these methods must be shown to be reproducible by other groups. To date, primer extension is the only method capable of detecting novel depurination sites, and we suggest an improvement that incorporates aniline cleavage and adapter ligation. Such a modification would allow widespread detection of abasic sites and minimize false-positive results due to RNA secondary structure.

5 Conclusion

Here, we have reviewed the latest information describing the effects of cellular and viral RNA depurination. Owing to its relative stability, abasic RNA likely persists in the cell for an extended period of time. However, the fate of abasic RNA in vivo is still being characterized. The recent discovery that a glycosylase mediating DNA repair also acts on RNA may indicate that specific mechanisms exist to detect and manage abasic RNA in the cell. Whether such RNA is repaired or degraded remains an existing question. From a wider perspective, it is unknown whether the formation of abasic RNA is a dynamic, responsive process that contributes to cellular adaptation.

It is well-established that RNA depurination has important and varied biological consequences, depending on the particular substrate. We have reviewed the effects of depurination on cellular and viral RNA. rRNA depurination leads to translational inhibition and activation of downstream responses including the unfolded protein response and ribotoxic stress response, both of which can trigger apoptosis. A link between depurination and the unfolded protein response has not yet been established; we suggest that it is the improper folding of proteins resulting from aberrant synthesis that triggers the response. Depurination of mRNA can inhibit or abolish translation and recent evidence suggests that depurinated mRNA may signal downstream apoptotic pathways. Less is known about depurinated tRNA, although the existence of antibiotics that cleave abasic RNA indicates that such damage is compromising to competitors. Depurination of viral RNA affects various steps of the viral life cycle. Although RNA glycosidases have promising medical applications with respect to antiviral treatments, an important limitation is that their use could inadvertently contribute to virus resistance. Methodical studies investigating the optimal threshold of RNA depurination leading to viral inhibition would be advantageous.

To encourage continued work that focuses on the characterization of RNA depurination, we have provided a comprehensive review of available detection methods. Generally, these methods fall into two categories: detecting the presence of an abasic site or the enrichment of free adenine in solution. Here, we also suggest an improvement to the primer extension method that would enable widespread detection of RNA depurination sites.

References

Ahlquist P (1992) Bromovirus RNA replication and transcription. Curr Opin Genet Dev 2:71–76

Allen T, Henschel EV, Coons T et al (1989) Purification and characterization of the adenine phosphoribosyltransferase and hypoxanthine-guanine phosphoribosyltransferase activities from Leishmania donovani. Mol Biochem Parasitol 33:273–281

An R, Jia Y, Wan B et al (2014) Non-enzymatic depurination of nucleic acids: factors and mechanisms. PLoS One 9, e115950

Bailly V, Verly WG (1988a) Importance of thiols in the repair mechanisms of DNA containing AP (apurinic or apyrimidinic) sites. Nucleic Acids Res 16:9489–9496

Bailly V, Derydt M, Verly WG (1989) Delta-elimination in the repair of AP (apurinic/apyrimidinic) sites in DNA. Biochem J 261:707–713

Bailly V, Verly WG (1988b) Possible roles of beta-elimination and delta-elimination reactions in the repair of DNA containing AP (apurinic/apyrimidinic) sites in mammalian cells. Biochem J 253:553–559

Barrio JR, Secrist JA, Leonard NJ (1972) Fluorescent adenosine and cytidine derivatives. Biochem Biophys Res Commun 46:597–604

Basu VP, Song M, Gao L et al (2008) Strand transfer events during HIV-1 reverse transcription. Virus Res 134:19–38

Bevilacqua VLH, Nilles JM, Rice JS et al (2010) Ricin activity assay by direct analysis in real time mass spectrometry: detection of adenine release. Anal Chem 82:798–800

Brigotti M, Alfieri R, Sestili P et al (2002) Damage to nuclear DNA induced by Shiga toxin 1 and ricin in human endothelial cells. FASEB J 16:365–372

Brookes P, Lawley P (1961) The reaction of mono- and di functional alkylating agents with nucleic acids. Biochem J 80:496–503

Browning KS, Humphreys J, Hobbs W et al (1990) Determination of the amounts of the protein synthesis initiation and elongation factors in wheat germ. J Biol Chem 265:17967–17973

Burton K, Petersen GB (1960) The frequencies of certain sequences of nucleotides in deoxyribonucleic acid. Biochem J 75:17–27

Calabretta A, Kupfer PA, Leumann CJ (2015) The effect of RNA base lesions on mRNA translation. Nucleic Acids Res 43:4713–4720

Chen XY, Berti PJ, Schramm VL (2000) Transition-state analysis for depurination of DNA by ricin A-chain. J Am Chem Soc 122:6527–6534

Chen XY, Link TM, Schramm VL (1998) Ricin A-chain: kinetics, mechanism, and RNA stem-loop inhibitors. Biochemistry 37:11605–11613

Cody RB, Laramée JA, Nilles JM et al (2005) Direct analysis in real time (DART) mass spectrometry. JEOL News 40:8–12

Colpoys WE, Cochran BH, Carducci TM et al (2005) Shiga toxins activate translational regulation pathways in intestinal epithelial cells. Cell Signal 17:891–899

Doma MK, Parker R (2006) Endonucleolytic cleavage of eukaryotic mRNAs with stalls in translation elongation. Nature 440:561–564

Drabløs F, Feyzi E, Aas PA et al (2004) Alkylation damage in DNA and RNA: repair mechanisms and medical significance. DNA Repair (Amst) 3:1389–1407

Endo Y, Mitsui K, Motizuki M et al (1987) The mechanism of action of ricin and related toxic lectins on eukaryotic ribosomes. J Biol Chem 262:5908–5912

Endo Y, Tsurugi K (1987) RNA N-glycosidase activity of ricin A chain. J Biol Chem 262:8128–8130

Endo Y, Tsurugi K (1988) The RNA N-glycosidase activity of ricin A-chain. Nucleic Acids Symp Ser 263:139–142

Endo Y, Tsurugi K (1986) Mechanism of action of ricin and related toxic lectins on eukaryotic ribosomes. Nucleic Acids Symp Ser 262:187–190

Fabris D (2000) Steady-state kinetics of ricin A-chain reaction with the sarcin: ricin Loop and with HIV-1 Ψ-RNA hairpins evaluated by direct infusion electrospray ionization mass spectrometry. J Am Chem Soc 122:8779–8780

Gandhi R, Manzoor M, Hudak KA (2008) Depurination of brome mosaic virus RNA3 in vivo results in translation-dependent accelerated degradation of the viral RNA. J Biol Chem 283:32218–32228

Gao L, Balakrishnan M, Roques BP et al (2006) Insights into the multiple roles of pausing in HIV-1 reverse transcriptase-promoted strand transfers. J Biol Chem 282:6222–6231

Garcia MA, Gil J, Ventoso I et al (2006) Impact of protein kinase PKR in cell biology: from antiviral to antiproliferative action. Microbiol Mol Biol Rev 70:1032–1060

Gates KS (2010) An overview of chemical processes that damage cellular DNA: spontaneous hydrolysis, alkylation, and reactions with radicals. Chem Res Toxicol 22:1747–1760

Gessner SL, Irvin JD (1980) Inhibition of elongation factor 2-dependent translocation by the pokeweed antiviral protein and ricin. J Biol Chem 255:3251–3253

Gray JS, Bae HK, Li JCB et al (2008) Double-stranded RNA-activated protein kinase mediates induction of interleukin-8 expression by deoxynivalenol, shiga toxin 1, and ricin in monocytes. Toxicol Sci 105:322–330

Heisler I, Keller J, Tauber R et al (2002) A colorimetric assay for the quantitation of free adenine applied to determine the enzymatic activity of ribosome-inactivating proteins. Anal Biochem 302:114–122

Hevesi L, Wolfson-Davidson E, Nagy JB et al (1972) Contribution to the mechanism of the acid-catalyzed hydrolysis of purine nucleosides. J Am Chem Soc 94:4715–4720

Hines HB, Brueggemann EE, Hale ML (2004) High-performance liquid chromatography-mass selective detection assay for adenine released from a synthetic RNA substrate by ricin A chain. Anal Biochem 330:119–122

Holmberg L, Melander Y, Nygård O (1994) Probing the structure of mouse Ehrlich ascites cell 5.8S, 18S and 28S ribosomal RNA in situ. Nucleic Acids Res 22:1374–1382

Horrix C, Raviv Z, Flescher E et al (2011) Plant ribosome-inactivating proteins type II induce the unfolded protein response in human cancer cells. Cell Mol Life Sci 68:1269–1281

Hoskins J, Scott Butler J (2008) RNA-based 5-fluorouracil toxicity requires the pseudouridylation activity of Cbf5p. Genetics 179:323–330

Iglesias R (2005) Molecular characterization and systemic induction of single-chain ribosome-inactivating proteins (RIPs) in sugar beet (Beta vulgaris) leaves. J Exp Bot 56:1675–1684

Iglesias R, Perez Y, Citores L et al (2008) Elicitor-dependent expression of the ribosome-inactivating protein beetin is developmentally regulated. J Exp Bot 59:1215–1223

Inada T (2013) Quality control systems for aberrant mRNAs induced by aberrant translation elongation and termination. Biochim Biophys Acta 1829:634–642

Iordanov MS, Pribnow D, Magun JL et al (1997) Ribotoxic stress response: activation of the stress-activated protein kinase JNK1 by inhibitors of the peptidyl transferase reaction and by sequence-specific RNA damage to the alpha-sarcin/ricin loop in the 28S rRNA. Mol Cell Biol 17:3373–3381

Irvin JD, Kelly T, Robertus JD (1980) Purification and properties of a second antiviral protein from Phytolacca Americana which inactivates eukaryotic ribosomes. Arch Biochem Biophys 200:418–425

Jandhyala DM, Thorpe CM, Magun B (2012) Ricin and shiga toxins: effects on host cell signal transduction. Curr Top Microbiol Immunol 357:41–65

Jobert L, Nilsen H (2014) Regulatory mechanisms of RNA function: emerging roles of DNA repair enzymes. Cell Mol Life Sci 71:2451–2465

Jobert L, Skjeldam HK, Dalhus B et al (2013) The human base excision repair enzyme SMUG1 directly interacts with DKC1 and contributes to RNA quality control. Mol Cell 49:339–345

Kang R, Tang D (2012) PKR-dependent inflammatory signals. Sci Signal 5:pe47

Kao CC, Sivakumaran K (2000) Brome mosaic virus, good for an RNA virologist's basic needs. Mol Plant Pathol 1:91–97

Karran RA, Hudak KA (2008) Depurination within the intergenic region of Brome mosaic virus RNA3 inhibits viral replication in vitro and in vivo. Nucleic Acids Res 36:7230–7239

Karran RA, Hudak KA (2011) Depurination of Brome mosaic virus RNA3 inhibits its packaging into virus particles. Nucleic Acids Res 39:7209–7222

Kim Y-J, Wilson DM III (2011) Overview of base excision repair biochemistry. Curr Mol Pharmacol 5:3–13

Kirk SR, Tor Y (1999) tRNA(Phe) binds aminoglycoside antibiotics. Bioorg Med Chem 7:1979–1991

Kochetkov N, Budovskii E (1972) Hydrolysis of N-glycosidic bonds in nucleosides, nucleotides, and their derivatives. In: Organic chemistry of nucleic acids part B. Plenum Press, New York, pp 425–448

Kong Q, Lin CLG (2010) Oxidative damage to RNA: mechanisms, consequences, and diseases. Cell Mol Life Sci 67:1817–1829

Krzyzosiak WJ, Marciniec T, Wiewiorowski M et al (1988) Characterization of the lead(II)-induced cleavages in tRNAs in solution and effect of the Y-base removal in yeast $tRNA^{Phe}$. Biochemistry 27:5771–5777

Küpfer PA, Crey-Desbiolles C, Leumann CJ (2007) Trans-lesion synthesis and RNaseH activity by reverse transcriptases on a true abasic RNA template. Nucleic Acids Res 35:6846–6853
Küpfer PA, Leumann CJ (2007) The chemical stability of abasic RNA compared to abasic DNA. Nucleic Acids Res 35:58–68
Lhomme J, Constant JF, Demeunynck M (1999) Abasic DNA structure, reactivity, and recognition. Biopolymers 52:65–83
Li J, Xia X, Ke Y et al (2007) Trichosanthin induced apoptosis in HL-60 cells via mitochondrial and endoplasmic reticulum stress signaling pathways. Biochim Biophys Acta 1770:1169–1180
Lindahl T, Nyberg B (1972) Rate of depurination of native deoxyribonucleic acid. Biochemistry 11:3610–3618
Lovell MA, Markesbery WR (2007) Oxidative DNA damage in mild cognitive impairment and late-stage Alzheimer's disease. Nucleic Acids Res 35:7497–7504
Manoharan M, Mazumder A, Wilde JA et al (1989) The characterization of abasic sites in DNA heteroduplexes by site specific labeling with C13. J Am Chem Soc 110:1620–1622
Mansouri S, Choudhary G, Sarzala PM et al (2009) Suppression of human T-cell leukemia virus I gene expression by pokeweed antiviral protein. J Biol Chem 284:31453–31462
Martinet W, de Meyer GRY, Herman AG et al (2004) Reactive oxygen species induce RNA damage in human atherosclerosis. Eur J Clin Invest 34:323–327
Masaoka A, Matsubara M, Hasegawa R et al (2003) Mammalian 5-formyluracil-DNA glycosylase: 2. Role of SMUG1 uracil-DNA glycosylase in repair of 5-formyluracil and other oxidized and deaminated base lesions. Biochemistry 42:5003–5012
Matsubara M, Tanaka T, Terato H et al (2004) Mutational analysis of the damage-recognition and catalytic mechanism of human SMUG1 DNA glycosylase. Nucleic Acids Res 32:5291–5302
McCann WP, Hall LM, Nonidez WK (1983) Preparation, titration, and storage of chloroacetaldehyde for fluorometric determination of adenine and its derivatives. Anal Chem 55:1454–1455
Moazed D, Robertson JM, Noller HF (1988) Interaction of elongation factors EF-G and EF-Tu with a conserved loop in 23S RNA. Nature 334:362–364
Morland I, Rolseth V, Luna L et al (2002) Human DNA glycosylases of the bacterial Fpg/MutM superfamily: an alternative pathway for the repair of 8-oxoguanine and other oxidation products in DNA. Nucleic Acids Res 30:4926–4936
Nakhoul H, Ke J, Zhou X et al (2014) Ribosomopathies: mechanisms of disease. Clin Med Insights Blood Disord 7:7–16
Obrig TG, Irvin JD, Hardesty B (1973) The effect of an antiviral peptide on the ribosomal reactions of the peptide elongation enzymes, EF-I and EF-II. Arch Biochem Biophys 155:278–289
Operario DJ, Balakrishnan M, Bambara RA et al (2006) Reduced dNTP interaction of human immunodeficiency virus type 1 reverse transcriptase promotes strand transfer. J Biol Chem 281:32113–32121
Ozawa A, Sawasaki T, Takai K et al (2003) RALyase; a terminator of elongation function of depurinated ribosomes. FEBS Lett 555:455–458
Pan X, Whitten DA, Wilkerson CG et al (2014) Dynamic changes in ribosome-associated proteome and phosphoproteome during deoxynivalenol-induced translation inhibition and ribotoxic stress. Toxicol Sci 138:217–233
Pan X, Whitten DA, Wu M et al (2013) Global protein phosphorylation dynamics during deoxynivalenol-induced ribotoxic stress response in the macrophage. Toxicol Appl Pharmacol 268:201–211
Parikh BA, Coetzer C, Tumer NE (2002) Pokeweed antiviral protein regulates the stability of its own mRNA by a mechanism that requires depurination but can be separated from depurination of the alpha-sarcin/ricin loop of rRNA. J Biol Chem 277:41428–41437
Pettersen HS, Visnes T, Vågbø CB et al (2011) UNG-initiated base excision repair is the major repair route for 5-fluorouracil in DNA, but 5-fluorouracil cytotoxicity depends mainly on RNA incorporation. Nucleic Acids Res 39:8430–8444

Pierce M, Kahn JN, Chiou J et al (2011) Development of a quantitative RT-PCR assay to examine the kinetics of ribosome depurination by ribosome inactivating proteins using *Saccharomyces cerevisiae* as a model. RNA 17:201–210

Roda RH, Balakrishnan M, Kim JK et al (2002) Strand transfer occurs in retroviruses by a pause-initiated two-step mechanism. J Biol Chem 277:46900–46911

Sato Y, Seino T, Nishizawa S et al (2007) Strong binding of naphthyridine derivatives to cytosine in an AP site-containing DNA duplex and their application to fluorescence detection of single nucleotide polymorphisms. Nucleic Acids Symp Ser (Oxf) 51:313–314

Sawasaki T, Morishita R, Ozawa A et al (1999) Mechanism of ribosome RNA apurinic site specific lyase. Nucleic Acids Symp Ser 42:257–258

Sawasaki T, Nishihara M, Endo Y (2008) RIP and RALyase cleave the sarcin/ricin domain, a critical domain for ribosome function, during senescence of wheat coleoptiles. Biochem Biophys Res Commun 370:561–565

Shapiro R, Danzig M (1972) Acidic hydrolysis of deoxycytidine and deoxyuridine derivatives: the general mechanism of deoxyribonucleoside hydrolysis. Biochemistry 11:23–29

Shifrin VI, Anderson P (1999) Trichothecene mycotoxins trigger a ribotoxic stress response that activates c-Jun N-terminal kinase and p38 mitogen-activated protein kinase and induces apoptosis. J Biol Chem 274:13985–13992

Shoemaker CJ, Green R (2011) Kinetic analysis reveals the ordered coupling of translation termination and ribosome recycling in yeast. Proc Natl Acad Sci 108:E1392–E1398

Smyth RP, Davenport MP, Mak J (2012) The origin of genetic diversity in HIV-1. Virus Res 169:415–429

Spahn CM, Gomez-Lorenzo MG, Grassucci RA et al (2004) Domain movements of elongation factor eEF2 and the eukaryotic 80S ribosome facilitate tRNA translocation. EMBO J 23:1008–1019

Stirpe F (2004) Ribosome-inactivating proteins. Toxicon 44:371–383

Stirpe F, Battelli MG (2006) Ribosome-inactivating proteins: progress and problems. Cell Mol Life Sci 63:1850–1866

Sturm MB, Schramm VL (2009) Detecting ricin: sensitive luminescent assay for ricin A-chain ribosome depurination kinetics. Anal Chem 81:2847–2853

Suzuki T, Ohsumi S, Makino K (1994) Mechanistic studies on depurination and apurinic site chain breakage in oligodeoxyribonucleotides. Nucleic Acids Res 22:4997–5003

Svarovskaia ES, Delviks KA, Hwang CK et al (2000) Structural determinants of murine leukemia virus reverse transcriptase that affect the frequency of template switching. J Virol 74:7171–7178

Szczepanik W, Dworniczek E, Ciesiołka J et al (2003) In vitro oxidative activity of cupric complexes of kanamycin A in comparison to in vivo bactericidal efficacy. J Inorg Biochem 94:355–364

Takeshita M, Chang CN, Johnson F et al (1987) Oligodeoxynucleotides containing synthetic abasic sites: model substrates for DNA polymerases and apurinic/apyrimidinic endonucleases. J Biol Chem 262:10171–10179

Tanaka M, Jaruga P, Küpfer PA et al (2012) RNA oxidation catalyzed by cytochrome c leads to its depurination and cross-linking, which may facilitate cytochrome c release from mitochondria. Free Radic Biol Med 53:854–862

Tanpure AA, Patheja P, Srivatsan SG (2012) Label-free fluorescence detection of the depurination activity of ribosome inactivating protein toxins. Chem Commun 48:501

Taylor S, Massiah A, Lomonossoff G et al (1994) Correlation between the activities of five ribosome-inactivating proteins in depurination of tobacco ribosomes and inhibition of tobacco mosaic virus infection. Plant J 5:827–835

Tesh VL (2012) The induction of apoptosis by shiga toxins and ricin. Curr Top Microbiol Immunol 357:137–178

Tsuboi T, Kuroha K, Kudo K et al (2012) Dom34:Hbs1 plays a general role in quality-control systems by dissociation of a stalled ribosome at the 3′ end of aberrant mRNA. Mol Cell 46:518–529

Venner H (1964) Studies on nucleic acids IX: stability of the N-glycosidic linkage in nucleosides. Hoppe Seylers Z Physiol Chem 339:14–27

Wang M, Hudak KA (2006) A novel interaction of pokeweed antiviral protein with translation initiation factors 4G and iso4G: a potential indirect mechanism to access viral RNAs. Nucleic Acids Res 34:1174–1181

Watanabe K, Kawasaki T, Sako N et al (1997) Actions of pokeweed antiviral protein on virus-infected protoplasts. Biosci Biotechnol Biochem 61:994–997

Wrzesiński J, Szczepanik W, Ciesiołka J et al (2005) $tRNA^{Phe}$ cleavage by aminoglycosides is triggered off by formation of an abasic site. Biochem Biophys Res Commun 331:267–271

Zamboni M, Brigotti M, Rambelli F et al (1989) High-pressure-liquid-chromatographic and fluorimetric methods for the determination of adenine released from ribosomes by ricin and gelonin. Biochem J 259:639–643

Zhabokritsky A, Mansouri S, Hudak KA (2014) Pokeweed antiviral protein alters splicing of HIV-1 RNAs, resulting in reduced virus production. RNA 20:1238–1247

Zhang J, Perry G, Smith MA et al (1999) Parkinson's disease is associated with oxidative damage to cytoplasmic DNA and RNA in substantia nigra neurons. Am J Pathol 154:1423–1429

Zhao L, Haslam DB (2005) A quantitative and highly sensitive luciferase-based assay for bacterial toxins that inhibit protein synthesis. J Med Microbiol 54:1023–1030

Zhou H-R (2003) Rapid, sequential activation of mitogen-activated protein kinases and transcription factors precedes proinflammatory cytokine mrna expression in spleens of mice exposed to the trichothecene vomitoxin. Toxicol Sci 72:130–142

Zhou H-R, He K, Landgraf J et al (2014) Direct activation of ribosome-associated double-stranded RNA-dependent protein kinase (PKR) by Deoxynivalenol, anisomycin and ricin: a new model for ribotoxic stress response induction. Toxins (Basel) 6:3406–3425

Zhou H-R, Lau AS, Pestka JJ (2003) Role of double-stranded RNA-activated protein kinase R (PKR) in deoxynivalenol-induced ribotoxic stress response. Toxicol Sci 74:335–344

Zoltewicz J, Clark D, Sharpless T, Grahe G (1970) Kinetics and mechanism of the acid-catalyzed hydrolysis of some purine nucleosides. J Am Chem Soc 92:1741–1749

Recognition of RNA Sequence and Structure by Duplex and Triplex Formation: Targeting miRNA and Pre-miRNA

Kiran M. Patil and Gang Chen

Contents

Abstract RNAs form complex structures containing both single-stranded (ss) and double-stranded (ds) regions for their diverse regulatory and catalytic functions. The emerging RNA sequence and structure databases provide the foundation for developing RNA-binding ligands for reprogramming RNA–RNA and RNA–protein interactions through the recognition of RNA sequence and structure. We choose miRNA biogenesis and gene regulation pathways as examples to summarize how chemically modified nucleic acid oligomers can be used to target specific RNA sequences and structures through duplex and triplex formation. We discuss the significant progress that has been made in using anti-miRNA oligonucleotides in targeting mature miRNA by duplex formation. The strategy of targeting dsRNA by triplex formation is relatively less explored. We summarize the recent results of developing nucleobase and backbone modifications in peptide nucleic acids (PNAs) to facilitate structure-specific and selective targeting of dsRNAs over ssRNA and dsDNA at physiological conditions. We briefly discuss how sequence-specific dsRNA-binding PNAs may be utilized to target disease-associated miRNA precursors and viral RNAs.

K.M. Patil • G. Chen (✉)
Division of Chemistry and Biological Chemistry, School of Physical and Mathematical Sciences, Nanyang Technological University, Singapore 637371, Singapore
e-mail: RNACHEN@ntu.edu.sg

S. Jurga et al. (eds.), *Modified Nucleic Acids in Biology and Medicine*, RNA Technologies, DOI 10.1007/978-3-319-34175-0_13

Keywords Chemically modified oligonucleotide • Peptide nucleic acid • Modified nucleobase • RNA recognition • RNA structure • dsRNA-binding ligands

1 Introduction

RNAs are involved in catalysing and/or regulating the gene expression processes at the levels of transcription, RNA splicing and processing, and translation (Cech and Steitz 2014). To perform the remarkably diverse biological functions, RNAs fold into complex secondary and three-dimensional structures and are often in complex with proteins and metabolites (Strobel and Cochrane 2007). Furthermore, RNAs can tune their structures and functions through the dynamic and reversible chemical modifications of the nucleobase and backbone (Ge and Yu 2013; Qi et al. 2014; Yue et al. 2015; Zipeto et al. 2015). The recent burst of the information of sequences, chemical modifications, and structures of the canonical and newly discovered functional RNAs provides the foundation for unravelling the molecular mechanisms of RNA biology and RNA-related diseases, including cancer and neurodegeneration. An appreciation of the essential functions of RNA has catalysed the development of numerous chemical probes and therapeutic ligands based on the sequences and dynamic structures of RNA (Li and Rana 2014; Shortridge and Varani 2015; Velagapudi et al. 2015).

Several nucleic acid oligomer-based compounds, e.g. fomivirsen, pegaptanib, and mipomersen, have reached the clinic (Boisguérin et al. 2015; Crooke and Geary 2013; Geary et al. 2002; Gelsinger et al. 2012; Jeker and Marone 2015; Ng et al. 2006). Many more are being evaluated in clinical trials for various applications (Li and Rana 2014; Sharma et al. 2014). Here we choose microRNAs (miRNAs or miRs) (Lee et al. 1993; Wightman et al. 1993) as examples to summarize how RNA sequences and structures can be probed and targeted by using chemically modified nucleic acid oligomers (Järver et al. 2014; Jeker and Marone 2015; Li and Rana 2014; van Rooij and Kauppinen 2014).

miRNAs serve as key players in many biological processes. Structural and biochemical studies have provided detailed insights into the molecular mechanisms of siRNA and miRNA biogenesis and gene regulation processes (Wilson and Doudna 2013). Mature miRNAs are double-stranded RNAs, but only one strand of which, usually 21–23 nucleotides (nt) in length, is functional in the RNA interference process. miRNA genes are first transcribed as primary miRNA (pri-miRNA) transcripts, which are processed into precursor miRNA (pre-miRNA) in the nucleus typically catalysed by double-stranded RNA-binding (dsRNA-binding) RNase Drosha and other protein factors. Pre-miRNAs are further processed into mature miRNA by dsRNA-binding RNase Dicer or Argonaute in the cytoplasm. Both pri-miRNA and pre-miRNA form stem-loop structures (hairpins). The processed double-stranded miRNAs contain the guide strand and the passenger strand, and are in complex with Argonaute proteins and other protein factors, forming the miRISC (miRNA-induced silencing complex). The passenger strand is removed from the miRISC complex by cleavage or unwinding by the miRISC complex. The 5′ side of the miRNA guide strand (residues 2–8, known as the seed region) in miRISC is preorganized and exposed for initiating the duplex formation with mRNA target. A miRNA–mRNA duplex formation in the miRISC complex results in the degradation of mRNA and repression of protein expression (van Rooij and Kauppinen 2014). The dysregulation of the miRNA levels results in many diseases such as cancers, neurodegeneration, and cardiovascular diseases. Thus, miRNAs may serve as biomarkers and drug targets for applications in diagnostics and therapeutics (Jayaraj et al. 2015; Li and Rana 2014; Penna et al. 2015; van Rooij and Kauppinen 2014).

The detailed mechanistic understanding of the miRNA biogenesis and gene regulation pathways paves the way for developing chemical probes and therapeutic ligands through reprogramming RNA–protein and RNA–RNA interactions in a sequence- and structure-specific manner (Jayaraj et al. 2015; Li and Rana 2014; van Rooij and Kauppinen 2014; Wilson and Doudna 2013). Small molecules are being developed to influence miRNA expression and function (Bose et al. 2012; Gumireddy et al. 2008; Velagapudi et al. 2015). Here we focus our summary on the reported work on (i) targeting mature miRNAs using chemically modified anti-miRNA oligonucleotides (anti-miRs) or antagomiRs through duplex formation and (ii) developing chemically modified peptide nucleic acids (PNAs) for targeting the double-stranded regions of pre-miRNAs or pri-miRNAs through triplex formation.

2 Targeting Mature miRNAs by Duplex Formation

Anti-miR oligonucleotides operate on the basis of sequence complementarity with the mature miRNA guide strand and can thus compete with an mRNA target in binding to the mature miRNA guide strand in the miRISC complex, resulting in the inhibition of the activity of miRNAs (Fabani and Gait 2008). Duplex formation between anti-miRs and miRNA guide strands is one of the most widely used strategies to inhibit the function of miRNAs.

Anti-miRs are usually chemically modified nucleic acid oligomers (Fig. 1). The chemical modifications in the artificial nucleic acids are often incorporated to enhance (i) the duplex formation between a miRNA and an anti-miR and (ii) the chemical stability and thus resistance against RNases including Argonaute. Most of the reported chemical modifications in anti-miRs are on the sugar phosphate backbone (Avitabile et al. 2014; Lennox et al. 2013; Oh et al. 2009; Simonson and Das 2015), some of which are discussed below.

Incorporation of 2′-*O*-methyl (2′-OMe) modification into oligonucleotides is known to improve the nuclease resistance and duplex formation stability with ssRNA (Cummins et al. 1995; Lamond and Sproat 1993; Majlessi et al. 1998; Verma and Eckstein 1998). In 2004, Zamore and Tuschl labs (Hutvagner

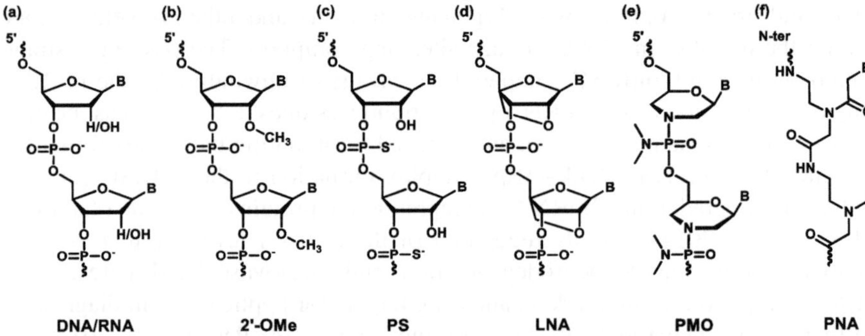

Fig. 1 Chemical structures of nucleic acids with unmodified and modified backbones. The letter "B" represents the nucleobase. (**a**) DNA/RNA. (**b**) 2′-*O*-methyl oligonucleotide (2′-OMe). The 2′-OME and 2′-F modifications have the 2′-OMe group replaced by 2′-*O*-methoxyethyl and 2′-fluorine, respectively. (**c**) Phosphorothioate (PS) linked RNA. (**d**) Locked Nucleic Acid (LNA). (**e**) Phosphorodiamidate morpholino oligonucleotide (PMO). (**f**) Peptide nucleic acid (PNA)

et al. 2004; Meister et al. 2004) showed that 21-nt 2′-OMe-modified oligonucleotides (Fig. 1b) can be used to inhibit siRNA and miRNA activities in human cell extract and cell culture. 2′-OMe-modified oligonucleotides, however, are not fully resistant against exonucleases. Replacement of one or more phosphate linkages by phosphorothioate linkages (PS, Fig. 1c) improves the nuclease resistance but results in a compromised binding affinity towards targeted miRNAs (Lennox and Behlke 2010). Optimization of the number and position of 2′-OMe modification and PS-linkage is needed to synergistically enhance the anti-miR activity (Krutzfeldt et al. 2005).

Similar to 2′-OMe, 2′-*O*-methoxyethyl (2′-OME) modification in an oligonucleotide stabilizes the C3′-*endo* sugar pucker conformation. Thus, compared to the 2′-OMe modification with PS-linkage, the 2′-OME modification combined with PS-linkage further improves the binding affinity of oligonucleotides with ssRNA targets and resistance against enzymatic degradation and thus anti-miR activity (Esau et al. 2006; Manoharan 1999). Insertion of PS-linkages into 2′-F-modified oligonucleotides in combination with other backbone modifications also enhances the nuclease resistance and improves anti-miR activity (Davis et al. 2006; Kawasaki et al. 1993).

Locked nucleic acid (LNA, Fig. 1d) is one of the most promising modifications in antisense technologies. Several LNA-incorporated oligonucleotides are being tested in clinical trials (Sharma et al. 2014). In 2005, LNA-incorporated oligonucleotides were used to probe the expression and activity of miR-21 in cell culture (Chan et al. 2005). The 23-mer LNA/2′-OMe mixmer antisense oligonucleotide was found to be more effective than those with 2′-OMe analogue alone in blocking miR-122 activity in human and rat liver cells (Fabani and Gait 2008). LNA exhibits excellent binding affinity, nuclease resistance, and anti-miR activity when in combination with other modifications such as 2′-F or 2′-OMe modifications

(Kierzek et al. 2005; Lennox and Behlke 2010). Oligonucleotides are usually delivered into cells using transfection agents or by conjugation with cell-penetrating moieties (Chan et al. 2005; Fabani and Gait 2008; Lennox and Behlke 2010; Stein et al. 2010). It is worth noting that relatively short seed-targeting 8-mer anti-miRs with full LNA modification and PS-linkages show uptake and activity in cell cultures and in mouse models without using transfection agents (Obad et al. 2011; Stein et al. 2010).

Charge-neutral phosphorodiamidate morpholino oligonucleotides (PMOs, Fig. 1e) and peptide nucleic acids (PNAs, Fig. 1f) show promising antisense activities in cell cultures and animal models (Ahn et al. 2002; Faria et al. 2001; Holmes et al. 2003; Suwanmanee et al. 2002) and in clinical trials (Gambari 2014; Sharma et al. 2014). For example, cell-penetrating peptide-PMO (CPP-PMO) conjugates were evaluated for their exon-skipping activity in mice and are now in clinical trials for Duchenne Muscular Dystrophy (Fletcher et al. 2007; Mendell et al. 2013; Voit et al. 2014). In 2007, PMOs of different sequences and lengths (21–25 mer) were shown to be able to knock down miRNA activities by targeting mature miRNA guide strands or miRNA precursors (through duplex formation between PMO anti-miRs and pre-miRNAs or pri-miRNAs via strand invasion) (Kloosterman et al. 2007). Cell culture studies show comparable anti-miR-155 and thus anticancer activities for 23-mer PMO and PNA delivered through encapsulation in biodegradable polymer nanoparticles coated with CPP (Cheng and Saltzman 2012).

PNA (Fig. 1f) (Nielsen and Egholm 1999) has a neutral backbone, which allows it to bind strongly to negatively charged RNA or DNA. The achiral and flexible backbone makes PNA to easily adopt the required conformation for strong and sequence-specific binding to the targeted sequences and structures (Nielsen and Egholm 1999). In addition, the relatively robust resistance towards chemical and enzymatic degradation makes PNAs promising candidates as chemical probes and therapeutic ligands. PNAs have been extensively applied to alter gene expression at different levels, including transcription, RNA processing, and translation (Oh et al. 2009; Wang and Xu 2004). For example, PNAs and CPP-PNA conjugates have high efficiency in the inhibition of reverse transcription (Koppelhus et al. 1997), splicing redirection (Abes et al. 2007; Cartegni and Krainer 2003; Turner et al. 2005), inhibition of HIV-1 Tat-dependent *trans*-activation (Das et al. 2012; Turner et al. 2005), and inhibition of bacterial mRNA translation (Huang et al. 2007). PNAs, however, have relatively low water solubility and cell permeability, which can be overcome by incorporating positively charged residues into PNAs (see below). Significant advances have been made to facilitate the cellular uptake of charge-neutral PMOs and PNAs by attaching small molecules, peptides, and nanoparticles or by novel liposome formulation (Avitabile et al. 2015; Boisguérin et al. 2015; Cheng et al. 2015; Das et al. 2012; Ma et al. 2014; Oh et al. 2010; Riguet et al. 2004; Shiraishi and Nielsen 2014).

Numerous reports demonstrated the application of PNA anti-miRs in targeting various miRNAs (Amato et al. 2014; Gaglione et al. 2011; Gambari 2014; Manicardi et al. 2012; Piva et al. 2013; Torres et al. 2011b). In 2008, Fabani and

Gait initiated the use of PNAs in blocking the function of miRNAs (Fabani and Gait 2008). The activity of the liver-specific miR-122 was shown to be inhibited by 23-mer PNA anti-miRs in human and rat liver cells. Remarkably, the anti-miR-122 PNA shows considerable activity upon conjugation with the CPP R_6-penetratin or simply by attachment of four lysine residues (Fabani and Gait 2008). An efficient inhibition of miR-155 function was also reported in vitro and in vivo by using a 23-mer PNA anti-miR with four lysine residues attached (Fabani et al. 2010). The tumour microenvironment has a relatively low pH, and mouse model studies demonstrated that tumour-targeted delivery of the same full-length 23-mer anti-miR-155 PNA can be facilitated by attaching a peptide with a low pH-induced transmembrane structure (Cheng et al. 2015). The presence of a free thiol group such as a cysteine at the end of an anti-miR-122 PNA enhances its anti-miR activity due to improved cellular uptake (Torres et al. 2012). Cell culture studies reveal that, compared to LNA-modified anti-miRs, a 15-mer PNA anti-miR conjugated with Tat peptide shows more effective and long-lasting anti-miR activity and is less toxic to the cell (Oh et al. 2009). The expression of miR-210 was found to be inhibited by 18-mer PNA anti-miRs with arginine residues attached on the PNA backbone or *N*-terminus (Manicardi et al. 2012).

PNA is known to be able to invade a preexisting dsDNA or dsRNA structure. Thus, PNA was used to invade the stem of a pre-miRNA through strand invasion, i.e. through PNA–RNA duplex formation by displacing one of the strands in the preexisting duplex of the pre-miRNA (Avitabile et al. 2012). Furthermore, Torres et al. demonstrated that decoy PNAs used in Northern blot assays can improve the detection of miRNAs upon targeting by anti-miRs (Torres et al. 2011a). Taken together, numerous anti-miR oligonucleotides and PNAs have been developed to successfully target mature miRNA sequences in cell cultures and in animal models through Watson–Crick duplex formation. The results suggest that miRNAs are excellent biomarkers and therapeutic targets, and nucleic acid oligomers are promising candidates as useful chemical probes and therapeutic ligands.

3 Targeting Pre-miRNA Duplex Regions by Triplex Formation

miRNA biogenesis and expression are sensitive to the pre-miRNA and pri-miRNA structures and chemical modifications (Feng et al. 2012; Li et al. 2013; Liu et al. 2013; Starega-Roslan et al. 2011; Yang et al. 2006), suggesting that pre-miRNA and pri-miRNA may be drug targets through the modulation of their structures and dynamics. Small molecules and peptides/peptoids have been developed to target the RNA duplex regions of pre-miRNA and pri-miRNA (Bose et al. 2015a; Chirayil et al. 2009; Jayaraj et al. 2015; Velagapudi et al. 2015). Since the major groove of an RNA duplex can accommodate a third nucleic acid strand to form a triplex structure, chemically modified triplex-forming

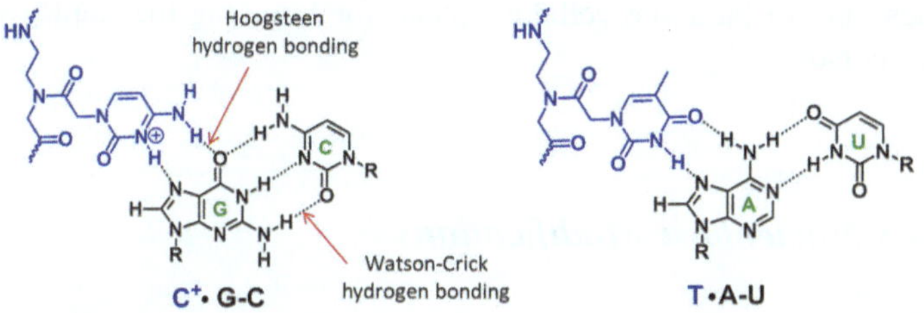

Fig. 2 Chemical structures of C^+·G-C and T·A-U base triples. The letter "R" represents the sugar phosphate backbone of RNA

oligonucleotides (TFOs) and PNAs have great potential in regulating RNA structures and interactions with proteins. For example, binding of triplex-forming PNAs to miRNA precursors may modulate the miRNA biogenesis process involving the actions on pri-miRNA and pre-miRNA by the RNase III type nucleases Drosha and Dicer, respectively. The advantage of targeting miRNA precursor structures through triplex formation is that the targeted sequences may not be limited to the seed regions of miRNAs. Sequence-specific recognition of dsRNAs in miRNA precursors also confers higher specificity compared to the traditional anti-miR strategy, which only recognizes the sequence of miRNA.

RNA triplexes are often found in a complex three-dimensional architecture essential for biologically functional RNAs (Conrad 2014; Devi et al. 2015). Naturally occurring major-groove RNA triplexes are stabilized by multiple base triples such as C^+·G-C and U·A-U or T·A-U (see Fig. 2), with the "·" and "-" representing a Hoogsteen (Nikolova et al. 2013) and a Watson–Crick pair, respectively. The main hurdles to the strategy of targeting RNA structures through triplex formation are that a relatively low pH is required for the stable C^+·G-C base triple formation (monomer $pK_a = 4.5$), and a stable triplex formation is often limited to a duplex target that carries a purine-rich strand.

In addition, to target RNA duplex regions through triplex formation, chemical modifications on the nucleobases and/or backbone are required to stabilize the Hoogsteen base pair formation but destabilize the Watson–Crick base pair formation. PNAs were originally designed to bind sequence specifically to the major groove of dsDNA (Nielsen et al. 1991). However, due to the fact that unmodified bases can form both Watson–Crick and Hoogsteen base pairs, an unmodified PNA often forms a strand invasion complex such as a PNA·DNA–PNA triplex, duplex invasion complex, or double duplex invasion complex (Buchardt et al. 1993; Gupta et al. 2012; Nielsen and Egholm 1999; Panyutin et al. 2012). PNA·DNA_2 triplex structures may form for G-C pair-rich duplexes (Hansen et al. 2009).

Here, we review the reported efforts on targeting RNA duplexes, including those present in pre-miRNAs, by chemically modified triplex-forming PNAs without breaking the preexisting RNA duplex structure. We summarize the initial efforts in designing and discovering chemically modified triplex-forming PNAs to enhance the recognition of dsRNA over ssRNA, to minimize the pH dependence of triplex formation, to recognize dsRNAs with a purine-rich strand interrupted by a few pyrimidines, and to facilitate cellular uptake for targeting the duplex regions of miRNA precursors.

3.1 PNA Nucleobase Modifications

Toulmé and coworkers (Aupeix et al. 1999) observed weak binding between a 12-mer unmodified PNA and an HIV-1 ribosomal frameshift-inducing RNA hairpin at pH 7.0 by gel shift assay. About one decade later, Rozners and coworkers (Li et al. 2010) characterized a stable PNA·RNA_2 triplex formation at pH 5.5 by biophysical methods, including UV absorbance-detected thermal melting, isothermal titration calorimetry (ITC), and circular dichroism (CD). The ITC data suggest that, compared to a homologous dsDNA, a dsRNA is more favourable for binding to a PNA to form a major-groove triplex (Li et al. 2010). In a major-groove PNA·RNA_2 triplex (see Fig. 3e, f), the PNA strand is in a parallel orientation relative to the purine-rich strand (*N*-terminus of PNA aligned to 5′ end of RNA) of the RNA duplex. It is significant that an unmodified short PNA with all-pyrimidine sequence can selectively bind to the major groove of dsRNA over dsDNA, which is opposite to the typical binding properties of a TFO (Devi et al. 2015; Zhou et al. 2013) and polyamides developed by the Dervan group (Chenoweth et al. 2013).

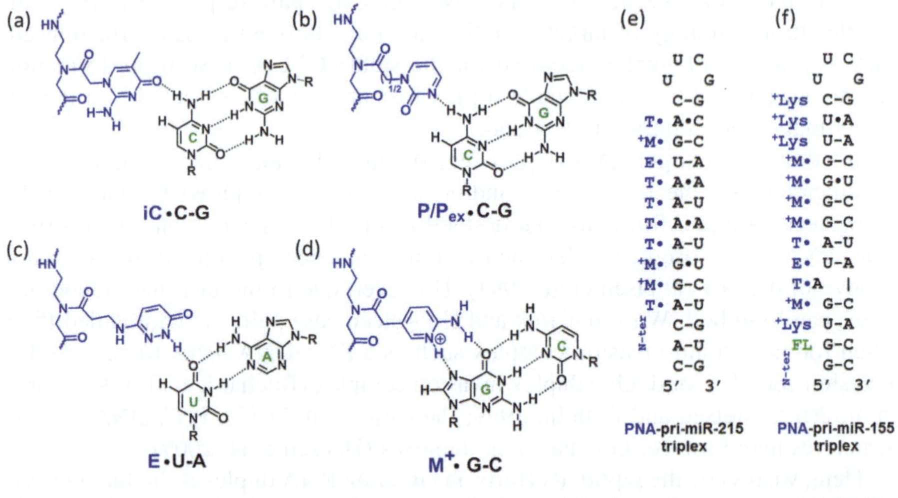

Fig. 3 (**a-d**) Proposed hydrogen bonding patterns for base triples formed between modified PNA bases and Watson–Crick pairs. (**a**) **iC•**C-G. (**b**) **P/P$_{ex}$•**C-G. (**c**) **E•**U-A. (**d**) **M$^+$•**G-C. The letter "R" represents the sugar phosphate backbone of RNA. Panels e and f show the binding of nucleobase-modified PNA to the duplex regions of (**e**) pri-miR-215 and (**f**) pri-miR-155. FL is the fluorescein tag along with the spacer ethylene glycol units attached to the *N*-terminus of the PNA (Muse et al. 2013)

Several modified triplex-forming PNAs have been reported (Rozners 2014). The 5-methylisocytosine (**iC,** see Fig. 3a) (Benner 2004) nucleobase modification was incorporated into a triplex-forming PNA to recognize a C-G inverted base pair by forming an **iC**·C-G base triple (Zengeya et al. 2011). The 2-pyrimidinone (**P,** see Fig. 3b) and its analogue (**P$_{ex}$,** see Fig. 3b) (Buchini and Leumann 2004; Ranasinghe et al. 2005) was incorporated into 9-mer PNAs to recognize a C-G inversion in RNA hairpins (Gupta et al. 2011). The 3-oxo-2,3-dihydropyridazine (**E,** see Fig. 3c) PNA monomer (Eldrup et al. 1997) was used to recognize a U-A base pair in an RNA hairpin (Gupta et al. 2011). An **E**-incorporated PNA (NH_2-Lys-CCCC**E**CTTC-$CONH_2$) was shown to bind to a bacterial ribosomal A-site RNA with a tight binding affinity ($K_D = 227$ nM) at pH 6.25 (Gupta et al. 2011).

To minimize the pH dependence in major-groove triplex formation, neutral or basic analogues of cytosine have been utilized in TFOs and PNAs to form triplexes with a parental DNA duplex (Egholm et al. 1995; Ono et al. 1991; von Krosigk and Benner 1995; Xiang et al. 1994; Xiang et al. 1996). A modified nucleobase pseudoisocytosine (**J**) (see Fig. 4a) was tested in modified TFOs and PNAs to target dsDNA with or without strand invasion (Ono et al. 1991) (Egholm et al. 1995; Hansen et al. 2009; Kurakin et al. 1998; Yaroslavsky and Smolina 2013). In 2012, Rozners and coworkers demonstrated (Zengeya et al. 2012) the recognition of dsRNAs at physiological pH with PNAs incorporating **J**, or 2-aminopyridine (abbreviated as **M**, see Fig.3d) (Cassidy et al. 1997; Chen and McLaughlin 2000; Hildbrand et al. 1997; Rusling et al. 2005). The pK_a values of the monomer forms

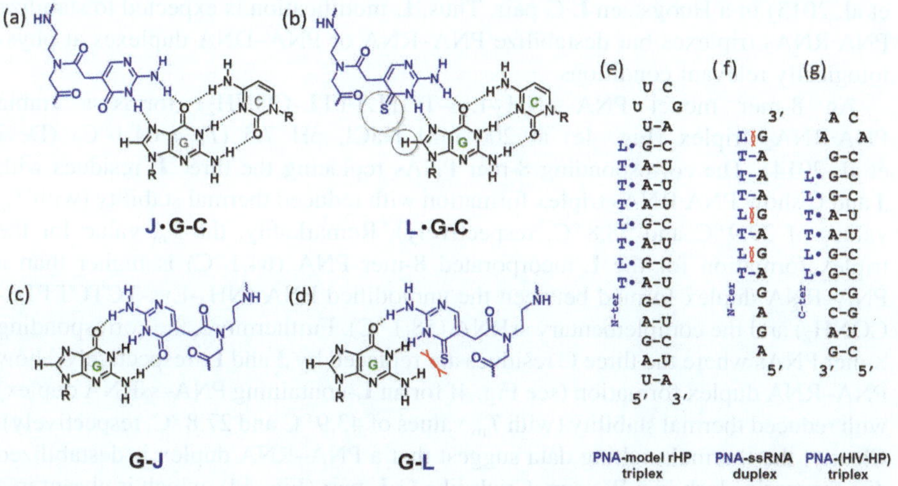

Fig. 4 **(a–d)** Proposed hydrogen bonding patterns for nucleobase triples and pairs involving modified PNA **J** and **L** bases. **(a) J•**G-C triple. **(b) L•**G-C triple. **(c)** Watson–Crick-like G-**J**. **(d)** Watson–Crick-like G-**L**. The letter "R" represents the sugar phosphate backbone of RNA. **(e–g)** Triplex and duplex structures formed involving PNA. **(e)** A triplex formed between an **L**-modified PNA and a model RNA hairpin (rHP). **(f)** A PNA–ssRNA in a parallel mode. The steric clash between **L** and G (shown in *red*) destabilizes the PNA–RNA duplex formation. **(g)** An **L**-modified PNA binds to the HIV-1 frameshift-inducing RNA hairpin (HIV-HP)

of C, **J**, and **M** are 4.5, 9.4, and 6.7, respectively (Hildbrand et al. 1997; Kan et al. 1999). Importantly, pri-miR-215 can be targeted by modified PNAs through triplex formation (Fig.3e) with a relatively tight binding affinity ($K_D = 83$ nM) at a near-physiological buffer condition (Zengeya et al. 2012). The fluorescently labelled PNAs (targeting pri-miR-155, see Fig. 3f) incorporating **M** and **E** nucleobases along with D-lysine residues exhibit significant cellular uptake in HEK293 cells (Muse et al. 2013).

To selectively target dsRNAs by triplex formation, a triplex-forming PNA is required to selectively bind to dsRNA over ssRNA. Through the **J**·G-C (Fig. 4a) base triple formation, **J**-incorporated PNAs can bind dsDNA (in most of the cases it forms a strand invasion complex) (Egholm et al. 1995; Hansen et al. 2009) and dsRNA (with minimal strand invasion) at physiologically relevant pH (Zengeya et al. 2012). However, **J** can recognize both the Hoogsteen face and Watson–Crick face of a G base, resulting in formation of stable Hoogsteen **J**·G and Watson–Crick-like G-**J** pairs, respectively (Fig. 4a, c). Thus, **J**-incorporated PNAs may bind to both dsRNA and ssRNA. To overcome this issue, in 2014 our laboratory designed thio-pseudoisocytosine (Cao et al. 2011) PNA monomer (**L**) and demonstrated its incorporation into PNAs for more selective and enhanced binding towards dsRNA over dsDNA, ssDNA, and ssRNA, compared to **J**-or C-incorporated PNAs (Devi et al. 2014). The oxygen atom of the **J** nucleobase is replaced by a relatively bulky sulphur atom in an **L** nucleobase (Fig. 4b). The **L** nucleobase in a PNA has steric repulsion with a G in a Watson–Crick-like G-**L** pair (Fig. 4d), but has an enhanced van der Waals interaction (Miyata et al. 2009; Okamoto et al. 2006; Zhou et al. 2013) in a Hoogsteen **L**·G pair. Thus, **L**-modification is expected to stabilize PNA·RNA_2 triplexes but destabilize PNA–RNA or PNA–DNA duplexes at physiologically relevant conditions.

An 8-mer model PNA (NH_2-Lys-**TLTLTTTL**-$CONH_2$) forms a stable PNA·RNA_2 triplex (Fig. 4e) at 200 mM NaCl, pH 7.5 ($T_m = 64.1$ °C) (Devi et al. 2014). The corresponding 8-mer PNAs replacing the three **L** residues with **J** and C show PNA·RNA_2 triplex formation with reduced thermal stability (with T_m values of 29.9 °C and 38.8 °C, respectively). Remarkably, the T_m value for the triplex formation for the **L**-incorporated 8-mer PNA (64.1 °C) is higher than a PNA–RNA duplex formed between the unmodified PNA (NH_2-Lys-TCTCTTTC-$CONH_2$) and the complementary ssRNA (58.1 °C). Furthermore, the corresponding 8-mer PNAs where the three C residues are replaced by **J** and **L**, respectively, show PNA–RNA duplex formation (see Fig. 4f for an **L**-containing PNA–ssRNA duplex) with reduced thermal stability (with T_m values of 43.9 °C and 27.8 °C, respectively). Clearly, the thermal melting data suggest that a PNA–RNA duplex is destabilized due the steric clash in a Watson–Crick-like G-**L** pair (Fig. 4d), which is absent in a Watson–Crick-like G-**J** (Fig. 4c) or Watson–Crick G-C pair.

Importantly, gel shift assays indicate that a hexamer PNA (NH_2-Lys-**LLTTLL**-$CONH_2$) binds to the stem of the HIV-1 programmed −1 ribosomal frame shift-stimulating RNA hairpin through a stable triplex (see Fig. 4g) formation with a K_D value of 1.1 μM at 200 mM NaCl, pH 7.5 (Devi et al. 2014). The two corresponding

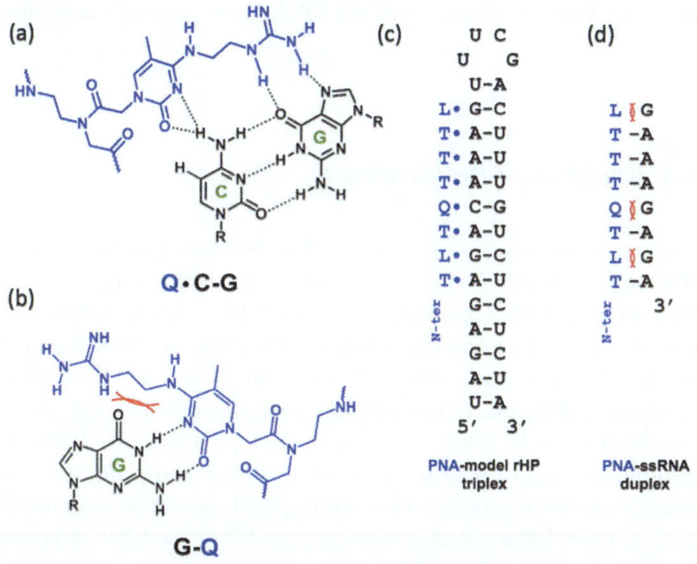

Fig. 5 (**a**) Proposed hydrogen bonding pattern for a base triple formed between modified PNA base **Q** with Watson–Crick C-G pair. (**b**) Steric clash occurs for a Watson–Crick-like G-**Q** pair. (**c**) An **L** and **Q** nucleobase-modified PNA binds to a model RNA hairpin with a C-G inversion (rHP). (**d**) The **L** and **Q** nucleobase-modified PNA shows no binding to a ssRNA due to the steric clash (shown in *red*) in Watson–Crick-like G-**Q** (see panel b) and G-**L** (see Fig. 4d) pairs

6-mer PNAs, where the four **L** residues are replaced by **J** and C, respectively, show no binding at similar conditions (Devi et al. 2014).

More recently, we demonstrated that a dsRNA with a C-G base pair (Fig. 5c) can be recognized by a short PNA (Toh et al. 2016) incorporating an alkylguanidine derivative of 5-methylcytosine (**Q**, see Fig. 5a) (Semenyuk et al. 2010). The gel electrophoresis studies reveal that an 8-mer PNA (NH_2-Lys-T**L**T**Q**TTT**L**-$CONH_2$) containing **L** and **Q** modified residues is capable of binding to the targeted model RNA hairpin with a K_D of 4.4 ± 0.5 μM by forming a PNA•RNA_2 triplex (Fig. 5c) in a near-physiological buffer (200 mM NaCl, pH 7.5). Importantly, no appreciable binding was observed for the 8-mer PNA upon changing the C-G base pair in the RNA hairpin to G-C, U-A, or A-U. In addition, due to the steric clash between a Watson–Crick-like G-**Q** pair (Fig. 5b), the 8-mer PNA shows no binding to a ssRNA (Fig. 5d) or to a homologous DNA hairpin with a C-G inversion. Confocal microscopic studies reveal that a Cy3-labelled PNA sequence with multiple **Q** residues (Cy3-LysNH-T**Q**T**Q**TTT**Q**-$CONH_2$) can pass into the HeLa cells, suggesting that the attached guanidine groups on the PNA nucleobase can facilitate the cellular uptake.

Taken together, the binding and cellular uptake studies reveal that it is possible to develop selective and bioactive dsRNA-binding PNAs with nucleobase modifications with minimized binding to dsDNA and ssRNA. Cell culture and in vivo studies are needed to see if the modified PNAs have anti-pre-miR and antiviral activities.

3.2 PNA Backbone Modifications

The backbone of PNA can also be modified at different positions (α, β, and/or γ position, Fig. 6a) to improve the water solubility, cellular uptake, and binding affinity (Bose et al. 2015b; Corradini et al. 2011; Dragulescu-Andrasi et al. 2005; Manicardi et al. 2012; Mitra and Ganesh 2012; Püschl et al. 1998; Sacui et al. 2015; Sahu et al. 2009; Zhou et al. 2003; Zhou et al. 2006). PNAs with guanidine-modified backbone (Fig. 6b) have been developed to enhance the cellular uptake (Dragulescu-Andrasi et al. 2005; Zhou et al. 2003; Zhou et al. 2006). It is worth noting that attaching multiple arginine residues to the anti-miR-210 PNAs significantly enhances cellular uptake, biostability, and anti-miR activity (Manicardi et al. 2012). A PNA incorporating α-alkyl guanidino-modified thymine forms a strand invasion complex with the HIV-1 trans-activation response element (TAR) hairpin (Gupta et al. 2012). HIV-1 TAR RNA may be a viral pre-miR (Moens 2009).

PNAs incorporating one or two benzyl-substituted thymine monomers (Fig. 6c) show enhanced mismatch discrimination and selective binding towards dsRNA over dsDNA (Zengeya et al. 2013). However, the benzyl modification slightly weakens the binding strength towards dsRNA (Zengeya et al. 2013). Clearly, PNA backbone and nucleobase modifications can improve pharmacodynamic and pharmacokinetic properties of PNA for biological and therapeutic applications.

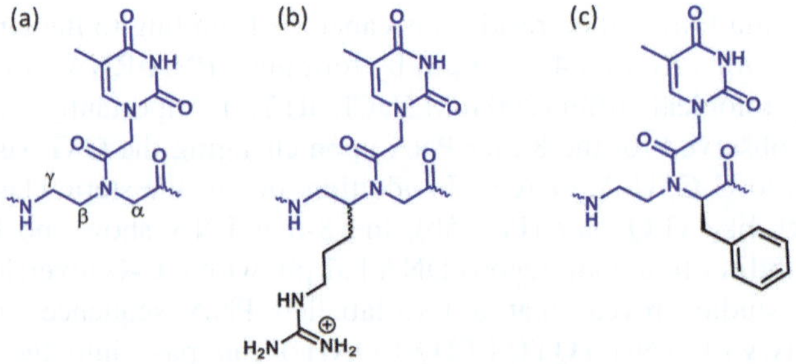

Fig. 6 Representative chemical structures of PNAs with (**a**) unmodified backbone, (**b**) β-modified PNA, and (**c**) α-modified PNA

4 Conclusions and Future Prospects

Clinical trials are underway in evaluating the therapeutic applications of anti-miR oligonucleotides (Shibata et al. 2013). Miravirsen, an LNA-based anti-miR drug candidate, shows promising results in clinical trials for targeting liver-specific miR-122 and controlling the hepatitis C virus (HCV) infection (Gebert et al. 2014; Lindow and Kauppinen 2012; Ottosen et al. 2015). Apart from the cellular miRNAs, the viral miRNAs may also serve as novel antiviral therapeutic targets (Moens 2009; Tycowski et al. 2015).

RNA structure-based design and discovery of novel chemically modified artificial nucleic acid oligomers such as LNA, PMO, and PNA will surely help further improve their binding affinity and selectivity. In parallel, the structural and mechanistic studies of miRNA biogenesis and gene regulation pathways can provide the guidelines for the design of anti-miRs (e.g. through duplex formation targeting seed region) and anti-pre-miRs (through selective triplex formation) in the context of complex RNA–protein structures and dynamics. Bioactivities of the nucleic acid oligomers can be enhanced by conjugation with small molecules (e.g. guanidine, amino sugar, and lipid), peptides, and nanoparticles (Boisguérin et al. 2015; Cheng et al. 2015; Das et al. 2012; Ma et al. 2014; Nair et al. 2014; Oh et al. 2010; Riguet et al. 2004; Shiraishi and Nielsen 2014; Wexselblatt et al. 2014), by novel liposome formulation (Avitabile et al. 2015), or by simply shortening the oligomer length (Obad et al. 2011; Stein et al. 2010). We expect that therapeutics based on the anti-miR and anti-pre-miR or anti-pri-miR compounds will be developed in the near future.

Acknowledgments We thank Prof Mike Gait, Prof Xavier Roca, Prof Ding Xiang Liu, Prof Sek-Man Wong, Dr Jiazi Tan, and Prof Souvik Maiti and his group members for critically reading the manuscript. This work was supported by Singapore Ministry of Education Tier 2 research grants MOE2013-T2-2-024 and MOE2015-T2-1-028 to GC.

References

Abes S, Turner JJ, Ivanova GD et al (2007) Efficient splicing correction by PNA conjugation to an R_6-Penetratin delivery peptide. Nucleic Acids Res 35:4495–4502

Ahn DG, Kourakis MJ, Rohde LA et al (2002) T-box gene tbx5 is essential for formation of the pectoral limb bud. Nature 417:754–758

Amato F, Tomaiuolo R, Nici F et al (2014) Exploitation of a very small peptide nucleic acid as a new inhibitor of miR-509-3p involved in the regulation of cystic fibrosis disease gene expression. Biomed Res Int 2014:610718

Aupeix K, Le Tinévez R, Toulmé J-J (1999) Binding of oligopyrimidines to the RNA hairpin responsible for the ribosome gag-pol frameshift in HIV-1. FEBS Lett 449:169–174

Avitabile C, Saviano M, D'Andrea L et al (2012) Targeting pre-miRNA by peptide nucleic acids: a new strategy to interfere in the miRNA maturation. Artif DNA PNA XNA 3:88–96

Avitabile C, Cimmino A, Romanelli A (2014) Oligonucleotide analogues as modulators of the expression and function of noncoding RNAs (ncRNAs): emerging therapeutics applications. J Med Chem 57:10220–10240

Avitabile C, Accardo A, Ringhieri P et al (2015) Incorporation of naked peptide nucleic acids into liposomes leads to fast and efficient delivery. Bioconjugate Chem 26:1533–1541
Benner SA (2004) Understanding nucleic acids using synthetic chemistry. Acc Chem Res 37:784–797
Boisguérin P, Deshayes S, Gait MJ et al (2015) Delivery of therapeutic oligonucleotides with cell penetrating peptides. Adv Drug Del Rev 87:52–67
Bose D, Jayaraj G, Suryawanshi H et al (2012) The tuberculosis drug streptomycin as a potential cancer therapeutic: inhibition of miR-21 function by directly targeting its precursor. Angew Chem Int Ed 51:1019–1023
Bose D, Nahar S, Rai MK et al (2015a) Selective inhibition of miR-21 by phage display screened peptide. Nucleic Acids Res 43:4342–4352
Bose T, Banerjee A, Nahar S et al (2015b) β,γ-Bis-substituted PNA with configurational and conformational switch: preferred binding to cDNA/RNA and cell-uptake studies. Chem Commun 51:7693–7696
Buchardt O, Egholm M, Berg RH et al (1993) Peptide nucleic acids and their potential applications in biotechnology. Trends Biotechnol 11:384–386
Buchini S, Leumann CJ (2004) Stable and selective recognition of three base pairs in the parallel triple-helical DNA binding motif. Angew Chem Int Ed 43:3925–3928
Cao S-Q, Okamoto I, Tsunoda H et al (2011) Synthesis and triplex-forming properties of oligonucleotides containing thio-substituted C-nucleoside 4-thiopseudoisocytidine. Tetrahedron Lett 52:407–410
Cartegni L, Krainer AR (2003) Correction of disease-associated exon skipping by synthetic exon-specific activators. Nat Struct Biol 10:120–125
Cassidy SA, Slickers P, Trent JO et al (1997) Recognition of GC base pairs by triplex forming oligonucleotides containing nucleosides derived from 2-aminopyridine. Nucleic Acids Res 25:4891–4898
Cech TR, Steitz JA (2014) The noncoding RNA revolution—trashing old rules to forge new ones. Cell 157:77–94
Chan JA, Krichevsky AM, Kosik KS (2005) MicroRNA-21 is an antiapoptotic factor in human glioblastoma cells. Cancer Res 65:6029–6033
Chen DL, McLaughlin LW (2000) Use of pK_a differences to enhance the formation of base triplets involving C−G and G−C base pairs. J Org Chem 65:7468–7474
Cheng CJ, Saltzman WM (2012) Polymer nanoparticle-mediated delivery of microRNA inhibition and alternative splicing. Mol Pharm 9:1481–1488
Cheng CJ, Bahal R, Babar IA et al (2015) MicroRNA silencing for cancer therapy targeted to the tumour microenvironment. Nature 518:107–110
Chenoweth DM, Meier JL, Dervan PB (2013) Pyrrole-imidazole polyamides distinguish between double-helical DNA and RNA. Angew Chem Int Ed 52:415–418
Chirayil S, Chirayil R, Luebke KJ (2009) Discovering ligands for a microRNA precursor with peptoid microarrays. Nucleic Acids Res 37:5486–5497
Conrad NK (2014) The emerging role of triple helices in RNA biology. Wiley Interdiscip Rev RNA 5:15–29
Corradini R, Sforza S, Tedeschi T et al (2011) Peptide nucleic acids with a structurally biased backbone: updated review and emerging challenges. Curr Top Med Chem 11:1535–1554
Crooke ST, Geary RS (2013) Clinical pharmacological properties of mipomersen (Kynamro), a second generation antisense inhibitor of apolipoprotein B. Br J Clin Pharmacol 76:269–276
Cummins LL, Owens SR, Risen LM et al (1995) Characterization of fully 2′-modified oligoribonucleotide hetero- and homoduplex hybridization and nuclease sensitivity. Nucleic Acids Res 23:2019–2024
Das I, Désiré J, Manvar D et al (2012) A peptide nucleic acid–aminosugar conjugate targeting transactivation response element of HIV-1 RNA genome shows a high bioavailability in human cells and strongly inhibits Tat-mediated transactivation of HIV-1 transcription. J Med Chem 55:6021–6032

Davis S, Lollo B, Freier S et al (2006) Improved targeting of miRNA with antisense oligonucleotides. Nucleic Acids Res 34:2294–2304

Devi G, Yuan Z, Lu Y et al (2014) Incorporation of thio-pseudoisocytosine into triplex-forming peptide nucleic acids for enhanced recognition of RNA duplexes. Nucleic Acids Res 42:4008–4018

Devi G, Zhou Y, Zhong Z et al (2015) RNA triplexes: from structural principles to biological and biotech applications. Wiley Interdiscip Rev RNA 6:111–128

Dragulescu-Andrasi A, Zhou P, He G et al (2005) Cell-permeable GPNA with appropriate backbone stereochemistry and spacing binds sequence-specifically to RNA. Chem Commun 2005:244–246

Egholm M, Christensen L, Dueholm KL et al (1995) Efficient pH-independent sequence-specific DNA binding by pseudoisocytosine-containing bis-PNA. Nucleic Acids Res 23:217–222

Eldrup AB, Dahl O, Nielsen PE (1997) A novel peptide nucleic acid monomer for recognition of thymine in triple-helix structures. J Am Chem Soc 119:11116–11117

Esau C, Davis S, Murray SF et al (2006) miR-122 regulation of lipid metabolism revealed by in vivo antisense targeting. Cell Metab 3:87–98

Fabani MM, Gait MJ (2008) miR-122 targeting with LNA/2′-O-methyl oligonucleotide mixmers, peptide nucleic acids (PNA), and PNA–peptide conjugates. RNA 14:336–346

Fabani MM, Abreu-Goodger C, Williams D et al (2010) Efficient inhibition of miR-155 function in vivo by peptide nucleic acids. Nucleic Acids Res 38:4466–4475

Faria M, Spiller DG, Dubertret C et al (2001) Phosphoramidate oligonucleotides as potent antisense molecules in cells and in vivo. Nat Biotechol 19:40–44

Feng Y, Zhang X, Graves P et al (2012) A comprehensive analysis of precursor microRNA cleavage by human Dicer. RNA 18:2083–2092

Fletcher S, Honeyman K, Fall AM et al (2007) Morpholino oligomer-mediated exon skipping averts the onset of dystrophic pathology in the mdx mouse. Mol Ther 15:1587–1592

Gaglione M, Milano G, Chambery A et al (2011) PNA-based artificial nucleases as antisense and anti-miRNA oligonucleotide agents. Mol Biosyst 7:2490–2499

Gambari R (2014) Peptide nucleic acids: a review on recent patents and technology transfer. Expert Opin Ther Pat 24:267–294

Ge J, Yu YT (2013) RNA pseudouridylation: new insights into an old modification. Trends Biochem Sci 38:210–218

Geary R, Henry S, Grillone L (2002) Fomivirsen. Clin Pharmacokinet 41:255–260

Gebert LF, Rebhan MA, Crivelli SE et al (2014) Miravirsen (SPC3649) can inhibit the biogenesis of miR-122. Nucleic Acids Res 42:609–621

Gelsinger C, Steinhagen-Thiessen E, Kassner U (2012) Therapeutic potential of Mipomersen in the management of familial hypercholesterolaemia. Drugs 72:1445–1455

Gumireddy K, Young DD, Xiong X et al (2008) Small-molecule inhibitors of microRNA miR-21 function. Angew Chem Int Ed 47:7482–7484

Gupta P, Zengeya T, Rozners E (2011) Triple helical recognition of pyrimidine inversions in polypurine tracts of RNA by nucleobase-modified PNA. Chem Commun 47:11125–11127

Gupta P, Muse O, Rozners E (2012) Recognition of double-stranded RNA by guanidine-modified peptide nucleic acids. Biochemistry 51:63–73

Hansen ME, Bentin T, Nielsen PE (2009) High-affinity triplex targeting of double stranded DNA using chemically modified peptide nucleic acid oligomers. Nucleic Acids Res 37:4498–4507

Hildbrand S, Blaser A, Parel SP et al (1997) 5-Substituted 2-aminopyridine C-nucleosides as protonated cytidine equivalents: increasing efficiency and selectivity in DNA triple-helix formation. J Am Chem Soc 119:5499–5511

Holmes SC, Arzumanov AA, Gait MJ (2003) Steric inhibition of human immunodeficiency virus type-1 Tat-dependent trans-activation in vitro and in cells by oligonucleotides containing 2′-O-methyl G-clamp ribonucleoside analogues. Nucleic Acids Res 31:2759–2768

Huang X-W, Pan J, An X-Y et al (2007) Inhibition of bacterial translation and growth by peptide nucleic acids targeted to domain II of 23S rRNA. J Pept Sci 13:220–226

Hutvagner G, Simard MJ, Mello CC et al (2004) Sequence-specific inhibition of small RNA function. PLoS Biol 2, E98
Järver P, Torres AG, Gait MJ (2014) Synthetic microRNA blocking agents. In: Arbuthnot P, Weinberg MS (eds) Applied RNAi: from fundamental research to therapeutic applications. Caister Academic Press, Norfolk, pp 105–126
Jayaraj GG, Nahar S, Maiti S (2015) Nonconventional chemical inhibitors of microRNA: therapeutic scope. Chem Commun 51:820–831
Jeker LT, Marone R (2015) Targeting microRNAs for immunomodulation. Curr Opin Pharmacol 23:25–31
Kan L-S, Lin WC, Yadav RD et al (1999) NMR studies of the tautomerism in pseudoisocytidine. Nucleosides Nucleotides 18:1091–1093
Kawasaki AM, Casper MD, Freier SM et al (1993) Uniformly modified 2′-deoxy-2′-fluoro phosphorothioate oligonucleotides as nuclease-resistant antisense compounds with high affinity and specificity for RNA targets. J Med Chem 36:831–841
Kierzek E, Ciesielska A, Pasternak K et al (2005) The influence of locked nucleic acid residues on the thermodynamic properties of 2′-O-methyl RNA/RNA heteroduplexes. Nucleic Acids Res 33:5082–5093
Kloosterman WP, Lagendijk AK, Ketting RF et al (2007) Targeted inhibition of miRNA maturation with morpholinos reveals a role for miR-375 in pancreatic islet development. PLoS Biol 5, e203
Koppelhus U, Zachar V, Nielsen PE et al (1997) Efficient in vitro inhibition of HIV-1 gag reverse transcription by peptide nucleic acid (PNA) at minimal ratios of PNA/RNA. Nucleic Acids Res 25:2167–2173
Krutzfeldt J, Rajewsky N, Braich R et al (2005) Silencing of microRNAs in vivo with 'antagomirs'. Nature 438:685–689
Kurakin A, Jakob Larsen H, Nielsen PE (1998) Cooperative strand displacement by peptide nucleic acid (PNA). Chem Biol 5:81–89
Lamond AI, Sproat BS (1993) Antisense oligonucleotides made of 2′-O-alkyl RNA: their properties and applications in RNA biochemistry. FEBS Lett 325:123–127
Lee RC, Feinbaum RL, Ambros V (1993) The C. elegans heterochronic gene lin-4 encodes small RNAs with antisense complementarity to lin-14. Cell 75:843–854
Lennox KA, Behlke MA (2010) A direct comparison of anti-microRNA oligonucleotide potency. Pharm Res 27:1788–1799
Lennox KA, Owczarzy R, Thomas DM et al (2013) Improved performance of anti-miRNA oligonucleotides using a novel non-nucleotide modifier. Mol Ther Nucleic Acids 2, e117
Li Z, Rana TM (2014) Therapeutic targeting of microRNAs: current status and future challenges. Nat Rev Drug Discov 13:622–638
Li M, Zengeya T, Rozners E (2010) Short peptide nucleic acids bind strongly to homopurine tract of double helical RNA at pH 5.5. J Am Chem Soc 132:8676–8681
Li N, You X, Chen T et al (2013) Global profiling of miRNAs and the hairpin precursors: insights into miRNA processing and novel miRNA discovery. Nucleic Acids Res 41:3619–3634
Lindow M, Kauppinen S (2012) Discovering the first microRNA-targeted drug. J Cell Biol 199:407–412
Liu YP, Schopman NC, Berkhout B (2013) Dicer-independent processing of short hairpin RNAs. Nucleic Acids Res 41:3723–3733
Ma X, Devi G, Qu Q et al (2014) Intracellular delivery of antisense peptide nucleic acid by fluorescent mesoporous silica nanoparticles. Bioconjugate Chem 25:1412–1420
Majlessi M, Nelson NC, Becker MM (1998) Advantages of 2′-O-methyl oligoribonucleotide probes for detecting RNA targets. Nucleic Acids Res 26:2224–2229
Manicardi A, Fabbri E, Tedeschi T et al (2012) Cellular uptakes, biostabilities and anti-miR-210 activities of chiral arginine-PNAs in leukaemic K562 cells. ChemBioChem 13:1327–1337
Manoharan M (1999) 2′-Carbohydrate modifications in antisense oligonucleotide therapy: importance of conformation, configuration and conjugation. Biochim Biophys Acta 1489:117–130

Meister G, Landthaler M, Dorsett Y et al (2004) Sequence-specific inhibition of microRNA- and siRNA-induced RNA silencing. RNA 10:544–550

Mendell JR, Rodino-Klapac LR, Sahenk Z et al (2013) Eteplirsen for the treatment of Duchenne muscular dystrophy. Ann Neurol 74:637–647

Mitra R, Ganesh KN (2012) Aminomethylene peptide nucleic acid (am-PNA): synthesis, regio-/stereospecific DNA binding, and differential cell uptake of (α/γ, R/S)am-PNA analogues. J Org Chem 77:5696–5704

Miyata K, Tamamushi R, Tsunoda H et al (2009) Synthesis and triplex formation of oligonucleotides containing 8-thioxodeoxyadenosine. Org Lett 11:605–608

Moens U (2009) Silencing viral microRNA as a novel antiviral therapy? J Biomed Biotechnol 2009:419539

Muse O, Zengeya T, Mwaura J et al (2013) Sequence selective recognition of double-stranded RNA at physiologically relevant conditions using PNA-peptide conjugates. ACS Chem Biol 8:1683–1686

Nair JK, Willoughby JL, Chan A et al (2014) Multivalent N-acetylgalactosamine-conjugated siRNA localizes in hepatocytes and elicits robust RNAi-mediated gene silencing. J Am Chem Soc 136:16958–16961

Ng EW, Shima DT, Calias P et al (2006) Pegaptanib, a targeted anti-VEGF aptamer for ocular vascular disease. Nat Rev Drug Discov 5:123–132

Nielsen PE, Egholm M (1999) An introduction to peptide nucleic acid. Curr Issues Mol Biol 1:89–104

Nielsen PE, Egholm M, Berg RH et al (1991) Sequence-selective recognition of DNA by strand displacement with a thymine-substituted polyamide. Science 254:1497–1500

Nikolova EN, Zhou H, Gottardo FL et al (2013) A historical account of Hoogsteen base-pairs in duplex DNA. Biopolymers 99:955–968

Obad S, dos Santos CO, Petri A et al (2011) Silencing of microRNA families by seed-targeting tiny LNAs. Nat Genet 43:371–378

Oh S, Ju Y, Park H (2009) A highly effective and long-lasting inhibition of miRNAs with PNA-based antisense oligonucleotides. Mol Cells 28:341–345

Oh SY, Ju Y, Kim S et al (2010) PNA-based antisense oligonucleotides for microRNAs inhibition in the absence of a transfection reagent. Oligonucleotides 20:225–230

Okamoto I, Seio K, Sekine M (2006) Triplex forming ability of oligonucleotides containing 2′-O-methyl-2-thiouridine or 2-thiothymidine. Bioorg Med Chem Lett 16:3334–3336

Ono A, Ts'o POP, Kan LS (1991) Triplex formation of oligonucleotides containing 2′-O-methylpseudoisocytidine in substitution for 2′-deoxycytidine. J Am Chem Soc 113:4032–4033

Ottosen S, Parsley TB, Yang L et al (2015) In vitro antiviral activity and preclinical and clinical resistance profile of miravirsen, a novel anti-hepatitis C virus therapeutic targeting the human factor miR-122. Antimicrob Agents Chemother 59:599–608

Panyutin IG, Onyshchenko MI, Englund EA et al (2012) Targeting DNA G-quadruplex structures with peptide nucleic acids. Curr Pharm Des 18:1984–1991

Penna E, Orso F, Taverna D (2015) miR-214 as a key hub that controls cancer networks: small player, multiple functions. J Invest Dermatol 135:960–969

Piva R, Spandidos DA, Gambari R (2013) From microRNA functions to microRNA therapeutics: novel targets and novel drugs in breast cancer research and treatment. Int J Oncol 43:985–994

Püschl A, Sforza S, Haaima G et al (1998) Peptide nucleic acids (PNAs) with a functional backbone. Tetrahedron Lett 39:4707–4710

Qi L, Chan TH, Tenen DG et al (2014) RNA editome imbalance in hepatocellular carcinoma. Cancer Res 74:1301–1306

Ranasinghe RT, Rusling DA, Powers VEC et al (2005) Recognition of CG inversions in DNA triple helices by methylated 3*H*-pyrrolo [2,3-d] pyrimidin-2(7*H*)-one nucleoside analogues. Chem Commun 2005:2555–2557

Riguet E, Tripathi S, Chaubey B et al (2004) A peptide nucleic acid-neamine conjugate that targets and cleaves HIV-1 TAR RNA inhibits viral replication. J Med Chem 47:4806–4809

Rozners E (2014) Sequence-selective recognition of double-stranded RNA. In: Erdmann VA, Markiewicz WT, Barciszewski J (eds) Chemical biology of nucleic acids, vol 10. Springer Berlin, Heidelberg, pp 167–180
Rusling DA, Powers VEC, Ranasinghe RT et al (2005) Four base recognition by triplex-forming oligonucleotides at physiological pH. Nucleic Acids Res 33:3025–3032
Sacui I, Hsieh W-C, Manna A et al (2015) Gamma peptide nucleic acids: as orthogonal nucleic acid recognition codes for organizing molecular self-assembly. J Am Chem Soc 137:8603–8610
Sahu B, Chenna V, Lathrop KL et al (2009) Synthesis of conformationally preorganized and cell-permeable guanidine-based γ-peptide nucleic acids (γGPNAs). J Org Chem 74:1509–1516
Semenyuk A, Darian E, Liu J et al (2010) Targeting of an interrupted polypurine:polypyrimidine sequence in mammalian cells by a triplex-forming oligonucleotide containing a novel base analogue. Biochemistry 49:7867–7878
Sharma VK, Sharma RK, Singh SK (2014) Antisense oligonucleotides: modifications and clinical trials. Med Chem Commun 5:1454–1471
Shibata C, Otsuka M, Kishikawa T et al (2013) Current status of miRNA-targeting therapeutics and preclinical studies against gastroenterological carcinoma. Mol Cell Ther 1:5
Shiraishi T, Nielsen PE (2014) Cellular delivery of peptide nucleic acids (PNAs). Methods Mol Biol 1050:193–205
Shortridge MD, Varani G (2015) Structure based approaches for targeting non-coding RNAs with small molecules. Curr Opin Struct Biol 30:79–88
Simonson B, Das S (2015) MicroRNA therapeutics: the next magic bullet? Mini Rev Med Chem 15:467–474
Starega-Roslan J, Koscianska E, Kozlowski P et al (2011) The role of the precursor structure in the biogenesis of microRNA. Cell Mol Life Sci 68:2859–2871
Stein CA, Hansen JB, Lai J et al (2010) Efficient gene silencing by delivery of locked nucleic acid antisense oligonucleotides, unassisted by transfection reagents. Nucleic Acids Res 38, e3
Strobel SA, Cochrane JC (2007) RNA catalysis: ribozymes, ribosomes, and riboswitches. Curr Opin Chem Biol 11:636–643
Suwanmanee T, Sierakowska H, Fucharoen S et al (2002) Repair of a splicing defect in erythroid cells from patients with beta-thalassemia/HbE disorder. Mol Ther 6:718–726
Toh D-FK, Devi G, Patil KM et al (2016) Incorporating a guanidine-modified cytosine base into triplex-forming PNAs for the recognition of a C-G pyrimidine-purine inversion site of an RNA duplex. Nucleic Acids Res, Submitted
Torres AG, Fabani MM, Vigorito E et al (2011a) MicroRNA fate upon targeting with anti-miRNA oligonucleotides as revealed by an improved Northern-blot-based method for miRNA detection. RNA 17:933–943
Torres AG, Threlfall RN, Gait MJ (2011b) Potent and sustained cellular inhibition of miR-122 by lysine-derivatized peptide nucleic acids (PNA) and phosphorothioate locked nucleic acid (LNA)/2′-O-methyl (OMe) mixmer anti-miRs in the absence of transfection agents. Artif DNA: PNA XNA 2:71–78
Torres AG, Fabani MM, Vigorito E et al (2012) Chemical structure requirements and cellular targeting of microRNA-122 by peptide nucleic acids anti-miRs. Nucleic Acids Res 40:2152–2167
Turner JJ, Ivanova GD, Verbeure B et al (2005) Cell-penetrating peptide conjugates of peptide nucleic acids (PNA) as inhibitors of HIV-1 Tat-dependent trans-activation in cells. Nucleic Acids Res 33:6837–6849
Tycowski KT, Guo YE, Lee N et al (2015) Viral noncoding RNAs: more surprises. Genes Dev 29:567–584
van Rooij E, Kauppinen S (2014) Development of microRNA therapeutics is coming of age. EMBO Mol Med 6:851–864
Velagapudi SP, Vummidi BR, Disney MD (2015) Small molecule chemical probes of microRNA function. Curr Opin Chem Biol 24:97–103

Verma S, Eckstein F (1998) Modified oligonucleotides: synthesis and strategy for users. Annu Rev Biochem 67:99–134

Voit T, Topaloglu H, Straub V et al (2014) Safety and efficacy of drisapersen for the treatment of Duchenne muscular dystrophy (DEMAND II): an exploratory, randomised, placebo-controlled phase 2 study. Lancet Neurol 13:987–996

von Krosigk U, Benner SA (1995) pH-independent triple helix formation by an oligonucleotide containing a pyrazine donor-donor-acceptor base. J Am Chem Soc 117:5361–5362

Wang G, Xu XS (2004) Peptide nucleic acid (PNA) binding-mediated gene regulation. Cell Res 14:111–116

Wexselblatt E, Esko JD, Tor Y (2014) On guanidinium and cellular uptake. J Org Chem 79:6766–6774

Wightman B, Ha I, Ruvkun G (1993) Posttranscriptional regulation of the heterochronic gene lin-14 by lin-4 mediates temporal pattern formation in C. elegans. Cell 75:855–862

Wilson RC, Doudna JA (2013) Molecular mechanisms of RNA interference. Annu Rev Biophys 42:217–239

Xiang G, Soussou W, McLaughlin LW (1994) A new pyrimidine nucleoside ($m^{5ox}C$) for the pH-independent recognition of G-C base pairs by oligonucleotide-directed triplex formation. J Am Chem Soc 116:11155–11156

Xiang G, Bogacki R, McLaughlin LW (1996) Use of a pyrimidine nucleoside that functions as a bidentate hydrogen bond donor for the recognition of isolated or contiguous G-C base pairs by oligonucleotide-directed triplex formation. Nucleic Acids Res 24:1963–1970

Yang W, Chendrimada TP, Wang Q et al (2006) Modulation of microRNA processing and expression through RNA editing by ADAR deaminases. Nat Struct Mol Biol 13:13–21

Yaroslavsky Anastasia I, Smolina Irina V (2013) Fluorescence imaging of single-copy DNA sequences within the human genome using PNA-directed padlock probe assembly. Chem Biol 20:445–453

Yue Y, Liu J, He C (2015) RNA N^6-methyladenosine methylation in post-transcriptional gene expression regulation. Genes Dev 29:1343–1355

Zengeya T, Li M, Rozners E (2011) PNA containing isocytidine nucleobase: synthesis and recognition of double helical RNA. Bioorg Med Chem Lett 21:2121–2124

Zengeya T, Gupta P, Rozners E (2012) Triple-helical recognition of RNA using 2-aminopyridine-modified PNA at physiologically relevant conditions. Angew Chem Int Ed 51:12593–12596

Zengeya T, Gindin A, Rozners E (2013) Improvement of sequence selectivity in triple helical recognition of RNA by phenylalanine-derived PNA. Artif DNA: PNA XNA 4:69–76

Zhou P, Wang M, Du L et al (2003) Novel binding and efficient cellular uptake of guanidine-based peptide nucleic acids (GPNA). J Am Chem Soc 125:6878–6879

Zhou P, Dragulescu-Andrasi A, Bhattacharya B et al (2006) Synthesis of cell-permeable peptide nucleic acids and characterization of their hybridization and uptake properties. Bioorg Med Chem Lett 16:4931–4935

Zhou Y, Kierzek E, Loo ZP et al (2013) Recognition of RNA duplexes by chemically modified triplex-forming oligonucleotides. Nucleic Acids Res 41:6664–6673

Zipeto MA, Jiang Q, Melese E et al (2015) RNA rewriting, recoding, and rewiring in human disease. Trends Mol Med 21:549–559

Modifications in Therapeutic Oligonucleotides Improving the Delivery

Ilya Dovydenko, Alya Venyaminova, Dmitrii Pyshnyi, Ivan Tarassov, and Nina Entelis

Contents

Abstract Oligonucleotides are increasingly used in clinical applications. RNA-based therapeutics include inhibitors of mRNA translation, agents of RNA interference, ribozymes, and aptamers targeting various molecular targets. Challenges with the delivery, specificity, and stability of these therapeutics have spawned the development of chemically modified oligonucleotides. In this chapter, we will describe modifications improving delivery and stability of RNA molecules in human cells. Because the most of the cell transfection methods using oligonucleotide complexes with cationic lipids revealed to be toxic, specific modifications and various conjugates have been recently developed to promote the carrier-free

I. Dovydenko
UMR Génétique Moléculaire, Génomique, Microbiologie (GMGM), Strasbourg University-CNRS, Strasbourg 67084, France

Institute of Chemical Biology and Fundamental Medicine SB RAS, Novosibirsk, Russia

A. Venyaminova • D. Pyshnyi
Institute of Chemical Biology and Fundamental Medicine SB RAS, Novosibirsk, Russia

I. Tarassov • N. Entelis (✉)
UMR Génétique Moléculaire, Génomique, Microbiologie (GMGM), Strasbourg University-CNRS, Strasbourg 67084, France
e-mail: n.entelis@unistra.fr

S. Jurga et al. (eds.), *Modified Nucleic Acids in Biology and Medicine*, RNA Technologies, DOI 10.1007/978-3-319-34175-0_14

cellular uptake of therapeutic oligonucleotides. Nucleic acids (NA) are relatively unstable in cytosol due to a plethora of nucleases; therefore, various modifications can be introduced to design nuclease-resistant molecules. These modifications should not interfere with the therapeutic activity and intracellular localization of the oligonucleotides. The influence of nucleotide modifications on the siRNA efficiency and on the anti-replicative activity of therapeutic RNA imported into human mitochondria is discussed.

Keywords Delivery • Oligonucleotide therapeutics • Lipophilic conjugates • Modified oligonucleotides • Mitochondrial diseases • Anti-replicative RNA

1 Introduction: Nucleic Acids' Delivery Systems

Synthetic oligonucleotides and their conjugates are widely used in various fields of molecular biology, nanobiotechnology, and medicine as tools for fundamental and applied research, as well as promising drugs for diagnosis and treatment of viral and genetic diseases, cancer, and other diseases of humans and animals (Ginn et al. 2013; Tan et al. 2011). Development of oligonucleotides as potential therapeutic agents is limited by low efficiency of penetration into target cells due to their large size, negative charge, and low stability. Many systems of gene targeting have been developed to overcome these barriers. Delivery vectors can be divided into the two major groups: viral and nonviral systems (Kay 2011). Each type has its own advantages and disadvantages. Systems based on viral vectors have efficient mechanisms for entering the cell, escaping endosomal entrapment, and translocating gene cargo to the nucleus. Despite of a high efficiency of targeting, several limitations are associated with viral systems: insertional mutagenesis, immune response to viral proteins, tumorigenesis, and cytotoxic effects (Thomas et al. 2003; Walther and Stein 2000).

The shortcomings in viral vectors stimulated development of nonviral delivery carriers, which can be readily synthesized and modified to facilitate biocompatibility. Improving of nonviral delivery systems relies on the detailed understanding of the barriers associated with the nucleic acids targeting into cells. The successful system for nucleic acids' (NA) delivery should meet a number of requirements: biocompatibility and low cytotoxicity, resistance to nuclease activity, possibility of endosomal escape, and capability of entering the appropriate cellular compartment.

Mammalian cells internalize extracellular macromolecules by the endocytosis leading to formation of vesicle-like structures that fuse with early endosomes (De Haes et al. 2012). Thus, the efficacy of transfection and the expected effect depend on both the ability of a carrier to efficiently deliver the NA cargo with minimal toxicity and its potential to overcome the endosomal compartmentalization (Huotari and Helenius 2011). To facilitate NA escape into the cytosol, various compounds have been used in combination with the delivery vectors.

Endosomolytic agents vary in type (natural or synthetic compounds) as well as in their mechanisms of action, which include the endosomal membrane destabilization (TAT HIV, KALA, or GALA peptides), pore formation (e.g., listeriolysin O toxin produced by *Listeria monocytogenes,* gp41HIV protein), and endosomal disruption via the "proton sponge" mechanism (e.g., PEI, ammonium chloride, chloroquine, methylamine) (Varkouhi et al. 2011).

Nonviral delivery systems can be carrier-mediated or carrier-free. The carrier-mediated NA targeting systems can be further subdivided into the three main groups:

(a) Polymeric systems, in which NA form complex with a polymer through charge interactions between the positive groups of the polymer and the negatively charged NA (Oliveira et al. 2015)
(b) Lipidic systems, in which cationic lipids interact with negatively charged NA and condensate or encapsulate them (Balazs and Godbey 2011)
(c) Inorganic carriers involving various materials such as gold nanoparticles, silica, and carbon nanotubes, which can bind NA through different mechanisms (Dizaj et al. 2014)

In this chapter we will describe another approach for delivery of oligonucleotides, which consists in conjugation of NA molecules with targeting ligands.

2 Carrier-Free Targeting Systems

2.1 Nucleic Acids' Conjugation with Targeting Ligands

The transporting molecule should be capable of binding to the cell surface or to specific receptors on it and inducing endocytosis (Juliano et al. 2013). The ligand molecule can be attached to NA directly or through a linker. Depending on the type of oligonucleotide and its purpose, the linker can be connected via the 2′-, 3′-, or 5′-terminus, the C5 atom of pyrimidine bases, the C8 atom of adenine, the exocyclic amino group of guanine, or an internucleoside phosphate (Winkler 2013). Conjugated NA cargo needs additional chemical modifications shielding it from nucleases (see Sect. 3 for details). For most carrier-free systems, the endosomal escape is passive, which reduces the efficiency of transfection; thus, there is a necessity to use additional compounds to promote the release of oligonucleotides from endosomes before they are degraded and recycled. However, the simplicity of design and the small size, in comparison with nanoparticles, ensure a lower toxicity and a better biodistribution of NA conjugates. Carriers with sizes larger than 5 nm can only be used for NA delivery to certain types of tumors and to normal tissues with fenestrated endothelia, such as the liver and spleen, whereas conjugates can also reach many other types of tissues (Juliano et al. 2009).

Various delivery systems through covalent attachment of addressing ligands to the NA cargo have been developed: carbohydrate–NA conjugates, peptide–NA conjugates, antibody–NA conjugates (Uckun et al. 2013), aptamer conjugates, and lipid–NA conjugates. Here, we will briefly characterize the main types of conjugates and provide a more detailed analysis of lipophilic molecules.

2.2 Conjugates with Carbohydrates

It had been demonstrated that the asialoglycoprotein receptor located on the surface of hepatocytes can bind diverse chemotherapeutic agents, including galactose glycoproteins, and helps their internalization by endocytosis (Stockert 1995). This permitted the use of carbohydrate-based vectors for addressed NA delivery. For instance, 5′-glycoconjugates of oligonucleotides have demonstrated excellent cell-type specificity and cellular uptake in the nanomolar concentration range (Biessen et al. 1999). Triantennary N-acetyl galactosamine conjugates (Fig. 1) facilitate the targeted delivery of siRNAs and antisense oligonucleotides to hepatocytes in vivo (Nair et al. 2014; Prakash et al. 2014).

2.3 Conjugates with Peptides and Aptamers

Peptides used for NA delivery can be divided into the two classes. The first group includes cell-targeting peptides, specific ligands for surface receptors overexpressed in diseased cells (Juliano et al. 2013; McGuire et al. 2014; Vives et al. 2008). For instance, bombesin peptide (a ligand for the gastrin-releasing peptide receptor) and a specific peptide for the IGF1R receptor overexpressed in

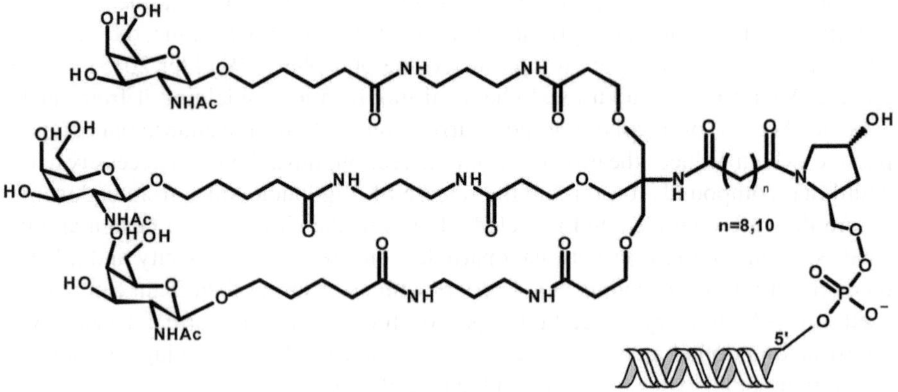

Fig. 1 Triantennary N-acetyl galactosamine-nucleic acid conjugate

breast cancer were conjugated with an siRNA and an antisense oligonucleotide, respectively, for targeted delivery (Cesarone et al. 2007; Ming et al. 2010). The second group comprises the cell-penetrating peptides (CPPs). These peptides are short, amphiphilic, and enriched with the basic amino acids. CPPs can enter cells by two ways, either via endocytosis through electrostatic interactions with negatively charged glycosaminoglycans, similar to cationic polymers (Juliano et al. 2008), or through membrane translocation. The choice of the entry pathway is dependent on the CPP sequence, concentration, and the temperature (Boisguerin et al. 2015). CPPs can also promote the endosomal escape.

Aptamers are small nucleic acids that fold into a well-defined structure, which determines their affinity and specificity for target molecules. They can be selected from pools of randomized sequences by SELEX (Systematic Evolution of Ligands by Exponential Enrichment) approach (Tuerk and Gold 1990). Aptamers can be evolved to bind small molecules but also nucleic acids, carbohydrates, and soluble or membrane proteins. For instance, so-called escort aptamers recognize cell surface receptors specific for the certain cell type and can be used for the targeted delivery of therapeutic agents. One of the most known aptamers of this type is the 2′-F-RNA aptamer against prostate-specific membrane antigene (PSMA). After binding to its target, the anti-PSMA aptamer can be internalized; therefore, this escort aptamer is widely used now as a delivery vehicle for a number of antitumor drugs, including siRNAs and shRNAs. A comprehensive overview of the therapeutic nucleic acids delivery strategies using aptamers can be found in recently published reviews (Aaldering et al. 2015; Davydova et al. 2011; Ming and Laing 2015; Tan et al. 2011; Zhou and Rossi 2011).

2.4 Lipid-Containing Conjugates

One of the most popular methods of delivery of conjugated oligonucleotides is the use of lipids. Various lipophilic molecules have been conjugated to oligonucleotides, including phospholipids, fatty acids, bile acids (e.g., cholic acid), cholesterol, and fat-soluble vitamins (as α-tocopherol, folic acid) (Bhat et al. 1999; Guzaev et al. 2002; Raouane et al. 2012). The structures of these compounds are shown in Fig. 2. Among those, cholesterol, studied by various groups for the past 25 years since the pioneering work of Letsinger et al. (1989), is by far the most extensively characterized addressing agent.

Cholesterol is an essential lipid of cell membranes of many eukaryotes, which make it attractive for delivery of various therapeutic molecules. Intravenous administration of cholesterol-containing siRNA conjugates resulted in significant levels of their accumulation in liver, heart, kidney, adipose, and lung tissues (Soutschek et al. 2004). In another study, accumulation of a cholesterol-conjugated siRNA in brain cells upon intrastriatal injection has been demonstrated (DiFiglia et al. 2007). These and other studies show that cholesterol conjugation significantly improves delivery of NA. Cellular uptake of cholesterol-conjugated oligonucleotides in vivo

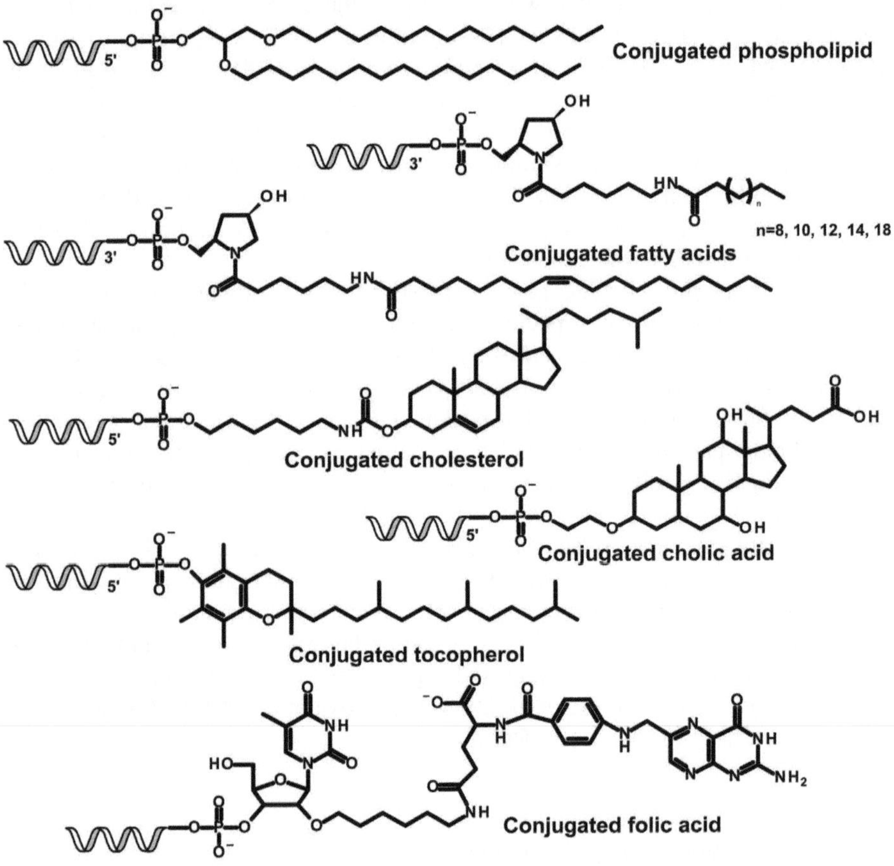

Fig. 2 Various lipophilic molecules used for conjugation with oligonucleotides

depends on the complex formation between the conjugates and the high-density or low-density lipoproteins circulating in the bloodstream. Binding of the complexes to the lipoprotein receptors leads to the uptake of the conjugates by the various tissues (Wolfrum et al. 2007).

Conjugates of oligonucleotides with cholesterol have been developed by many research groups, wherein attachment to an oligonucleotide was performed mainly through either the 5′- or the 3′-termini. Cholesterol can be attached through the unique hydroxyl group of the steroid either directly (MacKellar et al. 1992; Seo et al. 2006) or via various aliphatic linkers. Examples of linear linkers are molecules based on diamines (Letsinger et al. 1989), amino alcohols with various lengths of aliphatic chain (Lorenz et al. 2004; Petrova et al. 2011), and polyethylene glycol (Kubo et al. 2007) (Fig. 3).

Another approach to the synthesis of cholesterol-conjugated oligonucleotides is introducing the steroid residue at the 5′- or 3′-terminus of the oligonucleotide chain through branched linkers containing several reactive groups. The compounds used

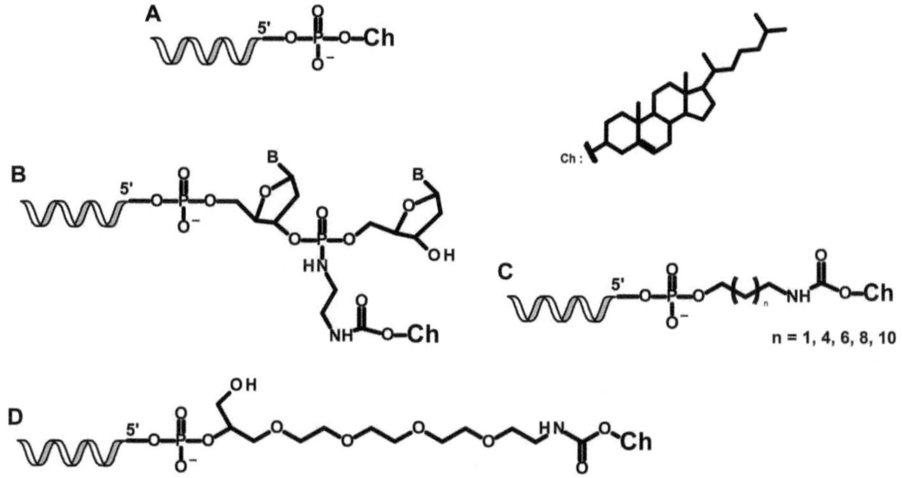

Fig. 3 Cholesterol conjugated with oligonucleotides directly (**a**), through linear linkers based on ethylenediamine (**b**), amino alcohols with various lengths of aliphatic chain (**c**), or polyethylene glycol (**d**)

as the branched linkers include glycerol (Ueno et al. 2008; Vu et al. 1993), 2-aminobutyl-1,3-propanediol (Rump et al. 1998), L-hydroxyproline (Manoharan et al. 2005; Reed et al. 1991), lysine (Stetsenko and Gait 2001), serine (Chaltin et al. 2005), and serinol (Manoharan et al. 2005) (Fig. 4).

A cholesterol residue may negatively affect the therapeutic effect by anchoring the NA cargo to the lipid bilayer membrane or by reducing the efficiency of annealing with the target molecule. To alleviate these effects, the cholesterol residue can be conjugated to the sense chain in siRNA, or it can be added through a cleavable arm, usually containing a disulfide bond (Boutorine and Kostina 1993; Chen et al. 2010; Manoharan et al. 2005; Moschos et al. 2007; Oberhauser and Wagner 1992) (Fig. 5).

The length of the linker can influence the cellular uptake; the optimal efficiency had been achieved for the RNA chain and the cholesterol residue spaced by 6–10 methylene units (Petrova et al. 2011). After penetration of the siRNA conjugates into cells, they affect the target gene expression, suggesting that they are able to escape from endosomes. So far, the mechanism of their endosomal release is still not understood, although it was hypothesized to be related to intracellular traffic of cholesterol (Maxfield and Wustner 2013).

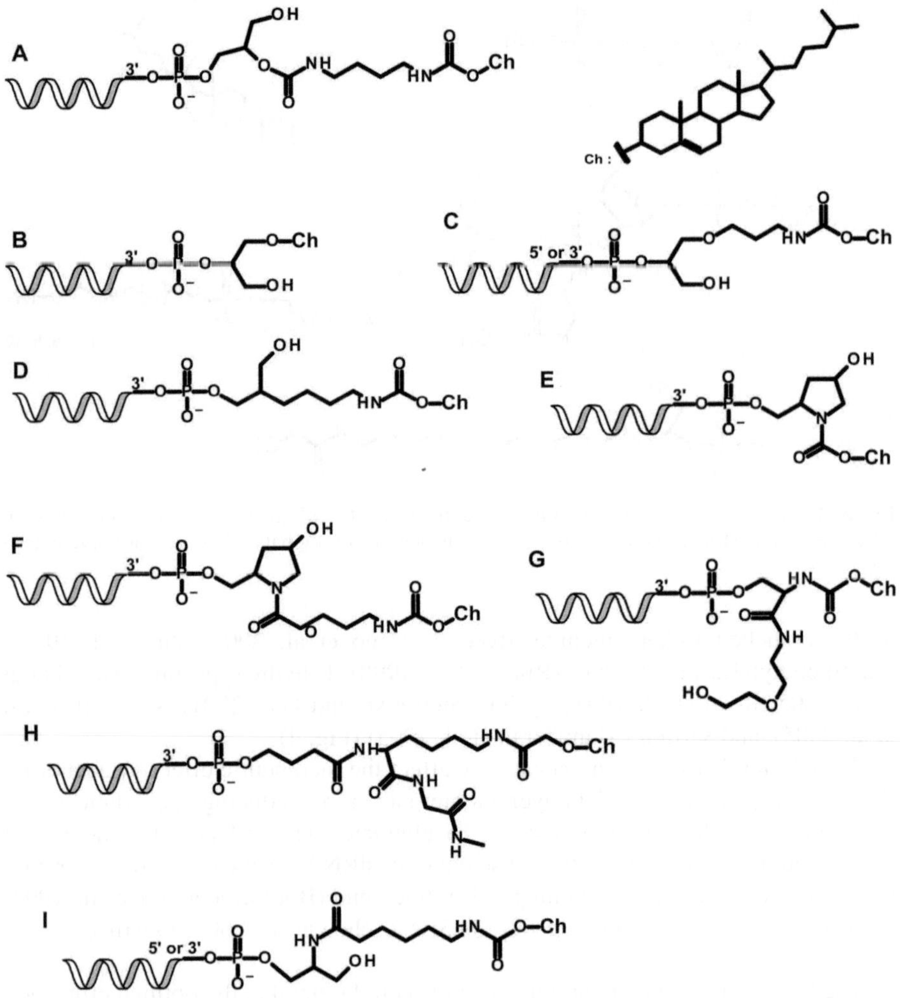

Fig. 4 Cholesterol conjugated with oligonucleotides through branched linkers containing the following compounds: (**a**–**c**) glycerol, (**d**) 2-aminobutyl-1,3-propanediol, (**e**, **f**) L-hydroxyproline, (**g**) lysine, (**h**) serine, (**i**) serinol

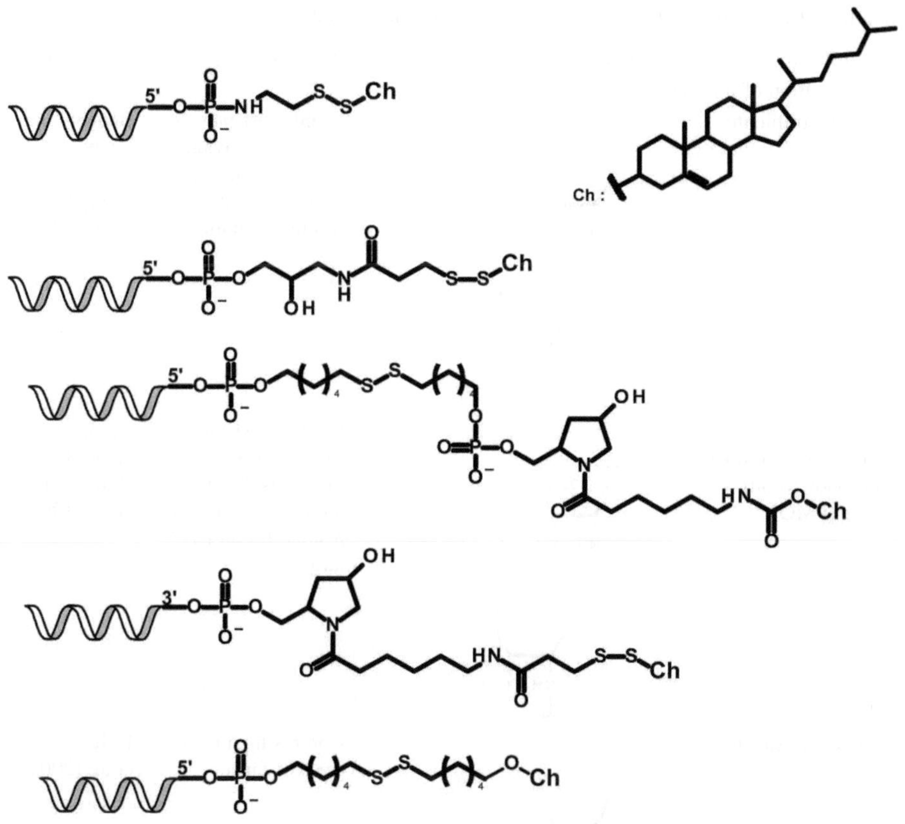

Fig. 5 Cholesterol conjugated with oligonucleotides through various cleavable linkers containing disulfide bond

3 Modifications Improving the Stability of Nucleic Acids

After penetration in the cell and release into the cytoplasm, NA become vulnerable to nuclease attack. This complication can be circumvented if the oligonucleotide cargo is chemically modified to improve its stability. Chemical modifications of nucleic acids can be classified into three distinct categories: internucleoside linkage modifications, sugar modifications, and base modifications (the latter type can affect the thermal stability of the duplex but is not used for NA stabilization and is not considered here). The different types of modifications may be used within the same molecule, depending on the desired effect. We present the structure and the properties of the most used and promising modifications (Table 1).

Among oligonucleotide derivatives listed in Table 1, there are modifications, which are well known and have been recommended for therapeutic use; one can mention phosphorothioates, phosphorodiamidate morpholino oligomers (PMO), peptide nucleic acids (PNA), and different types of 2′-modified oligonucleotides

Table 1 Chemical modifications improving nucleic acids' stability

Modification	Structure	Properties	References
Modification of internucleoside link			
Phosphorothioate	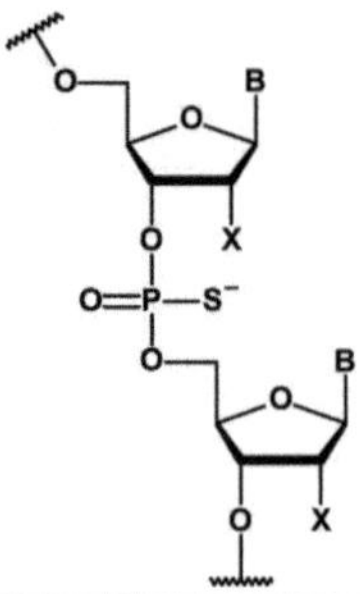	Highly resistant to nuclease cleavage; decreases the duplex stability; binds to serum albumin; toxic	Bennett and Swayze (2010), Milligan et al. (1993)
N3′ → P5′ Phosphorodiamidate morpholino oligomers (PMO)	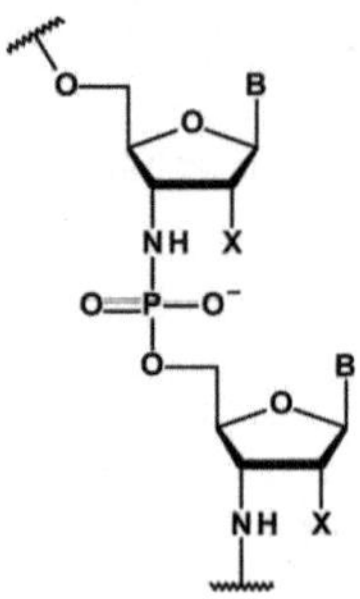	Confers resistance to nucleases; increases T_m of duplex by 2 °C per residue for an RNA target	Gryaznov et al. (1995), Heidenreich et al. (1997)
Boranophosphate	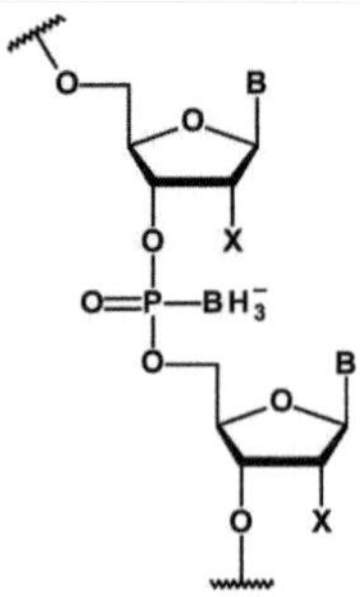	Confers high nuclease resistance	Hall et al. (2004)
Amide-internucleosidic linkage	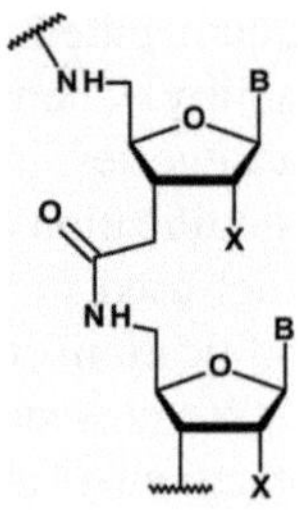	Confers nuclease resistance; leads to the duplex T_m change from −4 to +0.9 °C per residue for an RNA target	De Mesmaeker et al. (1994), Mutisya et al. (2014)

(continued)

Table 1 (continued)

Modification	Structure	Properties	References
Phosphonoacetate	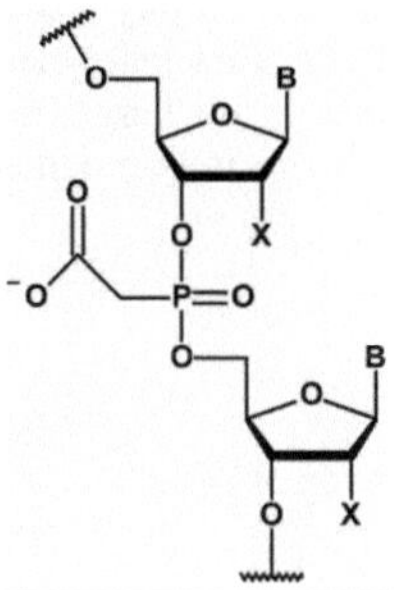	Confers nuclease resistance; uncharged; decreases T_m of duplex by 1.2 °C per residue for an RNA target	Sheehan et al. (2003)
Morpholino	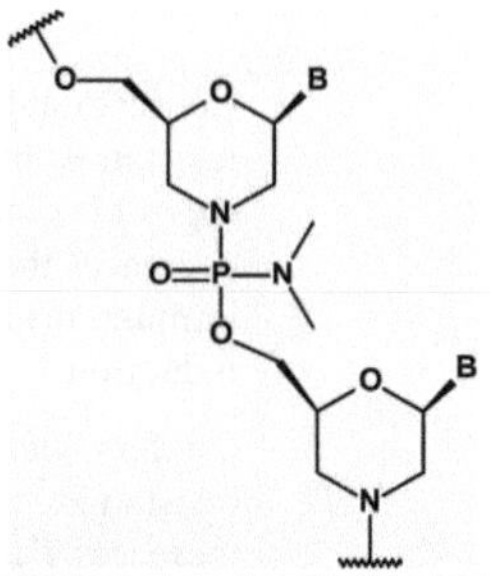	Highly resistant to nuclease cleavage; uncharged	Summerton and Weller (1997)
Peptide nucleic acid (PNA)	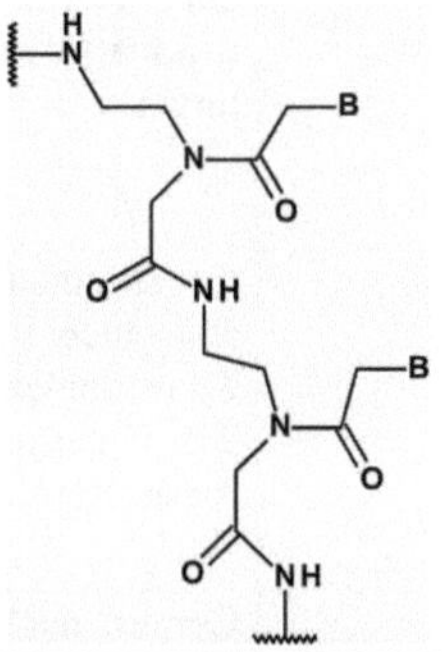	Highly resistant to nuclease cleavage; uncharged; increases T_m of duplex by 1 °C per residue for a DNA target	Shabi et al. (2006)
SATE (S-acyl-2-thioethyl-phosphotriesters)	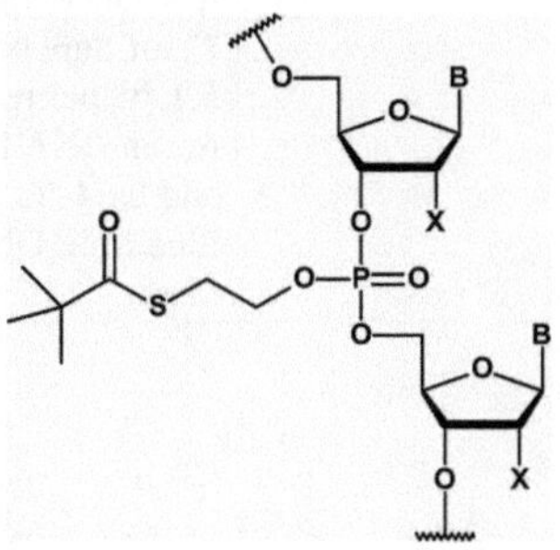	Highly resistant to nuclease cleavage; uncharged; reversible protecting group (cleaved off by thioesterase in the cytoplasm)	Meade et al. (2014)

(continued)

Table 1 (continued)

Modification	Structure	Properties	References
Phosphoryl guanidine (PG)	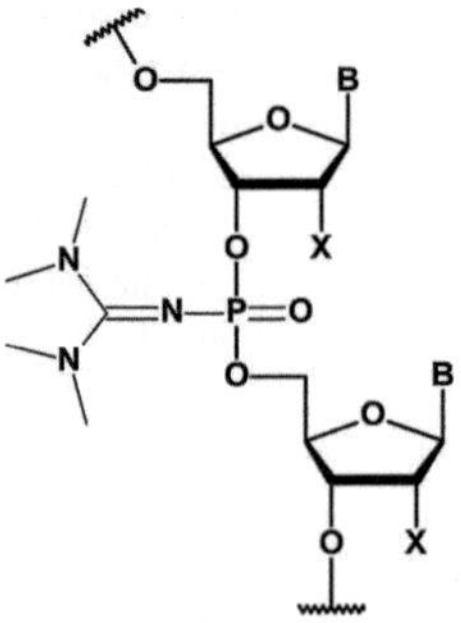	Highly resistant to nuclease cleavage; uncharged when fully modified	Kupryushkin et al. (2014), Lebedeva et al. (2015)
Sugar modifications			
2′-OMe-RNA	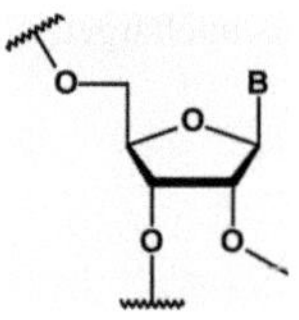	Confers nuclease resistance; increases the duplex stability; decreases the innate immune response induction	Bennett and Swayze (2010)
2′-OMOE-RNA	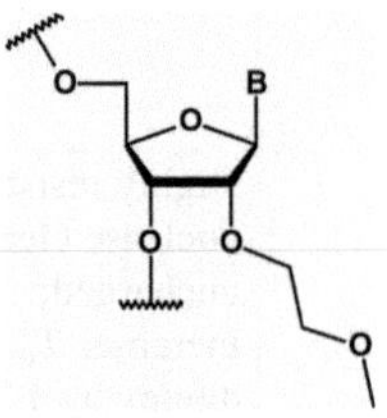	Confers nuclease resistance; increases T_m of duplex by 2 °C per residue for an RNA target	
2′F-RNA	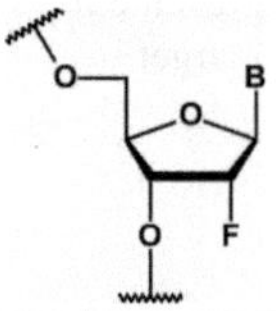	Confers nuclease resistance; increases T_m of duplex by 2.2 °C per residue for an RNA target	Kawasaki et al. (1993)
Locked nucleic acid (LNA)	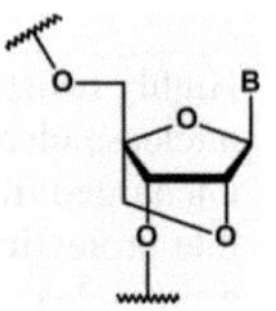	Confers nuclease resistance; increases T_m of duplex by 5.1 °C per residue for an RNA target and by 4 °C per residue for a DNA target	Koshkin et al. (1998)

(continued)

Table 1 (continued)

Modification	Structure	Properties	References
Unlocked nucleic acid (UNA)	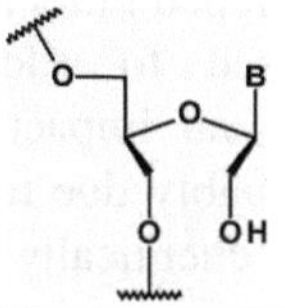	Decreases T_m of duplex by 5–10 °C per residue for an RNA target and by 7–10 °C per residue for a DNA target	Campbell and Wengel (2011)
Xylo nucleic acid (XNA)	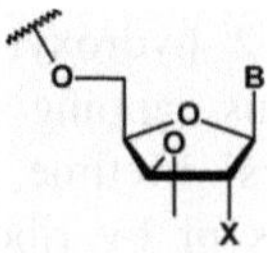	Confers nuclease resistance; decreases duplex stability	Poopeiko et al. (2003)

X—H (DNA) or OH (RNA); B—bases: A, C, G, T, U

(Wickstrom 2015). Another group consists of recently developed promising modifications of oligonucleotide structure whose potential should be studied in details in nearest future. Such modifications include amide-internucleosidic linkage, S-acyl-2-thioethyl-phosphotriesters (SATE), or phosphoryl guanidines (PG) incorporated instead of parent phosphodiester moieties. The derivatives and analogues listed in Table 1 can allow to design oligomers characterized by the electroneutral backbone, drastically increased nuclease resistance, and, therefore, enhanced therapeutic potential.

Modifications should be introduced with caution as they may change NA properties, such as toxicity and binding affinity for RNA/DNA targets. For instance, it was found that the increased amount of modified nucleotides enhanced the stability of siRNA in the presence of serum, but reduced its silencing activity. The targeted modification of nuclease-sensitive sites (mostly UpA, CpA, and UpG sites) improved the stability of siRNA and prolonged the silencing effect with minimal loss of silencing efficiency (Volkov et al. 2009). Moreover, 2′-O-methyl analogues of ribonucleotides introduced in the nuclease-sensitive sites of long dsRNA prevented the activation of innate immunity response without the loss of silencing efficiency and specificity (Gvozdeva et al. 2014).

4 Modifications of the Therapeutic RNA Imported into Human Mitochondria

Defects in human mitochondrial genome can cause a wide range of clinical disorders, mainly neuromuscular diseases. Most of deleterious mitochondrial mutations are heteroplasmic, meaning that wild-type and mutated forms of mtDNA coexist in the same cell (Pinto and Moraes 2014). Therefore, a shift in the proportion between mutant- and wild-type molecules could restore mitochondrial functions. The anti-replicative strategy aims to induce such a shift in heteroplasmy by mitochondrial targeting specifically designed molecules in order to inhibit replication of mutant mtDNA. Recently, we developed mitochondrial RNA vectors (Kolesnikova et al. 2011) that can be used to address anti-replicative oligoribonucleotides into human mitochondria and impact heteroplasmy level. The observed effect was however transient, probably due to a rapid degradation of RNA molecules (Comte et al. 2013). Various chemically modified nucleotides have then been introduced in anti-replicative oligoribonucleotides to improve their stability, namely, nucleotides substituted at the 2′-hydroxyl group with 2′-OMe, 2′-F, and 2′-deoxy in combination with terminus capping chemistry. The most important increase of anti-replicative molecules' lifetime can be achieved by using synthetic RNA–DNA chimeric molecules or by ribose 2′-O-methylation in nuclease-sensitive sites. The presence of "inverted" 3′-3′ thymidine at the 3′-terminus and modifications of 2′-OH ribose moiety did not prevent the mitochondrial uptake of the recombinant molecules. Nevertheless, the modified oligonucleotides did not cause a significant effect on the heteroplasmy level in transfected *trans*mitochondrial cybrid cells bearing a pathogenic mtDNA deletion, proving to be less efficient than nonmodified RNA molecules (Tonin et al. 2014). One can hypothesize that the C3′-endo sugar conformation and 3′-3′ inverted nucleotides might be recognized by the replisome or by other mitochondrial nucleoid proteins as nonnatural and quickly eliminated.

To decrease the toxicity of the cell transfection procedure and create an approach of carrier-free targeting of various anti-replicative RNAs into living human cells, we designed conjugates containing a cholesterol residue. Because cholesterol could stall the mitochondrial import of therapeutic anti-replicative RNA due to attachment to the mitochondrial membranes, we developed the protocol of chemical synthesis of oligoribonucleotides conjugated with cholesterol residue through cleavable covalent bonds. Conjugates containing pH-triggered hydrazone bond (Fig. 6a) were shown to be stable during the cell transfection procedure and rapidly cleaved in acidic endosomal cellular compartments. RNAs conjugated to cholesterol through a hydrazone bond were characterized by efficient carrier-free cellular uptake and partial co-localization with mitochondrial network. Moreover, the imported oligoribonucleotide designed to target a pathogenic point mutation in mitochondrial DNA was able to induce a decrease in the proportion of mutant mitochondrial genomes (Dovydenko et al. 2015).

We suppose that anti-replicative RNA conjugated to cholesterol can be internalized by the endocytosis pathway (Fig. 6b). Thereafter, the hydrazone bond between RNA and cholesterol moieties would be cleaved in the acidic conditions of the late endosomes, and the endosomal escape can be induced by destabilization of endosomal lipid bilayer, due to the positively charged hydrazide group which is formed by the conjugate hydrolysis. Released RNA molecules can be partially degraded in the cytoplasm, but still partially targeted into mitochondria due to the presence of a structural determinant for mitochondrial import.

To improve the in vivo delivery of cholesterol-RNA conjugates, we are planning to design and synthesize conjugated molecules containing various nucleotide and

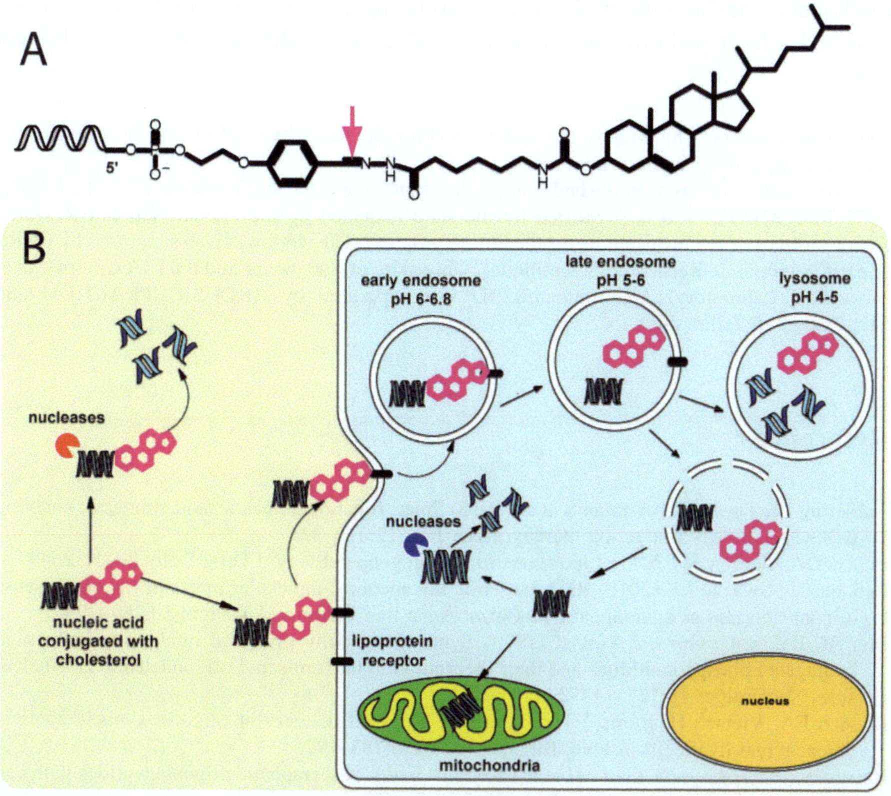

Fig. 6 Cell delivery of RNA conjugated to cholesterol through cleavable linker. (**a**) Oligoribonucleotide conjugated with cholesterol residue through pH-triggered hydrazone bond. The point of the cleavage in acidic conditions is shown by an *arrow*. (**b**) Schematic representation of the cell delivery of RNA conjugated to cholesterol through cleavable linker. RNA is represented by a *helix*; cholesterol residue is highlighted in *purple*. See text for the details

internucleotide bond modifications, which can improve the stability of anti-replicative RNA moieties and promote their tissue distribution and cellular uptake.

5 Conclusion and Future Prospects

As one can deduce from the examples discussed above, there are various nucleotide modifications that can protect RNA molecules introduced into human cells against nucleolytic degradation. Another type of modification, the conjugation of oligonucleotides to the ligands, which can be internalized into the cell by natural transport mechanisms, is a promising approach to overcome the problem of their inefficient delivery to target cells and tissues. Further development of the oligonucleotide modification technology will allow creating novel therapeutic molecules characterized by high stability, low toxicity, and efficient delivery to various human tissues.

Acknowledgments The authors are grateful to Prof. R. Afasizhev and Dr. A. Smirnov for the critical reading of the manuscript.

This work has been published under the framework of the LABEX ANR-11-LABX-0057_MITOCROSS and is supported by the state managed by the French National Research Agency as part of the Investments for the Future program. The work was also supported by CNRS (Centre National de Recherche Scientifique), University of Strasbourg and the LIA (International Associated Laboratory) ARN-mitocure. ID was supported by ARCUS–SUPRACHEM and MitoCross PhD fellowships.

References

Aaldering LJ, Tayeb H, Krishnan S et al (2015) Smart functional nucleic acid chimeras: enabling tissue specific RNA targeting therapy. RNA Biol 12:412–425

Balazs DA, Godbey W (2011) Liposomes for use in gene delivery. J Drug Deliv 2011:326497

Bennett CF, Swayze EE (2010) RNA targeting therapeutics: molecular mechanisms of antisense oligonucleotides as a therapeutic platform. Annu Rev Pharmacol Toxicol 50:259–293

Bhat B, Balow G, Guzaev A et al (1999) Synthesis of fully protected nucleoside- folic acid conjugated phosphoramidites and their incorporation into antisense oligonucleotides. Nucleosides Nucleotides 18:1471–1472

Biessen EA, Vietsch H, Rump ET et al (1999) Targeted delivery of oligodeoxynucleotides to parenchymal liver cells in vivo. Biochem J 340(3):783–792

Boisguerin P, Deshayes S, Gait MJ et al (2015) Delivery of therapeutic oligonucleotides with cell penetrating peptides. Adv Drug Deliv Rev 87:52–67

Boutorine AS, Kostina EV (1993) Reversible covalent attachment of cholesterol to oligodeoxyribonucleotides for studies of the mechanisms of their penetration into eukaryotic cells. Biochimie 75:35–41

Campbell MA, Wengel J (2011) Locked vs. unlocked nucleic acids (LNA vs. UNA): contrasting structures work towards common therapeutic goals. Chem Soc Rev 40:5680–5689

Cesarone G, Edupuganti OP, Chen CP et al (2007) Insulin receptor substrate 1 knockdown in human MCF7 ER+ breast cancer cells by nuclease-resistant IRS1 siRNA conjugated to a disulfide-bridged D-peptide analogue of insulin-like growth factor 1. Bioconjug Chem 18:1831–1840

Chaltin P, Margineanu A, Marchand D et al (2005) Delivery of antisense oligonucleotides using cholesterol-modified sense dendrimers and cationic lipids. Bioconjug Chem 16:827–836

Chen Q, Butler D, Querbes W et al (2010) Lipophilic siRNAs mediate efficient gene silencing in oligodendrocytes with direct CNS delivery. J Control Release 144:227–232

Comte C, Tonin Y, Heckel-Mager AM et al (2013) Mitochondrial targeting of recombinant RNAs modulates the level of a heteroplasmic mutation in human mitochondrial DNA associated with Kearns Sayre Syndrome. Nucleic Acids Res 41:418–433

Davydova AS, Vorobjeva MA, Venyaminova AG (2011) Escort aptamers: new tools for the targeted delivery of therapeutics into cells. Acta Nat 3:12–29

De Haes W, Van Mol G, Merlin C et al (2012) Internalization of mRNA lipoplexes by dendritic cells. Mol Pharm 9:2942–2949

De Mesmaeker A, Lebreton J, Waldner A et al (1994) Replacement of the phosphodiester linkage in oligonucleotides: comparison of two structural amide isomers. Bioorg Med Chem Lett 4:873–878

DiFiglia M, Sena-Esteves M, Chase K et al (2007) Therapeutic silencing of mutant huntingtin with siRNA attenuates striatal and cortical neuropathology and behavioral deficits. Proc Natl Acad Sci USA 104:17204–17209

Dizaj SM, Jafari S, Khosroushahi AY (2014) A sight on the current nanoparticle-based gene delivery vectors. Nanoscale Res Lett 9:252

Dovydenko I, Tarassov I, Venyaminova A et al (2015) Method of carrier-free delivery of therapeutic RNA importable into human mitochondria: lipophilic conjugates with cleavable bonds. Biomaterials 76:408–417

Ginn SL, Alexander IE, Edelstein ML et al (2013) Gene therapy clinical trials worldwide to 2012: an update. J Gene Med 15:65–77

Gryaznov SM, Lloyd DH, Chen JK et al (1995) Oligonucleotide N3′→P5′ phosphoramidates. Proc Natl Acad Sci USA 92:5798–5802

Guzaev AP, Cook PD, Manoharan M et al (2002) Nucleosidic and non-nucleosidic folate conjugates. ISIS Pharmaceuticals Inc. US Patent 6,335,434 B1: 84

Gvozdeva OV, Dovydenko IS, Venyaminova AG et al (2014) 42- and 63-bp anti-MDR1-siRNAs bearing 2′-OMe modifications in nuclease-sensitive sites induce specific and potent gene silencing. FEBS Lett 588:1037–1043

Hall AH, Wan J, Shaughnessy EE et al (2004) RNA interference using boranophosphate siRNAs: structure-activity relationships. Nucleic Acids Res 32:5991–6000

Heidenreich O, Gryaznov S, Nerenberg M (1997) RNase H-independent antisense activity of oligonucleotide N3′→P5′ phosphoramidates. Nucleic Acids Res 25:776–780

Huotari J, Helenius A (2011) Endosome maturation. Embo J 30:3481–3500

Juliano R, Alam MR, Dixit V et al (2008) Mechanisms and strategies for effective delivery of antisense and siRNA oligonucleotides. Nucleic Acids Res 36:4158–4171

Juliano R, Bauman J, Kang H et al (2009) Biological barriers to therapy with antisense and siRNA oligonucleotides. Mol Pharm 6:686–695

Juliano RL, Carver K, Cao C et al (2013) Receptors, endocytosis, and trafficking: the biological basis of targeted delivery of antisense and siRNA oligonucleotides. J Drug Target 21:27–43

Kawasaki AM, Casper MD, Freier SM et al (1993) Uniformly modified 2′-deoxy-2′-fluoro phosphorothioate oligonucleotides as nuclease-resistant antisense compounds with high affinity and specificity for RNA targets. J Med Chem 36:831–841

Kay MA (2011) State-of-the-art gene-based therapies: the road ahead. Nat Rev Genet 12:316–328

Kolesnikova O, Kazakova H, Comte C et al (2011) Selection of RNA aptamers imported into yeast and human mitochondria. RNA 16:926–941

Koshkin AA, Singha SK, Nielsen P et al (1998) LNA (Locked Nucleic Acids): synthesis of the adenine, cytosine, guanine, 5-methylcytosine, thymine and uracil bicyclonucleoside monomers, oligomerisation, and unprecedented nucleic acid recognition. Tetrahedron 54:3607–3630

Kubo T, Zhelev Z, Ohba H et al (2007) Modified 27-nt dsRNAs with dramatically enhanced stability in serum and long-term RNAi activity. Oligonucleotides 17:445–464

Kupryushkin MS, Pyshnyi DV, Stetsenko DA (2014) Phosphoryl guanidines: a new type of nucleic acid analogues. Acta Nat 6:116–118

Lebedeva NA, Anarbaev RO, Kupryushkin MS et al (2015) Design of a new fluorescent oligonucleotide-based assay for a highly specific real-time detection of apurinic/apyrimidinic site cleavage by tyrosyl-DNA phosphodiesterase 1. Bioconjug Chem 26:2046–2053

Letsinger RL, Zhang GR, Sun DK et al (1989) Cholesteryl-conjugated oligonucleotides: synthesis, properties, and activity as inhibitors of replication of human immunodeficiency virus in cell culture. Proc Natl Acad Sci USA 86:6553–6556

Lorenz C, Hadwiger P, John M et al (2004) Steroid and lipid conjugates of siRNAs to enhance cellular uptake and gene silencing in liver cells. Bioorg Med Chem Lett 14:4975–4977

MacKellar C, Graham D, Will DW et al (1992) Synthesis and physical properties of anti-HIV antisense oligonucleotides bearing terminal lipophilic groups. Nucleic Acids Res 20:3411–3417

Manoharan M, Kesavan V, Rajeev KG (2005) Modified iRNA agents. Fish and Richardson Paper Corporation Boston MA. 20050107325 A1, USA, pp 245

Maxfield FR, Wustner D (2013) Analysis of cholesterol trafficking with fluorescent probes. Methods Cell Biol 108:367–393

McGuire MJ, Gray BP, Li S et al (2014) Identification and characterization of a suite of tumor targeting peptides for non-small cell lung cancer. Sci Rep 4:4480

Meade BR, Gogoi K, Hamil AS et al (2014) Efficient delivery of RNAi prodrugs containing reversible charge-neutralizing phosphotriester backbone modifications. Nat Biotechnol 32:1256–1261

Milligan JF, Matteucci MD, Martin JC (1993) Current concepts in antisense drug design. J Med Chem 36:1923–1937

Ming X, Laing B (2015) Bioconjugates for targeted delivery of therapeutic oligonucleotides. Adv Drug Deliv Rev 87:81–89

Ming X, Alam MR, Fisher M et al (2010) Intracellular delivery of an antisense oligonucleotide via endocytosis of a G protein-coupled receptor. Nucleic Acids Res 38:6567–6576

Moschos SA, Jones SW, Perry MM et al (2007) Lung delivery studies using siRNA conjugated to TAT (48–60) and penetratin reveal peptide induced reduction in gene expression and induction of innate immunity. Bioconjug Chem 18:1450–1459

Mutisya D, Selvam C, Lunstad BD et al (2014) Amides are excellent mimics of phosphate internucleoside linkages and are well tolerated in short interfering RNAs. Nucleic Acids Res 42:6542–6551

Nair JK, Willoughby JL, Chan A et al (2014) Multivalent N-acetylgalactosamine conjugated siRNA localizes in hepatocytes and elicits robust RNAi-mediated gene silencing. J Am Chem Soc 136:16958–16961

Oberhauser B, Wagner E (1992) Effective incorporation of 2′-O-methyl-oligoribonucleotides into liposomes and enhanced cell association through modification with thiocholesterol. Nucleic Acids Res 20:533–538

Oliveira C, Silveira I, Veiga F et al (2015) Recent advances in characterization of nonviral vectors for delivery of nucleic acids: impact on their biological performance. Expert Opin Drug Deliv 12:27–39

Petrova NS, Chernikov IV, Meschaninova MI et al (2011) Carrier-free cellular uptake and the gene-silencing activity of the lipophilic siRNAs is strongly affected by the length of the linker between siRNA and lipophilic group. Nucleic Acids Res 40:2330–2344

Pinto M, Moraes CT (2014) Mitochondrial genome changes and neurodegenerative diseases. Biochim Biophys Acta 1842:1198–1207

Poopeiko NE, Dahl BM, Wengel J (2003) Xylo-configured oligonucleotides (XNA, xylo nucleic acids): synthesis and hybridization studies. Nucleosides Nucleotides Nucleic Acids 22:1147–1149

Prakash TP, Graham MJ, Yu J et al (2014) Targeted delivery of antisense oligonucleotides to hepatocytes using triantennary N-acetyl galactosamine improves potency 10-fold in mice. Nucleic Acids Res 42:8796–8807

Raouane M, Desmaele D, Urbinati G et al (2012) Lipid conjugated oligonucleotides: a useful strategy for delivery. Bioconjug Chem 23:1091–1104

Reed MW, Adams AD, Nelson JS et al (1991) Acridine- and cholesterol-derivatized solid supports for improved synthesis of 3′-modified oligonucleotides. Bioconjug Chem 2:217–225

Rump ET, de Vrueh RL, Sliedregt LA et al (1998) Preparation of conjugates of oligodeoxynucleotides and lipid structures and their interaction with low-density lipoprotein. Bioconjug Chem 9:341–349

Seo YJ, Jeong HS, Bang EK et al (2006) Cholesterol-linked fluorescent molecular beacons with enhanced cell permeability. Bioconjug Chem 17:1151–1155

Shabi S, Sajjad K, Arif A (2006) Peptide nucleic acid (PNA): a review. J Chem Technol Biotechnol 81:892–899

Sheehan D, Lunstad B, Yamada CM et al (2003) Biochemical properties of phosphonoacetate and thiophosphonoacetate oligodeoxyribonucleotides. Nucleic Acids Res 31:4109–4118
Soutschek J, Akinc A, Bramlage B et al (2004) Therapeutic silencing of an endogenous gene by systemic administration of modified siRNAs. Nature 432:173–178
Stetsenko DA, Gait MJ (2001) A convenient solid-phase method for synthesis of 3′-conjugates of oligonucleotides. Bioconjug Chem 12:576–586
Stockert RJ (1995) The asialoglycoprotein receptor: relationships between structure, function, and expression. Physiol Rev 75:591–609
Summerton J, Weller D (1997) Morpholino antisense oligomers: design, preparation, and properties. Antisense Nucleic Acid Drug Dev 7:187–195
Tan W, Wang H, Chen Y et al (2011) Molecular aptamers for drug delivery. Trends Biotechnol 29:634–640
Thomas CE, Ehrhardt A, Kay MA (2003) Progress and problems with the use of viral vectors for gene therapy. Nat Rev Genet 4:346–358
Tonin Y, Heckel AM, Dovydenko I et al (2014) Characterization of chemically modified oligonucleotides targeting a pathogenic mutation in human mitochondrial DNA. Biochimie 100:192–199
Tuerk C, Gold L (1990) Systematic evolution of ligands by exponential enrichment: RNA ligands to bacteriophage T4 DNA polymerase. Science 249:505–510
Uckun FM, Qazi S, Dibirdik I et al (2013) Rational design of an immunoconjugate for selective knock-down of leukemia-specific E2A-PBX1 fusion gene expression in human Pre-B leukemia. Integr Biol (Camb) 5:122–132
Ueno Y, Kawada K, Naito T et al (2008) Synthesis and silencing properties of siRNAs possessing lipophilic groups at their 3′-termini. Bioorg Med Chem 16:7698–7704
Varkouhi AK, Scholte M, Storm G et al (2011) Endosomal escape pathways for delivery of biologicals. J Control Release 151:220–228
Vives E, Schmidt J, Pelegrin A (2008) Cell-penetrating and cell-targeting peptides in drug delivery. Biochim Biophys Acta 1786:126–138
Volkov AA, Kruglova NS, Meschaninova MI et al (2009) Selective protection of nuclease-sensitive sites in siRNA prolongs silencing effect. Oligonucleotides 19:191–202
Vu H, Singh P, Joyce N et al (1993) Synthesis of cholesteryl supports and phosphoramidites containing a novel peptidyl linker for automated synthesis of triple-helix forming oligonucleotides (TFOs). Nucleic Acids Symp Ser:19–20
Walther W, Stein U (2000) Viral vectors for gene transfer: a review of their use in the treatment of human diseases. Drugs 60:249–271
Wickstrom E (2015) DNA and RNA derivatives to optimize distribution and delivery. Adv Drug Deliv Rev 87:25–34
Winkler J (2013) Oligonucleotide conjugates for therapeutic applications. Ther Deliv 4:791–809
Wolfrum C, Shi S, Jayaprakash KN et al (2007) Mechanisms and optimization of in vivo delivery of lipophilic siRNAs. Nat Biotechnol 25:1149–1157
Zhou J, Rossi JJ (2011) Cell-specific aptamer-mediated targeted drug delivery. Oligonucleotides 21:1–10

Interstrand Cross-Linking of Nucleic Acids: From History to Recent and Future Applications

Ellen Gyssels, Nathalie De Laet, Emily Lumley, and Annemieke Madder

Contents

E. Gyssels • N. De Laet • E. Lumley • A. Madder (✉)
Department of Organic and Macromolecular Chemistry, Organic and Biomimetic Chemistry Research Group, Ghent University, Krijgslaan 281, S4, 9000 Ghent, Belgium
e-mail: annemieke.madder@ugent.be

S. Jurga et al. (eds.), *Modified Nucleic Acids in Biology and Medicine*, RNA Technologies, DOI 10.1007/978-3-319-34175-0_15

Abstract This chapter contains an elaborate overview of various methods that have been developed for interstrand cross-linking of nucleic acids. The existing cross-link methodologies can be subcategorized in different groups, each interesting depending on the requirements of the envisaged application. An endogenous or exogenous bifunctional compound can react with two different nucleophilic groups on the nucleobases, resulting in fast and high-yielding cross-linking. However, when site selectivity of the formed cross-link is desired, other approaches are required. Therefore, a series of methodologies are at hand where functionalities are introduced in oligonucleotide probes, which can be intrinsically reactive or can be triggered and activated at a selected time upon target recognition. In a third class, both strands are modified with orthogonal groups. After reaction of the introduced functionalities, a cross-link is formed at a specific and fixed position. The chapter ends with an overview of and outlook to present and future applications. It is thus shown that cross-linking agents can be exploited as useful therapeutic or diagnostic tools and have further proven their utility in the stabilization or structure elucidation of cross-linked systems aimed at studying and understanding the repair systems in a cell.

Keywords Nucleic acids • Oligonucleotide probes • Interstrand cross-linking • Oligonucleotide therapeutics • Exogenous cross-linkers

1 Introduction

An interstrand cross-link (ICL), a covalent ligation between two complementary nucleic acid strands, can be introduced by a variety of exogenous or endogenous agents. The presence of these ICLs in DNA can prevent cell replication, transcription, or translation. After their formation, ICL repair mechanisms are initiated which act to repair the strands; however, due to the need for both strands to be repaired, interstrand cross-linking agents are still genotoxic and, therefore, popular as anticancer therapeutics (Noll et al. 2006). Due to the impact of ICLs on cell viability, a wealth of researchers have devoted their attention to the preparation and analysis of interstrand cross-linked duplexes, and a series of methods have been developed to access such structures in an efficient and site-selective way. Consequently, applications of cross-linkable oligonucleotides in therapeutics and nanotechnology or to study or reveal cellular pathways have been established.

Here, we present an overview of methods to achieve nucleic acid interstrand cross-linking developed to date. We have opted for a systematic overview of methodologies structured according to the conceptual mechanistic difference for obtaining the interstrand cross-linked (ICL) species, as represented by the cartoon in Fig. 1.

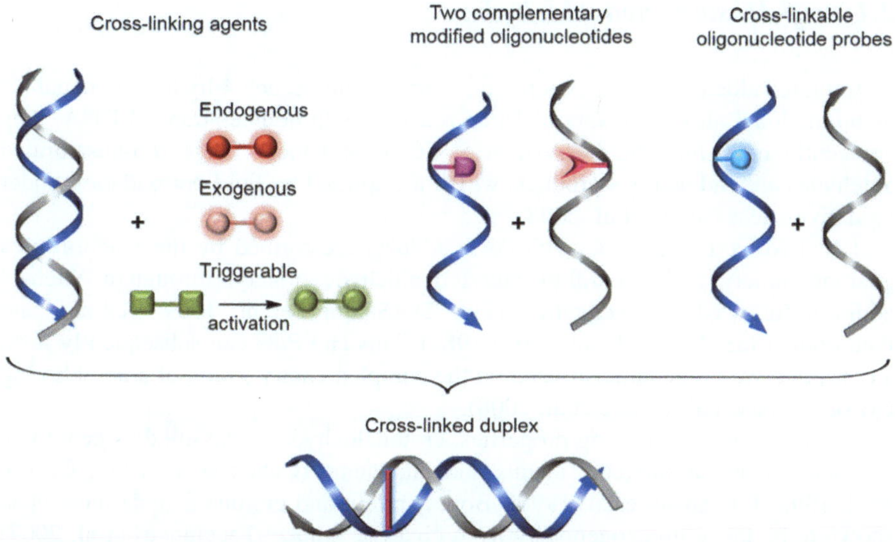

Fig. 1 Schematic overview of different cross-link strategies

In addition to endogenously present cross-linking agents (Sect. 2), reactive bifunctional species can be externally added to duplexes to achieve ICL formation (Sect. 3). Triggerable or activatable agents have been developed which are able to fine-tune and augment selectivity. Furthermore, introduction of those moieties into single-stranded probes allows targeting specific nucleic acid targets as discussed in Sect. 4. In view of material science applications, we review methods for the covalent connection of complementary oligonucleotides modified with matching chemical probes (Sect. 5) or the construction and solid-phase-based incorporation of cross-linked nucleoside building blocks (Sect. 6). Finally an overview of potential applications is given in Sect. 7. For a more in-depth discussion on mechanistic aspects and specific applications, we would like to refer to important reviews on this topic (Rajski and Williams 1998; Noll et al. 2006; Efimov and Fedyunin 2010). Due to the extensive, ongoing research in this area, our aim is to provide a broad overview of cross-linking methodology in nucleic acids, developed to date.

2 Endogenous Cross-Linking Agents

Compounds possessing double functionalities can react with two corresponding groups positioned nearby. When these groups are present on complementary oligonucleotide strands, a covalent bond can be formed: an interstrand cross-link. Compounds of this type that are found naturally or generated inside the body, able to induce ICL formation, are categorized as endogenous cross-linking agents.

2.1 *α,β-Unsaturated Aldehydes*

Unsaturated aldehydes such as acrolein, crotonaldehyde, and 4-hydroxynonenal act as bifunctional alkylating agents. Upon reaction with double-stranded DNA, they covalently cross-link the DNA in a 5′-CG-3′ sequence. These α,β-unsaturated aldehydes are endogenous products which are formed by lipid peroxidation under oxidative stress (Stone et al. 2008).

1,N^2-hydroxypropano-dG (HO-PdG) adducts are formed by the reaction of a guanine moiety with an α,β-unsaturated aldehyde moiety, through a Michael addition followed by cyclization (Fig. 2) (Shapiro et al. 1969; Galliani and Pantarotto 1983; Leonard and Barrio 1984). This HO-PdG can subsequently form a cross-link with a guanine situated on the complementary strand of a neighboring CG or GC base pair (Stone et al. 2008).

Due to these cross-linking properties, crotonaldehyde is classified as genotoxic and acrolein is carcinogenic in rats and mutagenic (Cohen et al. 1992; Czerny et al. 1998; Kawanishi et al. 1998). Both acrolein and crotonaldehyde have been shown to be taken up exogenously from cigarette smoke (Facchinetti et al. 2007) and automobile exhaust gases (Nishikawa et al. 1987).

2.2 *Nitric Oxide*

Nitric oxide is formed endogenously during nitric oxide synthase-catalyzed oxidation of L-arginine, and its metabolites can react with nucleic acids (Berka et al. 2004; Beda and Nedospasov 2006). One of the major modifications which is found after reaction of N_2O_3 with DNA is the deamination of bases with an exocyclic amino group (A, G, and C), resulting in mutations and transitions in the sequence. DNA interstrand cross-links, intrastrand cross-links, and DNA-protein cross-links have been found after reaction with nitrous acid (Shapiro et al. 1977;

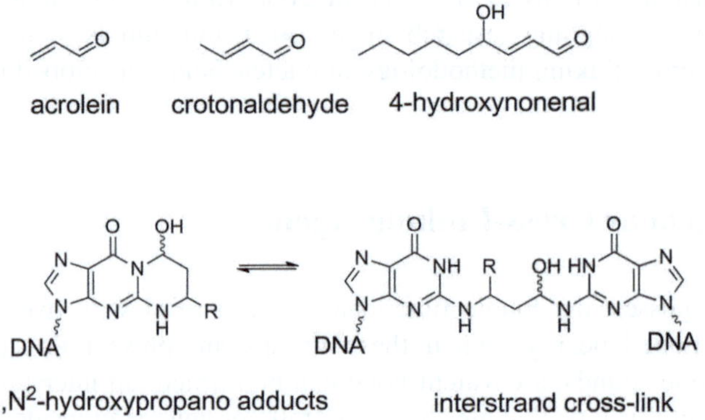

Fig. 2 Examples of α, β-unsaturated aldehydes and their cross-link formation

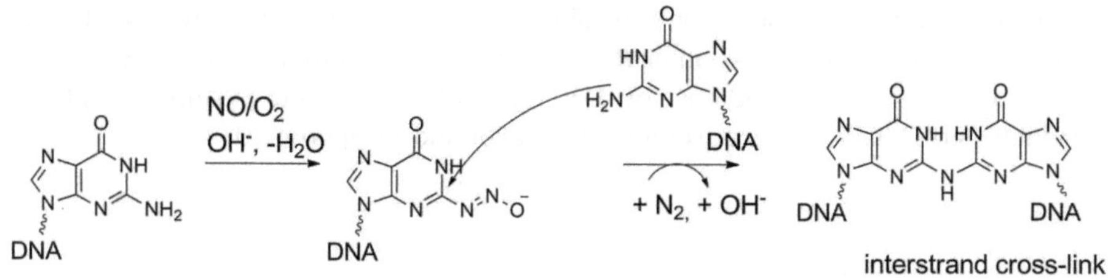

Fig. 3 G-G cross-link formed after nitric oxide treatment

Suzuki et al. 2000). However, G-G cross-links (Fig. 3) are also formed after reaction with endogenous nitric oxide in aerobic conditions (Caulfield et al. 2003; Beda and Nedospasov 2007).

Inspired by this natural toxicity of nitrous acid, a solid-phase synthesis route was developed for the production of nitrous acid cross-linked DNA by Hopkins et al. (Harwood et al. 1999b).

2.3 *Formaldehyde*

Formaldehyde is a widely used carcinogenic chemical and is present in cigarette smoke, automobile exhaust gases, and smog (Swenberg et al. 2013). Next to this exogenous occurrence of formaldehyde, it also has many endogenous sources (metabolite of amino acids, lipid peroxidation, metabolite of methanol, demethylation with P450). Formaldehyde can react with the nucleophilic groups present on amino acids (mainly lysine, cysteine, and tryptophan) (Klockenbusch et al. 2012) and on nucleotides (mainly dG) (Swenberg et al. 2013). Upon further reaction of the formed adducts, cross-linked adducts can be formed, however, mainly encompassing DNA-protein, RNA-protein, and protein-protein cross-links rather than DNA interstrand cross-links. As a result, formaldehyde treatment is routinely used in chromatin immunoprecipitation (ChIP) to investigate the interaction between proteins and DNA (Orlando et al. 1997).

2.4 *Pyrimidine Dimers*

Under influence of UV light, pyrimidine CC and TT dimers can be formed in DNA. Due to the presence of sufficiently energetic UVB radiation in sunlight, photochemically induced [2 + 2] dimerization of pyrimidines can be initiated. When a CC dimer is mispaired to two adenines, a CC to TT mutation is introduced in replication of the dimer (Goodsell 2001). These mutations are most often found in the p53 tumor suppressor gene in skin cancers. UVA is also photo-carcinogenic and is involved in photoaging but is only weakly absorbed by DNA and proteins. Examples of DNA photoproducts, other than the most commonly occurring pyrimidine dimers that are directly induced by UV light, are pyrimidine-pyrimidone photoproducts and Dewar valence isomers (El-Yazbi and Loppnow 2014).

2.5 Abasic Sites

Abasic sites are prevalent lesions lacking a nucleobase on the 1′-position of the sugar moiety (Greenberg 2014). They are formed as an intermediate during DNA repair and are produced in every cell 2000–10,000 times per day (Lindahl 1993). Abasic sites lack hydrogen bonding, causing destabilization of the duplex structure and imparting an electrophilic character to them. The reactivity of these lesions can be accredited to the formation of an aldehyde moiety, present only in a minor amount (1 %) through the equilibrium between the closed and open ring form. This aldehyde is then available to react with the N^2-position of dG to form an interstrand cross-link (Fig. 4) (Dutta et al. 2007; Johnson et al. 2013; Catalano et al. 2015). The interstrand cross-link is generally only formed when the abasic site follows a C in the sequence (5′-C-abasic-3′). Very recently, however, it has been shown that adenine was also able to attack the aldehyde, resulting in interstrand cross-link formation (Price et al. 2014).

Inspired by the natural reactivity of these abasic sites, the group of Gates recently developed a high-yielding synthetic route to cross-linked duplexes. Here, two complementary non-modified oligonucleotide strands are post-synthetically modified, one to yield an abasic site and the other an N^4-amino-2′-deoxycytidine. Attack of the exocyclic amine functionality of this modified cytidine moiety on the abasic site yields the formation of an ICL (Gamboa Varela and Gates 2015). An alternative approach taken by Komatsu and co-workers exploits the formation of two abasic sites on opposite positions of complementary strands. When adding a

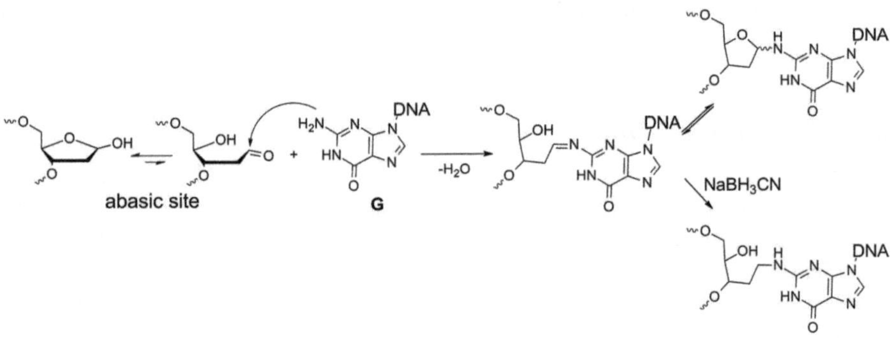

Fig. 4 Interstrand cross-link formation with abasic sites

bifunctional aminooxy derivative, reaction with both abasic sites leads to the formation of an ICL (Ichikawa et al. 2012).

3 Exogenous Cross-Linking Agents

In the following, an overview of exogenously added agents which cause cross-linking is discussed. Often, these agents can be or have been applied as chemotherapeutics as it has been shown that such cross-linking treatment can interfere with cell replication and ultimately cause cell death (Rajski and Williams 1998).

3.1 *Inherently Reactive, Non-Triggerable Reagents*

3.1.1 Nitrogen Mustards and Derivatives

Nitrogen mustards comprise reactive agents characterized by the presence of chloroethyl side chains on a nitrogen (Fig. 5a). By intramolecular alkylation the chloroethyl group is transformed into a reactive aziridinium ion, susceptible to nucleophilic attack of the N^7 of a guanine residue. Formation of a second aziridinium ion can then occur, which is capable of reacting with a guanine of the complementary strand, thus forming an interstrand cross-link (Fig. 5b) (Brookes and Lawley 1961).

Mustard gases were used as a toxin in the First World War and have since been found to possess chemotherapeutic properties. The use of mustine as an anticancer agent was first proposed in 1946 (Gilman and Philips 1946) and is still used in chemotherapy today. Some other important nitrogen mustards which are used as drugs are presented in Fig. 5a.

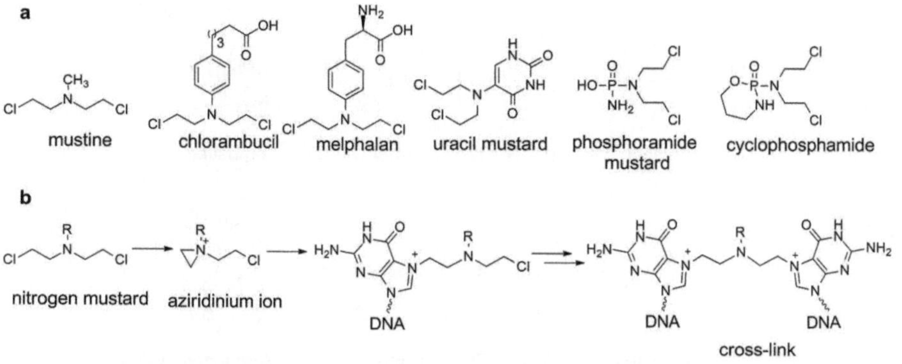

Fig. 5 (**a**) Some important nitrogen mustards and (**b**) their general mechanism of interstrand cross-link formation

3.1.2 Chloroethylating Agents

Unlike the alkylating agents discussed above, chloroethylating agents introduce only a small ethylene linkage between two bases. Initially, a single base of the DNA strand is modified with a chloroethyl group, and although mechanistic uncertainties exist, it is widely assumed that, upon internal cyclization, interstrand cross-linking then occurs, involving a base of the opposite DNA strand as represented in Fig. 6.

3.1.3 Nitrosoureas

Chloroethylnitrosoureas or methylnitrosoureas are commonly used to treat brain tumors. These agents can form an interstrand cross-link in a CG Watson-Crick base pair (see Fig. 6) (Fischhaber et al. 1999). The chemical structures of the most important nitrosoureas are presented in Fig. 7. N,N′-Bis(2-chloroethyl)-nitrosourea (BCNU) or carmustine is a drug approved to treat brain cancer, multiple myeloma, and lymphoma.

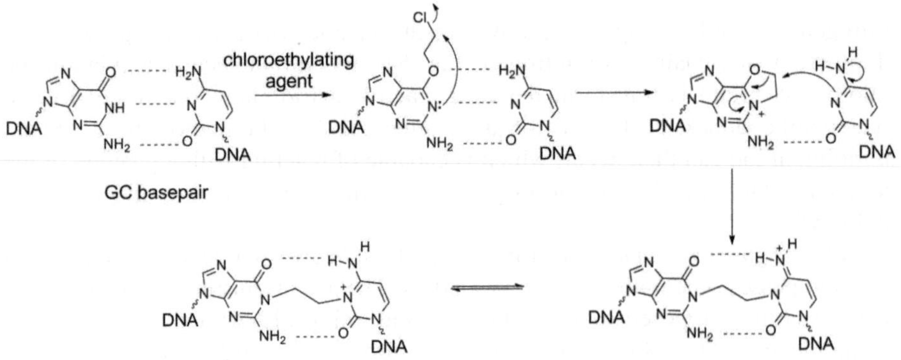

Fig. 6 Proposed mechanism of interstrand cross-linking through reaction with a chloroethylating agent

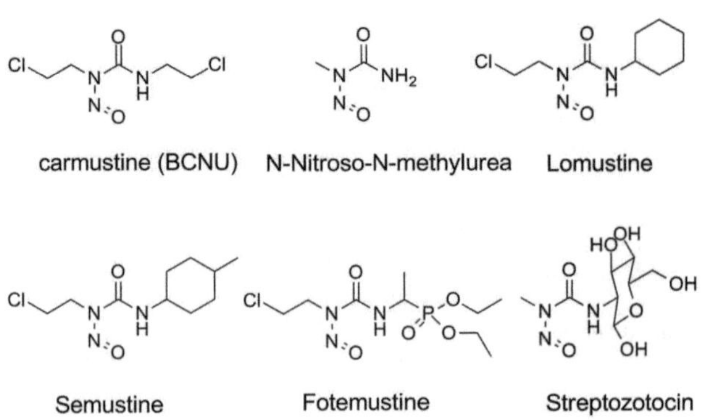

Fig. 7 Some examples of nitrosoureas

3.1.4 Triazene Derivatives

Triazenes consist of an amine directly connected to an azo group ($R^1R^2N\text{-}N{=}NR^3$). Triazenes are known to alkylate the N^7- or O^6-position of a guanosine. Formation of N^7-chloroethylguanosine can lead to formation of a GG intrastrand cross-link, whereas O^6-chloroethylguanosine can generate a CG interstrand cross-link (Tong and Ludlum 1981; Smith et al. 1996). Two commonly used triazene chemotherapeutics are temozolomide and dacarbazine (Fig. 8). Both drugs act through methylation of the guanosine of DNA at N^7 or O^6, thus resulting in apoptosis without interstrand cross-link formation (Sanada et al. 2007).

3.1.5 Epoxides

Epoxides are capable of reacting with DNA and follow a mechanism similar to the nitrogen mustards (Millard and White 1993). Introduction of an interstrand cross-link in a 5′-GC-3′ or 5′-CG-3′ sequence is realized through a reaction with the N^7-position of guanosine. Diepoxybutane is the most simple epoxide cross-linking agent (Fig. 9). Carzinophilin, also called azinomycin B, is a more complex chemotherapeutic agent combining an epoxide and a strained aziridine for reaction with DNA.

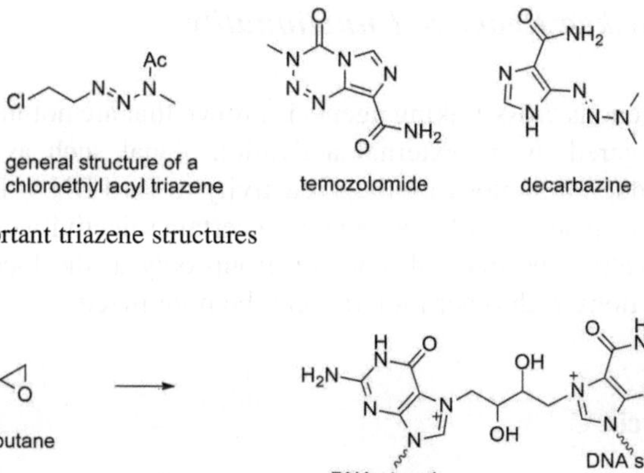

Fig. 8 Some important triazene structures

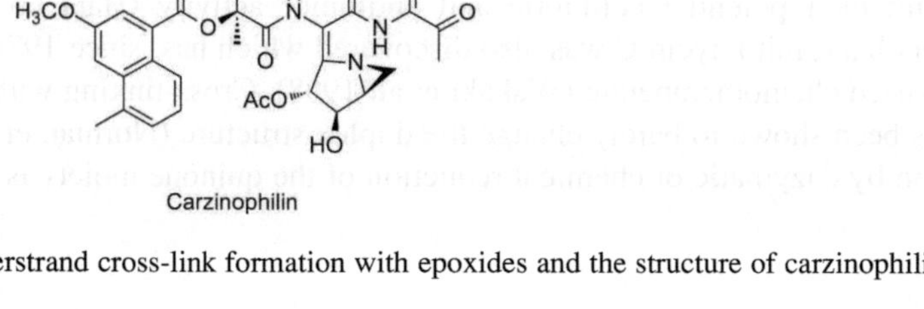

Fig. 9 Interstrand cross-link formation with epoxides and the structure of carzinophilin

Fig. 10 Structure of commonly used platinum complexes as cross-linking agents

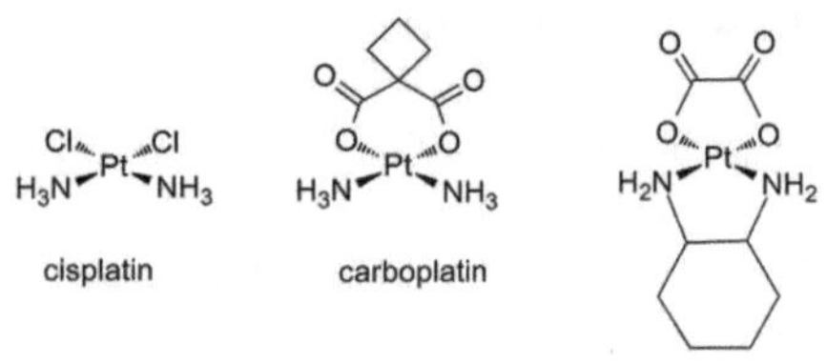

3.1.6 Cisplatin and Other Transition Metal Complexes

Cisplatin was the first transition metal complex to be used as a chemotherapeutic agent in 1979 and is used to date to treat various cancers (e.g., bladder cancer, cervical cancer, ovarian cancer, and testicular cancer, among others). It is capable of forming both intrastrand and interstrand cross-links (5′-GC-3′). The chloride atoms can be displaced by water, followed by nucleotide binding at the N^7-position of guanine (Reishus and Martin 1961). The stereoisomer, transplatin, also shows toxicity but has no antitumor activity. To diminish the side effects of cisplatin chemotherapy, various derivatives have been developed and approved, the most important of which are presented in Fig. 10.

3.2 Triggerable Exogenous Cross-Linking Agents with Masked Reactive Functionality

A series of exogenous cross-linking agents is known that are not inherently reactive but can be triggered by an external activation signal such as UV irradiation, oxidation, or reduction, unlocking their reactivity toward DNA. Due to the possibility to induce the reactivity of these agents, a certain selectivity can be introduced. Indeed, by activating the masked reactive group only at the location of interest, nonspecific reactions with other moieties can be minimized.

3.2.1 Mitomycin C

Mitomycins are natural products, containing an aziridine moiety, which are isolated from *Streptomyces caespitosus*. Mitomycins A and B were discovered in 1956 along with their potential antibiotic and antitumor activity (Hata et al. 1956). Two years later, mitomycin C was also discovered which has, since 1974, become a widely used chemotherapeutic (Wakaki et al. 1958). Cross-linking with mitomycin C has been shown to barely change the duplex structure (Norman et al. 1990). Activation by enzymatic or chemical reduction of the quinone moiety is needed to induce reaction with DNA (Fig. 11). Subsequently, a cascade of reactions takes place, ultimately leading to interstrand cross-link formation. Although the interstrand cross-link is most abundant, the intrastrand cross-link product or the monoadduct can also be formed (Tomasz 1995). Modifications of these mitomycins have been considered for the development of more selective or efficient cross-linkers. Among these, dimers of mitomycin-like structures have been proven to impart regioselectivity (Paz et al. 2004).

3.2.2 Psoralens

Psoralens are natural products of the furocoumarin family, structurally related to the coumarins. The most common psoralens are presented in Fig. 12. Due to their flat tricyclic structure, they can intercalate in dsDNA and upon activation by UV irradiation produce a cross-link by a [2+2] cycloaddition between two thymine moieties in a 5′-TA-3′ or a 5′-AT-3′ region. Psoralen-induced cross-linking can be reversible by tuning the wavelength of irradiation (Decout and Lhomme 1992).

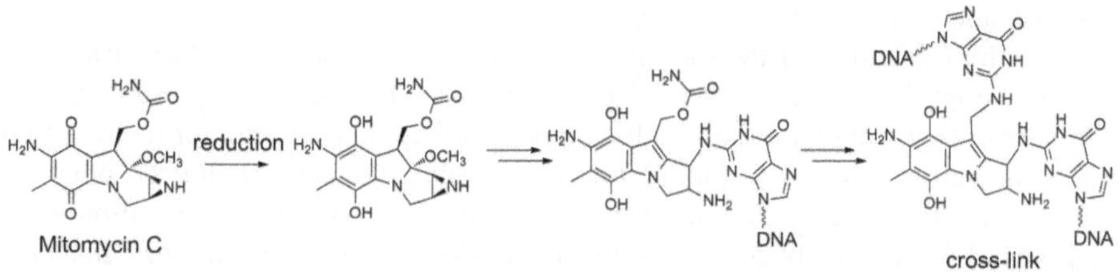

Fig. 11 Cross-link formation with mitomycin C

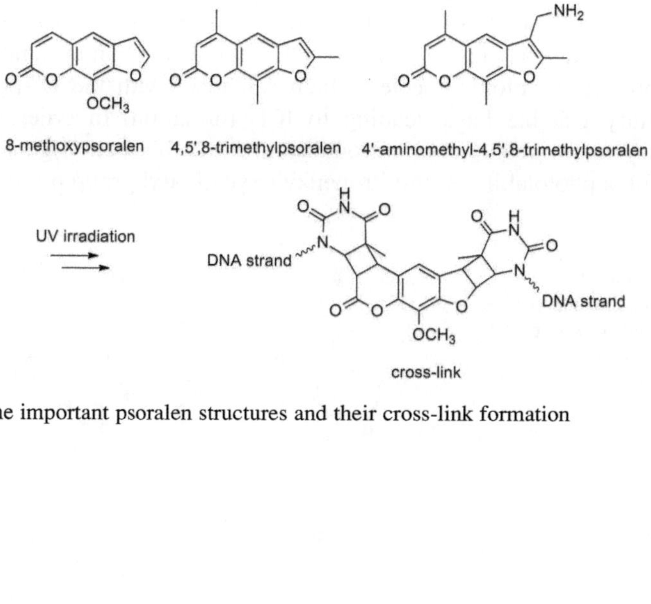

Fig. 12 Some important psoralen structures and their cross-link formation

4 Cross-Linkable Oligonucleotide Probes

Introduction of reactive moieties within the oligonucleotide sequence through the synthesis of unnatural nucleotides allows the formation of cross-links between complementary pairs. In the following section, various approaches to this methodology are presented, classified according to various chemical reactivities.

4.1 *Oligonucleotide Probes with Intrinsic S_N2 Reactivity*

4.1.1 Oligonucleotides Modified with Halogenated Building Blocks

The presence of a halomethyl moiety within an oligonucleotide creates an electrophilic center sensitive to nucleophilic attack. Coleman and co-workers developed a method for modifying an oligonucleotide sequence with 4-thio-2′-deoxyuridine. After reacting this moiety with a bifunctional electrophilic group containing an α-halo carbonyl function, an electrophilic center is generated. Various building blocks were synthesized (Coleman and Kesicki 1995; Coleman and Pires 1997), of which a selection is presented in Fig. 13.

After hybridization of the modified sequence with a complementary strand, an S_N2 nucleophilic attack of the N^7-position of a guanine base, present on the complementary strand, on the α-halocarbonyl moiety, leads to the formation of an interstrand cross-link. Cross-linking not only involves the guanine directly opposite the modified building block, but other guanine moieties present on the complementary strand have also been shown to be involved in ICL formation. This lack of selectivity, next to the ICL yield, was shown to depend on the bifunctional linker used and the exact oligonucleotide sequence.

Incorporation of a chloroethyl-modified thymine base in an oligonucleotide sequence was described by Schärer et al. (Alzeer and Schärer 2006). The proposed mechanistic pathway, as presented in Fig. 14, is initiated by an intramolecular S_N2 reaction, forming a cationic species which can react with the O^6-position of a complementary guanine base, leading to ICL formation. In order to obtain a triggerable system, Luedtke et al. introduced a O^6-(2-chloroethyl)guanine moiety, modified with a photolabile ortho-nitrobenzyloxycarbonyl group present on the N^2

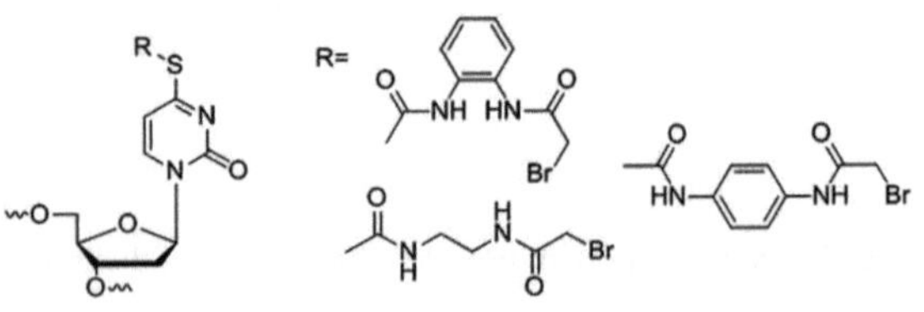

Fig. 13 Selection of building blocks developed by Coleman and co-workers for cross-linking of oligonucleotides

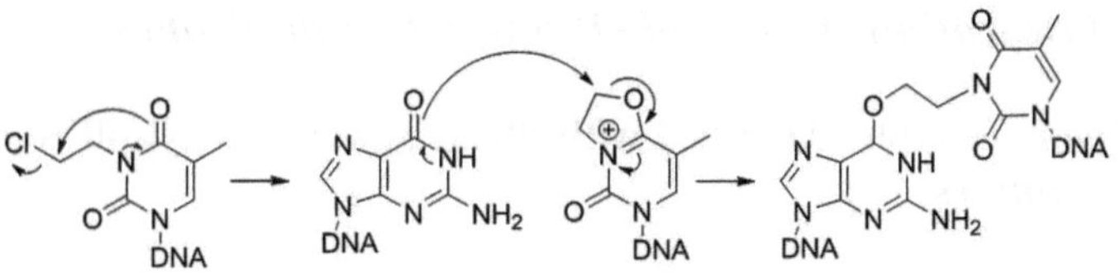

Fig. 14 Reaction of chloroethyl thymine with a complementary guanine base leading to ICL formation

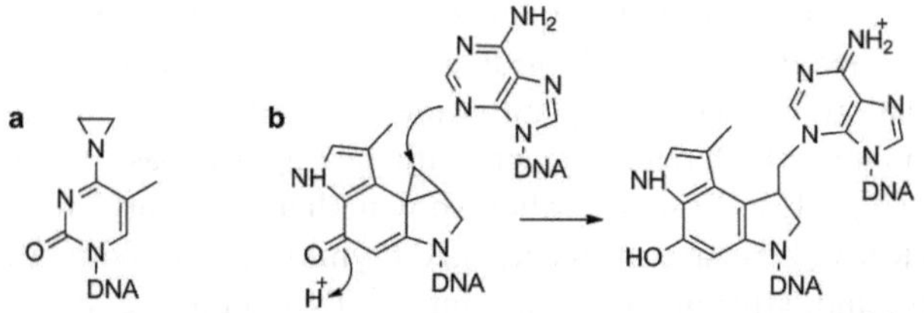

Fig. 15 (**a**) 5-Methyl cytidine modified with an aziridine moiety, (**b**) reaction of the N^7-position of an adenine base with the cyclopropyl unit on a cyclopropapyrroloindole

nitrogen. Upon UV irradiation, the protecting group is cleaved, allowing the nitrogen to undergo an intramolecular S_N2 reaction (Hentschel et al. 2012).

4.1.2 Oligonucleotides Modified with Strained Ring Systems

Oligonucleotides modified with an electrophilic moiety in the form of strained ring systems or heterocycles are capable of inducing interstrand cross-links, through attack of nucleophilic sites present on the opposite strand. It was shown that cross-linking to the opposite strand occurred when 5-methyl-cytosine (Fig. 15a) was modified with an aziridine ring (Webb and Matteucci 1986a, b). Incorporation of a more complex strained ring system has also been demonstrated by conjugating an oligonucleotide sequence with a cyclopropapyrroloindole unit which contains an electrophilic cyclopropyl moiety. After hybridization with the complementary strand, nucleophilic attack of the N^7-position of an adenosine moiety leads to the formation of a cross-link (Fig. 15b) (Lukhtanov et al. 1996, 1997).

The previously described alkylating properties of mitomycin C also resulted in the exploration of its conjugation ability to an oligonucleotide sequence to improve sequence selectivity. Tomasz and co-workers incorporated a mitomycin derivative at the 5′-end of a synthetic oligonucleotide sequence by attack of a nearby 2-amino group of guanine, present in the minor groove. Reduction of the mitomycin by NADPH cytochrome C reductase or NADH cytochrome C reductase (vide supra) resulted in successful ICL formation with complementary strands (Maruenda and Tomasz 1996).

4.2 Hybridization-Activated Oligonucleotide Probes

4.2.1 Oligonucleotides Modified with Phenylsulfide/Phenylsulfoxide Nucleotides

The group of Sasaki reported on the synthesis of phenylsulfide derivatives of purines and their incorporation in oligonucleotide sequences. Such derivatives have been shown to selectively form cross-links with an opposite cytosine present on the hybridized complementary strand (Nagatsugi et al. 1999; Kawasaki et al. 2005). These modified sequences were activated by oxidation to the corresponding phenylsulfoxide. Hybridization with the complementary strand stimulates elimination of the phenylsulfoxide and formation of the 2-amino-6-vinylpurine moiety, which readily reacts with the opposite base, showing selectivity for cytosine (Fig. 16). It was further shown that this vinyl methodology was applicable in RNA chemistry. A 4-vinylated thymine and a 4-vinylated deoxyuracil were designed, and, after incorporation into a DNA sequence, cross-linking to the complementary RNA sequence was observed. Preference for cross-linking to an opposite uracil base in case of the modified thymine (and competitive cross-linking to flanking adenine bases when present as neighboring bases of the uracil moiety) and virtually complete selectivity for a uracil base in case of the modified deoxyuracil was observed.

Furthermore, Nagatsugi and co-workers described the synthesis and incorporation of 6-vinylpurine in an oligonucleotide sequence, demonstrating that removal of the 2-amino group of the purine moiety enabled faster cross-linking (Imoto et al. 2012). Recently, the same group synthesized a novel nucleobase, a 2-amino-6-(1-ethylthiovinyl) purine, which could be oxidized using the biologically relevant hydrogen peroxide and $FeCl_2$, forming an ICL with a complementary RNA sequence (Kusano et al. 2014).

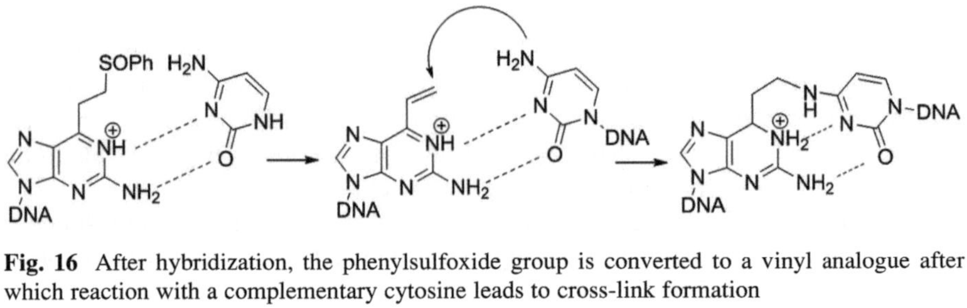

Fig. 16 After hybridization, the phenylsulfoxide group is converted to a vinyl analogue after which reaction with a complementary cytosine leads to cross-link formation

4.3 *Photo-Activatable Oligonucleotide Probes*

4.3.1 Oligonucleotides Modified with Halogenated Building Blocks

Inspired by the applicability of 5-brominated deoxyuridine as a radiosensitizing agent in cells, the group of Hunting investigated the effect of this moiety on the formation of interstrand cross-links between oligonucleotides. After irradiation of this halogenated nucleoside with UV or γ-irradiation, a very reactive σ-radical is formed (Fig. 17), which was observed to react with the complementary strand when present in a bulge region, resulting in the formation of an ICL (Cecchini et al. 2005; Dextraze et al. 2009). To our knowledge, the exact structure of the formed cross-link has not yet been specified.

The group of Greenberg demonstrated that the elimination of hydrogen bonding to the complementary base, through replacement of the uridine by a substituted phenyl functionality, made ICL formation possible in regions that lacked the bulge area (Hou and Greenberg 2014). The different aryl halide building blocks are presented in Fig. 18. Cross-linking to all four complementary bases was observed in competition with cross-linking to flanking thymidine bases. The obtained ICL yields were dependent on the substitution pattern of the building block and the method of activation (i.e., UV or γ-irradiation).

4.3.2 Oligonucleotides Modified with Phenyl Selenium Nucleotides

When cells are subjected to oxidative stress, various sorts of DNA lesions can arise, including the formation of a 5-(2′-deoxyuracil) methyl radical. The group of

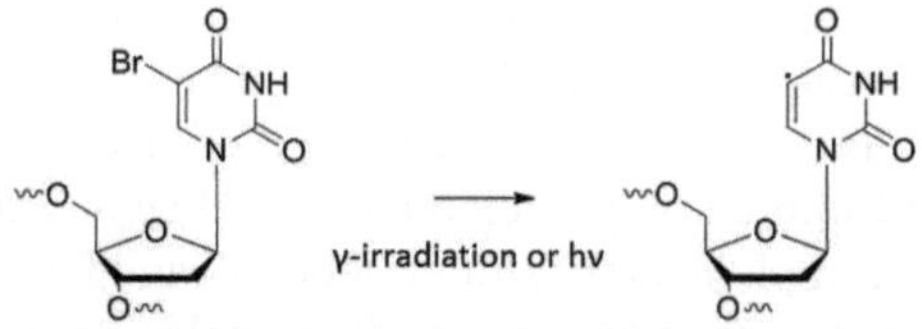

Fig. 17 Irradiation of 5′-brominated deoxyuridine leads to the formation of a very reactive σ-radical

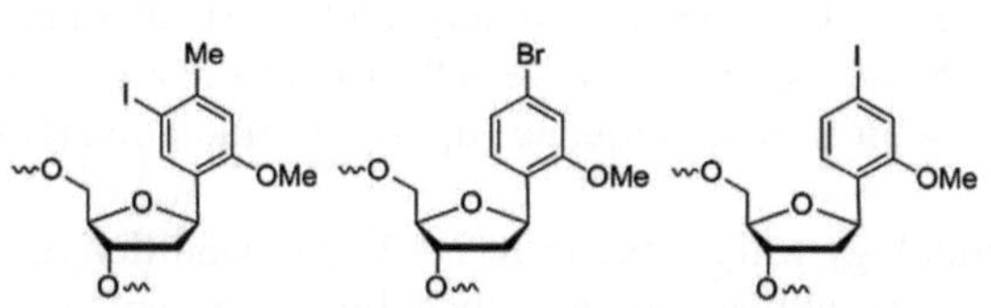

Fig. 18 Aryl halide building blocks, designed by the group of Greenberg, for cross-linking of oligonucleotides

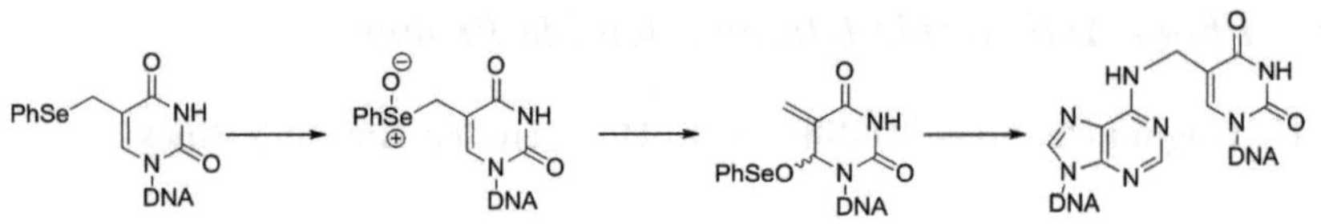

Fig. 19 Phenyl selenium cross-linking strategy for cross-linking to adenine

Greenberg showed that this lesion could induce an ICL with the complementary hybridized strand by reaction with an opposite adenine base. A phenyl selenide nucleotide precursor was used for the synthetic in situ generation of this radical moiety (Fig. 19). Additionally, also cross-linking to a guanine was achieved by modification of 5-methyl-2′-deoxycytidine to the phenyl selenide analogue (Peng et al. 2008). Alternative activation methods for these precursors were also described including UV irradiation and thermal activation in the presence of glutathione, oxygen, and azo initiator (Hong and Greenberg 2005a). Non-radical pathways for the formation of ICLs after the incorporation of phenyl selenium nucleotides have also been demonstrated, by oxidation with hydrogen peroxide, singlet oxygen, or $NaIO_4$ (Hong and Greenberg 2005b; Hong et al. 2006a, b).

Greenberg's phenyl selenide strategy was further shown to be applicable for RNA cross-linking after incorporation of a phenyl selenide modified 5-methyl uridine in an RNA sequence (Sloane and Greenberg 2014).

4.3.3 Oligonucleotides Modified with Derivatives Reacting Through [2 + 2] Cycloaddition

As described above, psoralens have been widely used as externally added cross-linking agents. Yet, when sequence selectivity is required, incorporation of the psoralen derivative into an oligonucleotide sequence allows limiting the reaction to cross-link formation between the modified strand and its complement. The group of Essigmann successfully synthesized a psoralen-modified thymine moiety which was then incorporated into an oligonucleotide sequence and proven to selectively form an ICL with the complementary strand (Kobertz and Essigmann 1997). Miller and co-workers reported on the attachment of a psoralen derivative to the 5′-end of an oligodeoxynucleotide methylphosphonate sequence and demonstrated its cross-linking abilities to DNA sequences (Lee et al. 1988) and RNA sequences (Kean et al. 1988). Furthermore, Murakami and co-workers described the synthesis of a psoralen-derivatized 2′-O-adenosine moiety and, through internal incorporation into a phosphorothioate sequence, were able to selectively cross-link a complementary uridine base in an RNA sequence upon UV irradiation (Higuchi et al. 2007, 2009).

Another methodology which utilizes a [2 + 2] cycloaddition reaction as a tool for oligonucleotide cross-linking was described by the group of Fujimoto. Here, a 3-cyanovinylcarbazole nucleoside was incorporated in an oligonucleotide sequence, and reversible cross-linking was observed through irradiation at different

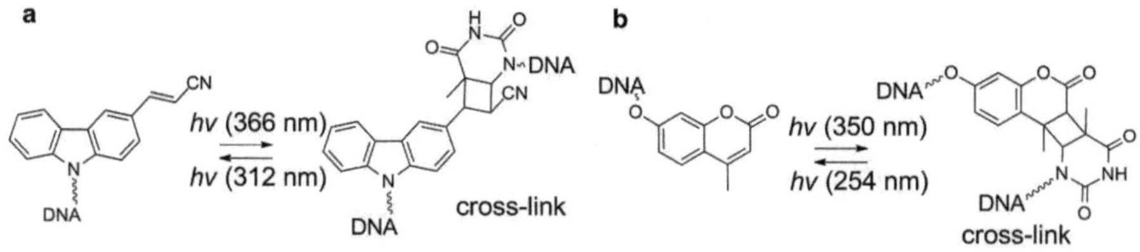

Fig. 20 (**a**) [2 + 2] cycloaddition with 3-cyanovinylcarbazole-modified DNA, (**b**) [2 + 2] cycloaddition with coumarin-modified DNA

wavelengths of light. Initially, fast and efficient cross-linking was observed with a pyrimidine base on the complementary strand upon irradiation at 366 nm (Fig. 20a). Subsequent irradiation at 312 nm was shown to break the cross-link, reforming the original single strands (Yoshimura and Fujimoto 2008; Sakamoto et al. 2015a, b). They also extended their method to cross-linking of RNA (Yoshimura et al. 2009). Editing the RNA sequence was demonstrated, after cross-linking to a cytosine, by the reverse reaction which led to substitution of this cytosine with a uracil base (Fujimoto et al. 2010).

Recently, Peng et al. described the incorporation of a coumarin moiety in an oligonucleotide sequence (Fig. 20b). Cross-linking was observed when irradiating the duplex with 350 nm, inducing a [2 + 2] cycloaddition with thymine (and in much lesser extent with cytosine and adenosine) present on the complementary strand. By irradiating the duplex at 254 nm, complete reversibility was observed (Haque et al. 2014; Sun et al. 2014).

4.3.4 Oligonucleotides Modified with a Furan Moiety

Inspired by the natural toxicity of furan in the human body, our group established an efficient cross-link methodology through incorporation of a furan moiety in an oligonucleotide sequence (Stevens and Madder 2009; Jawalekar et al. 2011; Op de Beeck and Madder 2011). After oxidation of the furan group, a reactive keto-enal is formed which is prone to react with an exocyclic amine present on the opposite base and directed in the major groove, leading to ICL formation. This indicates cytidine and adenine as cross-linking partners. Oxidation of the furan unit can be achieved by using N-bromosuccinimide or singlet oxygen (Op De Beeck and Madder 2012), which is generated by irradiation of a photosensitizer with a characteristic wavelength of the visible spectrum (Fig. 21).

Various furan building blocks were synthesized (Fig. 22) and evaluated for their cross-link abilities. ICL yields were shown to depend upon the furan moiety used.

In addition to cross-linking DNA, the furan methodology shown can be extended to the field of RNA. It was shown that incorporation of a furan moiety in a 2′-OMe RNA sequence lead to cross-link formation to a target unmodified RNA sequence (Carrette et al. 2014).

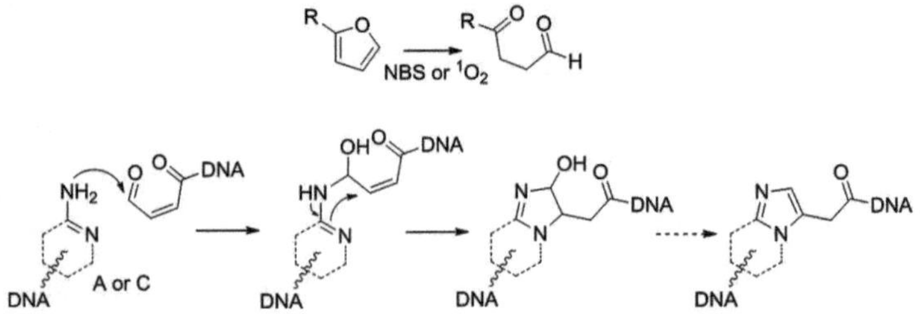

Fig. 21 Oxidation of the furan unit followed by nucleophilic attack of exocyclic amine with the formation of an ICL

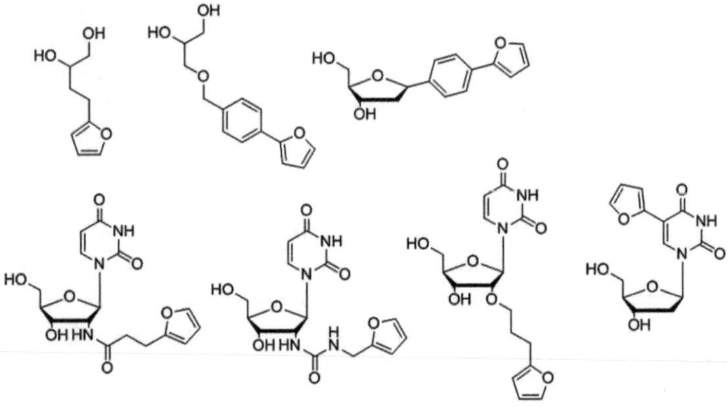

Fig. 22 Various furan building blocks developed for oligonucleotide cross-linking

5 Cross-Linking Between Two Complementary Modified Oligonucleotides

Modification of two complementary bases in the oligonucleotide strands can result in formation of a cross-link through simple and well-established chemical reactions.

5.1 Disulfide Formation

Modification with sulfide derivatives, of two positions opposite one another on complementary oligonucleotide strands, can lead to disulfide bond formation upon oxidation, leading to the formation of a stable cross-linked duplex (Fig. 23).

The sulfide moiety can be incorporated during oligonucleotide synthesis providing a sulfide-modified nucleoside which can be displayed at a given position. This

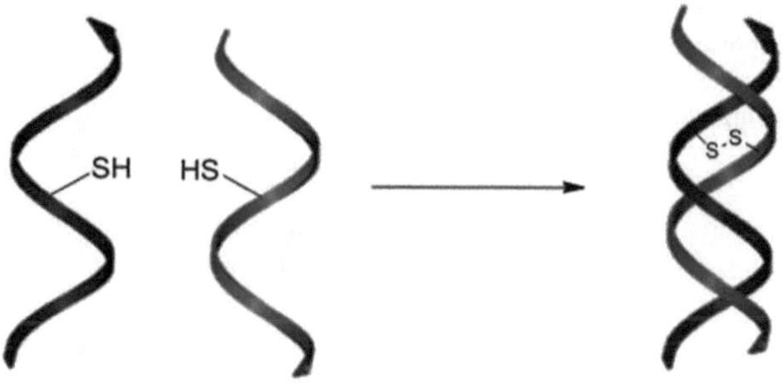

Fig. 23 Schematic representation of cross-linking of oligonucleotides by disulfide bond formation

was described by the group of Jones who showed internal incorporation of sulfide-linked (deoxy)guanine nucleosides in DNA and RNA sequences (Xiaorong et al. 2009). Glick et al. incorporated various sulfide-modified thymine residues with the ability to form cross-links upon oxidation by air when situated at the 5′-strand end and the 3′-end of the complementary sequence in an oligonucleotide duplex (Osborne et al. 1996; Glick 1998). Additionally, Hatano and co-workers described the internal incorporation of a building block containing a sulfide group directly attached to the sugar ring. Stable cross-links were observed that were reversible under reducing conditions (Hatano et al. 2004, 2012).

5.2 *Click Chemistry*

A well-known click reaction is the copper-mediated reaction between azide and alkyne functionalities. Due to advantages such as efficiency and regioselectivity, this click methodology has been utilized for many applications, and it has also been found to be useful for forming nucleic acid interstrand cross-links. This technique involves the modification of one strand with an azide function, while the opposite base on the complementary strand holds the alkyne derivative. Copper is then used to catalyze the formation of a stable triazole function, yielding an interstrand cross-link (Fig. 24) (Kocalka et al. 2008). An alternative approach was taken by the group of Seela to use this click chemistry. They modified both oligonucleotides with an alkyne function and added an external bifunctional azide moiety (Pujari et al. 2010). They have subsequently used this methodology to form cross-links in various DNA configurations showing the versatility of the approach (Xiong and Seela 2011, 2012; Pujari and Seela 2012, 2013; Ingale and Seela 2013; Pujari et al. 2014).

Due to the high toxicity of copper, developing a method which has no need for copper catalysis is of interest for biological applications. The group of Bertozzi initially determined that a ring-strained alkyne, in most cases cyclooctyn, with electron withdrawing substituents, can from triazole compounds with azides even in

Fig. 24 Schematic representation of cross-linking of oligonucleotides using click chemistry

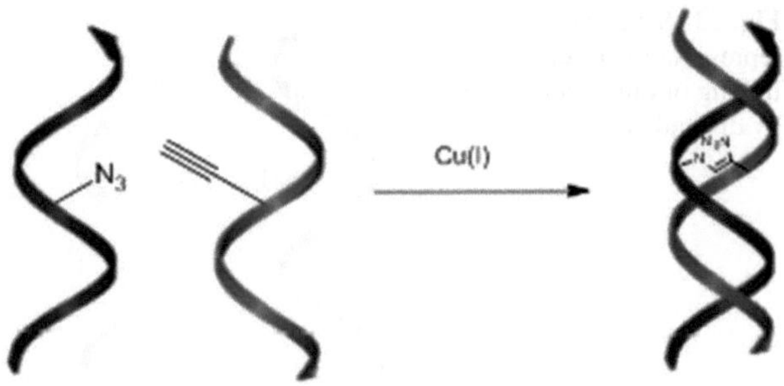

the absence of a copper catalyst (Baskin et al. 2007). This method has been demonstrated in DNA cross-links by the group of Seela and the group of Brown (Gerrard et al. 2012; Shelbourne et al. 2012). Another alternative reaction occurs between an azide and a nitrile oxide, which is electronically similar to the nitrile, and was demonstrated by Heaney in collaboration with our group (Singh et al. 2009, 2012).

5.3 [2 + 2] Photocycloaddition Reactions

As previously described in Sect. 4.3, photo-activatable oligonucleotide probes have been used for the cross-linking of DNA strands through modification of one strand and reaction with a base pair on the other strand. The group of Asanuma have chosen to modify both DNA strands with stilbene derivatives which become active under light irradiation of 340 nm to undergo photocycloaddition reactions (Kashida et al. 2013; Doi et al. 2015).

5.4 Adding an External Reagent to Two Modified Oligonucleotides

Introduction of electrophilic moieties in oligonucleotide sequences allows formation of reactive centers which are susceptible to nucleophilic attack. Schärer et al. developed a methodology using this theory, in which two aldehyde precursors were introduced on opposite strands. The corresponding aldehydes were formed upon oxidation, after which an external bifunctional compound was added, covalently bonding these two moieties together and leading to the formation of an ICL (Angelov et al. 2009; Mukherjee et al. 2014).

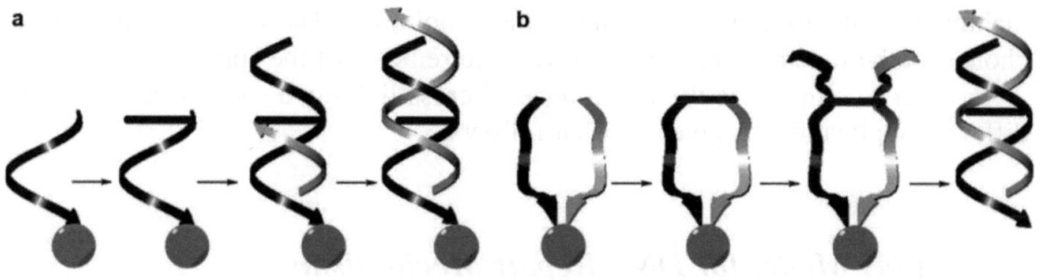

Fig. 25 Solid-phase ICL synthesis. (**a**) First strategy, (**b**) second strategy

6 Solid-Phase ICL Synthesis

To obtain insight into the characteristics of ICLs formed by treatment of DNA with nitrous acid, Hopkins and Sigurdsson developed an approach for the stepwise generation of cross-linked duplexes (Harwood et al. 1999a). Here, cross-links were introduced during solid-phase synthesis, which was done following two strategies. The first approach introduced a cross-linked phosphoramidite during elongation of an oligonucleotide chain. When the synthesis proceeded, both the original and the introduced nucleosides were extended simultaneously toward the 5′-end, forming an identical base pair sequence. After end capping of these growing sequences, elongation of the 3′-end of the introduced chain, preceded by its deprotection, was completed (Fig. 25a). The second strategy involved introduction of coupled phosphoramidites simultaneously on two growing oligonucleotide chains, both attached to solid support (Fig. 25b).

These strategies were applied by the group of Kishi for synthesis of various duplexes (Li et al. 2001). The cross-link, which was represented by a G-T or A-T coupled base pair, was introduced internally, at the terminus of a duplex, or by covalently coupling an internal position of the oligonucleotide chain with the 3′-end of its complement.

To obtain insight in the repair mechanism active upon application of various chemotherapeutics, the group of Miller, Noronha, and Wilds synthesized different cross-linked duplexes relevant in this context. By introducing a C-C (Noll et al. 2001; Noronha et al. 2002) or a T-T cross-link (Wilds et al. 2004), a system is formed which resembles the lesions found after treatment with mechloretamine. Alternatively, a G-G cross-link resembles the lesions upon treatment with hepsulfam (Wilds et al. 2006).

7 Applications

A wide variety of techniques for cross-link formation has been presented here, and this results from the diverse array of possible applications for this procedure. Subject to the application, cross-link yield, selectivity, instant reactivity or reactivity on demand, and cross-link trigger are all factors which can be optimized by making use of different cross-linking methodologies. The cross-link method of choice can be tuned in order to meet the requirements of the application.

In order to emphasize the importance of cross-linking, a short overview of the different applications is given in what follows.

7.1 *Study Model for DNA Repair Mechanisms*

As described above, nucleic acid cross-links in cells are a key cause of DNA damage, formed under the influence of certain endo- and exogenous agents. These highly toxic lesions prevent DNA replication which can ultimately lead to cell apoptosis (Schärer 2003). In addition to this, DNA mutations and rearrangements can also result from cross-link formation, with the possibility of tumor growth as a consequence (Dronkert and Kanaar 2001). Therefore, organisms have found a way to protect themselves in the form of DNA repair enzymes. Yet, whereas this repair is essential in healthy cells, it can compromise the effect of chemotherapeutics of cancer cells, where cell death is the main goal (Chaney and Sancar 1996). Therefore, elucidation of the repair mechanism of these lesions is of great importance to improve the quality of certain chemotherapeutics, where resistance of the cancerous cells to the treatment is interfering with the desired therapeutic effect (Guainazzi and Scharer 2010). Due to the high complexity of the repair of these lesions in mammalian cells, the exact working mechanism is of much interest. Therefore, there is a need for high accessibility of well-defined cross-linked oligonucleotide duplexes.

7.2 *Nanotechnology*

The high specificity with which an oligonucleotide sequence recognizes its complement contributes to the applicability in the field of nanotechnology. Indeed, this specificity guides the assembly of DNA strands to a highly controlled nanoconstruct. DNA origami constructs were first introduced by Seeman in 1982 and were further developed by Rothemund (Seeman 1982; Rothemund 2006). The use of long single-stranded DNA was combined with many short sequences to make folded constructs. Yet, due to the electrostatic character of the bonds holding the DNA helices together, working temperatures had to remain under 55°C to prevent melting of the DNA duplexes and thus damage to the material (Castro et al. 2011). To expand the applicability of the DNA composing materials, a higher temperature resistance is required. It was shown that the introduction of a cross-link between the DNA strands could improve the temperature stability of the DNA-based materials (Lundberg et al. 2010; Rajendran et al. 2011; Tagawa et al. 2011; Nakamura and Fujimoto 2014). Furthermore by making use of a cross-link, another nucleic acid strand or other moiety can be conjugated onto the nanomaterial at a specific position.

7.3 Secondary and Tertiary Structural Information

Cross-linking can be used for the structural elucidation of 3D folding of nucleic acid structures. RNA, for example, forms numerous secondary and tertiary structures (e.g., helices, hairpins, loops, etc.). These can be analyzed using classical methods such as X-ray crystallography or NMR; however, the results from these studies are difficult to interpret because many of the structures involved are formed in a reversible fashion. The close proximity between certain functional groups can be proven through cross-links which also helps in revealing the secondary and tertiary structures. Furthermore, an installed cross-link helps in stabilizing and rendering formation of these structures irreversible so that further analysis is possible (Harris and Christian 2009). One example is the discovery of a DNA-RNA hybrid-type G-quadruplex through fixation of the structure by making use of the copper-based cycloaddition cross-linking between the 5′-end and the 3′-end of the sequence (Xu et al. 2009). A second example involves the use of platinum(II) cross-linking to a phosphorothioate substitution in the hammerhead ribozyme (HHRz) (Chapman and DeRose 2012).

7.4 Cross-Linking as a Tool for Therapeutics

Applications in the cellular environment are more challenging due to numerous requirements resulting from the complex mixture in which cross-linking needs to occur. Due to the fact that the cross-link target is a natural sequence, evidently only small exogenous agents or oligonucleotide probes, in which the reactive functionality is present in a single strand, can be applied. To improve the selectivity in this context, it is beneficial that the reactivity of the used probe can be triggered, preferably with a non-detrimental activation method. Additionally, cell stability and cell permeability are also factors of importance when applying the technique in a cellular context.

7.4.1 Chemotherapeutics

As previously discussed, many cross-linking agents with intrinsic reactivity are used as chemotherapeutics. After the Second World War, autopsies made clear that sulfur mustard had specifically attacked the white blood cells, and it was hypothesized that this could help in the treatment of leukemia. At present, a long list of chemotherapeutic agents is used to cure or treat cancer; however, the cross-linking agents, especially bifunctional alkylating agents and platinum-based agents, represent a large proportion of this list. It is not our intention to list them all here, but the most important chemotherapeutic cross-linking agents (e.g., cyclophosphamide, mustine, melphalan, cisplatin, oxaliplatin, etc.) are discussed in Sect. 3. The major disadvantage of chemotherapeutics is the lack of selectivity resulting in a debilitating number of side effects for all these drugs. To circumvent these side effects, the fields of antisense therapy, decoy DNA, and RNA interference have been developed to create therapeutics with more selectivity.

7.4.2 Antisense

The working principle behind the majority of therapeutics is an interaction with the proteins that play a key role in certain diseases. Antisense therapeutics, on the other hand, reduce protein expression by interfering with the translation of mRNA sequencing that codes for this specific protein. Although discussion about the specific mechanisms of action is still ongoing, two types of working principle can be described (Dias and Stein 2002). The first mechanism exploits the function of the RNAse H endonucleases. After hybridization of the synthetic DNA with the mRNA strand, the formed DNA-RNA duplex is recognized by RNase H which selectively hydrolyzes the RNA sequence, preventing its translation. Both mipomersen and fomivirsen are antisense therapeutics, approved by the FDA, which induce degradation of the specific mRNA by recruiting RNAse H. The second mechanistic pathway involves sterical blocking of the mRNA sequence, rendering it unavailable for splicing or binding of the ribosome, and in this way the production pathway of the corresponding protein is blocked. By introducing a covalent bond between the mRNA sequence and its externally added complement, a more rigid and sustainable blocking motif can be formed, thus improving the potential of the antisense technology. Indeed, Sasaki and co-workers demonstrated that their vinyl purine methodology was applicable in the antisense methodology. They synthesized PEG-conjugated analogues of sulfide-protected 2-amino-6-alkyl purine, complementary to the firefly luciferase mRNA sequence. Intracellular application of these modified strands led to successful inhibition of luciferase fluorescence. They also proved that more inhibition was obtained when cross-linking was applied than when using the non-modified antisense sequence (Ali et al. 2006). The group of Fujimoto successfully demonstrated the possibility of reducing GFP (green fluorescent protein) expression in HeLa cells, by the use of photo-activatable phosphorothioate oligonucleotides modified with a 3-cyanovinylcarbazole nucleoside, complementary to a specified region in the mRNA sequence coding for this protein. They showed a diminished fluorescence after UV irradiation of the probe-containing cells, proving a reduced GFP concentration (Sakamoto et al. 2014).

7.4.3 Transcription Factor Decoy Oligonucleotides

An important step in the transcription of genes is the binding of transcription factors (TFs) to their binding motifs located on several positions near the transcription initiation site. The binding of these TFs and their mutual interactions have a high influence on the expression of the gene to its corresponding proteins. Therefore, they have become a relevant target for therapeutic applications. By introducing "decoy" DNA, a short duplex sequence containing the binding motif of the transcription factor, into the cell, competition arises between the binding motif located on the gene and on the decoy DNA, resulting in suppression of TF gene binding. Yet, an important limitation of the transcription factor decoy is the stability of the decoy DNA in cellular context due to the presence of nucleases. By introducing a cross-link between the strands in the decoy duplex or by closing the decoy duplex on both sides rendering a cyclic decoy or dumbbell, a higher stability toward endonucleases can be achieved (Gao et al. 1994; Murakami et al. 2001). When this is utilized, successful inhibition of protein expression can be observed (Amoah-Apraku et al. 2000; Murakami et al. 2001). Furthermore, work carried out by El-Sagheer and Brown illustrated the use of dumbbell DNA, cyclized using the copper-catalyzed azide-alkyne cycloaddition reaction, for targeting transcription factors. This cyclic decoy has greater stability toward degradation in comparison with the noncyclic hairpin and shows smooth cell uptake (El-Sagheer et al. 2008; El-Sagheer and Brown 2008, 2012).

7.4.4 RNA Interference

Interference with mRNA can occur by other means than the antisense strategy as a tool to prevent its translation. Endogenously, this is done in the process of RNA interference through natural RNA molecules which inhibit or stimulate gene expression. These processes and mechanisms are still to be fully revealed; however, it is known that both microRNA (miRNA) and small interfering RNA (siRNA) can downregulate protein translation from mRNA by binding to the RNA-induced silencing complex (RISC) (Ipsaro and Joshua-Tor 2015). Upregulation however is also possible by targeting promoter regions. Artificial siRNA or anti-miRNA can be delivered into the cell to interfere with this posttranscriptional gene expression (Chabot et al. 2015). This is an emerging field since the discovery of RNA interference was first described by Fire and Mello (Fire et al. 1998). Cross-linking can be useful in this field, as was demonstrated by the use of a psoralen-modified oligonucleotide which inhibits the RISC function better than the similar non-psoralen-modified analogue suggesting that cross-linking indeed is useful for regulation of RNA interference (Matsuyama et al. 2014).

References

Ali MM, Oishi M, Nagatsugi F et al (2006) Intracellular inducible alkylation system that exhibits antisense effects with greater potency and selectivity than the natural oligonucleotide. Angew Chem Int Ed Engl 45:3136–3140

Alzeer J, Schärer OD (2006) A modified thymine for the synthesis of site-specific thymine-guanine DNA interstrand crosslinks. Nucleic Acids Res 34:4458–4466

Amoah-Apraku B, Fang MZ, Guzman NJ (2000) A non-nucleotide-bridged DNA decoy inhibits renal epithelial nitric oxide synthase expression. Kidney Int 57:83–91
Angelov T, Guainazzi A, Schärer OD (2009) Generation of DNA interstrand cross-links by post-synthetic reductive amination. Org Lett 11:661–664
Baskin JM, Prescher J a, Laughlin ST et al (2007) Copper-free click chemistry for dynamic in vivo imaging. Proc Natl Acad Sci USA 104:16793–16797
Beda NV, Nedospasov AA (2006) Inorganic nitric oxide metabolites participating in no-dependent modifications of biopolymers. Russ J Bioorg Chem 32:1–22
Beda NV, Nedospasov AA (2007) NO-dependent modifications of nucleic acids. Russ J Bioorg Chem 33:181–212
Berka V, Yeh H-C, Gao D et al (2004) Redox function of tetrahydrobiopterin and effect of L-arginine on oxygen binding in endothelial nitric oxide synthase. Biochemistry 43:13137–13148
Brookes P, Lawley PD (1961) The alkylation of guanosine and guanylic acid. J Chem Soc 3923–3928
Carrette LLG, Gyssels E, Loncke J, Madder A (2014) A mildly inducible and selective cross-link methodology for RNA duplexes. Org Biomol Chem 12:931–935
Castro CE, Kilchherr F, Kim D-N et al (2011) A primer to scaffolded DNA origami. Nat Methods 8:221–229
Catalano MJ, Liu S, Andersen N et al (2015) Chemical structure and properties of interstrand cross-links formed by reaction of guanine residues with abasic sites in duplex DNA. J Am Chem Soc 137:3933–3945
Caulfield JL, Wishnok JS, Tannenbaum SR (2003) Nitric oxide-induced interstrand cross-links in DNA. Chem Res Toxicol 16:571–574
Cecchini S, Massen C, La Madeleine C et al (2005) Interstrand cross-link induction by UV radiation in bromodeoxyuridine- substituted DNA: dependence on DNA conformation. Biochemistry 44:16957–16966
Chabot S, Teissié J, Golzio M (2015) Targeted electro-delivery of oligonucleotides for RNA interference: siRNA and antimiR. Adv Drug Deliv Rev 81:161–168
Chaney SG, Sancar A (1996) DNA repair: enzymatic mechanisms and relevance to drug response. J Natl Cancer Inst 88:1346–1360
Chapman EG, DeRose VJ (2012) Site-specific platinum(II) cross-linking in a ribozyme active site. J Am Chem Soc 134:256–262
Cohen SM, Garland EM, St John M et al (1992) Acrolein initiates rat urinary bladder carcinogenesis. Cancer Res 52:3577–3581
Coleman RS, Kesicki EA (1995) Template-directed cross-linking of oligonucleotides: site-specific covalent modification of dG-N7 within duplex DNA. J Org Chem 60:6252–6253
Coleman RS, Pires RM (1997) Covalent cross-linking of duplex DNA using 4-thio-2′-deoxyuridine as a readily modifiable platform for introduction of reactive functionality into oligonucleotides. Nucleic Acids Res 25:4771–4777
Czerny C, Eder E, Rünger TM (1998) Genotoxicity and mutagenicity of the α, β-unsaturated carbonyl compound crotonaldehyde (butenal) on a plasmid shuttle vector. Mutat Res 407:125–134
Decout J-L, Lhomme J (1992) Reversible photocycloaddition of a 4′,5′-dihydropsoralen derivative with thymine. Photochem Photobiol 56:431–440
Dextraze ME, Cecchini S, Bergeron F et al (2009) Reaching for the other side: generating sequence-dependent interstrand cross-links with 5-bromodeoxyuridine and γ-rays. Biochemistry 48:2005–2011
Dias N, Stein C (2002) Antisense oligonucleotides: basic concepts and mechanisms minireview antisense oligonucleotides: basic concepts and mechanisms. Mol Cancer Ther 1:347–355
Doi T, Kashida H, Asanuma H (2015) Efficiency of [2 + 2] photodimerization of various stilbene derivatives within DNA duplex scaffold. Org Biomol Chem 13:4430–4437
Dronkert ML, Kanaar R (2001) Repair of DNA interstrand cross-links. Mutat Res 486:217–247

Dutta S, Chowdhury G, Gates KS (2007) Interstrand cross-links generated by abasic sites in duplex DNA. J Am Chem Soc 129:1852–1853

Efimov VA, Fedyunin SV (2010) Cross-linked nucleic acids: isolation, structure, and biological role. Biochemistry 75:1606–1627

El-Sagheer AH, Brown T (2008) Synthesis, serum stability and cell uptake of cyclic and hairpin decoy oligonucleotides for TCF/LEF and GLI transcription factors. Int J Pept Res Ther 14:367–372

El-Sagheer AH, Brown T (2012) Click nucleic acid ligation: applications in biology and nanotechnology. Acc Chem Res 45:1258–1267

El-Sagheer AH, Kumar R, Findlow S et al (2008) A very stable cyclic DNA miniduplex with just two base pairs. Chembiochem 9:50–52

El-Yazbi AF, Loppnow GR (2014) Detecting UV-induced nucleic-acid damage. TrAC Trends Anal Chem 61:83–91

Facchinetti F, Amadei F, Geppetti P et al (2007) Alpha, beta-unsaturated aldehydes in cigarette smoke release inflammatory mediators from human macrophages. Am J Respir Cell Mol Biol 37:617–623

Fire A, Xu S, Montgomery MK et al (1998) Potent and specific genetic interference by double-stranded RNA in *Caenorhabditis elegans*. Nature 391:806–811

Fischhaber P, Gall A, Duncan J et al (1999) Direct demonstration in synthetic oligonucleotides that N, N′-bis(2-chloroethyl)-nitrosourea cross-links N-1 of deoxyguanosine to N-3 of deoxycytidine on opposite strands of duplex DNA. Cancer Res 59:4363–4368

Fujimoto K, Konishi-Hiratsuka K, Sakamoto T et al (2010) Site-specific photochemical RNA editing. Chem Commun (Camb) 46:7545–7547

Galliani G, Pantarotto C (1983) The reaction of guanosine and 2′-deoxyguanosine with acrolein. Tetrahedron Lett 24:4491–4492

Gamboa Varela J, Gates KS (2015) A simple, high-yield synthesis of DNA duplexes containing a covalent, thermally cleavable interstrand cross-link at a defined location. Angew Chem Int Ed 54:7666–7669

Gao H, Chidambaram N, Chen BC et al (1994) Double-stranded cyclic oligonucleotides with non-nucleotide bridges. Bioconjug Chem 5:445–453

Gerrard SR, Hardiman C, Shelbourne M et al (2012) A new modular approach to nanoassembly: stable and addressable DNA nanoconstructs via orthogonal click chemistries. ACS Nano 6:9221–9228

Gilman A, Philips FS (1946) The biological actions and therapeutic applications of the b-chloroethyl amines and sulfides. Science 103:409–436

Glick GD (1998) Design, synthesis, and analysis of conformationally constrained nucleic acids. Biopolymers 48:83–96

Goodsell DS (2001) The molecular perspective: ultraviolet light and pyrimidine dimers. Oncologist 6:298–299

Greenberg MM (2014) Abasic and oxidized abasic site reactivity in DNA: enzyme inhibition, cross-linking, and nucleosome catalyzed reactions. Acc Chem Res 47:646–55

Guainazzi A, Scharer O (2010) Using synthetic DNA interstrand crosslinks to elucidate repair pathways and identify new therapeutic targets for cancer chemotherapy. Cell Mol Life Sci 67:3683–3697

Haque MM, Sun H, Liu S et al (2014) Photoswitchable formation of a DNA interstrand cross-link by a coumarin-modified nucleotide. Angew Chem Int Ed 53:7001–7005

Harris ME, Christian EL (2009) RNA crosslinking methods. Methods Enzymol 468:127–146

Harwood E a, Sigurdsson ST, Edfeldt NBF et al (1999a) Chemical synthesis and preliminary structural characterization of a nitrous acid interstrand cross-linked duplex DNA. J Am Chem Soc 121:5081–5082

Harwood EA, Sigurdsson ST, Edfeldt NBF et al (1999b) Chemical synthesis and preliminary structural characterization of a nitrous acid interstrand cross-linked duplex DNA. J Am Chem Soc 121:5081–5082

Hata T, Hoshi T, Kanamori K et al (1956) Mitomycin, a new antibiotic from streptomyces. I J Antibiot (Tokyo) 9:141–146

Hatano A, Makita S, Kirihara M (2004) Synthesis and characterization of a DNA analogue stabilized by mercapto C-nucleoside induced disulfide bonding. Bioorganic Med Chem Lett 14:2459–2462

Hatano A, Okada M, Kawai G (2012) Solution structure of S-DNA formed by covalent base pairing involving a disulfide bond. Org Biomol Chem 10:7327–7333

Hentschel S, Alzeer J, Angelov T et al (2012) Synthesis of DNA interstrand cross-links using a photocaged nucleobase. Angew Chem Int Ed 51:3466–3469

Higuchi M, Kobori A, Yamayoshi A et al (2009) Synthesis of antisense oligonucleotides containing 2′-O-psoralenylmethoxyalkyl adenosine for photodynamic regulation of point mutations in RNA. Bioorg Med Chem 17:475–483

Higuchi M, Yamayoshi A, Yamaguchi T et al (2007) Selective photo-cross-linking of 2′-O-psoralen-conjugated oligonucleotide with RNAs having point mutations. Nucleosides Nucleotides Nucleic Acids 26:277–290

Hong IS, Ding H, Greenberg MM (2006a) Oxygen independent DNA interstrand cross-link formation by a nucleotide radical. J Am Chem Soc 128:485–491

Hong IS, Ding H, Greenberg MM (2006b) Radiosensitization by a modified nucleotide that produces DNA interstrand cross-links under hypoxic conditions. J Am Chem Soc 128:2230–2231

Hong IS, Greenberg MM (2005a) Efficient DNA interstrand cross-link formation from a nucleotide radical. J Am Chem Soc 127:3692–3693

Hong IS, Greenberg MM (2005b) DNA interstrand cross-link formation initiated by reaction between singlet oxygen and a modified nucleotide. J Am Chem Soc 127(30):10510–10511

Hou D, Greenberg MM (2014) DNA interstrand cross-linking upon irradiation of aryl halide C-nucleotides. J Org Chem 79:1877–1884

Ichikawa K, Kojima N, Hirano Y et al (2012) Interstrand cross-link of DNA by covalently linking a pair of abasic sites. Chem Commun 48:2143–2145

Imoto S, Chikuni T, Kansui H et al (2012) Fast DNA interstrand cross-linking reaction by 6-vinylpurine. Nucleosides Nucleotides Nucleic Acids 31:752–762

Ingale SA, Seela F (2013) Stepwise click functionalization of DNA through a bifunctional azide with a chelating and a nonchelating azido group. J Org Chem 78:3394–3399

Ipsaro JJ, Joshua-Tor L (2015) From guide to target: molecular insights into eukaryotic RNA-interference machinery. Nat Struct Mol Biol 22:20–28

Jawalekar AM, Op de Beeck M, van Delft FL et al (2011) Synthesis and incorporation of a furan-modified adenosine building block for DNA interstrand crosslinking. Chem Commun (Camb) 47:2796–2798

Johnson KM, Price NE, Wang J et al (2013) On the formation and properties of interstrand DNA-DNA cross-links forged by reaction of an abasic site with the opposing guanine residue of 5′-CAp sequences in duplex DNA. J Am Chem Soc 135:1015–1025

Kashida H, Doi T, Sakakibara T et al (2013) p-Stilbazole moieties as artificial base pairs for photo-cross-linking of DNA duplex. J Am Chem Soc 135:7960–7966

Kawanishi M, Matsuda T, Nakayama A et al (1998) Molecular analysis of mutations induced by acrolein in human fibroblast cells using supF shuttle vector plasmids. Mutat Res 417:65–73

Kawasaki T, Nagatsugi F, Ali MM et al (2005) Hybridization-promoted and cytidine-selective activation for cross-linking with the use of 2-amino-6-vinylpurine derivatives. J Org Chem 70:14–23

Kean JM, Murakami A, Blake KR et al (1988) Photochemical cross-linking of psoralen-derivatized oligonucleoside methylphosphonates to rabbit globin messenger RNA. Biochemistry 27:9113–9121

Klockenbusch C, O'Hara JE, Kast J (2012) Advancing formaldehyde cross-linking towards quantitative proteomic applications. Anal Bioanal Chem 404:1057–1067

Kobertz WR, Essigmann JM (1997) Solid-phase synthesis of oligonucleotides containing a site-specific psoralen derivative. J Am Chem Soc 119:5960–5961

Kocalka P, El-Sagheer AH, Brown T (2008) Rapid and efficient DNA strand cross-linking by click chemistry. Chembiochem 9:1280–1285

Kusano S, Haruyama T, Ishiyama S et al (2014) Development of the crosslinking reactions to RNA triggered by oxidation. Chem Commun (Camb) 50:3951–3954

Lee BL, Murakami A, Blake KR et al (1988) Interaction of psoralen-derivatized oligodeoxyribonucleoside methylphosphonates with single-stranded DNA. Biochemistry 27:3197–3203

Leonard NJ, Barrio JR (1984) Etheno-substituted nucleotides and coenzymes: fluorescence and biological activity. Crit Rev Biochem Mol Biol 15:125–199

Li HY, Qiu YL, Moyroud E et al (2001) Synthesis of DNA oligomers possessing a covalently cross-linked Watson–Crick base pair model. Angew Chem Int Ed Engl 113:1519–1523

Lindahl T (1993) Instability and decay of the primary structure of DNA. Nature 362:709–715

Lukhtanov EA, Mills AG, Kutyavin IV et al (1997) Minor groove DNA alkylation directed by major groove triplex forming oligodeoxyribonucleotides. Nucleic Acids Res 25:5077–5084

Lukhtanov EA, Podyminogin MA, Kutyavin IV et al (1996) Rapid and efficient hybridization-triggered crosslinking within a DNA duplex by an oligodeoxyribonucleotide bearing a conjugated cyclopropapyrroloindole. Nucleic Acids Res 24:683–687

Lundberg EP, El-Sagheer AH, Kocalka P et al (2010) A new fixation strategy for addressable nano-network building blocks. Chem Commun 46:3714–3716

Maruenda H, Tomasz M (1996) Antisense sequence-directed cross-linking of DNA oligonucleotides by mitomycin C. Bioconjug Chem 7:541–544

Matsuyama Y, Yamayoshi A, Kobori A et al (2014) Functional regulation of RNA-induced silencing complex by photoreactive oligonucleotides. Bioorg Med Chem 22:1003–1007

Millard JT, White MM (1993) Diepoxybutane cross-links DNA at 5′-GNC sequences. Biochemistry 32:2120–2124

Mukherjee S, Guainazzi A, Schärer OD (2014) Synthesis of structurally diverse major groove DNA interstrand crosslinks using three different aldehyde precursors. Nucleic Acids Res 42:7429–7435

Murakami A, Yamamoto Y, Namba M et al (2001) Photo-cross-linked oligonucleotide duplex as a decoy-DNA for inhibition of restriction endonuclease activity. Bioorg Chem 29:223–233

Nagatsugi F, Kawasaki T, Usui D et al (1999) Highly efficient and selective cross-linking to cytidine based on a new strategy for auto-activation within a duplex. J Am Chem Soc 121:6753–6754

Nakamura S, Fujimoto K (2014) Creation of DNA array structure equipped with heat resistance by ultrafast photocrosslinking. J Chem Technol Biotechnol 89:1086–1090

Nishikawa H, Hayakawa T, Sakai T (1987) Determination of acrolein and crotonaldehyde in automobile exhaust gas by gas chromatography with electron-capture detection. Analyst 112:859–862

Noll DM, Mason TM, Miller PS (2006) Formation and repair of interstrand cross-links in DNA. Chem Rev 106:277–301

Noll DM, Noronha AM, Miller PS (2001) Synthesis and characterization of DNA duplexes containing an N4C-ethyl-N4C interstrand cross-link. J Am Chem Soc 123:3405–3411

Norman D, Live D, Sastry M et al (1990) NMR and computational characterization of mitomycin cross-linked to adjacent deoxyguanosines in the minor groove of the d(T-A-C-G-T-A).cntdot.d (T-A-C-G-T-A) duplex. Biochemistry 29:2861–2875

Noronha AM, Noll DM, Wilds CJ et al (2002) N4C-ethyl-N4C cross-linked DNA: synthesis and characterization of duplexes with interstrand cross-links of different orientations. Biochemistry 41:760–771

Op de Beeck M, Madder A (2011) Unprecedented C-selective interstrand cross-linking through in situ oxidation of furan-modified oligodeoxynucleotides. J Am Chem Soc 133:796–807

Op De Beeck M, Madder A (2012) Sequence specific DNA cross-linking triggered by visible light. J Am Chem Soc 134:10737–10740
Orlando V, Strutt H, Paro R (1997) Analysis of chromatin structure by in vivo formaldehyde cross-linking. Methods 11:205–214
Osborne SE, Völker J, Stevens SY et al (1996) Design, synthesis, and analysis of disulfide cross-linked DNA duplexes. J Am Chem Soc 118:11993–12003
Paz MM, Kumar GS, Glover M et al (2004) Mitomycin dimers: polyfunctional cross-linkers of DNA. J Med Chem 47:3308–3319
Peng X, In SH, Li H et al (2008) Interstrand cross-link formation in duplex and triplex DNA by modified pyrimidines. J Am Chem Soc 130:10299–10306
Price NE, Johnson KM, Wang J et al (2014) Interstrand DNA-DNA cross-link formation between adenine residues and abasic sites in duplex DNA. J Am Chem Soc 136:3483–3490
Pujari SS, Leonard P, Seela F (2014) Oligonucleotides with "clickable" sugar residues: synthesis, duplex stability, and terminal versus central inter-strand cross-linking of 2′-O-propargylated 2-aminoadenosine with a bifunctional azide. J Org Chem 79:4423–4437
Pujari SS, Seela F (2012) Cross-linked DNA: propargylated ribonucleosides as "click" ligation sites for bifunctional azides. J Org Chem 77:4460–4465
Pujari SS, Seela F (2013) Parallel stranded DNA stabilized with internal sugar cross-links: synthesis and click ligation of oligonucleotides containing 2′-propargylated isoguanosine. J Org Chem 78:8545–8561
Pujari SS, Xiong H, Seela F (2010) Cross-linked DNA generated by "bis-click" reactions with bis-functional azides: site independent ligation of oligonucleotides via nucleobase alkynyl chains. J Org Chem 75:8693–8696
Rajendran A, Endo M, Katsuda Y et al (2011) Photo-cross-linking-assisted thermal stability of DNA origami structures and its application for higher-temperature self-assembly. J Am Chem Soc 133:14488–14491
Rajski SR, Williams RM (1998) DNA cross-linking agents as antitumor drugs. Chem Rev 98:2723–2796
Reishus JW, Martin DS (1961) cis-Dichlorodiammineplatinum(ii): acid hydrolysis and isotopic exchange of the chloride ligands 1. J Am Chem Soc 83:2457–2462
Rothemund PWK (2006) Folding DNA to create nanoscale shapes and patterns. Nature 440:297–302
Sakamoto T, Ooe M, Fujimoto K (2015a) Critical effect of base pairing of target pyrimidine on the interstrand photo-cross-linking of DNA via 3-cyanovinylcarbazole nucleoside. Bioconjug Chem 26:1475–1478
Sakamoto T, Shigeno A, Ohtaki Y et al (2014) Photo-regulation of constitutive gene expression in living cells by using ultrafast photo-cross-linking oligonucleotides. Biomater Sci 2:1154–1157
Sakamoto T, Tanaka Y, Fujimoto K (2015b) DNA photo-cross-linking using 3-cyanovinylcarbazole modified oligonucleotide with threoninol linker. Org Lett 17:3–6
Sanada M, Hidaka M, Takagi Y et al (2007) Modes of actions of two types of anti-neoplastic drugs, dacarbazine and ACNU, to induce apoptosis. Carcinogenesis 28:2657–2663
Schärer OD (2003) Chemistry and biology of DNA repair. Angew Chem Int Ed Engl 42:2946–2974
Seeman NC (1982) Nucleic acid junctions and lattices. J Theor Biol 99:237–247
Shapiro R, Cohen BI, Shiuey S-J et al (1969) Reaction of guanine with glyoxal, pyruvaldehyde, and kethoxal, and the structure of the acylguanines. Synthesis of N2-alkylguanines. Biochemistry 8:238–245
Shapiro R, Dubelman S, Feinberg AM et al (1977) Isolation and identification of cross-linked nucleosides from nitrous acid treated deoxyribonucleic acid. J Am Chem Soc 99:302–303
Shelbourne M, Brown T, El-Sagheer AH et al (2012) Fast and efficient DNA crosslinking and multiple orthogonal labelling by copper-free click chemistry. Chem Commun 48:11184–11186
Singh I, Vyle JS, Heaney F (2009) Fast, copper-free click chemistry: a convenient solid-phase approach to oligonucleotide conjugation. Chem Commun 3276–3278

Singh I, Freeman C, Madder A, Vyle JS, Heaney F (2012) Fast RNA conjugations on solid phase by strain-promoted cycloadditions. Org Biomol Chem 10:6633–6639

Sloane JL, Greenberg MM (2014) Interstrand cross-link and bioconjugate formation in RNA from a modified nucleotide. J Org Chem 79:9792–9798

Smith MB, Schmidt BF, Czerwinski G et al (1996) Specificity of DNA alkylation by 1-(2-chloroethyl)-3-alkyl-3-acyltriazenes depends on the structure of the acyl group: kinetic and product studies. Chem Res Toxicol 9:466–475

Stevens K, Madder A (2009) Furan-modified oligonucleotides for fast, high-yielding and site-selective DNA inter-strand cross-linking with non-modified complements. Nucleic Acids Res 37:1555–1565

Stone MP, Cho Y-J, Huang H et al (2008) Interstrand DNA cross-links induced by alpha, beta-unsaturated aldehydes derived from lipid peroxidation and environmental sources. Acc Chem Res 41:793–804

Sun H, Fan H, Peng X (2014) Quantitative DNA interstrand cross-link formation by coumarin and thymine: structure determination, sequence effect, and fluorescence detection. J Org Chem 79:11359–11369

Suzuki T, Yamada M, Nakamura T et al (2000) Products of the reaction between a diazoate derivative of 2′-deoxycytidine and l -lysine and its implication for DNA—nucleoprotein cross-linking by NO or HNO 2. Chem Res Toxicol 13:1223–1227

Swenberg JA, Moeller BC, Lu K et al (2013) Formaldehyde carcinogenicity research: 30 years and counting for mode of action, epidemiology, and cancer risk assessment. Toxicol Pathol 41:181–189

Tagawa M, Shohda K, Fujimoto K et al (2011) Stabilization of DNA nanostructures by photo-cross-linking. Soft Matter 7:10931–10934

Tomasz M (1995) Mitomycin C: small, fast and deadly (but very selective). Chem Biol 2:575–579

Tong WP, Ludlum DB (1981) Formation of the cross-linked base, diguanylethane, in DNA treated with N, N′-Bis(2-chloroethyl)-N-nitrosourea. Cancer Res 41:380–382

Wakaki S, Marumo H, Tomioka K et al (1958) Isolation of new fractions of antitumor mitomycins. Antibiot Chemother 8:228–240

Webb TR, Matteucci MD (1986a) Sequence-specific cross-linking of deoxyoligonucleotides via hybridization-triggered alkylation. J Am Chem Soc 108:2764–2765

Webb TR, Matteucci MD (1986b) Hybridization triggered cross-linking of deoxyoligonucleotides. Nucleic Acids Res 14:7661–7674

Wilds CJ, Booth JD, Noronha AM (2006) Synthesis of oligonucleotides containing an O6-G-alkyl-O6-G interstrand cross-link. Tetrahedron Lett 47:9125–9128

Wilds CJ, Noronha AM, Robidoux S et al (2004) Mispair-aligned N3T-alkyl-N3T interstrand cross-linked DNA: synthesis and characterization of duplexes with interstrand cross-links of variable lengths. J Am Chem Soc 126:9257–9265

Xiaorong H, Gang W, Gaffney BL et al (2009) Synthesis of guanosine and deoxyguanosine phosphoramidites with cross-linkable thioalkyl tethers for direct incorporation into RNA and DNA. Nucleosides Nucleotides Nucleic Acids 28:1076–1094

Xiong H, Seela F (2011) Stepwise "click" chemistry for the template independent construction of a broad variety of cross-linked oligonucleotides: influence of linker length, position, and linking number on DNA duplex stability. J Org Chem 76:5584–5597

Xiong H, Seela F (2012) Cross-linked DNA: site-selective "click" ligation in duplexes with bis-azides and stability changes caused by internal cross-links. Bioconjug Chem 23:1230–1243

Xu Y, Suzuki Y, Komiyama M (2009) Click chemistry for the identification of G-quadruplex structures: discovery of a DNA-RNA G-quadruplex. Angew Chem Int Ed Engl 48:3281–3284

Yoshimura Y, Fujimoto K (2008) Ultrafast reversible photo-cross-linking reaction: toward in situ DNA manipulation. Org Lett 10:3227–3230

Yoshimura Y, Ohtake T, Okada H et al (2009) A new approach for reversible RNA photocrosslinking reaction: application to sequence-specific RNA selection. ChemBioChem 10:1473–1476

Chemical Synthesis of Lesion-Containing Oligonucleotides for DNA Repair Studies

Rémy Lartia

Contents

R. Lartia (✉)
Département de Chimie Moléculaire, UMR CNRS 5250, Université Grenoble Alpes, BP 53, 38041 Grenoble Cedex 9, France
e-mail: remy.lartia@ujf-grenoble.fr

S. Jurga et al. (eds.), *Modified Nucleic Acids in Biology and Medicine*, RNA Technologies, DOI 10.1007/978-3-319-34175-0_16

Abstract DNA is continuously exposed to various damaging conditions (oxidizing agents, UV, ionizing radiations, or endogenous species) that produce strand breaks, cross-links, or damaged nucleotides. In most cases, several kinds of DNA damages are simultaneously produced, often in low yield. Consequently, this hampered isolation and study of a well-defined damaged oligonucleotide (ODN). In order to get deeper insights into biological properties of a given lesion, chemistry enables preparation of large quantity of ODN containing a single and site-specifically located lesion. In addition, chemistry allows fine-tuning of lesions properties: influence of stereochemistry or of the neighboring nucleobase.

This book chapter will emphasize on the methods aimed to chemical introduction of damaged nucleotides into a synthetic ODN. Phosphoramidite chemistry will be highlighted as it enables high-yielding, automatable and user-friendly site-specific incorporation of the DNA lesions.

Keywords Nucleic acids • DNA lesions • Oligonucleotide • Phosphoramidite

1 Introduction

Chemical introduction of damaged nucleobase by phosphoramidite approach was previously reviewed (Butenandt 1999; Iwai 2006), and nomenclature rules were recently suggested for clarity purposes (Cooke et al. 2010; Cadet et al. 2012).The aim of this review is to recall earlier reported methods and focus on new synthetic advances. Biochemical investigation of the lesions will not be detailed.

Two main chemical pathways toward damaged ODN are known: (1) the pioneering approach based on post-synthetic generation of damage in an ODN or (2) the synthetic incorporation of a suitably protected damaged nucleotide into the target ODN.

The post-synthetic methods consisted in submitting a short ODN containing a single sensitive target nucleobase to the exogenous conditions (UV, γ-irradiation, chemicals, etc.) known to transform it. Damaged ODN had then to be purified following extensive and tedious separative methods. Although simple, this method is restricted to short ODN and will not be extensively discussed. Finally, some mixed methods are based on the incorporation of a reactive nucleoside during elongation, followed by ODN post-synthetic modification.

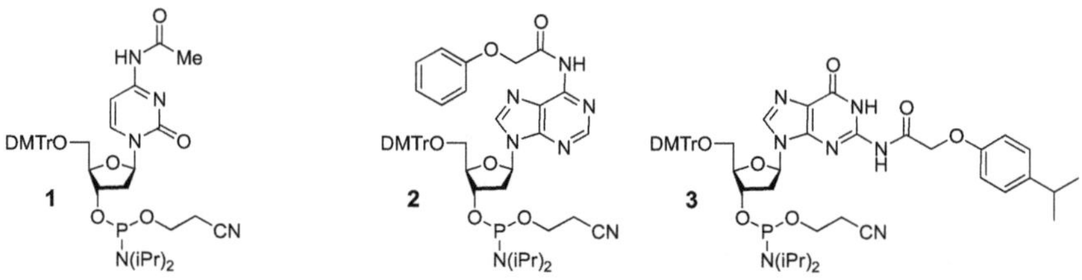

Fig. 1 Ultrasensitive phosphoramidites. **2** and **3** are protected with pac (phenoxyacetic) groups

So far, the phosphoramidite approach is the most appropriate method for site-specific incorporation of a lesion within an ODN irrespective of its length and sequence context.

A common feature shared by an impressive number of lesions is their base sensitivity, either due to facile epimerization or to the presence of groups sensitive to β-elimination. Unfortunately, usual final step at the end of ODN synthesis requires harsh alkaline conditions (30 % aqueous solution, 55 °C for 16 h for dG-containing ODN). An important breakthrough was achieved with the development of the commercially available "pac phosphoramidite" **1–3** (Fig. 1) (Virta 2009) based on a new set of protecting groups removable in mild conditions (methanolic 50 mM K_2CO_3 for 4 h at RT or concentrated ammonia for 2 h at RT). Phosphoramidites **1–3** were then commonly used for synthesis of lesion-containing ODNs.

2 Abasic Sites

2.1 "True" Abasic Site (4, 5, Ap)

This lesion is mainly present as hemiacetal **4** (99 %) in equilibrium with unstable aldehyde **5** (1 %) prone to strand breakage at physiological pH and lower via a β-elimination reaction. To overcome this instability, Ap site was first chemically incorporated in ODN as a stable analogue lacking the 1′-hydroxyl group using **6a** or **6b**, as a suitably protected aldehyde precursor **7** or as a protected hemiacetalic derivative **8** and **9** (Fig. 2).

The phosphoramidite of the stable tetrahydrofuran analogue **6a** (Eritja et al. 1987) was the first to be synthesized with a methyl group for transient protection of the P^{III} atom. The more user-friendly cyanoethyl derivative **6b** (Iyer et al. 1990) was next developed and is still commercially available. Stable analogues have also been synthesized for determining the structural feature of the Ap site (Huang and Greenberg 2008).

Synthesis of the "true" abasic site **4** is more challenging due to its intrinsic instability. Earlier post-synthetic methods relied on heat deglycosylation of a short oligomer containing a single acid-sensitive nucleobase/pyrimidine (Iocono

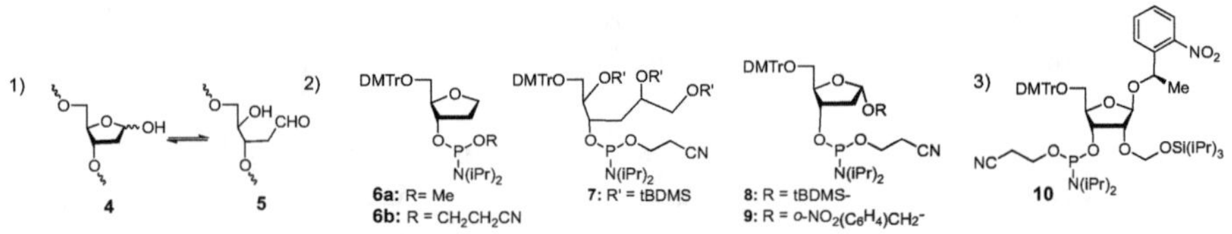

Fig. 2 (1) Equilibrium between hemiacetal (**4**) and aldehyde (**5**) forms of Ap site; (2) structures of the phosphoramidite derivatives of the stable Ap site analogues (**6a** and **6b**) of the protected aldehyde (**7**) or of the hemiacetal (**8** and **9**); (3) phosphoramidite of RNA abasic site **10**

et al. 1990), 8-alkylthiopurine (Laayoun et al. 1994), or 1-deazaguanine (Kojima et al. 2005). Direct incorporation of suitably protected precursors of abasic site was performed by using **8** (Groebke and Leumann 1990) or **9** (Péoc'h et al. 1991). In the former strategy, 1′-hydroxy group was protected as a tBDMS derivative and released under mild acidic conditions. In the second strategy: the 1′-hydroxy was protected with a photosensitive *o*-nitrobenzyl group derivative. The phosphoramidite **9** was introduced during ODN elongation, and the free Ap site was released quantitatively after UV illumination of the purified ODN.

A more sophisticated procedure was also described (Coleman and Pires 1999): a dG-containing target ODN was hybridized with a complementary strand bearing a bromoacetyl residue opposite a guanine residue. Upon hybridization, N7 position of the target dG was alkylated and became more prone to depurination.

A further improvement was carried out by the introduction of the aldehydic form of the abasic site **5** with a phosphoramidite precursor **7** introduced as a tBDMS protected triol (Shishkina and Johnson 2000). Unmasking was performed by DMT removal and triol desilylation by aqueous AcOH, followed by periodate treatment (glycol oxidation) and spontaneous ring closure.

The more recent works on Ap site chemistry focused on synthesis of its RNA counterpart (Trzupek and Sheppard 2005; Küpfer and Leumann 2005, 2007). These teams used photolabile precursors such as **10**.

2.2 The 4-Dioxo-2-Phosphorylbutane Lesion (11, 12, DOB)

The DOB precursor **15** was obtained as a mixture of four isomers. The hydroxyl groups in 1′ and 4′ positions were protected as photolabile nitroveratrol derivatives (Kodama and Greenberg 2005; Guan et al. 2010) (Fig. 3).

Interestingly, when opposite dA in a duplex, DOB reacts with exocyclic amino group to form interstrand cross-link by formation of a pyrrolidine tether (Guan and Greenberg 2009, 2011). A DOB-dA cross-linked analogue was introduced by phosphoramidite chemistry following either normal or reverse phosphoramidite. Authors elegantly obtained an isomerically pure cross-linked duplex (Ghosh and Greenberg 2014).

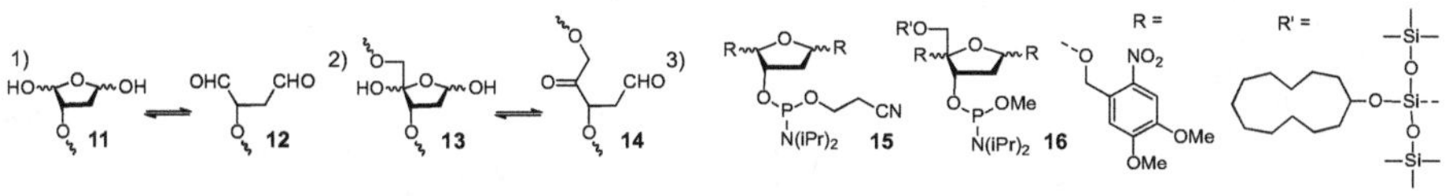

Fig. 3 Structures of (1) the DOB lesion (**11, 12**), (2) the C4-AP lesion (**13, 14**), (3) phosphoramidite derivatives for the introduction of the DOB lesion (**15**), or the C4-AP lesion (**16**)

2.3 C4′-Oxidized Abasic Site (13, 14, C4-AP)

Chemical incorporation of C4-AP site (**13, 14**) was performed in 2003 using a modified phosphoramidite (Kim et al. 2003). The 5′-alcohol in phosphoramidite **16** (and in the upcoming phosphoramidites) was silylated and the 1′- and 4′-hydroxyl groups protected by the photolabile nitroveratryl group. Stable analogues were synthesized following similar strategies (Kim et al. 2005).

Other approaches relying on incorporation of a caged precursor of the C4′-oxidized abasic site were described (Aso et al. 2006; Usui et al. 2008).

2.4 The 2′-Deoxyribonolactone Lesion (17, 1-AP)

Post-synthetic light-triggered generation of deoxyribonolactone **17** was achieved by illumination of an ODN containing a 7-nitroindole ring as a nucleobase analogue **18**. Under UV illumination the nitro group leads to radical H-abstraction at the 1′ position. The resulting radical reacts with water leading to the expected ribonolactone and free 7-nitrosoindole as by-product (Kotera et al. 1998). Other strategies, all relying on a key photochemical reaction to produce radical intermediates, were reported, either using 3-deaza-3-nitro-2′-deoxyadenosine **19** (Berthet et al. 2009; Crey-Desbiolles and Kotera 2006), 2-nitrotoluene **20a** or 2-nitroveratrol **20b** as sacrificial base analogue (Zheng and Sheppard 2004; Lenox et al. 2001), or **21** (Tronche et al. 1998) or 5-halouridine (Tashiro et al. 2008) for H1′ radical generation (Fig. 4).

2.5 The Erythrose Lesion (25, C2-AP)

Erythrose lesion **25** was incorporated in an ODN as a protected triol phosphoramidite **22**. The aldehyde function was released by post-synthetic mild oxidation of the glycol moiety **24** (Kim et al. 2004) (Fig. 5).

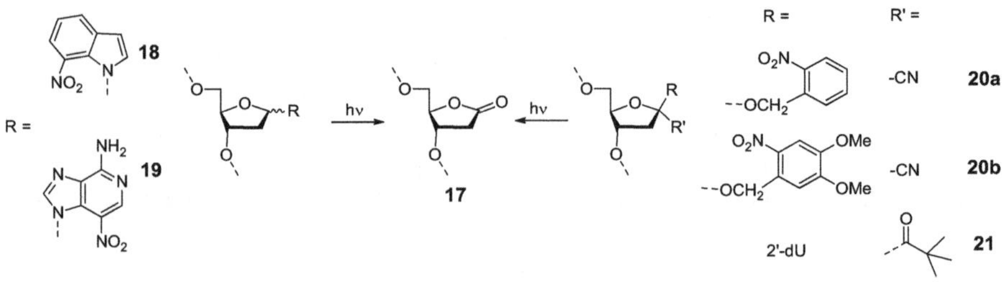

Fig. 4 Nonnatural nucleobase analogues used for the post-synthetic generation of 2′-deoxyribonolactone lesion

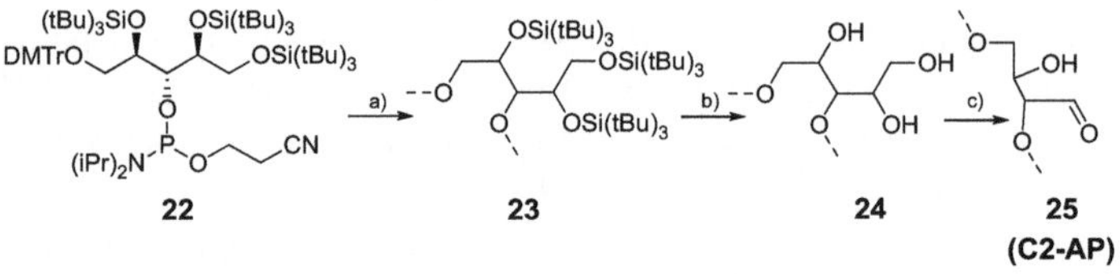

Fig. 5 Structures of the C2-AP lesion **25**. (a) Oligonucleotide elongation, (b) NMP/Et_3N/$Et_3N{\cdot}3HF$, 65 °C, 3 h, (c) $NaIO_4$, pH = 6, 30 min, RT

3 Oxidized Purines

3.1 7,8-Dihydro-8-oxo-2′-Deoxyguanosine (8-OxodG)

Phosphoramidite **31** was prepared from 2′-deoxyguanosine (dG) after bromination of the 8-position, followed by substitution by a benzyloxy group and its hydrogenation (Einolf et al. 1998). Substitution of the bromine atom (**26** to **27**) is the bottleneck of the reaction (28 % yield and large amount of by-products). Increasing the BnOH/DMSO ratio from 1:3 to 3:2 led to an 80 % yield and easier purification (Nampalli and Kumar 2000).

Direct incorporation of 8-oxodG by phosphoramidite chemistry was achieved in 1992 using **31**. Importantly, mercaptoethanol had to be added in the deprotection mixture as 8-oxoG is sensitive to aerial oxidation (60 % degradation yield was reported in the absence of reducing agent) (Bodepudi et al. 1992) (Fig. 6).

Other biologically relevant 8-oxopurines were also reported: *N*2-alkyl 8-oxoG (Kannan and Burrows 2011), 2-amino, 8-oxopurine (namely, 8-oxoG lacking the 6-oxo group) (Cadena-Amaro et al. 2005), and the 6*O* methyl derivative (Varaprasad et al. 1996).

Synthesis of 8-oxoA with a 70 % yield relied on a similar strategy for the 8-position oxidation (Bodepudi et al. 1992). The direct conversion of the 8-bromo derivative into 8-oxoA was further reported with a 95 % yield (Chatgilialoglu et al. 2006).

The phosphoramidite derivatives of the putative (4*R*) and (4*S*) diastereoisomers of 4-hydroxy-8-oxodG (formally 4,8-dihydro-4-hydroxy-8-oxo-2′-deoxyguanosine)

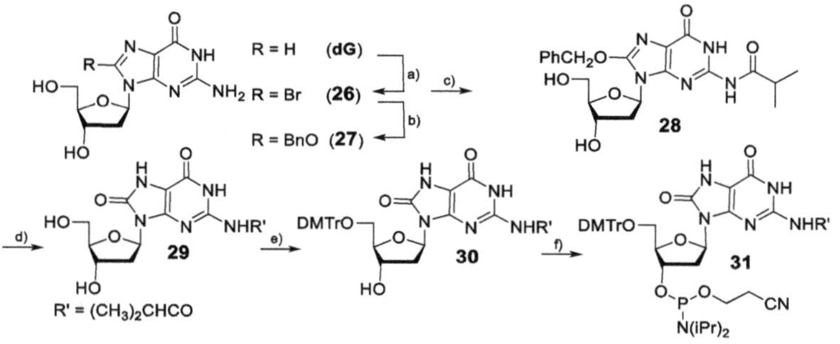

Fig. 6 Synthesis of 8-oxodG phosphoramidite **31**. (a) Br_2, H_2O, 0°C; (b) BnONa, DMSO, 65 °C, 24 h; (c) i-Me_3SiCl, pyr.; ii-isobutyric anhydride, RT, 4 h.; iii-NH_3 (aq), 15 min, 0 °C; (d) H_2 (60 psi), Pd/C, *n*-BuOH/H_2O, 55 °C, 12 h; (e) DMTrCl, pyr., 0 °C → RT, 15 min; (f) $(iPr)_2NP(Cl)OCH_2CH_2CN$, TEA, CH_2Cl_2, 20 min, RT

were also synthesized. This lesion was supposed to arise from 8-oxodG oxidation by singlet oxygen ($^1O_2$). Key step of the synthesis was a 30 h photosensitized oxidation of an oxygen-saturated aqueous solution of dG yielding ca. 45 % of the expected diastereoisomer pair (Romieu et al. 1999a). Further studies revealed that the right structure was a spiranic compound **127** (*vide infra*) (Adam et al. 2002).

3.2 The 2,6-Diamino-4-Hydroxy-5-Formamidopyrimidine Lesion (32, Fapy-dG) and 4,6-Diamino-5-Formamidopyrimidine (Fapy-dA) Lesions

Fapy-dG was incorporated by phosphoramidite chemistry as a lesion-containing dimer (Haraguchi and Greenberg 2001). An important structural feature of the Fapy-dG nucleoside is the easy conversion between furanoid and pyranoid isomers and their fast epimerization. As this conversion cannot happen when Fapy-dG is embedded in an oligonucleotidic scaffold (5′-alcohol being linked to a phosphate residue), the chosen chemical pathway relied on direct incorporation of a suitably protected dinucleotide. On the other hand, facile epimerization at the C1′ position precludes the need to design stereospecific synthetic pathway (Fig. 7).

The nitro Fapy-dG precursor **34** was used as a racemic mixture and coupled with **33** to afford a Fapy-dG-T dimer **35**. Internucleotidic phosphate was protected as a methyl triester as the classical β-cyanoethyl group did not withstand the subsequent transformations. Finally, the formamido moiety in **37** was obtained through reduction of the nitro group **35** and formylation of the resulting amino group **36**. Phosphoramidite **38** was incorporated with a 70 % yield using a reverse strategy (i.e., chain elongation from 5′ to 3′ end). Pivalic anhydride/lutidine mixture was used for capping step (usual acetylation conditions lead to deformylation). Four years later, the same authors presented a new dimer phosphoramidite **39** which was incorporated by the usual 3′ to 5′ chain elongation (Jiang et al. 2005). Improvements

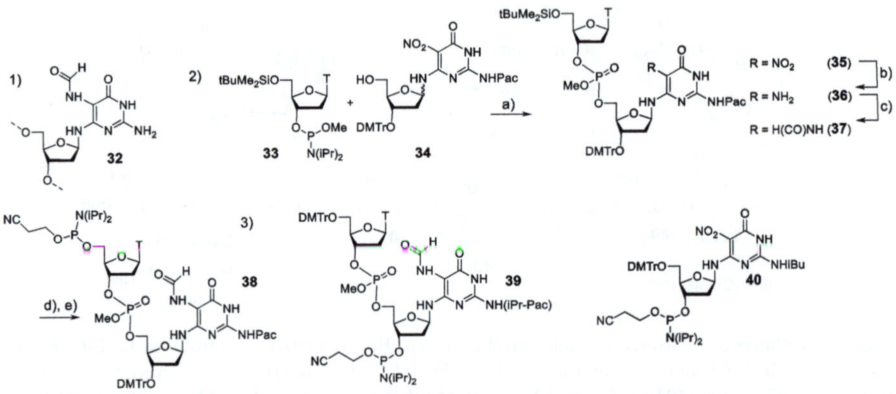

Fig. 7 (1) Structure of the Fapy-dG lesion **32**, (2) synthesis of the phosphoramidite **38** for the introduction of the Fapy-dG lesion, (a) i-tetrazole, ii-tBuOOH; (b) H_2, Pd/C, MeOH; (c) H(C=O)OAc, pyr., THF; (d) Bu_4N^+, F^-, AcOH, THF; (e) $(iPr)_2NP(Cl)OCH_2CH_2CN$, TEA, CH_2Cl_2, 20 min, RT. (3) Phosphoramidite **39** and **40**

of the synthesis of the nitro Fapy-dG precursor were also reported (Haraguchi et al. 2002).

A post-synthetic generation of Fapy-dG was described: nitro phosphoramidite **40** was incorporated in a short trinucleotide (TXT), which was in turn submitted to the reduction and formylation steps. Formylation was performed using *N*-formylimidazole. No over-formylation was reported (Lukin et al. 2011).

New syntheses of the α and β anomers of Fapy-dG as well as the RNA counterpart (Fapy-G) were reported (Burgdorf and Carell 2002). Fapy-dGTP was also prepared and its incorporation by DNA polymerase appeared to be roughly 1000 times less effective than normal dGTP (Imoto et al. 2006). A close analogue, Me-Fapy-dG, was also synthesized. This lesion is supposed to arise from *N*7 alkylation of dG followed by imidazole ring opening (Christov et al. 2008).

3.3 The Imidazolone (41, dIz) and Oxazolone Lesions (42, dZ)

The dIz **41** and dZ **4** lesions (Fig. 8) were first isolated and characterized after UV illumination of an aqueous solution of 2′-deoxyguanosine in the presence of riboflavin as photosensitizer. Deoxyguanosine is partially converted into the dIz lesion **41** which in turn is quantitatively converted into dZ **42**. Unfortunately, a rapid degradation into guanidine was observed at pH = 10 for dZ and dIZ (Cadet et al. 1994) as well as for the dinucleotide d(TpIz) (Buchko et al. 1995). This instability precluded their chemical incorporation in ODN.

First attempt to obtain dIZ and dZ lesions was achieved post-synthetically by illumination of a single-G-containing ODN in aerated aqueous solution in the presence of riboflavin (Kino and Sugiyama 2001; Gasparutto et al. 1998).

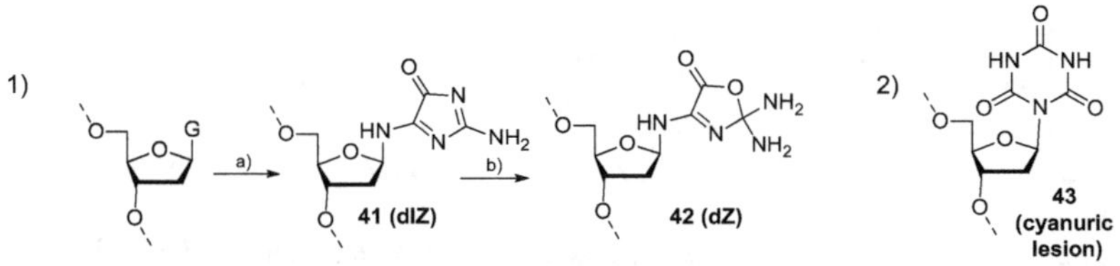

Fig. 8 Structures of the (1) dIZ, dZ and (2) cyanuric lesions. (a) Irradiation of a riboflavin aerated aqueous solution; (b) H_2O, pH = 7, RT, 20 h

ODN containing the dIZ lesion was obtained as the major photoproduct, but spontaneous hydrolysis of the dIZ lesion in neutral conditions leads to a final 15 to 38 % yield.

Direct chemical incorporation of dZ or dIz by phosphoramidite chemistry has not been reported. However, a stabilized dZ lesion analogue was post-synthetically obtained by illumination of an ODN containing the 1′*C*-carbanucleoside of 8-oxodG (Mueller et al. 2008).

3.4 *The Cyanuric Acid Lesion (43)*

Cyanuric acid lesion (Fig. 8) is not alkali-labile. Its incorporation in an ODN was performed by automated synthesis using phosphoramidite precursor **43**. The β-anomer was purified by silica gel chromatography at the nucleoside step (Gasparutto et al. 1999).

4 Damaged Pyrimidine: Oxidized and Reduced Ring (Fig. 9)

4.1 *The 5,6-Dihydrothymine Lesion (dDHT, 44)*

The poor stability of dDHT **44** in concentrated ammonia at RT is the main bottleneck for successful chemical incorporation of this lesion in synthetic ODN (a less than 2 h alkaline treatment is recommended). The first incorporation of **51** was achieved in a synthetic poly(dT) (Molko et al. 1985), and its incorporation in an ODN containing the four nucleobases could be achieved using the pac protecting groups (Schulhof et al. 1988). Synthesis of **51** started from rhodium-catalyzed hydrogenation of the thymine double bond yielding the 5*S* isomer as a major product.

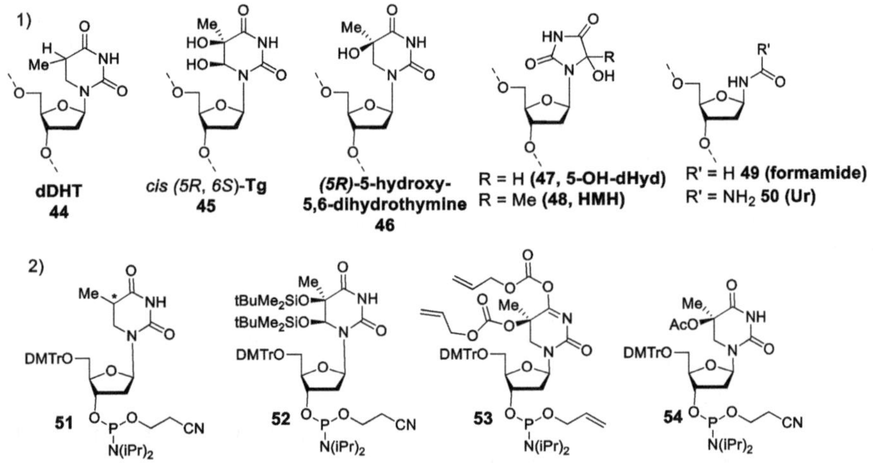

Fig. 9 (1) Structures of the main thymine lesions (**44–50**) and (2) corresponding phosphoramidites **51–54**

4.2 *The 5,6-Dihydroxythymine Lesion (Tg, 45)*

Chemical synthesis of Tg **45** was achieved by incorporation of **52**. This phosphoramidite was obtained by osmylation of the thymine double bond leading to the formation of both *cis* isomers. The synthesis was pursued only on the major 5*R*, 6*S* isomer obtained in a 6:1 ratio. Both 5- and 6-hydroxy groups were protected as TBDMS ether, and chain elongation was performed using pac groups for the exocyclic amines protection. Finally, diol was freed by a fluoride treatment (Iwai 2000). Synthesis of the other main diastereoisomer 5*S*, 6*R* was undertaken after chromatographic separation of both isomers. Interestingly, tertiary alcohol in position 5 did not need to be protected (Iwai 2001). A strong improvement of the synthesis of **52** was obtained by using ionic liquid as a cosolvent for the osmylation reaction (Shimizu et al. 2006). Under such conditions, the 5*S*, 6*R* isomer is predominantly obtained in a 1:2 mixture. Diastereoisomeric mixture was also incorporated in synthetic oligomers following a close chemical pathway: 5- and 6-hydroxyl groups were protected as levulinyl ester (Gasparutto et al. 2005).

4.3 *The 5-Hydroxy-5,6-Dihydrothymine Lesion (46)*

This lesion does not withstand the usual final alkaline cleavage. A precursor **53** protected by allyloxy and allyloxycarbonyl groups was designed. Final ODN was linked to the support by a photocleavable tether. After chain elongation, the deprotection and release of the free oligomer were achieved by Pd(0) treatment and photoillumination (Bernadou et al. 1994). However, amino groups' deprotection appeared to be incomplete (up to four allyl remaining residues). Moreover, illumination step precluded the use of ODN containing contiguous thymine in the sequence (Greenberg and Matray 1997).The design of phosphoramidite **54** based on acetyl protection of the 5-hydroxy group and the use of pac protecting groups circumvented these drawbacks (Sambandam and Greenberg 1999).

4.4 The 5-Methylhydantoin Lesions (5-OH-dHyd 47) and 5-Hydroxy, 5-Methylhydantoin (HMH 48)

The 5-OH-dHyd **47** lesion is the major decomposition product of ozone-mediated oxidation of 2′-deoxycytidine. Its phosphoramidite derivative was prepared in 2001 as a racemic mixture. The key step is the direct conversion of dC into **47** derivative by ozonolysis, followed by 5′-DMT protection (28 % yield, two steps) and protection of the secondary alcohol in position 5 as a levulinic ester (Muller et al. 2001). This lesion appeared to be more alkali-labile than its methylated counterpart HMH **48** produced by ionizing radiation. Thymine was converted in one step into HMH by reaction with a $KMnO_4/Pb(OAc)_4$ aqueous solution in 35–41 % yield (Guy et al. 1993). The resulting HMH derivatives were obtained as a racemic mixture and converted into phosphoramidite without protection of the tertiary 5-hydroxy group. Although HMH lesion is prone to C1′ racemization, its alkali lability occurs only at high temperature, and it could be inserted in ODN by using pac strategy (Gasparutto et al. 2000).

4.5 The Ribosylformamide (49) and Ribosylurea (Ur, 50) Lesions

The ribosylformamide lesion **49** is often produced along with the Ur lesion **50** during oxidation of thymine. It is neither alkali-labile nor prone to anomerization, but it exists as *cis* and *trans* forms in a 3:7 ratio supposed to display different hybridization properties once embedded into DNA; lesion **49** was easily introduced by phosphoramidite chemistry (Baillet and Behr 1995).

A post-synthetic procedure was reported starting from oligomer containing a Tg lesion **45**. Overnight mild oxidation with sodium periodate leads to the expected lesion **49**-containing ODN in a 43 % yield (Toga et al. 2009).

The phosphoramidite synthon of ribosylurea was obtained by $KMnO_4/Pb(OAc)_4$ double bond oxidation of a suitably protected thymine derivative leading to a mixture containing urea derivative (16 %) and *N*-formylurea (18 %). The urea moiety was inserted in synthetic oligomer without protection using pac strategy (Guy et al. 1990). Others claimed that the oxidation step is poorly reproducible and proposed a two-step strategy: firstly permanganate-mediated oxidation of T into alkali-labile thymidine glycol derivative and secondly degradation into a 3:1 ratio of α- and β-ribosylurea isomers in sodium hydroxide solution in a 57 % yield (Dubey et al. 2001).

5 Pyrimidine Dimers

5.1 *Cyclobutane Pyrimidine Dimer (CPD, 55)*

CPD **55** results from photocross-linking of two contiguous thymines. Earlier synthesis of this damage was based on direct UV illumination of a short oligomer (Kemmink et al. 1987) affording CPD-containing oligomer in 20–30 % yield with *cis-syn* stereochemistry. This method is still of synthetic importance and was employed for generating TpC and Tp^{Me}C CPD-containing oligomers (Song et al. 2012) as well as CPD-containing ODN with thiophosphate internucleotidic bond (Othuka et al. 1992) (Fig. 10).

First, chemical incorporation of a *cis-syn*-cyclobutane dimer into synthetic oligomer by phosphoramidite **58** was reported in 1987 (Taylor et al. 1987). The key step was the photoillumination of a TpT dimer **56a** leading to a mixture of four column-separable compounds displaying R_P or S_P isomerism and *cis-syn* or *trans-syn* stereochemistry. The most abundant (27 %) *cis-syn* isomer **57** was purified and

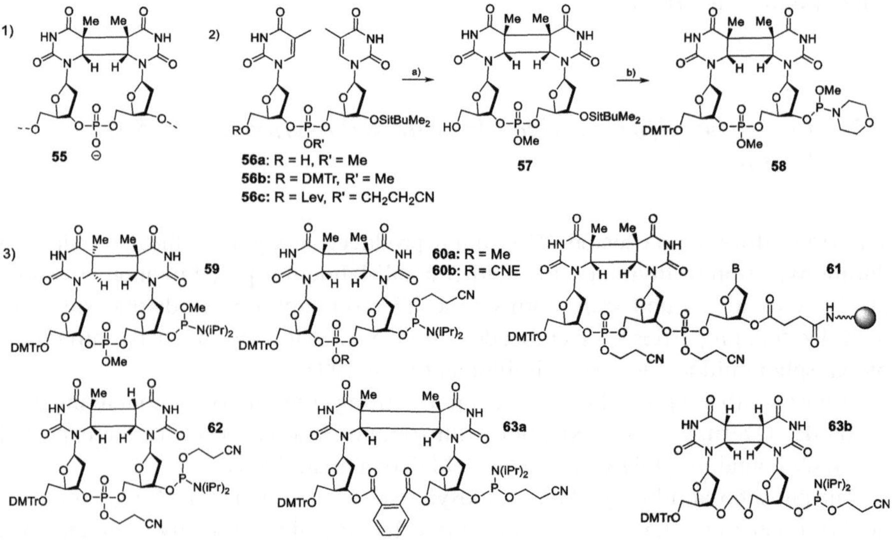

Fig. 10 (1) CPD lesion **55**, (2) synthesis of *cis-syn* phosphoramidite **58**, (a) hν, (b) i-DMTrCl, pyr., ii-TBAF, THF, AcOH; (3) CPD phosphoramidites **59**, **60a,b** and **62**, support **61** and analogues **63a,b**

converted to a ready-to-use phosphoramidite **58**. Incorporation of **58** into a tetramer was achieved in a 50 % yield. Although C3-N4 bond of the imide moiety of CPD is known to rapidly hydrolyze under alkaline conditions, CPD lesion withstands final ammonolysis conditions (Blackburn and Davies 1965). An intractable mixture was obtained when performing photoillumination of **56b**. Trityl group was claimed to be detrimental to [2+2] cycloaddition (Taylor et al. 1987) although others reported opposite results (Nadji et al. 1992). Levulinate derivative **56c** was also successfully used (30 % conversion) (Ortiz Mayo et al. 2003).

Phosphoramidite **59** was obtained after tedious chromatography steps from *trans-syn* stereoisomer (Taylor and Brockie 1988). More user-friendly derivatives **60a** and, especially, **60b** (Kosmoski and Smerdon 1999) were further developed with an internucleotidic phosphate group protected as a β-cyanoethyl derivative (Tommasi et al. 1996; Murata et al. 1990). Others reported that this group is not compatible with photoillumination (Yamamoto et al. 2014). The resulting CPD-containing oligomer was deprotected in concentrated ammonia at 55 °C for 5 h. A phosphorodithioate analogue (Murata et al. 1990) and a [3-^{15}N]-labeled CPD-containing ODN (Bdour et al. 2006) were also synthesized following this approach.

Synthesis of a d(TpU) CPD phosphoramidite **62** was also described (Takasawa et al. 2004) and appeared to be more alkali-sensitive than its TpT counterpart (Ortiz Mayo et al. 2003). Successful incorporation of TpC CPD lesion analogue by phosphoramidite approach was achieved following similar chemical pathway. Surprisingly, illumination of the TpC dimer leads to the *trans-syn* CPD as the major product, and cytosine moiety is rapidly converted into U by deamination (Cannistraro and Taylor 2009). Consequently, the authors introduced cytosine as its *N*4-methylated counterpart. The electron-donating ability of the methyl group is supposed to reduce its hydrolysis susceptibility (Yamamoto et al. 2014).

A TpT dimer analogue phosphoramidite **63a** with a transient phthalate internucleotidic linkage was devised to unravel the influence of the natural phosphodiester bond (Nadji et al. 1992). Ester bonds of the phthalate tether were hydrolyzed during the final deprotection. A formacetal dimer **63b** was also devised to introduce isostere analogue of UpU CPD-containing lesion (Butenandt et al. 1998). Although CPD-containing oligomers can be successfully synthesized by this approach, large excess of phosphoramidite was required, and little was said on incorporation yields. To extend the usefulness of these phosphoramidite building blocks, the solid support **61** was synthesized and ODNs were further extended firstly chemically (from the 5′ end) and then enzymatically (Ordoukhanian and Taylor 1997).

5.2 The (6-4)-Photoproducts (64) and the Dewar Photolesion (65)

The synthesis and incorporation of the T-(6-4)-T photoproduct phosphoramidite **70** were achieved in 1996 (Iwai et al. 1996). A suitably modified TpT dimer **66** was UV illuminated to form the (6-4)-photoproduct **67** in a 16 % yield. Attempts to protect the tertiary alcohol in **67** with acetic anhydride only lead to *N*3-acetylation. Consequently, the tertiary alcohol in **68** was left unprotected. The T-(6-4)-T photoproduct revealed to be stable in AcOH 80 % RT and in concentrated ammonia but only at RT. In harsh alkaline conditions, it decomposed by hydrolysis of the N3 and C4 bond of the 5′-pyrimidine component (Higurashi et al. 2003) by subsequent hydrolysis of the glycosidic bond of the 3′ moiety (Arichi et al. 2012) and by deamination (Lin et al. 2014). Consequently, the pac protection scheme was used to assemble the T-(6-4)-T photoproduct-containing ODN. Several slower-eluting by-products were observed after deprotection, suggesting that branching occurred by over-phosphitylation of the imide *N*3 atom of the (6-4) moiety. The expected oligomer was isolated as the major product by switching from tetrazole to benzimidazolium triflate as activator thus dramatically enhancing the *O*- vs. *N*-phosphitylation (Iwai et al. 1999). The T-(6-4)-C photoproduct lesion was also incorporated using the phosphoramidite **71** (Mizukoshi et al. 1998) synthesized by irradiation of the TpC dimer in a 3.9 % yield. The T-(6-4)-C lesion being alkali-sensitive, the pac strategy was required. However, capping steps after incorporation of **71** were omitted to avoid acylation of the tertiary amine. Similarly to what was noticed for CPD syntheses, the 5′-DMTr group was reported to be unstable when irradiated at 254 nm. Consequently, it was removed prior to photoillumination and reintroduced after.

The thio analogue **72** was also reported and successfully incorporated in ODN (Matus et al. 2003), although the chosen protecting group for the thiol moiety did not fulfill all the synthetic requirement for incorporation in ODN (Guérineau et al. 2004). The phosphoramidite of the Dewar lesion **73** was also prepared with an unprotected tertiary alcohol and was incorporated in synthetic ODN. The use of benzimidazolium triflate as activator instead of the usual tetrazaole leads to the formation of lower amount of branched oligomers. This lesion is compatible with the ultramild deprotection conditions, although small amount of by-product formation was reported after ammonia deprotection (Yamamoto et al. 2006) (Fig. 11).

5.3 The Spore Photoproduct (SP, 74)

Spore photoproduct dimer synthesis was achieved in 1995. The key step is the bromine substitution in **75a** by protected hydrogenated thymine unit **75b** affording the SP dimer isomers **76** in 1:1.2 ratio (Kim et al. 1995). Synthesis was further improved (Chandra et al. 2009) and analogues were developed for crystallographic

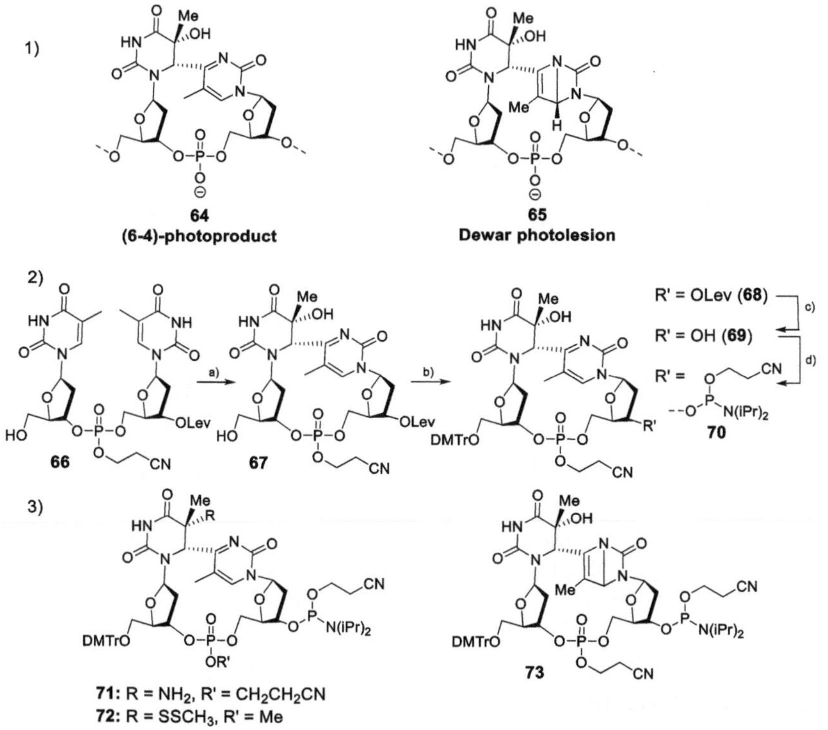

Fig. 11 (1) Structures of the (6-4)-photoproduct **64** and the Dewar photolesion **65**, (2) synthesis of the phosphoramidite derivative of the (6-4)-photoproduct **70**: (a) UV 254 nm irradiation, (b) DMTrCl, pyr., (c) $N_2H_4 \cdot H_2O$, pyr./AcOH, (d) $(iPr)_2NP(Cl)OCH_2CH_2CN$, DIEA. (3) Phosphoramidites **71, 72** and **73**

(Lin et al. 2011) or enzymatic studies (Friedel et al. 2006). Incorporation of SP analogues was next achieved with phosphoramidite lacking the internucleotidic phosphate bond **78** (Bürckstümmer and Carell 2008).

The chemical incorporation of the "true" SP lesion was recently achieved (Fig. 12, phosphoramidites **81a,b**) (Jian and Li 2013). Coupling of both thymidine units **75a** and **75b** by S_N reaction afforded the *R* and *S* isomers **76** in 28 and 23 % yields, respectively. After RP-HPLC separation of the biologically relevant *R* isomer, the 3′ and 5′ alcohols of both thymine units were bridged by a *o*-chlorophenyl phosphate triester link (**80**). However, incorporation of the resulting amidite **81a** into growing ODN chain proceeded poorly (15 % yield). The phosphate chlorophenyl group was thus converted into the less bulky and more reactive (90 % yield) methyl ester (phosphoramidite **81b**).

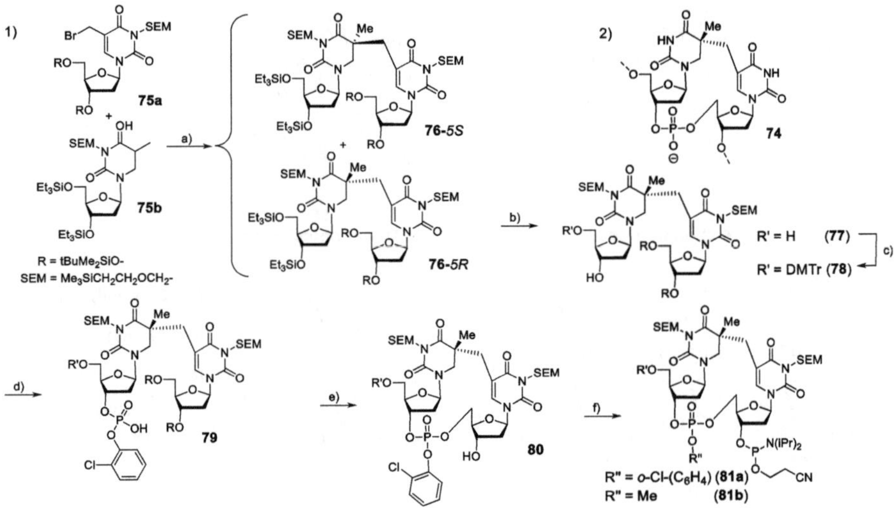

Fig. 12 (1) Synthesis of phosphoramidite **81a** (a) LDA, THF, −78 °C; (b) HF- pyr., −20 °C, MeCN; (c) DMTrCl, pyr.; (d) *o*-$ClC_6H_4POCl_2$, 1,2,4-triazole, THF, pyr.; (e) i-TBAF, THF; ii-1-mesitylene-sulfonyl-3-nitrotriazole, pyr.; (f) $(iPr)_2NP(Cl)OCH_2CH_2CN$, DIEA; (2) spore photoproduct **74**

6 Cyclonucleosides

The cyclonucleoside lesions (e.g., cyclo-dA **82**) are formed by ionizing radiation following a radical process. Synthesis of (5′*S*)-cyclo-dA phosphoramidite **91** (Romieu et al. 1998) is depicted in Fig. 13. The suitable protected adenosine derivative was turned into thioether **83**, which was converted in 51 % yield in anhydronucleoside **84** after illumination at 254 nm. The 3′-position in **85** was then easily protected. The chirality of the 5′-position was further introduced by an oxidation/reduction sequence. Steric hindrance led to the sole formation of the 5′*S* isomer **87**. The phosphoramidite **91** was stable and conveniently incorporated in synthetic oligomers. The synthesis of the phosphoramidite derivative of (5′*S*)-cyclo-dG was performed following similar chemical pathway (Romieu et al. 1999b). The configuration at the 5′-position could be inverted by either Mitsunobu reaction (Chatgilialoglu et al. 2014) or Corey's method (Romieu et al. 1999b) leading to 5′*R* derivatives of cyclo-dA and cyclo-dG. An alternative strategy was developed relying on 5′ → 3′ elongation by incorporating the reverse phosphoramidite **98** whose 3′ position was protected by silyl ether and further deprotected by 1 M TBAF (Brooks et al. 2000).

Synthesis of suitably protected cyclopyrimidine was more difficult because of the alkali lability of the hydrogenated pyrimidine and the formation of three chiral centers at the 5′, 5, and 6 positions. Moreover, rapid epimerization at position 5 precludes the need to incorporate a single 5*R* or 5*S* isomer. Cyclothymidine **93** was obtained as a mixture of both (5′*S*, 5*R*, 6*S*) and (5′*S*, 5*S*, 6*S*) isomers in a 2.7:1

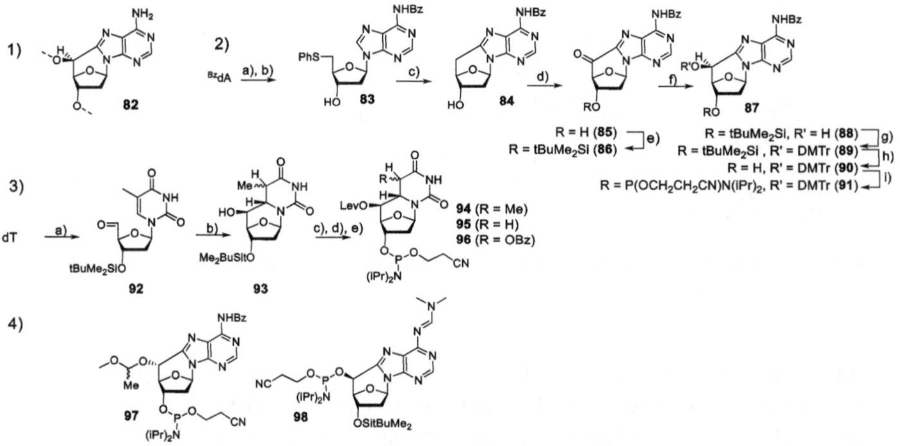

Fig. 13 (1) Structure of the cyclo-dA lesion **82**; (2) synthesis of phosphoramidite **91**. (a) TsCl, pyr., −20 °C; (b) PhSH, MeONa, MeOH, reflux; (c) 254 nm irradiation, $(EtO)_3P$, MeCN; (d) $tBuMe_2SiCl$, imidazole, DMF; (e) SeO_2, dioxane, reflux, 30 min; (f) $NaBH_4$, MeOH; (g) DMTrCl, pyr.; (h) TBAF, THF; (i) $(iPr)_2NP(Cl)OCH_2CH_2CN$, DIEA; (3) synthesis of phosphoramidites **94–96**, (a) Dess-Martin periodinane, (b) Bu_3SnH, AIBN, benzene, (c) levulinic acid, DCC, DMAP; (d) TEA·3HF, THF; (e) $(iPr)_2NP(Cl)OCH_2CH_2CN$, DIEA. (4) Structures of phosphoramidites **97–98**

ratio by intramolecular radical reaction of the aldehydic derivative **92**. The hindered 5′-secondary alcohol **93** appeared to be reluctant to tritylation, and the incorporation of the less bulky levulinate group was retained instead in phosphoramidite **94** (Romieu et al. 1999c). After ODN synthesis, the levulinate group was removed by hydrazine treatment. Reaction of the unprotected 5′-secondary alcohol with the next upcoming phosphoramidite required extended coupling time (Romieu et al. 1999a). The preparation and incorporation of (5′*S*,6*S*)-cyclo-5,6-dihydro-2′-deoxyuridine phosphoramidite **95** was also reported (Muller et al. 2000). Alternative introduction of cyclo-dA (phosphoramidite **97**) and unnatural cyclo-dU derivative using a different protection scheme was also reported (Yueh et al. 2012).

Recently, the cyclo-dA phosphoramidite synthesis has been improved by photoillumination of a 8-bromopurine derivative affording 54% of the expected cyclonucleoside in a 3:7 5′*S*/5′*R* ratio (Chatgilialoglu et al. 2014). Following a similar chemical pathway, the synthesis of the nonnatural cyclo-(5-hydroxy-5,6-dihydrouridine) phosphoramidite **96** was also reported (Muller et al. 2002).

7 Intrastrand Cross-Link

In order to achieve incorporation of a site-specific intrastrand cross-link, the alkali-sensitive photoactivable nucleoside phosphoramidite **99a** was devised (Fig. 14). Once embedded in oligomer **100**, and under UV exposure, the modified nucleobase

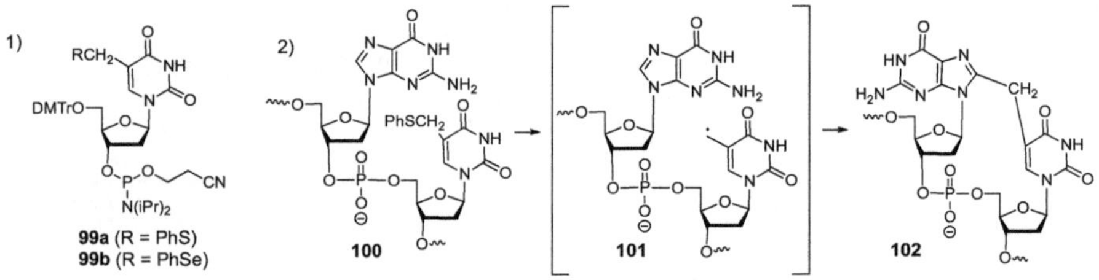

Fig. 14 (1) Structure of the phosphoramidite **99a** and **99b**; (2) post-synthetic generation of intrastrand cross-link

yielded a highly reactive radical **101** prone to cross-link with adjacent nucleobase in deaerated aqueous conditions. The d(T^G) and d(G^T) (**102**) lesions were post-synthetically generated (Bellon et al. 2006)as well as the d(A^T) and d(T^A) (Bellon et al. 2002).

Similarly, post-synthetic UV illumination of the ODN prepared from selenium-containing nucleoside **99b** led to interstrand cross-link (Hong and Greenberg 2005; Sloane and Greenberg 2014). UV irradiation of 5-bromo-dU/dC containing ODN produced U/C^G and A^U/C lesions (Zeng and Wang 2006; Hong and Wang 2005).

8 Strand-Break Lesions

One major lesion is the 5′-aldehyde formation **103** (Fig. 15). An important feature of this lesion is its versatility as it exists under several forms (A, T, G, C lesions). The thymine derivative was obtained by multistep synthesis. A suitably protected diol moiety **104** was incorporated in synthetic oligomers. The aldehyde function was released by post-synthetic periodate oxidation (Kodama and Greenberg 2005; Jung et al. 2011). A more universal method was devised later. After chain elongation, the 5′ unprotected alcohol **105** was mildly oxidized whatever the sequence context (**106**) (Sallamand et al. 2012) and protected as a *N*,*N*′-diphenylethylenediamine derivative **107**. Subsequent ammonolysis, purification, and aldehyde deprotection with usual AcOH 80 % RT 5 min treatment yield the expected 5′-aldehyde-containing ODN (Lartia and Constant 2015). Other strand-break lesion, DOB **11**, is described in Sect. 2.2.

9 Small Molecules Adducts

DNA is a known target for numerous pollutants widely present in the environment and their metabolites (polycyclic aromatic hydrocarbons, aromatic amines, drugs, aldehyde).

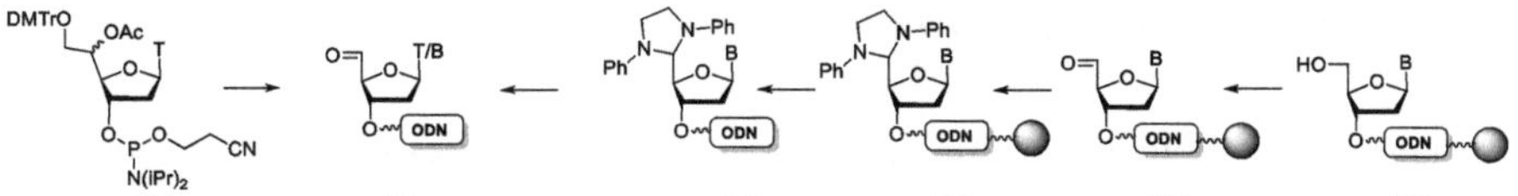

104 103 108 107 106 105

Fig. 15 Generation of the aldehyde lesion **103** by phosphoramidite approach (*left*) or by post-synthetic approach (*right*, B = A, T, G or C). *Gray ball*: CPG support

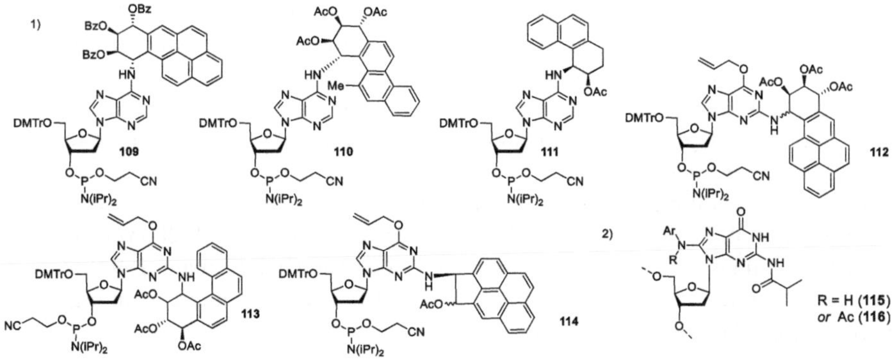

Fig. 16 (1) Some phosphoramidites used for chemical incorporation of nucleobase-PAH and (2) nucleobase-amine adducts lesions **115** and **116**

9.1 Polycyclic Aromatic Hydrocarbon (PAH) Adducts

Many adducts of PAH metabolites with purine have been reported so far and have been introduced into synthetic oligomers using phosphoramidite chemistry (Fig. 16). This is the case for adenosine adducts of benzo[*a*]pyrene **109** (Johnson et al. 2002; Ling et al. 2004; Yagi et al. 2007), 1-methylpyrene (Lee et al. 1990), and phenanthrene **111** (Lakshman et al. 1992a, b), as well as guanosine adducts of benzo[*a*]pyrene **112** (Kroth et al. 2001; Johnson et al. 2002; Yagi et al. 2007), benzo [*c*]phenanthrene **113** (Kroth et al. 2001), fluorene (Zhou et al. 1994; Bonala et al. 2005), naphthalene (Bonala et al. 2005), pyrene (Lee et al. 1990; Kirouac et al. 2013), nitropyrene (Colis et al. 2009), and cyclopenta[*c,d*]pyrene **114** (Yagi et al. 2007). Some were incorporated using H-phosphonate chemistry (Iyer et al. 2007). Another strategy relied on the incorporation of a fluorinated purine followed by post-synthetic displacement of the fluorine atom by an aminated PAH derivatives (Kim et al. 1992).

9.2 Aromatic Amines Adducts

Arylamine metabolites constitute a known class of reactive carcinogenic chemicals leading to formation of 8-arylaminated purines **115** as well as their *N*-acetyl counterparts **116** (Fig. 16). Phosphoramidite derivatives of aniline, *p*-methylaniline, *p*-cyanoaniline, 3,5-dimethylaniline, *p*-anisidine, 4-aminobiphenyl, and 2-aminofluorene (Böge et al. 2007, 2008) and their *N*-acetyl counterparts (Zhou and Romano 1993; Gillet and Schärer 2002; Gillet et al. 2005; Böge et al. 2006; Szombati et al. 2012; Krüger and Meier 2013) were synthesized and incorporated during chain elongation of synthetic ODNs. Alkylamination of the position 8 was performed by Buchwald-Hartwig amination of the corresponding 8-bromoderivative by aromatic amines. No further reactions took place on the resulting secondary amine. The final ODN had to be deprotected in hot ammonia in the presence of β-mercaptoethanol to avoid oxidative side reactions. *N*-acetyl derivatives are alkali-labile, and mild deprotection conditions such as 5 % methanolic DIEA at 55 °C for 14 h were required (Szombati et al. 2012). Others used the pac strategy (Gillet et al. 2005) or unusual exocyclic amines protecting groups such as Fmoc (Zhou and Romano 1993). A post-synthetic method, relying on Suzuki coupling of 8-bromoguanine-containing ODN, was reported (Omumi et al. 2011).

9.3 Aldehyde Adducts

The phosphoramidite derivative **117** of the adduct formed by the reaction of malondialdehyde with dG was synthesized by enzymatic transribosylation and incorporated in synthetic oligomers (Schnetz-Boutaud et al. 2000). Adducts formed with dA and dC were further reported (Wang et al. 2004), but their high alkali lability precludes their incorporation by phosphoramidite chemistry. Post-synthetic incorporation was then achieved for site-specific incorporation of nucleobase damaged by α,β-insaturated aldehydes (Wang et al. 2006; Maddukuri et al. 2010). Briefly, a 2-fluoropurine phosphoramidite derivative was incorporated in ODN **118**. After deprotection (0.1 M NaOH, 10 h, RT), ODN **118** was reacted with an amino derivative containing a diol precursor of aldehyde function, leading to the expected adduct. In some cases (synthesis of **120**), further reaction was required (Fig. 17).

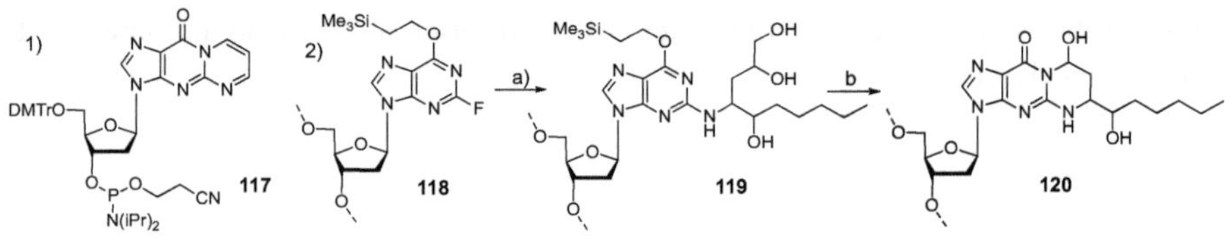

Fig. 17 (1) Phosphoramidite **117** and (2) post-synthetic generation of the-4-hydroxynonanal adduct **120**

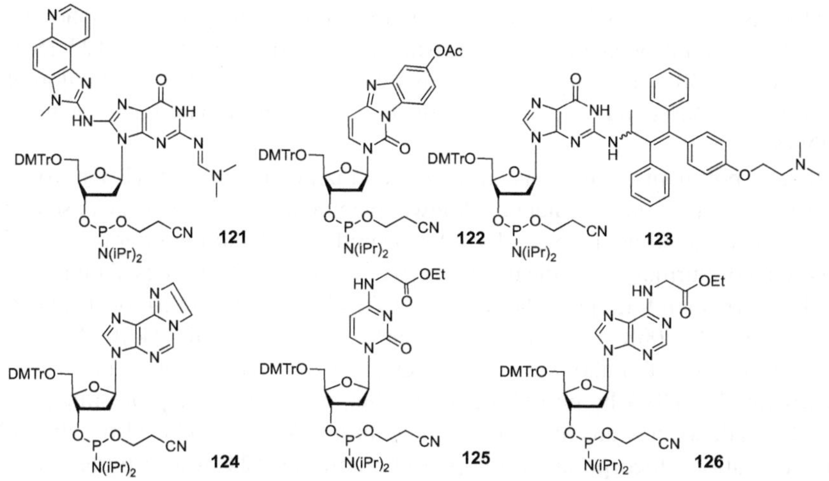

Fig. 18 Some adducts formed between nucleobase and various mutagen chemicals

Following similar strategies, other damages provoked by methylglyoxal (Cao et al. 2007), acrolein, crotonaldehyde, *trans*-4-hydroxy-nonenal **119** (Wang et al. 2003), or even styrene (DeCorte et al. 1996) could be introduced at site-specific positions.

9.4 Dietary Mutagen Adducts

Overcooking of meat produces a variety of mutagenic compounds. These compounds share common structural features: extended polycyclic core, endocyclic nitrogen atoms, and reactive exocyclic amines. Some phosphoramidite derivatives of nucleobases adducts formed with these complex heterocyclic amines were synthesized and incorporated in ODNs: 2-amino-3-methylimidazo[4,5-*f*]quinolone (Fig. 18, IQ, **121**) (Elmquist et al. 2004; Stover et al. 2006), 2-amino-1-methyl-6-phenylimidazo[4,5-*b*]pyridine (PhIP) (Bonala et al. 2006; Stover and Rizzo 2007), and 2-amino-3,8-dimethylimidazo[4,5-*f*]quinoxaline (MeIQx) (Takamura-Enya et al. 2006).

9.5 Miscellaneous Adducts

The benzene metabolite *p*-benzoquinone is known to react with DNA nucleobases. The phosphoramidite derivatives of such adducts with dC (**122**), dA (Chenna and Singer 1995), and dG (Chenna et al. 2008) were synthesized and incorporated in synthetic ODN.

DNA is also a target for numerous alkylating antitumor agents. Phosphoramidite derivatives of nitrogen mustard (Christov et al. 2014), tamoxifen **123** (Santosh Laxmi et al. 2002), mitomycin C (Champeil et al. 2008), or 1,3-bis(2-chloroethyl) nitrosourea (Maruenda et al. 1998) adducts have been prepared and incorporated in ODN.

Some pollutants, such as dimethylhydrazine or *N*-methyl-*N*-nitrosourea, encountered in the environment are known to be methylating agents. Consequently, various methylated nucleobases have been synthesized and incorporated in ODNs. Position 8 of purines being the most reactive, synthetic effort focused on the design of the corresponding phosphoramidite (Virgilio et al. 2005) or triphosphate (Cahová et al. 2008) derivatives. Similarly, derivatives of *O*4-alkyl dT (Vyle et al. 1998) and *O*6-alkyl dG (Smith et al. 1990) have been synthesized and incorporated in DNA by phosphoramidite chemistry using pac protecting groups.

Ethenoadenosines are fluorescent nucleobase adducts produced by vinyl chloride alkylation. Incorporation by phosphoramidite **124** was reported in 1994 (Srivastava et al. 1994). Recently, post-synthetic conversion of dA into etheno-dA was performed through hybridization of the target strand with a reactive strand bearing an acetaldehyde precursor (Egloff et al. 2015).

Last but not the least, the carboxymethylation of nucleobase by known carcinogenic *N*-nitroso metabolites was also investigated. The phosphoramidite derivatives of dC alkylated in *N*4 **125** as well as dA in *N*6 **126** (Wang and Wang 2010) thymine alkylated in *N*3 or *O*4 (Wang and Wang 2009; Xu and Swann 1990) and guanine alkylated in *O*6 (Xu 2000) position have been prepared and incorporated. In this case, the ester moiety is hydrolyzed by NaOH 0.4 M prior to usual ammonolysis (Wang and Wang 2010).

10 Conclusion

Numerous other lesions have been reported so far. However, their successful chemical incorporation in synthetic DNA has not been reported yet (Fig. 19).

Spiroiminodihydantoin (**127**, dSp), for example, is a photoproduct of dG oxidation (Adam et al. 1996). Its chemical structure was first wrongly assigned to be a 4-hydroxyderivative of 8-oxoG (cf. Sect. 3.1). NMR experiments further

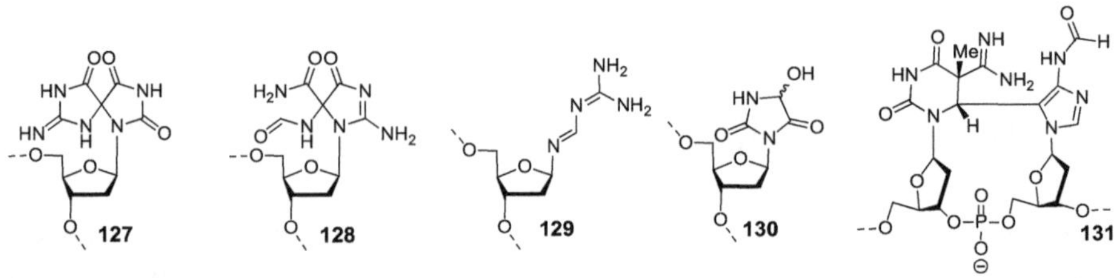

Fig. 19 Some known lesions remaining to be chemically incorporated into synthetic ODNs

demonstrated its spiranic skeleton (Adam et al. 2002). 5-Carboxamido-5-formamido-2-iminohydantoin **128** was recently reported to be formed under X-ray irradiation of dG (Alshykhly et al. 2015). Guanidinoformimine **129** (dGfg) is a lesion produced after decarboxylation of the open form of oxazolone lesion (dZ) (Stathis et al. 2012). Isodialuric acid lesion **130** (Iso-4-ohHyd) is formed by oxidation of 5-hydroxy-uracil (Simon et al. 2006). Similarly, the [2+2] cycloaddition photoproduct of thymine with adenosine was first reported in 1983 (Bose et al. 1983; Zhao et al. 1995) as well as its main degradation product **131** (Perrier et al. 2006). In all cases, the lack of straightforward chemical synthesis toward site-specific incorporation of these lesions is a major bottleneck to get further insights into their biological implication.

Phosphoramidite derivatives of the first discovered DNA lesions are now commercially available and routinely incorporated in synthetic ODN. Undoubtedly, chemists will provide in the near future new synthetic tools to access these demanding lesions as well as the ones to be discovered.

References

Adam W, Saha-Möller CR, Schönberger A (1996) Photooxidation of 8-oxo-7,8-dihydro-2′-deoxyguanosine by thermally generated triplet-excited ketones from 3-(hydroxymethyl)-3,4,4-trimethyl-1,2-dioxetane and comparison with Type I and Type II photosensitizers. J Am Chem Soc 118:9233–9238

Adam W, Arnold MA, Grüne M et al (2002) Spiroiminodihydantoin is a major product in the photooxidation of 2′-deoxyguanosine by the triplet states and oxyl radicals generated from hydroxyacetophenone photolysis and dioxetane thermolysis. Org Lett 4:537–540

Alshykhly OR, Fleming AM, Burrows CJ (2015) 5-Carboxamido-5-formamido-2-iminohydantoin, in addition to 8-oxo-7,8-dihydroguanine, is the major product of the Iron-Fenton or X-ray radiation-induced oxidation of guanine under aerobic reducing conditions in nucleoside and DNA contexts. J Org Chem 80:6996–7007

Arichi N, Inase A, Eto S et al (2012) Mechanism of the alkali degradation of (6-4) photoproduct-containing DNA. Org Biomol Chem 10:2318–2325

Aso M, Usui K, Fukuda M et al (2006) Photochemical generation of C4′-oxidized abasic site containing oligodeoxynucleotide and its efficient amine modification. Org Lett 8:3183–3186

Baillet S, Behr J (1995) Deoxyribosyl urea and deoxyribosylformamide oligonucleotides. Tetrahedron Lett 36:8981–8984

Bdour HM, Kao JL, Taylor J (2006) Synthesis and characterization of a [3-^{15}N]-labeled cis-syn thymine dimer-containing DNA duplex. J Org Chem 71:1640–1646

Bellon S, Ravanat J, Gasparutto D et al (2002) Cross-linked thymine-purine base tandem lesions: synthesis, characterization, and measurement in γ-irradiated isolated DNA. Chem Res Toxicol 15:598–606

Bellon S, Gasparutto D, Saint-Pierre C et al (2006) Guanine-thymine intrastrand cross-linked lesion containing oligonucleotides: from chemical synthesis to in vitro enzymatic replication. Org Biomol Chem 4:3831–3837

Bernadou J, Fabiano A, Robert A et al (1994) “Redox tautomerism” in high-valent metal-oxo-aquo complexes. Origin of the oxygen atom in epoxidation reactions catalyzed by water-soluble metalloporphyrins. J Am Chem Soc 116:12135

Berthet N, Crey-Desbiolles C, Kotera M et al (2009) Chemical synthesis, DNA incorporation and biological study of a new photocleavable 2′-deoxyadenosine mimic. Nucleic Acids Res 37:5237–5245

Blackburn GM, Davies RJH (1965) The structure of thymine photo-dimer. Chem Commun 11:215–216

Bodepudi V, Shibutani S, Johnson F (1992) Synthesis of 2′-deoxy-7,8-dihydro-8-oxoguanosine and 2′-deoxy-7,8-dihydro-8-oxoadenosine and their incorporation into oligomeric DNA. Chem Res Toxicol 5:608–617

Böge N, Gräsl S, Meier C (2006) Synthesis and properties of oligonucleotides containing C8-deoxyguanosine arylamine adducts of borderline carcinogens. J Org Chem 71:9728–9738

Böge N, Szombati Z, Meier C (2007) Synthesis of DNA-oligonucleotides damaged by arylamine-modified 2′-deoxyguanosine. Nucleoside Nucleotide Nucleic Acids 26:705–708

Böge N, Jacobsen MI, Szombati Z et al (2008) Synthesis of DNA strands site-specifically damaged by c8-arylamine purine adducts and effects on various DNA polymerases. Chem Eur J 14:11194–11208

Bonala RR, Torres MC, Attaluri S et al (2005) Incorporation of N2-deoxyguanosine metabolic adducts of 2-aminonaphthalene and 2-aminofluorene into oligomeric DNA. Chem Res Toxicol 18:457–465

Bonala R, Torres MC, Iden CR et al (2006) Synthesis of the PhIP adduct of 2′-deoxyguanosine and its incorporation into oligomeric DNA. Chem Res Toxicol 19:734–738

Bose S, Davies R, Sethi S et al (1983) Formation of an adenine-thymine photoadduct in the deoxydinucleoside monophosphate d(TpA) and in DNA. Science 220:723–725

Brooks PJ, Wise DS, Berry DA et al (2000) The oxidative DNA lesion 8,5′-(S)-cyclo-2′-deoxyadenosine is repaired by the nucleotide excision repair pathway and blocks gene expression in mammalian cell. J Biol Chem 275:22355–22362

Buchko GW, Cadet J, Morin B et al (1995) Photooxidation of d(TpG) by riboflavin and methylene blue. Isolation and characterization of thymidylyl-(3′,5′)-2-amino-5-[(2-deoxy-β-D-erythro-pentofuranosyl)amino]-4 H -imidazol-4-one and its primary decomposition product thymidylyl-(3′, 5′)-2, 2-diamino-4-[(2-deoxy-β-D- erythro -pentofuranosyl)amino]-5(2H)-oxazolone. Nucleic Acids Res 23:3954–3961

Bürckstümmer E, Carell T (2008) Synthesis and properties of DNA containing a spore photoproduct analog. Chem Commun 34:4037–4039

Burgdorf LT, Carell T (2002) Synthesis, stability, and conformation of the formamidopyrimidine G DNA lesion. Chem Eur J 8:293–301

Butenandt J (1999) Synthesis of DNA lesions and DNA-lesion-containing oligonucleotides for DNA-repair studies. Synthesis 1999(7):1085–1105

Butenandt J, Eker APM, Carell T (1998) Synthesis, crystal structure, and enzymatic evaluation of a DNA-photolesion isostere. Chem Eur J 4:642–654

Cadena-Amaro C, Delepierre M, Pochet S (2005) Synthesis and incorporation into DNA fragments of the artificial nucleobase, 2-amino-8-oxopurine. Bioorg Med Chem Lett 15:1069–1073

Cadet J, Berger M, Buchko GW et al (1994) 2,2-Diamino-4-[(3,5-di-O-acetyl-2-deoxy-beta-D-erythro- pentofuranosyl)amino]-5-(2H)-oxazolone: a novel and predominant radical oxidation product of 3′,5′-di-O-acetyl-2′-deoxyguanosine. J Am Chem Soc 116:7403–7404

Cadet J, Loft S, Olinski R et al (2012) Biologically relevant oxidants and terminology, classification and nomenclature of oxidatively generated damage to nucleobases and 2-deoxyribose in nucleic acids. Free Radic Res 46:367–381

Cahová H, Pohl R, Bednárová L et al (2008) Synthesis of 8-bromo-, 8-methyl- and 8-phenyl-dATP and their polymerase incorporation into DNA. Org Biomol Chem 6:3657–3660

Cannistraro VJ, Taylor J (2009) Acceleration of 5-methylcytosine deamination in cyclobutane dimers by G and its implications for UV-induced C-to-T mutation hotspots. J Mol Biol 392:1145–1157

Cao H, Jiang Y, Wang Y (2007) Stereospecific synthesis and characterization of oligodeoxyribonucleotides containing an N2-(1-carboxyethyl)-2′-deoxyguanosine. J Am Chem Soc 129:12123–12130

Champeil E, Paz MM, Ladwa S et al (2008) Synthesis of an oligodeoxyribonucleotide adduct of mitomycin C by the postoligomerization method via a triamino mitosene. J Am Chem Soc 130:9556–9565

Chandra T, Broderick WE, Broderick JB (2009) Chemoselective deprotection of triethylsilyl ethers. Nucleoside Nucleotide Nucleic Acids 28:1016–1029

Chatgilialoglu C, Navacchia ML, Postigo A (2006) A facile one-pot synthesis of 8-oxo-7,8-dihydro-(2′-deoxy)adenosine in water. Tetrahedron Lett 47:711–714

Chatgilialoglu C, Ferreri C, Masi A et al (2014) A problem solving approach for the diastereoselective synthesis of (5′, S)- and (5′, R)-5′,8-cyclopurine lesions. Org Chem Front 1:698–702

Chenna A, Singer B (1995) Large scale synthesis of p-benzoquinone-2′-deoxycytidine and p-benzoquinone-2′-deoxyadenosine adducts and their site-specific incorporation into DNA oligodeoxyribonucleotides. Chem Res Toxicol 8:865–874

Chenna A, Gupta RC, Bonala RR et al (2008) Synthesis of the fully protected phosphoramidite of the benzene-DNA adduct, N2-(4-hydroxyphenyl)-2′-deoxyguanosine and incorporation of the later into DNA oligomers. Nucleoside Nucleotide Nucleic Acids 27:979–991

Christov PP, Brown KL, Kozekov ID et al (2008) Site-specific synthesis and characterization of oligonucleotides containing an N6-(2-deoxy-D-erythro-pentofuranosyl)-2,6-diamino-3,4-dihydro-4-oxo-5-N-methylformamidopyrimidine lesion, the ring-opened product from N7-methylation of deoxyguanosine. Chem Res Toxicol 21:2324–2333

Christov PP, Son K, Rizzo CJ (2014) Synthesis and characterization of oligonucleotides containing a nitrogen mustard formamidopyrimidine monoadduct of deoxyguanosine. Chem Res Toxicol 27:1610–1618

Coleman RS, Pires RM (1999) Site-specific formation of abasic lesions in DNA. Nucleoside Nucleotide Nucleic Acids 18:2141–2146

Colis LC, Chakraborti D, Hilario P et al (2009) Synthesis of oligonucleotides containing 2′-deoxyguanosine adducts of nitropyrenes. Nucleoside Nucleotide Nucleic Acids 28:67–77

Cooke MS, Loft S, Olinski R et al (2010) Recommendations for standardized description of and nomenclature concerning oxidatively damaged nucleobases in DNA. Chem Res Toxicol 23:705–707

Crey-Desbiolles C, Kotera M (2006) Synthesis of 3-deaza-3-nitro-2′-deoxyadenosine. Bioorg Med Chem 14:1935–1941

DeCorte BL, Tsarouhtsis D, Kuchimanchi S et al (1996) Improved strategies for postoligomerization synthesis of oligodeoxynucleotides bearing structurally defined adducts at the N2 position of deoxyguanosine. Chem Res Toxicol 9:630–637

Dubey I, Pratviel G, Robert A et al (2001) Convenient method for the preparation of 2′-deoxyribosylurea by thymidine oxidation and NMR study of both anomers. Nucleoside Nucleotide Nucleic Acids 20:1463–1471

Egloff D, Oleinich IA, Freisinger E (2015) Sequence-specific generation of 1, N(6)-ethenoadenine and 3, N(4)-ethenocytosine in single-stranded unmodified DNA. ACS Chem Biol 10:547–553

Einolf HJ, Schnetz-Boutaud N, Guengerich FP (1998) Steady-state and pre-steady-state kinetic analysis of 8-oxo-7,8-dihydroguanosine triphosphate incorporation and extension by replicative and repair DNA polymerases. Biochemistry 37:13300–13312

Elmquist CE, Stover JS, Wang Z et al (2004) Site-specific synthesis and properties of oligonucleotides containing C8-deoxyguanosine adducts of the dietary mutagen IQ. J Am Chem Soc 126:11189–11201

Eritja R, Walker PA, Randall SK et al (1987) Synthesis of oligonucleotides containing the abasic site model compound 1,4-anhydro-2-deoxy-D-ribitol. Nucleoside Nucleotide Nucleic Acids 6:803–814

Friedel MG, Berteau O, Pieck JC et al (2006) The spore photoproduct lyase repairs the 5S- and not the 5R-configured spore photoproduct DNA lesion. Chem Commun 445–447

Gasparutto D, Ravanat J, Gérot O et al (1998) Characterization and chemical stability of photooxidized oligonucleotides that contain 2,2-diamino-4-[(2-deoxy-β-d-erythro-pentofuranosyl)amino]-5(2H)-oxazolone. J Am Chem Soc 120:10283–10286

Gasparutto D, Da Cruz S, Bourdat AG et al (1999) Synthesis and biochemical properties of cyanuric acid nucleoside-containing DNA oligomers. Chem Res Toxicol 12:630–638

Gasparutto D, Ait-Abbas M, Jaquinod M et al (2000) Repair and coding properties of 5-hydroxy-5-methylhydantoin nucleosides inserted into DNA oligomers. Chem Res Toxicol 13:575–584

Gasparutto D, Cognet S, Roussel S et al (2005) Synthesis of a convenient thymidine glycol phosphoramidite monomer and its site-specific incorporation into DNA fragments. Nucleoside Nucleotide Nucleic Acids 24:1831–1842

Ghosh S, Greenberg MM (2014) Synthesis of cross-linked DNA containing oxidized abasic site analogues. J Org Chem 79:5948–5957

Gillet LCJ, Schärer OD (2002) Preparation of C8-amine and acetylamine adducts of 2′-deoxyguanosine suitably protected for DNA synthesis. Org Lett 4:4205–4208

Gillet LCJ, Alzeer J, Schärer OD (2005) Site-specific incorporation of N-(deoxyguanosin-8-yl)-2-acetylaminofluorene (dG-AAF) into oligonucleotides using modified 'ultra-mild' DNA synthesis. Nucleic Acids Res 33:1961–1969

Greenberg MM, Matray TJ (1997) Inhibition of Klenow fragment (exo-) catalyzed DNA polymerization by (5R)-5,6-dihydro-5-hydroxythymidine and structural analogue 5,6-dihydro-5-methylthymidine. Biochemistry 36:14071–14079

Groebke K, Leumann C (1990) A method for preparing oligodeoxynucleotides containing an apurinic site. Helv Chim Acta 73:608–617

Guan L, Greenberg MM (2009) DNA interstrand cross-link formation by the 1,4-dioxobutane abasic lesion. J Am Chem Soc 131:15225–15231

Guan L, Greenberg MM (2011) An oxidized abasic lesion as an intramolecular source of DNA adducts. Aust J Chem 64:438–442

Guan L, Bebenek K, Kunkel TA et al (2010) Inhibition of short patch and long patch base excision repair by an oxidized abasic site. Biochemistry 49:9904–9910

Guérineau V, Matus SKA, Halgand F et al (2004) Studies on the chemical synthesis of oligodeoxynucleotides containing the s5T(6-4)T photoproduct: side reactions derived from the methylsulfenyl thiol protection elucidated by MALDI mass spectrometry. Org Biomol Chem 2:899–907

Guy A, Ahmad S, Téoule R (1990) Insertion of the fragile 2′-deoxyribosylurea residue into oligodeoxynucleotides. Tetrahedron Lett 31:5745–5748

Guy A, Dubet J, Téoule R (1993) The synthesis of 5-hydroxy-5-methylhydantoin nucleoside and its insertion into oligodeoxyribonucleotides. Tetrahedron Lett 34:8101–8102

Haraguchi K, Greenberg MM (2001) Synthesis of oligonucleotides containing Fapy·dG (N6 - (2-Deoxy-α, β- d - erythro-pentofuranosyl)-2,6-diamino-4-hydroxy-5-formamidopyrimidine). J Am Chem Soc 123:8636–8637

Haraguchi K, Delaney MO, Wiederholt CJ et al (2002) Synthesis and characterization of oligodeoxynucleotides containing formamidopyrimidine lesions and nonhydrolyzable analogues. J Am Chem Soc 124:3263–3269

Higurashi M, Ohtsuki T, Inase A et al (2003) 530 - Degradation photoproducts. J Biol Chem 278:51968–51973

Hong IS, Greenberg MM (2005) DNA interstrand cross-link formation initiated by reaction between singlet oxygen and a modified nucleotide. J Am Chem Soc 127:10510–10511

Hong H, Wang Y (2005) Formation of intrastrand cross-link products between cytosine and adenine from UV irradiation of d((Br)CA) and duplex DNA containing a 5-bromocytosine. J Am Chem Soc 127:13969–13977

Huang H, Greenberg MM (2008) Synthesis and analysis of oligonucleotides containing abasic site analogues. J Org Chem 73:2695–2703

Imoto S, Patro JN, Jiang YL et al (2006) Synthesis, DNA polymerase incorporation, and enzymatic phosphate hydrolysis of formamidopyrimidine nucleoside triphosphates. J Am Chem Soc 128:14606–14611

Iocono JA, Gildea B, McLaughlin LW (1990) Mild acid hydrolysis of 2-pyrimidinone-containing DNA fragments generates apurinic/apyrimidinic sites. Tetrahedron Lett 31:175–178

Iwai S (2000) Synthesis of thymine glycol containing oligonucleotides from a building block with the oxidized base. Angew Chem Int Ed 39:3874–3876

Iwai S (2001) Synthesis and thermodynamic studies of oligonucleotides containing the two isomers of thymine glycol. Chem Eur J 7:4343–4351

Iwai S (2006) Chemical synthesis of oligonucleotides containing damaged bases for biological studies. Nucleoside Nucleotide Nucleic Acids 25:561–582

Iwai S, Shimizu M, Kamiya H et al (1996) Synthesis of a phosphoramidite coupling unit of the pyrimidine (6−4) pyrimidone photoproduct and its incorporation into oligodeoxynucleotides. J Am Chem Soc 118:7642–7643

Iwai S, Mizukoshi T, Fujiwara Y et al (1999) Benzimidazolium triflate-activated synthesis of (6-4) photoproduct-containing oligonucleotides and its application. Nucleic Acids Res 27:2299–2303

Iyer RP, Uznanski B, Boal J et al (1990) Abasic oligodeoxyribonucleoside phosphorothioates: synthesis and evaluation as anti-HIV-1 agents. Nucleic Acids Res 18:2855–2859

Iyer PC, Yagi H, Sayer JM et al (2007) 3′-H-phosphonate synthesis of chiral benzo[*a*]pyrene diol epoxide adducts at N(2) of deoxyguanosine in oligonucleotides. Chem Res Toxicol 20:311–315

Jian Y, Li L (2013) Chemical syntheses of oligodeoxyribonucleotides containing spore photoproduct. J Org Chem 78:3021–3029

Jiang YL, Wiederholt CJ, Patro JN et al (2005) Synthesis of oligonucleotides containing Fapy.dG (N(6)-(2-deoxy-alpha, beta-D-erythropentofuranosyl)-2,6-diamino-4-hydroxy-5-formamidop yrimidine) using a 5′-dimethoxytrityl dinucleotide phosphoramidite. J Org Chem 70:141–149

Johnson F, Bonala R, Tawde D et al (2002) Efficient synthesis of the benzo[*a*]pyrene metabolic adducts of 2′-deoxyguanosine and 2′-deoxyadenosine and their direct incorporation into DNA. Chem Res Toxicol 15:1489–1494

Jung K, Kodama T, Greenberg MM (2011) Repair of the major lesion resulting from C5′-oxidation of DNA. Biochemistry 50:6273–6279

Kannan A, Burrows CJ (2011) Synthesis of N2-alkyl-8-oxo-7,8-dihydro-2′-deoxyguanosine derivatives and effects of these modifications on RNA duplex stability. J Org Chem 76:720–723

Kemmink J, Boelens R, Koning TMG et al (1987) Conformational changes in the oligonucleotide duplex d(GCGTTGCG). d (CGCAACGC) induced by formation of a cis-syn thymine dimer. A two-dimensional NMR study. Eur J Biochem 162:37–43

Kim SJ, Stone MP, Harris CM et al (1992) A postoligomerization synthesis of oligodeoxynucleotides containing polycyclic aromatic hydrocarbon adducts at the N6 position of deoxyadenosine. J Am Chem Soc 114:5480–5481

Kim SJ, Lester C, Begley TP (1995) Synthesis of the dinucleotide spore photoproduct. J Org Chem 60:6256–6257

Kim J, Gil JM, Greenberg MM (2003) Synthesis and characterization of oligonucleotides containing the C4′-oxidized abasic site produced by bleomycin and other DNA damaging agents. Angew Chem Int Ed Eng 42:5882–5885

Kim J, Weledji YN, Greenberg MM (2004) Independent generation and characterization of a C2′-oxidized abasic site in chemically synthesized oligonucleotides. J Org Chem 69:6100–6104

Kim J, Kreller CR, Greenberg MM (2005) Preparation and analysis of oligonucleotides containing the C4′-oxidized abasic site and related mechanistic probes. J Org Chem 70:8122–8129

Kino K, Sugiyama H (2001) Possible cause of G·C→C·G transversion mutation by guanine oxidation product, imidazolone. Chem Biol 8:369–378

Kirouac KN, Basu AK, Ling H (2013) Replication of a carcinogenic nitropyrene DNA lesion by human Y-family DNA polymerase. Nucleic Acids Res 41:2060–2071

Kodama T, Greenberg MM (2005) Preparation and analysis of oligonucleotides containing lesions resulting from C5′-oxidation. J Org Chem 70:9916–9924

Kojima N, Sugino M, Mikami A et al (2005) Generation of an abasic site in an oligonucleotide by using acid-labile 1-deaza-2′-deoxyguanosine and its application to postsynthetic modification. Org Lett 7:709–712

Kosmoski JV, Smerdon MJ (1999) Synthesis and nucleosome structure of DNA containing a UV photoproduct at a specific site. Biochemistry 38:9485–9494

Kotera M, Bourdat A, Defrancq E et al (1998) A highly efficient synthesis of oligodeoxyribonucleotides containing the 2′-deoxyribonolactone lesion. J Am Chem Soc 120:11810–11811

Kroth H, Yagi H, Sayer JM et al (2001) O6-allyl protected deoxyguanosine adducts of polycyclic aromatic hydrocarbons as building blocks for the synthesis of oligonucleotides. Chem Res Toxicol 14:708–719

Krüger S, Meier C (2013) Synthesis of site-specific damaged DNA strands by 8-(acetylarylamino)-2′-deoxyguanosine adducts and effects on various DNA polymerases. Eur J Org Chem 2013:1158–1169

Küpfer PA, Leumann CJ (2005) RNA abasic sites: preparation and trans-lesion synthesis by HIV-1 reverse transcriptase. Chembiochem 6:1970–1973

Küpfer PA, Leumann CJ (2007) The chemical stability of abasic RNA compared to abasic DNA. Nucleic Acids Res 35:58–68

Laayoun A, Décout J, Defrancq E et al (1994) Hydrolysis of oligonucleotides containing 8-substituted purine nucleosides. A new route for preparing abasic oligodeoxynucleotides. Tetrahedron Lett 35:4991–4994

Lakshman MK, Sayer JM, Jerina DM (1992a) Synthesis and site-specific incorporation of a bay-region cis ring-opened tetrahydro epoxide-deoxyadenosine adduct into a DNA oligomer. J Org Chem 57:3438–3443

Lakshman MK, Sayer JM, Yagi H et al (1992b) Synthesis and duplex-forming properties of a nonanucleotide containing an N6-deoxyadenosine adduct of a bay-region diol epoxide. J Org Chem 57:4585–4590

Lartia R, Constant J (2015) Synthetic access to the chemical diversity of DNA and RNA 5′-aldehyde lesions. J Org Chem 80:705–710

Lee H, Hinz M, Stezowski JJ et al (1990) Syntheses of polycyclic aromatic hydrocarbon-nucleoside and oligonucleotide adducts specifically alkylated on the amino functions of deoxyguanosine and deoxyadenosine. Tetrahedron Lett 31:6773–6776

Lenox HJ, McCoy CP, Sheppard TL (2001) Site-specific generation of deoxyribonolactone lesions in DNA oligonucleotides. Org Lett 3:2415–2418

Lin G, Chen C, Pink M et al (2011) Chemical synthesis, crystal structure and enzymatic evaluation of a dinucleotide spore photoproduct analogue containing a formacetal linker. Chem Eur J 17:9658–9668

Lin G, Jian Y, Ouyang H, Li L (2014) An unexpected deamination reaction after hydrolysis of the pyrimidine (6-4) pyrimidone photoproduct. Org Lett 16:5076–5079

Ling H, Sayer JM, Plosky BS et al (2004) Crystal structure of a benzo[*a*]pyrene diol epoxide adduct in a ternary complex with a DNA polymerase. Proc Nat Acad Sci USA 101:2265–2269

Lukin M, Minetti, Conceição ASA et al (2011) Novel post-synthetic generation, isomeric resolution, and characterization of Fapy-dG within oligodeoxynucleotides: differential anomeric impacts on DNA duplex properties. Nucleic Acids Res 39:5776–5789

Maddukuri L, Eoff RL, Choi J et al (2010) In vitro bypass of the major malondialdehyde- and base propenal-derived DNA adduct by human Y-family DNA polymerases κ, ι, and Rev1. Biochemistry 49:8415–8424

Maruenda H, Chenna A, Liem L et al (1998) Synthesis of 1, N6-ethano-2′-deoxyadenosine, a metabolic product of 1,3-Bis(2-chloroethyl)nitrosourea, and its incorporation into oligomeric DNA. J Org Chem 63:4385–4389

Matus SKA, Fourrey J, Clivio P (2003) Synthesis of the TT pyrimidine (6:4) pyrimidone photoproduct thio analogue phosphoramidite building block. Org Biomol Chem 1:3316

Mizukoshi T, Hitomi K, Todo T et al (1998) Studies on the chemical synthesis of oligonucleotides containing the (6−4) photoproduct of thymine−cytosine and its repair by (6−4) photolyase. J Am Chem Soc 120:10634–10642

Molko D, Delort A, Guy A et al (1985) Synthèse d'un fragment de DNA comportant la dihydro-5,6 thymine. Biochimie 67:801–809

Mueller H, Hopfinger M, Carell T (2008) Synthesis of a stabilized version of the imidazolone DNA lesion. Chembiochem 9:1617–1622

Muller E, Gasparutto D, Jaquinod M et al (2000) Chemical and biochemical properties of oligonucleotides that contain (5′S,6S)-Cyclo-5,6-dihydro-2′-deoxyuridine and (5′S,6S)-cyclo-5,6-dihydrothymidine, two main radiation-induced degradation products of pyrimidine 2′-deoxyribonucleosides. Tetrahedron 56:8689–8701

Muller E, Gasparutto D, Lebrun C et al (2001) Site-specific insertion of the (5, R) and (5, S) diastereoisomers of 1-[2-deoxy-β-D-erythro-pentofuranosyl]-5-hydroxyhydantoin into oligodeoxyribonucleotides. Eur J Org Chem 2001:2091–2099

Muller E, Gasparutto D, Cadet J (2002) Chemical synthesis and biochemical properties of oligonucleotides that contain the (5′S,5S,6S)-5′,6-cyclo-5-hydroxy-5,6-dihydro-2′-deoxyuridine DNA lesion. ChemBioChem 3:534–542

Murata T, Iwai S, Ohtsuka E (1990) Synthesis and characterization of a substrate for T4 endonuclease V containing a phosphorodithioate linkage at the thymine dimer site. Nucleic Acids Res 18:7279–7286

Nadji S, Wang CI, Taylor JS (1992) Photochemically and photoenzymatically cleavable DNA. J Am Chem Soc 114(24):9266–9269

Nampalli S, Kumar S (2000) Efficient synthesis of 8-Oxo-dGTP: a mutagenic nucleotide. Bioorg Med Chem Lett 10:1677–1679

Omumi A, Beach DG, Baker M et al (2011) Postsynthetic guanine arylation of DNA by Suzuki-Miyaura cross-coupling. J Am Chem Soc 133:42–50

Ordoukhanian P, Taylor J (1997) Solid phase-supported thymine dimers for the construction of dimer-containing DNA by combined chemical and enzymatic synthesis: a potentially general method for the efficient incorporation of modified nucleotides into DNA. Nucleic Acids Res 25:3783–3786

Ortiz Mayo JU, Thomas M, Saintomé C et al (2003) Facile synthesis of a cis–syn thymine dimer building block and its incorporation into oligodeoxynucleotides. Tetrahedron 59:7377–7383

Othuka E, Murata T, Iwai S et al (1992) Synthesis of a dodecadeoxyribooligonucleotide containing a 3′-thio analogue of thymidine photodimer. Heterocycles 33:529

Péoc'h D, Meyer A, Imbach J et al (1991) Efficient chemical synthesis of oligodeoxynucleotides containing a true abasic site. Tetrahedron Lett 32:207–210

Perrier S, Hau J, Gasparutto D et al (2006) Characterization of lysine-guanine cross-links upon one-electron oxidation of a guanine-containing oligonucleotide in the presence of a trilysine peptide. J Am Chem Soc 128:5703–5710

Romieu A, Gasparutto D, Molko D et al (1998) Site-specific introduction of (5′, S)-5,8-cyclo-2′-deoxyadenosine into oligodeoxyribonucleotides. J Org Chem 63:5245–5249

Romieu A, Gasparutto D, Molko D et al (1999a) Synthesis of oligonucleotides containing the (4R) and (4S) diastereoisomers of 4,8-dihydro-4-hydroxy-8-oxo-2′-deoxyguanosine. Eur J Org Chem 1999:49–56

Romieu A, Gasparutto D, Cadet J (1999b) Synthesis and characterization of oligonucleotides containing 5′,8-cyclopurine 2′-deoxyribonucleosides: (5′R)-5′,8-cyclo-2′-deoxyadenosine, (5′S)-5′,8-cyclo-2′-deoxyguanosine, and (5′R)-5′,8-cyclo-2′-deoxyguanosine. Chem Res Toxicol 12:412–421

Romieu A, Gasparutto D, Cadet J (1999c) Synthesis and characterization of oligodeoxynucleotides containing the two 5R and 5S diastereomers of (5′S,6S)-5′,6-cyclo-5,6-dihydrothymidine; radiation-induced tandem lesions of thymidine. J Chem Soc Perkin Trans 1:1257–1264

Sallamand C, Miscioscia A, Lartia R et al (2012) Phosphorylating reagent-free synthesis of 5′-phosphate oligonucleotides by controlled oxidative degradation of their 5′-end. Org Lett 14:2030–2033

Sambandam A, Greenberg MM (1999) The effects of 5R-5,6-dihydro-5-hydroxythymidine on duplex DNA stability and structure. Nucleic Acids Res 27:3597–3602

Santosh Laxmi YR, Suzuki N, Dasaradhi L et al (2002) Preparation of oligodeoxynucleotides containing a diastereoisomer of α-(N^2-2′-deoxyguanosinyl)tamoxifen by phosphoramidite chemical synthesis. Chem Res Toxicol 15:218–225

Schnetz-Boutaud NC, Mao H, Stone MP et al (2000) Synthesis of oligonucleotides containing the alkali-labile pyrimidopurinone adduct, M1G. Chem Res Toxicol 13:90–95

Schulhof JC, Molko D, Teoule R (1988) Synthesis of DNA fragments containing 5,6-dihydrothymine, a major product of thymine gamma radiolysis. Nucleic Acids Res 16:319–326

Shimizu T, Manabe K, Yoshikawa S et al (2006) Preferential formation of (5S,6R)-thymine glycol for oligodeoxyribonucleotide synthesis and analysis of drug binding to thymine glycol-containing DNA. Nucleic Acids Res 34:313–321

Shishkina IG, Johnson F (2000) A new method for the postsynthetic generation of abasic sites in oligomeric DNA. Chem Res Toxicol 13:907–912

Simon P, Gasparutto D, Gambarelli S et al (2006) Formation of isodialuric acid lesion within DNA oligomers via one-electron oxidation of 5-hydroxyuracil: characterization, stability and excision repair. Nucleic Acids Res 34:3660–3669

Sloane JL, Greenberg MM (2014) Interstrand cross-link and bioconjugate formation in RNA from a modified nucleotide. J Org Chem 79:9792–9798

Smith CA, Xu Y, Swann PF (1990) Solid-phase synthesis of oligodeoxynucleotides containing O6-alkylguanine. Carcinogenesis 11:811–816

Song Q, Sherrer SM, Suo Z et al (2012) Preparation of site-specific T=mCG cis-syn cyclobutane dimer-containing template and its error-free bypass by yeast and human polymerase η. J Biol Chem 287:8021–8028

Srivastava SC, Raza SK, Misra R (1994) 1, N6 -etheno deoxy and ribo adenoGine and 3, N4 -etheno deoxy and ribo cytidine phosphoramidites. Strongly fluorescent structures for selective introduction in defined sequence DNA and RNA molecules. Nucleic Acids Res 22:1296–1304

Stathis D, Lischke U, Koch SC et al (2012) Discovery and mutagenicity of a guanidinoformimine lesion as a new intermediate of the oxidative deoxyguanosine degradation pathway. J Am Chem Soc 134:4925–4930

Stover JS, Rizzo CJ (2007) Synthesis of oligonucleotides containing the N2-deoxyguanosine adduct of the dietary carcinogen 2-amino-3-methylimidazo[4,5-f]quinoline. Chem Res Toxicol 20:1972–1979

Stover JS, Chowdhury G, Zang H et al (2006) Translesion synthesis past the C8- and N2-deoxyguanosine adducts of the dietary mutagen 2-amino-3-methylimidazo[4,5-f]quinoline in the NarI recognition sequence by prokaryotic DNA polymerases. Chem Res Toxicol 19:1506–1517

Szombati Z, Baerns S, Marx A et al (2012) Synthesis of C8-arylamine-modified 2′-deoxyadenosine phosphoramidites and their site-specific incorporation into oligonucleotides. Chembiochem 13:700–712

Takamura-Enya T, Ishikawa S, Mochizuki M et al (2006) Chemical synthesis of 2′-deoxyguanosine-C8 adducts with heterocyclic amines: an application to synthesis of oligonucleotides site-specifically adducted with 2-amino-1-methyl-6-phenylimidazo[4,5-b]pyridine. Chem Res Toxicol 19:770–778

Takasawa K, Masutani C, Hanaoka F et al (2004) Chemical synthesis and translesion replication of a cis-syn cyclobutane thymine-uracil dimer. Nucleic Acids Res 32:1738

Tashiro R, Nakamura K, Sugiyama H (2008) Photoreaction of iodouracil in DNA duplex; C–I bond is cleaved via two different pathways 'homolysis and heterolysis'. Tetrahedron Lett 49:428–431

Taylor J, Brockie IR (1988) Synthesis of a trans-syn thymine dimer building block. Solid phase synthesis of CGTAT[t, s]TATGC. Nucleic Acids Res 16:5123–5136

Taylor JS, Brockie IR, O'Day CL (1987) A building block for the sequence-specific introduction of cis-syn thymine dimers into oligonucleotides. Solid-phase synthesis of TpT[c, s]pTpT. J Am Chem Soc 109:6735–6742

Toga T, Yamamoto J, Iwai S (2009) Efficient conversion of thymine glycol into the formamide lesion in oligonucleotides. Tetrahedron Lett 50:723–726

Tommasi S, Swiderski PM, Tu Y et al (1996) Inhibition of transcription factor binding by ultraviolet-induced pyrimidine dimers. Biochemistry 35:15693–15703

Tronche C, Goodman BK, Greenberg MM (1998) DNA damage induced via independent generation of the radical resulting from formal hydrogen atom abstraction from the C1′-position of a nucleotide. Chem Biol 5:263–271

Trzupek JD, Sheppard TL (2005) Photochemical generation of ribose abasic sites in RNA oligonucleotides. Org Lett 7:1493–1496

Usui K, Aso M, Fukuda M et al (2008) Photochemical generation of oligodeoxynucleotide containing a C4′-oxidized abasic site and its efficient amine modification: dependence on structure and microenvironment. J Org Chem 73:241–248

Varaprasad CV, Bulychev N, Grollman AP et al (1996) Synthesis of 8-oxo-7,8-dihydro-6-O-methyl-2′-deoxyguanosine and its use as a probe to study DNA-base excision by MutY enzyme. Tetrahedron Lett 37:9–12

Virgilio A, Esposito V, Randazzo A et al (2005) Effects of 8-methyl-2′-deoxyadenosine incorporation into quadruplex forming oligodeoxyribonucleotides. Bioorg Med Chem 13:1037–1044

Virta P (2009) Solid-phase synthesis of base-sensitive oligonucleotides. Arkivoc 40:54–83

Vyle JS, Young KJ, Grasby JA (1998) Incorporation of methylated pyrimidine analogues into RNA. Tetrahedron Lett 39:5093–5096

Wang J, Wang Y (2009) Chemical synthesis of oligodeoxyribonucleotides containing N^3- and O^4-carboxymethylthymidine and their formation in DNA. Nucleic Acids Res 37:336–345

Wang J, Wang Y (2010) Synthesis and characterization of oligodeoxyribonucleotides containing a site-specifically incorporated N6-carboxymethyl-2′-deoxyadenosine or N4-carboxymethyl-2′-deoxycytidine. Nucleic Acids Res 38:6774–6784

Wang H, Kozekov ID, Harris TM et al (2003) Site-specific synthesis and reactivity of oligonucleotides containing stereochemically defined 1, N2-deoxyguanosine adducts of the lipid peroxidation product trans-4-hydroxynonenal. J Am Chem Soc 125:5687–5700

Wang H, Marnett LJ, Harris TM et al (2004) A novel synthesis of malondialdehyde adducts of deoxyguanosine, deoxyadenosine, and deoxycytidine. Chem Res Toxicol 17:144–149

Wang H, Kozekov ID, Kozekova A et al (2006) Site-specific synthesis of oligonucleotides containing malondialdehyde adducts of deoxyguanosine and deoxyadenosine via a postsynthetic modification strategy. Chem Res Toxicol 19:1467–1474

Xu Y (2000) Synthesis and characterization of DNA containing O6-carboxymethylguanine. Tetrahedron 56:6075–6081

Xu Y, Swann PF (1990) A simple method for the solid phase synthesis of oligodeoxynucleotides containing O^4 -alkylthymine. Nucleic Acids Res 18:4061–4065

Yagi H, Frank H, Seidel A et al (2007) Synthesis and absolute configuration of cis- and trans-opened cyclopenta[cd]pyrene 3,4-oxide N2-deoxyguanosine adducts: conversion to phosphoramidites for oligonucleotide synthesis. Chem Res Toxicol 20:650–661

Yamamoto J, Hitomi K, Todo T et al (2006) Chemical synthesis of oligodeoxyribonucleotides containing the Dewar valence isomer of the (6-4) photoproduct and their use in (6-4) photolyase studies. Nucleic Acids Res 34:4406–4415

Yamamoto J, Oyama T, Kunishi T et al (2014) A cyclobutane thymine-N4-methylcytosine dimer is resistant to hydrolysis but strongly blocks DNA synthesis. Nucleic Acids Res 42:2075–2084

Yueh H, Yu H, Theile CS et al (2012) Synthesis and properties of DNA containing cyclonucleosides. Nucleoside Nucleotide Nucleic Acids 31:661–679

Zeng Y, Wang Y (2006) Sequence-dependent formation of intrastrand crosslink products from the UVB irradiation of duplex DNA containing a 5-bromo-2′-deoxyuridine or 5-bromo-2′-deoxycytidine. Nucleic Acids Res 34:6521–6529

Zhao X, Kao JL, Taylor J (1995) Preparation and characterization of a deoxyoligonucleotide 49-mer containing a site-specific thymidylyl-3′,5′-adenosine photoproduct. Biochemistry 34:1386–1392

Zheng Y, Sheppard TL (2004) Half-life and DNA strand scission products of 2-deoxyribonolactone oxidative DNA damage lesions. Chem Res Toxicol 17:197–207

Zhou Y, Romano LJ (1993) Solid-phase synthesis of oligonucleotides containing site-specific N-(2′-deoxyguanosin-8-yl)-2-(acetylamino)fluorene adducts using 9-fluorenylmethoxycarbonyl as the base-protecting group. Biochemistry 32:14043–14052

Zhou Y, Chladek S, Romano LJ (1994) Synthesis of oligonucleotides containing site-specific carcinogen adducts. Preparation of the 2-cyanoethyl N, N-diisopropylphosphoramidite of N-(2′-deoxyguanosin-8-yl)-2-(acetylamino)fluorene with Fmoc as the base-protecting group. J Org Chem 59:556–563

Single-Molecule Visualization of Biomolecules in the Designed DNA Origami Nanostructures Using High-Speed Atomic Force Microscopy

Masayuki Endo

Contents

Abstract Visualization of biomolecules is one of the straightforward ways to elucidate the physical properties of individual molecules and their reaction processes. Atomic force microscopy (AFM) enables direct imaging of biomolecules in the physiological environment. Because AFM visualizes the molecules at nanometer-scale spatial resolution, a versatile observation scaffold should be

M. Endo (✉)
Institute for Integrated Cell-Material Sciences (WPI-iCeMS), Kyoto University, Yoshida-ushinomiyacho, Sakyo-ku, Kyoto 606-8501, Japan
e-mail: endo@kuchem.kyoto-u.ac.jp

S. Jurga et al. (eds.), *Modified Nucleic Acids in Biology and Medicine*, RNA Technologies, DOI 10.1007/978-3-319-34175-0_17

required for the precise imaging of molecules in the reactions. The emergence of DNA origami technology allows the precise placement of target molecules in the designed nanostructures and enables molecules to be detected at the single-molecule level. The DNA origami is applied for visualizing the detailed motions of target molecules in the reaction using high-speed AFM (HS-AFM), which enables the analysis of dynamic motions of biomolecules in a subsecond time resolution.

In this review article, I describe the combination of the DNA origami system and HS-AFM for imaging various biochemical reactions including enzyme reactions and DNA structural changes. For observation of the enzyme reactions including the DNA methylation, base-excision repair, recombination, and transcription, the substrate DNA strands were placed in DNA nanostructures and their reactions were observed by HS-AFM. In addition, DNA structural changes including G-quadruplex formation and disruption, the hybridization and dehybridization of photoresponsive oligonucleotides, B–Z transition, and G-quadruplex/i-motif formation were visualized using this observation system. The stepwise movement of mobile DNA molecule along the DNA track was visualized on the DNA origami surface. Furthermore, the dynamic assembly/disassembly of photoresponsive DNA origami structures and formation of micrometer-sized DNA origami assemblies were directly visualized on the lipid bilayer. Catalytic reactions and RNA interactions were imaged by HS-AFM. These target-orientated observation systems should contribute to the detailed analysis of biomolecule motions in real time and at molecular resolution.

Keywords DNA nanotechnology • DNA origami • Single-molecule observation • High-speed atomic force microscopy • DNA nanodevice

1 Introduction

Direct imaging of target biomolecules is one of the straightforward ways to study physical properties of molecules in various phenomena involved in living systems. Single-molecule imaging using a probe microscope is a practical approach for investigating the motions of biomolecules during reactions. Atomic force microscopy (AFM) enables the direct observation of biomolecules at nanoscale spatial resolution, and the imaging can be performed under the physiological conditions. To facilitate the observation of single biomolecules, a versatile observation scaffold is needed for the precise analysis of interactions and reactions (Torring et al. 2011; Rajendran et al. 2012; Endo et al. 2013). Based on well-established DNA nanotechnology, DNA origami (Rothemund 2006), a new DNA self-assembly system, has recently been developed for the construction of a wide variety of two-dimensional and three-dimensional nanostructures (Fig. 1). DNA origami can

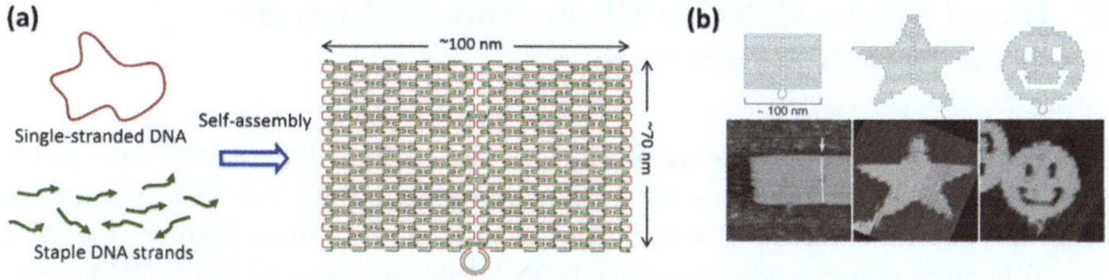

Fig. 1 DNA origami structures. (**a**) Assembly of DNA origami structure by annealing single-stranded long DNA and complementary short DNA strands (*staples*). (**b**) Various 2D structures including rectangle, *star shape*, and *smile mark* can be designed and constructed by DNA origami methods

be used as scaffolds for incorporation of various functional molecules and nanoparticles at desired positions. The DNA origami system is also expanded for single-molecule detection of target molecules and for the analysis of single chemical reactions, which are imaged by AFM (Rajendran et al. 2012; Torring et al. 2011). The various states of individual molecules can be imaged on the DNA origami surface if the DNA nanostructures are used as a scaffold for AFM observation. Therefore, the detailed dynamics of the molecules can also be visualized if the single-molecule imaging is performed on the DNA origami nanostructure. In the past decade, visualization of molecular movement during biological reactions in a subsecond time scale has been achieved using high-speed AFM (HS-AFM) (Ando et al. 2001; Ando and Kodera 2012; Uchihashi et al. 2012; Rajendran et al. 2014). The HS-AFM realizes imaging at 5–20 frames/s rate (depending on the scanning dimension and the number of data points), which enables imaging of the dynamic movements of biomolecules in real time at molecular resolution (Walters et al. 1996; Schitter et al. 2007). Combining the DNA origami system and HS-AFM, the dynamic movement of mobile molecules can be imaged when the substrate double-stranded DNA (dsDNA) is attached to the robust origami structure. The DNA origami system can be expanded to visualize the dynamic movement of various biochemical reactions including enzymatic reactions, DNA structural changes, DNA photoreactions, DNA catalytic reactions, and RNA interactions at the single-molecule level (Rajendran et al. 2012; Endo et al. 2010a). In addition, the lipid bilayer was used to expand the HS-AFM imaging for direct observation of the dynamic formation of DNA origami assemblies. Using the designed DNA origami scaffolds and improving the HS-AFM imaging technique, these observation systems can be extensively used to elucidate the physical properties of individual molecules involved in various biological and nonbiological phenomena.

2 Direct Single-Molecule Observation of Enzymes in the DNA Nanostructure

To directly observe the dynamic motions of the enzymes interacting with a substrate dsDNA, several studies have been performed using HS-AFM (Crampton et al. 2007; Gilmore et al. 2009; Suzuki et al. 2011). In these former studies, it was difficult to obtain a homogeneous dsDNA substrate because the dsDNA forms various random shapes. To overcome this problem, we created an observation scaffold based on the DNA origami structure carrying substrate dsDNAs. We designed and prepared a DNA origami scaffold called a "DNA frame" for this purpose. This robust DNA frame can accommodate two different dsDNA fragments in its cavity (ca. 40 nm × 40 nm) to control physical properties such as the tensions of incorporated dsDNAs (Fig. 2) (Endo et al. 2010a).

2.1 DNA Methylation

DNA-modifying enzymes often require the bending of specific DNA strands to facilitate reactions. DNA methylation enzyme *Eco*RI methyltransferase (M.*Eco*RI) bends dsDNA by 55–59°, enabling the methyl-transfer reaction to proceed

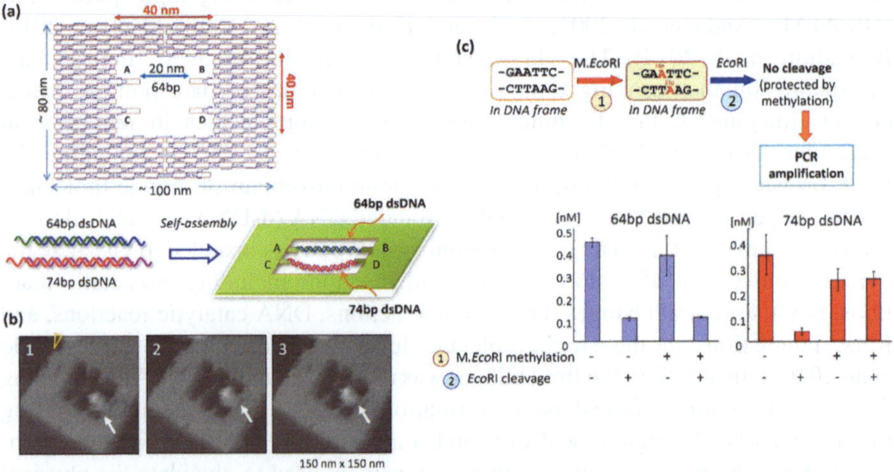

Fig. 2 Regulation of the methylation with M.*Eco*RI in the DNA origami frame and HS-AFM images of M.*Eco*RI movement. (**a**) DNA frame structure designed to incorporate two different dsDNAs, tense 64 bp dsDNA and relaxed 74 bp dsDNA, which contain a specific sequence for M. *Eco*RI at the center. (**b**) HS-AFM images of M.*Eco*RI moving on the 64 bp dsDNA in the DNA frame. Successive AFM images were taken at a scanning rate of 1.0 frame/s. (**c**) Quantification of DNA methylation in the DNA frame. When methylation occurs, the target sequence is protected from the subsequent cleavage by restriction enzyme *Eco*RI, resulting in amplification of the target sequence by PCR. The graphs show the quantification of methylation of the target site in 64 bp and 74 bp dsDNA by real-time PCR amplification

(Youngblood and Reich 2006). To examine the DNA bending effect on methylation with M.*Eco*RI, two different lengths of dsDNA fragments, 64 base pair (bp) (short-strand) and 74 bp (long-strand) dsDNA, were incorporated into the DNA frame structure (Endo et al. 2010a). Because the length between connectors in the DNA frame cavity is 64 bp, 64 bp dsDNA fits to the DNA frame as a tense form in which its conformational flexibility should be suppressed, while the 74 bp dsDNA is more relaxed, allowing the local dsDNA bending. The dynamic movements of the dsDNAs and the formation of M.*Eco*RI complexes with dsDNAs were observed by HS-AFM (Fig. 2b). After treatment of the dsDNAs in the DNA frame with M. *Eco*RI and subsequent digestion with restriction enzyme *Eco*RI, AFM analysis revealed that the 74 bp dsDNA was less effectively cleaved compared with the 64mer dsDNA. This indicates that the methylation occurred preferentially in the relaxed 74 bp dsDNA rather than in the tense 64 bp dsDNA. Biochemical analysis of the methylation to the target sequence was performed using real-time PCR. After methylation with M.*Eco*RI and subsequent *Eco*RI digestion, the 74 bp dsDNA substrate in the DNA frame was efficiently amplified, indicating that the preferential methylation occurred in the target sequence of the 74 bp dsDNA (Fig. 2c). These results suggest the importance of structural flexibility for the bending of dsDNA during the methyl-transfer reaction with M.*Eco*RI. Therefore, DNA methylation required the bending and can be regulated using the tension-controlled dsDNAs constructed in the DNA frame.

2.2 DNA Base-Excision Repair

DNA repair is an indispensable biological function to preserve genetic information from mutations such as transversion during replication (David et al. 2007). DNA base-excision repair enzymes 8-oxoguanine glycosylase (hOGG1) and T4 pyrimidine dimer glycosylase (PDG) were employed for the analysis of repair reactions in the DNA frame (Endo et al. 2010b). These enzymes have glycosylase/AP lyase activity, which removes damaged nucleobases and cleaves the DNA strand. Repair enzymes often require local structural changes in the target DNA strands, such as DNA bending, for the reaction to proceed (Nash et al. 1996; Vassylyev and Morikawa 1997). hOGG1 bends dsDNA about 70° to flip out the oxoG base to proceed the reaction (Bruner et al. 2000). PDG also bends the double helix by 60° to flip out the 3′-side of adenine in the opposite strand of the thymine dimer (Morikawa et al. 1992; Vassylyev et al. 1995). Various dsDNAs containing a damaged base were incorporated into a DNA frame as dsDNA cassettes, and the repair reactions were analyzed at the single-molecule level. To examine the structural effect on glycosylase/AP lyase activity including cleavage of the DNA strand and the trapping of reaction intermediates, two different lengths of substrate dsDNAs, tense 64 bp and relaxed 74 bp dsDNAs, were incorporated to the DNA frame. The enzymes more favorably cleaved the relaxed dsDNA and were covalently trapped after $NaBH_4$ reduction compared with the tense dsDNA. In addition, enzyme movement and the DNA repair reaction were directly observed on the DNA frame using HS-AFM. The DNA frame system serves to analyze the detailed repair process by directly observing the events involved in DNA repair reaction such as binding, sliding, catalytic reaction, and dissociation.

2.3 DNA Recombination

DNA recombination plays an important role including the generation of genetic diversity and mediating DNA integration into the host genome (Grindley et al. 2006). Cre recombinase recognizes the 34 bp *lox*P sequence and forms a synaptic complex with two *lox*P dsDNAs as a tetramer form (Guo et al. 1997; Van Duyne 2001). Here, substrate dsDNAs with the *lox*P sequence were placed in the DNA frame, and the recombination events were analyzed at the single-molecule level (Fig. 3) (Suzuki et al. 2014b).

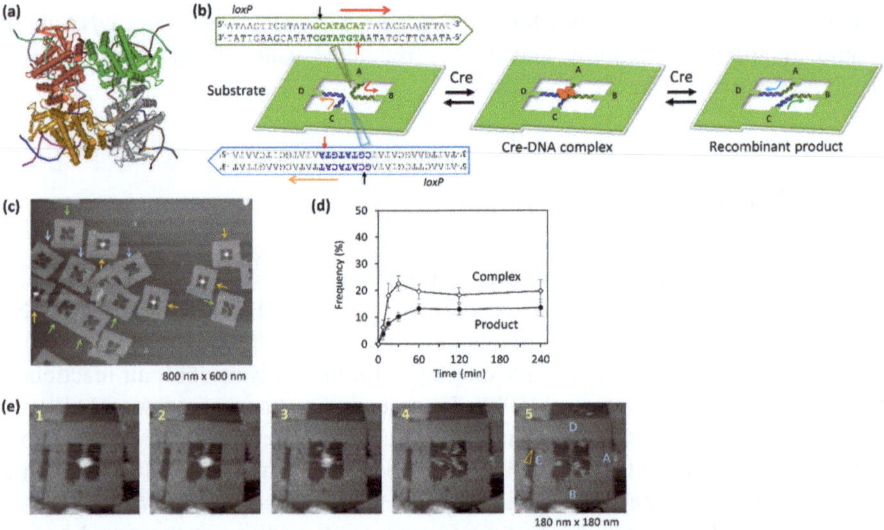

Fig. 3 Regulation and single-molecule observation of Cre-mediated recombination in the DNA frame. (**a**) Crystal structure of the Cre tetramer bound to the Holliday junction intermediate (PDB: 3CRX). (**b**) Substrate dsDNAs having the *loxP* sequence in the antiparallel orientation were placed onto the DNA frame. In the recombination, the Cre tetramer forms a synaptic complex and subsequently generates a recombinant product, which is easily identified by the topological arrangement of two dsDNAs as different loop structures and cross shape. (**c**) AFM image of the substrate in the DNA frame after incubation with Cre. The Cre–DNA complex, starting substrate, and recombinant product are indicated by *orange*, *green*, and *blue arrows*, respectively. (**d**) Time-course analysis of the formation of the Cre–DNA complex and recombinant product. (**e**) Synaptic complex structure with the Cre tetramer and two dsDNAs. HS-AFM images of the dissociation of the Cre tetramer from the dsDNAs into four Cre monomers and the appearance of a recombinant product. Successive AFM images were taken at a scanning rate of 1.0 frame/s

The DNA frame can control the orientation of two *lox*P sequences in the antiparallel and parallel arrangements. First, recombination was directly observed using the antiparallel arrangement placed in the DNA frame (Fig. 3b). Cre performs two-step cleavage/strand exchange reactions to form the Holliday junction (HJ) intermediate and the subsequent recombinant product. After incubation with Cre, the Cre–DNA complex and recombinant products were clearly observed in the DNA frame, showing that recombination occurred even using the substrates placed in the nanospace (Fig. 3c). In the reaction time course, the Cre–DNA complex formed first, followed by the recombinant product (Fig. 3d). During observation of the Cre–DNA complex by HS-AFM, the tetrameric Cre that formed the synaptic complex dissociated into four monomers, and the recombinant product simultaneously appeared (Fig. 3e). In the parallel arrangement, the Cre–*lox*P complex formed, while the recombinant product was not observed.

The DNA scaffold can also control the angle of the HJ and impose structural stress. To regulate the direction of recombination, the HJ intermediates crossing at 90° and 60° were prepared using different DNA frames (Suzuki et al. 2014b). By adjusting the directions of the *lox*P sequences, the HJ intermediates crossing at 90° were resolved to give products in a usual fashion (Guo et al. 1999; Gopaul et al. 1998). On the other hand, the reaction with HJ intermediates crossing at 60° changed the formation of the resolution products by different proportions. Therefore, the structural stress imposed on the HJ intermediates in the DNA frame can regulate the direction of recombination. The desired geometric arrangements of the substrate dsDNAs using the DNA frames are valuable for studying recombination events, which are controlled by the orientation of substrate dsDNAs and the angle of HJ intermediates. The HJ-containing DNA frame was used for the observation of the resolution of HJ by RecU resolvase (Suzuki et al. 2014a). The resolved products were observed in the DNA frame, indicating that the HJ in the DNA frame can also be a substrate for the resolvase.

2.4 Transcription

Transcription, including the RNA synthesis by RNA polymerase (RNAP), is one of the most important common biological phenomena in the living system. Transcription involves a series of the RNAP behaviors, including binding to double-stranded DNA (dsDNA), sliding along the dsDNA, RNA synthesis, and dissociation from the dsDNA. We directly observed the sliding of RNAP and transcription by creating the direct observation system with a template dsDNA (1.0 kbp) containing the T7 promoter which was attached to the designed DNA nanostructure (Fig. 4) (Endo et al. 2012a).

For the preparation of the observation system, a 400 nm-sized rectangular DNA origami scaffold was designed, and a PCR-amplified template dsDNA was incorporated between the connection sites in the DNA scaffold via corresponding staple strands attached to its ends (Fig. 4a). The purified template dsDNA-attached

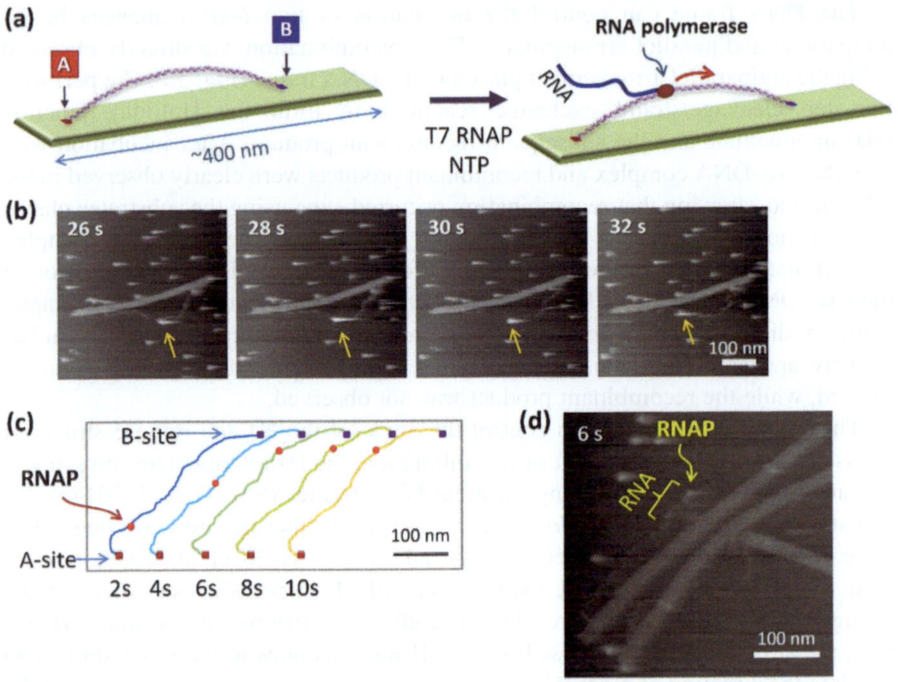

Fig. 4 AFM observation of the movement and reaction of RNA polymerase using the DNA nanoscaffold. (**a**) Structure of a template dsDNA-attached scaffold and transcription by addition of RNAP and NTP. (**b**) Successive AFM images of RNAP sliding along the template dsDNA attached to the scaffold. (**c**) The traces of RNAP movement on the template dsDNA during transcription. (**d**) HS-AFM image of RNAP in transcription

scaffold was observed in the buffer solution using HS-AFM. From the successive AFM images, the DNA scaffold part was tightly fixed on the mica surface; on the other hand, the template dsDNA chain moved flexibly. Most of the dsDNA chain was visualized by HS-AFM, because the template dsDNA was connected to the robust DNA nanoscaffold at the two positions.

To observe the sliding of RNAP, RNAP was added to the template dsDNA-attached scaffold. Successive images of RNAP sliding on the template dsDNA were obtained. From the analysis of the shape of the template dsDNA and the position of RNAP, RNAP moved in the area where the template dsDNA was also mobile (Fig. 4b). The one-dimensional diffusion coefficient ($D = <\Delta l^2>/2t$) obtained from the analysis of the AFM images was $5.1 \pm 0.7 \times 10^{-12}$ cm^2/s. Next, the direct observation of transcription including RNA synthesis was examined in the template dsDNA-attached scaffold (Fig. 4c,d). In the presence of NTPs, the target transcript was synthesized from the template dsDNA-attached scaffold. From the HS-AFM images, the RNAP movement from the promoter region and dissociation from the template dsDNA were observed. In a series of the reactions, single RNAP appeared around the promoter region (time = 2 s). The RNAP started to move downstream on the template dsDNA, with some unclear AFM scanning traces (time = 4–10 s), and continued to move to the middle of the template dsDNA. The RNAP finally dissociated from the dsDNA. In the expanded image (time = 6 s), the unclear AFM scanning trace would be the image of the RNA transcript synthesized by RNAP. The speed of the RNAP movement in the transcription was not constant during the time scale of the AFM measurement, because of the contact of template dsDNA to the mica surface.

We also successfully visualized transcription including RNAP sliding and RNA synthesis using a template dsDNA-attached nanoscaffold and HS-AFM. The observation system can be used for observation of a wide variety of events with DNA-binding proteins and enzymes that move along dsDNA during the reaction. A combination of the DNA origami nanoscaffold and HS-AFM is an intelligent system for observing enzyme reactions and relevant events including complex formation, catalysis, and dissociation at the single-molecule resolution.

3 Direct Observation of DNA Structural Changes in the DNA Nanostructure

Structural variations and conformational changes in the DNA involved in living systems are closely linked to the regulation of their biological functions, such as gene expression (Bacolla and Wells 2004, 2009). The DNA frame system allows the introduction of various dsDNAs with unrestricted sequences for observation of reactions, and DNA frames can also control the physical properties of dsDNAs such as the tensions and rotations of dsDNAs. Here, the DNA frame system is applied for the visualization of DNA structural changes: (1) formation/disruption of the G-quadruplex, (2) formation/dissociation of dsDNA into GQ/i-motif structures, (3) hybridization/dissociation of photoresponsive duplex, and (4) helical rotation in the B–Z transition.

3.1 G-Quadruplex Formation and Disruption

In the field of structural and molecular biology, G-quadruplex (GQ), a four-stranded helix structure, is of great interest because of its structural variations and biological functions (Shirude et al. 2007). For the detection of the formation and disruption of a GQ structure, we employed our observation system by monitoring the shape of dsDNA chains in the DNA frame (Sannohe et al. 2010). To place the G-rich sequences, two dsDNA chains containing single-stranded triple guanine (GGG) overhangs at the center were prepared for the interstrand GQ formation (Xu et al. 2008). Three G-tracts were placed in the upper dsDNA chain, whereas the lower dsDNA chain had a single G-tract (Fig. 5a). Initially, the two dsDNA

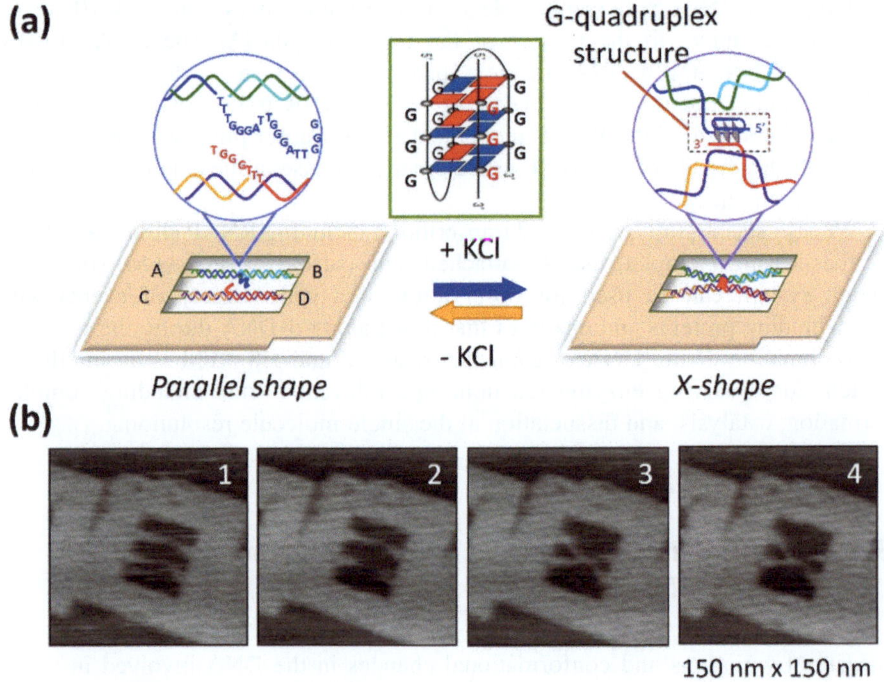

Fig. 5 Visualization of G-quadruplex (GQ) formation using the structural change of two dsDNA chains placed in the DNA frame. (**a**) In the presence of KCl, the separated state changes to the X-shape by connection at the center of two dsDNA chains via GQ formation. (**b**) HS-AFM images for the formation of an X-shape structure by the association of two dsDNA chains via GQ formation. Four successive AFM images were taken at a scanning rate of 0.2 frames/s

chains introduced did not contact each other. In the presence of K^+, the two dsDNA chains in the DNA frame clearly showed an X-shaped structure with a 44 % yield, demonstrating the formation of an interstrand GQ. The dynamic formation of the GQ was further examined in real time by HS-AFM. During scanning of the sample in the presence of K^+, the two dsDNA chains with G-tracks maintained a separated state for a given period, and then they suddenly formed the X-shaped structure (Fig. 5b). In a similar fashion, the disruption of the preformed GQ was directly observed in the absence of K^+. The X-shape remained unchanged for a while, and then it reverted to the separated state during AFM scanning.

Furthermore, four DNA strands containing various G-tracts were assembled in the DNA frame to form a synaptic G-quadruplex using the same system (Rajendran et al. 2013a, c). Thus, the single-molecule observation of the dynamic formation and disruption of GQ was successfully achieved by monitoring the global structural changes of the two incorporated dsDNA chains in the DNA frame using HS-AFM.

3.2 Hybridization and Dissociation of Photoresponsive Oligonucleotides

The direct observation of hybridization and dissociation of a single duplex DNA at molecular resolution is quite challenging. The dynamic hybridization and dissociation of single DNA strands were examined on the DNA origami scaffold using fluorescence microscopy by tracking fluorescence-labeled DNA strands (Jungmann et al. 2010). AFM-based single-molecule imaging can visualize whole nanostructures by directly monitoring the shape of DNA strands. Photoresponsive DNAs containing azobenzene moieties were employed to observe the hybridization and dissociation of duplex DNA (Asanuma et al. 2007; Liang et al. 2009). A pair of photoresponsive DNAs was connected to the individual dsDNA chains, which were then placed in the cavity of the DNA frame (Fig. 6a) (Endo et al. 2012b). The photoresponsive domain can hybridize with the corresponding counterpart in the *trans*-form of the azobenzene moiety and dissociate in the *cis*-form UV irradiation (Liang et al. 2009). The dissociated DNAs in the *cis*-form hybridize again upon visible light (Vis) irradiation (Liang et al. 2009). Two dsDNA chains possessing photoresponsive domains at the center were placed in the cavity of the DNA frame. The hybridized photoresponsive duplex in the center of the supporting dsDNA chains in the DNA frame was clearly visualized (Fig. 6b). Using this origami system, hybridization and dissociation can be identified by the global structural change of the two supporting dsDNA chains as an X-shape and as a separated shape in the DNA frame, respectively. Hybridization and dissociation were observed directly using HS-AFM. The dissociation of the two dsDNA chains in contact at the center (X-shape) was imaged during UV irradiation. The contact of the two separated dsDNA chains in the center was then imaged again during Vis irradiation (Fig. 6c). Successive switching of dissociation and hybridization of the photoresponsive domains was visualized at the single-molecule level by observing the global change of two dsDNA chains, the X-shape and the separated shape, in the DNA frame.

We also incorporated two switching motifs, a pair of photoresponsive DNAs and another pair of G-telomeric repeats for GQ, which were incorporated together into three parallel dsDNAs assembled in the DNA frame with six connectors (Yang et al. 2014). Photoirradiation and K^+ were used as input stimuli to switch the interactions among three dsDNAs in a logical manner. Cascading transformation from photoinduced dissociation to G-quadruplex formation in bulk solution was achieved successfully. In addition, a series of dual-switching configuration conversions were also observed in a single DNA frame.

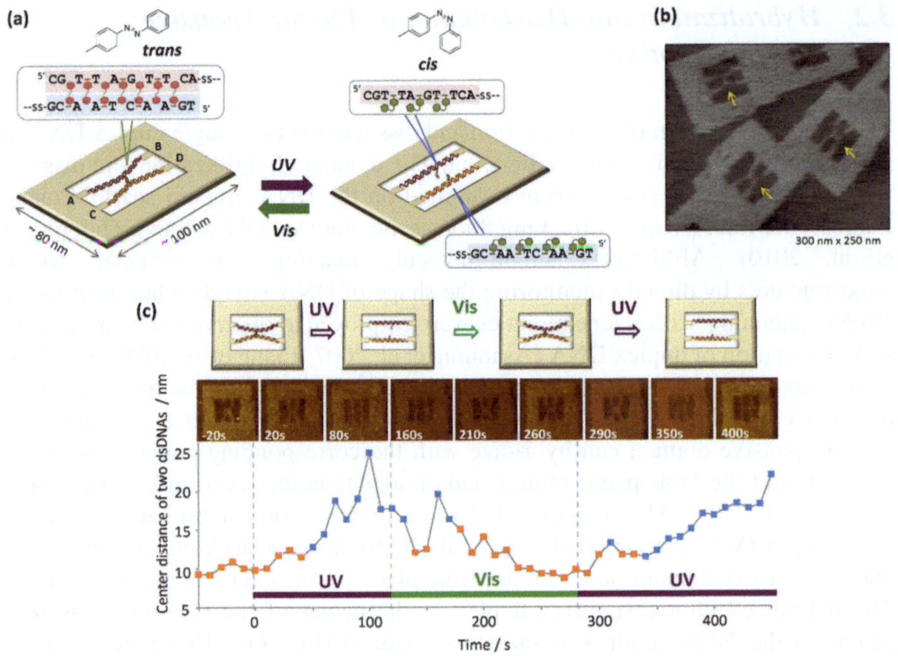

Fig. 6 Direct observation of the hybridization and dissociation of a pair of photoresponsive DNAs. (**a**) Single-molecule observation system. In the *trans*-form of an azobenzene moiety, two photoresponsive domains hybridize to form a duplex. In the *cis*-form induced by UV irradiation, the two domains dissociate. Two dsDNA chains containing photoresponsive DNAs were placed in the DNA frame structure to observe the dissociation and hybridization after UV and Vis irradiation, respectively. Two different dsDNA chains containing different photoresponsive DNAs were connected between the specific sites in the DNA frame via the corresponding overhanging ssDNAs. (**b**) AFM image of the photoresponsive duplex supported by two dsDNA chains in the DNA frame (*orange arrow*). (**c**) Photoswitching activity of photoresponsive DNAs in the DNA frame. The repeating dissociation and hybridization were visualized by HS-AFM during alternative UV/Vis photoirradiation. The distance between the centers of two dsDNA chains was plotted. The appearance of an X-shape and a separated shape is shown as an orange and blue rectangle in the graph, respectively

3.3 B–Z Transition in the Equilibrium State

A right-handed B-form dsDNA containing the CG repeat sequence is known to transit to the left-handed Z-form structure by increasing the salt concentration (Jovin et al. 1987). A nanomechanical device employing the rotation of the B–Z transition was created on the DNA nanostructure, and the rotation of the nanostructure was investigated by FRET (Mao et al. 1999). Using our observation system, a rotary motion of the B–Z conformational transition of double-helix DNA was directly visualized in the DNA frame structure (Rajendran et al. 2013b). To visualize the B–Z transition, a dsDNA with a 5-methyl-CG six-repeat sequence

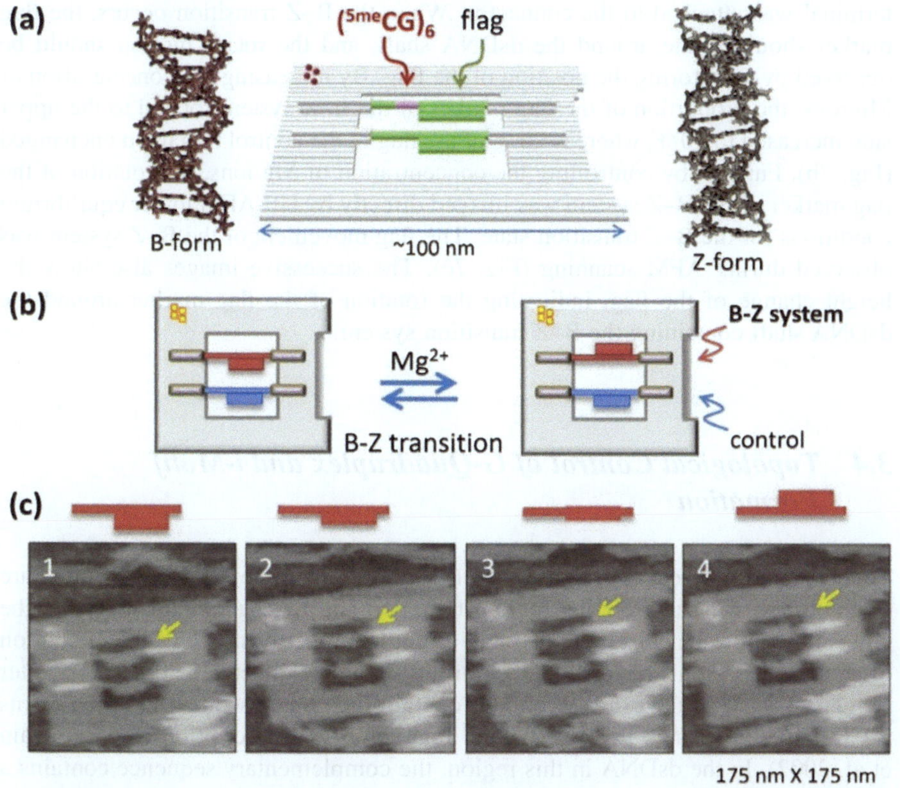

Fig. 7 Direct observation of B–Z transition in the DNA frame. (**a**) Single-molecule observation system for B–Z transition and B- and Z-form dsDNA structures. (**b**) Two dsDNAs having a (^{5me}CG)$_6$ sequence (B–Z system) and a random sequence (control) were introduced to the top and bottom site in the DNA frame, respectively. For observation of the dsDNA rotation, both of the ssDNA linkers in the left terminal of the dsDNA were fixed to the connector, while one ssDNA linker in the right terminal was attached to the connector. The dsDNAs have a flag marker for observation of the rotation of the dsDNA shaft during the B–Z transition. (**c**) HS-AFM images of the flipping motion of the flag marker at the top site (*yellow arrow*). Four successive images are presented at a scanning rate of 0.2 frames/s

(mCG)$_6$ and a flag marker containing three bundled dsDNAs connected by crossovers was introduced to the DNA frame structure (Fig. 7). The (mCG)$_6$ repeat can promote the formation of a Z-form even at low salt concentrations (Behe and Felsenfeld 1981). One dsDNA with a (^{me}CG)$_6$ sequence and a flag was introduced to the top as a B–Z transition system, and the other dsDNA with a random sequence and a flag was introduced to the bottom as a control. To allow rotation during the B–Z transition, four connectors in the DNA frame were designed to lift both dsDNAs from the scaffold surface (Fig. 7a). In addition, for observation of the rotation of double helix, both of the single-stranded DNA (ssDNA) linkers in the left terminal of the dsDNA were fixed to the connector, while one ssDNA linker in the right terminal was attached to the connector. When the B–Z transition occurs, the flag marker should rotate around the dsDNA shaft, and the rotary motion should be observed by monitoring the position of the flag. By increasing the concentration of Mg ions, the proportion of the flag marker of the B–Z system rotated to the upper side increased to 70 %, whereas 76 % of the flag in the control remained unchanged (Fig. 7b). Further, by controlling the concentration of Mg ions, the rotation of the flag marker of the B–Z system was imaged directly by HS-AFM under equilibrium conditions for the B–Z transition state. The flag movement of the B–Z system was observed during AFM scanning (Fig. 7c). The successive images also show the height change of the flag, indicating the rotation of the flag marker around the dsDNA shaft containing the B–Z transition system.

3.4 Topological Control of G-Quadruplex and i-Motif Formation

Tandem G-rich repeat sequences, which form G-quadruplex (GQ) structures, are often observed in the promoter regions (Huppert and Balasubramanian 2007). The GQ is thought to be closely involved in biological functions such as the regulation of gene expression and cell fate control (Siddiqui-Jain et al. 2002). Promoter sequences such as the *c-myc* promoter contain the G-rich sequence, which forms a GQ structure that is involved in the regulation of transcription (Siddiqui-Jain et al. 2002). In the dsDNA in this region, the complementary sequence contains a C-rich repeat sequence, which forms i-motif structure (Gehring et al. 1993). A number of studies have reported on the formation of the GQ and/or i-motif in the promoter regions containing GQ- and i-motif-forming sequences (Xu and Sugiyama 2006; Sun and Hurley 2009; Dhakal et al. 2010, 2012). The i-motif structure is physically induced in an acidic condition because the i-motif formation requires a hemiprotonated cytosine dimer formation (Gehring et al. 1993). The i-motif formation at neutral pH has also been observed in dsDNA with negative superhelicity (Sun and Hurley 2009) and in a molecular crowding environment (Miyoshi et al. 2004).

We demonstrate the single-molecule operation and observation of the formation and resolution of dsDNA containing a GQ forming and counterpart i-motif-forming sequence in the DNA nanostructure (Endo et al. 2015b). The formation and dissociation of the GQ/i-motif complementary sequence in the insulin-linked polymorphic region (ILPR) core sequence were manipulated in the DNA frame structure. The ILPR promoter region is considered to separate into the GQ and i-motif structures in the regulation of transcription (Dhakal et al. 2012). The GQ/i-motif sequence was introduced into the DNA frame for observation of the GQ and i-motif formation under the various conditions. In this system, topologically controlled dsDNA was prepared using the sequential manipulation of the interaction of DNA strands in a series of programmed operations. Using the strand displacement and the addition and removal of K^+, the topologically controlled GQ/i-motif dsDNA in the DNA frame was obtained in high yield. We performed the resolution of dsDNA containing the GQ/i-motif sequence by controlling the pH and K^+ conditions which were changed individually or together during the formation of the GQ and i-motif. Furthermore, these structural changes were directly observed by HS-FM at the single-molecule level. The dissociation of the dsDNA under the GQ and i-motif formation condition was monitored by HS-AFM. The results indicate that the dsDNA containing the GQ- and i-motif sequence is effectively dissolved when the duplex is helically loosened in the DNA nanoscaffold.

Using the DNA origami scaffold and HS-AFM system, important DNA conformational changes including G-quadruplex formation, B–Z transition, and G-quadruplex/i-motif formation from the dsDNA were successfully imaged. In addition, the dissociation and hybridization of photoresponsive DNA, which were controlled by photoirradiation using different wavelengths, were directly visualized in the origami structure. The observation system in these experiments can be used as a general strategy for investigating various DNA structural changes and molecular switches working as a single molecule. It can also be applied to the single-molecule imaging of chemical reactions such as bond formation and cleavage.

4 Direct Observation of a Mobile DNA Nanomachine on the DNA Origami Surface

Considering the programmability of DNA sequences and the manipulation of association and dissociation of duplexes, DNA has been used for various molecular machines (Bath and Turberfield 2007). DNA nanomachines such as DNA tweezers and various walking devices have been developed by manipulating duplex formation using strand displacement (Bath and Turberfield 2007). DNA motor systems powered by enzyme cleavage and photocleavage were designed and created on the DNA tile, and their walking motions were directly observed by HS-AFM.

4.1 DNA Motor System Created on the DNA Origami Tile

A DNA transportation system was constructed with a mobile DNA nanomachine moving along a designed track on the DNA origami surface (Fig. 8). The track on the DNA scaffold was constructed to observe the multistep movement of a specific DNA strand (Wickham et al. 2011). Multiple ssDNAs (stators) were introduced onto the rectangular DNA origami tile as a track to hybridize a complementary mobile DNA strand (motor strand). As shown in Fig. 8a, when a DNA motor strand hybridizes to the specific stator strand, the stator/motor duplex is subsequently cleaved by the nicking enzyme Nt.BbvCI, which removes the short ssDNA from the

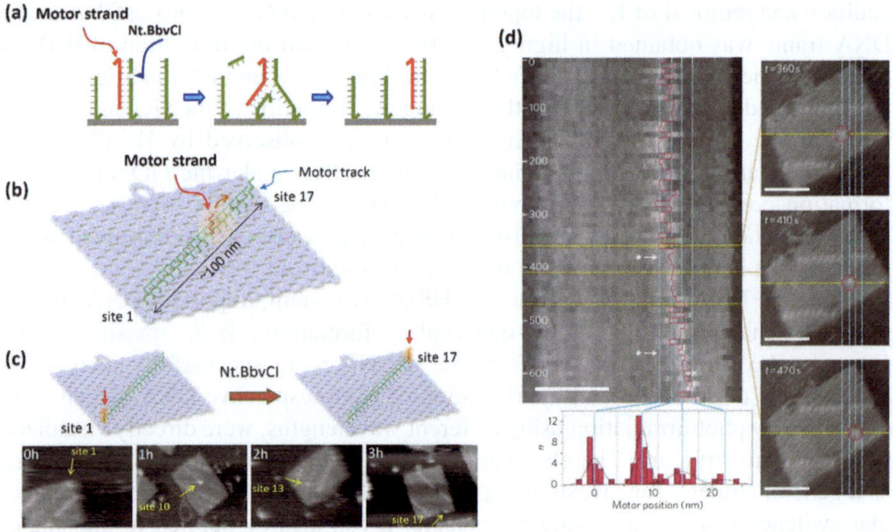

Fig. 8 Direct observation of DNA motor movement on the DNA origami surface. (**a**) Mechanism of the DNA motor movement. When a stator ssDNA (green ssDNA) forming a duplex with a DNA motor strand (red ssDNA) is cleaved by nicking enzyme Nt.BbvCI, the motor strand moves to the adjacent stator strand by branch migration. (**b**) A motor track consisting of 17 stators (green ssDNAs) was constructed on the DNA origami surface, and the movement of the DNA motor (red ssDNA) was examined. (**c**) AFM images of time-dependent movement of a DNA motor. A duplex containing a DNA motor strand was visualized as a *white dot*. (**d**) Stepwise movement of a DNA motor observed by HS-AFM and a kymograph and distribution of the motor positions. Scale bar 50 nm

stator (Bath et al. 2005). The motor strand then binds to the neighboring intact stator by branch migration and finally steps forward. The DNA tile carrying the motor strand at the initial position was incubated with Nt.BbvCI to examine the migration of the motor strand along the DNA motor track (Fig. 8b). The motor strand was imaged as a single spot of the duplex on the DNA origami scaffold, which was easily distinguished from the invisible single-stranded stators. Time-dependent movement of the motor strand along the motor track was observed (Fig. 8c). Furthermore, the movement of the motor strand was directly visualized by HS-AFM. The stator/motor duplex spot showed back and forth movement along the motor track and finally moved forward during AFM scanning. From the kymograph analysis, the distance of the motor-strand movement corresponded to the distance between the adjacent stators, indicating that the movement occurred stepwise on the track (Fig. 8d).

A programmed DNA motor system was constructed using the predesigned DNA track on the DNA origami scaffold. Using HS-AFM, the detailed motion of the DNA motor was directly observed and analyzed. The method was further applied to regulate the transportation of a DNA motor on a branched motor track. In this system, the direction of the DNA motor movement was precisely controlled by selectively blocking and releasing strands with predefined instructions (Wickham et al. 2012). Related works, such as a DNA spider (Lund et al. 2010) and a programmed assembly line (Gu et al. 2010), which also used a mobile DNA nanomachine and DNA origami scaffold, have been reported for the construction of a nanoscale transportation system.

4.2 *Photocontrolled DNA Motor System on the DNA Origami Tile*

The DNA motor system was created using a pyrene-attached DNA strand and disulfide-containing stator strands on the DNA origami tile. A single-stranded DNA carrying two pyrene molecules employed as the photocontrollable DNA motor was assembled on the DNA tile carrying four stator strands as the linear track (You et al. 2012). The stator/motor duplex was located on the surface of DNA tile. The excited pyrene molecules ($\lambda_{ex} = 350$ nm) of the motor can induce the cleavage of disulfide bond in stator strands by reduction via a photoinduced electron transfer. The DNA motor migrating from one cleaved stator to the next intact one on the DNA tile moved continuously until reaching the final stator during UV irradiation. The entire walking process of the motor was determined by characterizing the distribution ratios of stator/motor duplex at four anchorage sites on the tile separately under different irradiation times. The observed rate constants for the stepping of the motor were calculated from these data. Finally, the one-step movements of the motor during UV irradiation were observed in real time using HS-AFM. The photonic modification of DNA-based nanomachines can be used for biological applications such as cargo transport and manual configuration change of biomolecules in mesoscopic systems.

5 Direct Observation of Assembly and Disassembly of DNA Origami Structures

5.1 *Photocontrolled Assembly and Disassembly of DNA Origami*

The advantage for use of the photoreaction is that the target reactions can be initiated and controlled by just irradiating the light from outside. We have observed the single-molecule dissociation and hybridization of photoresponsive DNA strands containing azobenzene molecules in the DNA origami scaffold using HS-AFM (Endo et al. 2012b). The method is also applied for the assembly and disassembly of the DNA origami nanostructures. Using the hexagonally shaped DNA origami carrying photoresponsive DNAs, the hexagonal dimer in the initial state (*trans-*

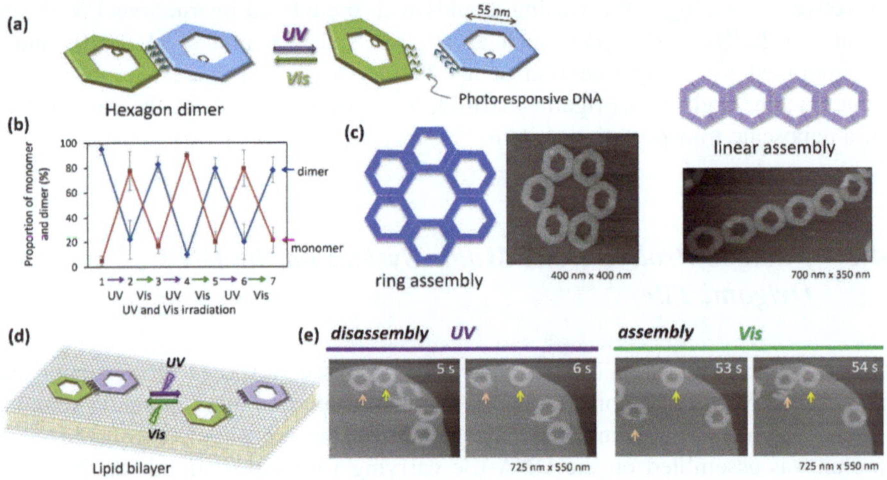

Fig. 9 Direct observation of assembly and disassembly of photoresponsive hexagonal origami on the lipid bilayer surface. (**a**) Assembly and disassembly of hexagonal origami having four photoresponsive DNAs by UV/Vis irradiation. (**b**) Proportion of dimer formation and dissociation by repeating UV/Vis irradiation. (**c**) Formation of ring-shape and linear assemblies and the AFM images. (**d**) Observation of dynamic dimer formation and dissociation on the lipid bilayer with UV/Vis irradiation. (**e**) Direct observation of dissociation of dimer with UV irradiation (*left*) and assembly with Vis irradiation (*right*) using HS-AFM

azobenzene) was dissociated into monomers (*cis*-azobenzene) by UV irradiation and then reassembled to form dimer by visible light (Vis) irradiation (Fig. 9a) (Yang et al. 2012). The assembly and disassembly are reversibly controlled by UV/Vis irradiation in solution. The dynamic dimer formation and dissociation in solution were examined by gel analysis (Fig. 9b) and monitored by fluorescence quenching. The repeated assembly and disassembly of the hexagons were observed in high yield by alternative UV/Vis irradiation. Various shapes such as linear and ring-shaped assemblies were prepared by programming the positions of the photoresponsive DNAs (Fig. 9c).

In addition, we observed the dynamic assembly and disassembly of photoresponsive hexagonal origami structures using a lipid bilayer surface by HS-AFM (Suzuki et al. 2014c). Because of the fluid property of a lipid bilayer, a lipid bilayer can be used for the dynamic assembly of DNA origami structures on the surface (Suzuki et al. 2015b). Cholesterol moieties were introduced to the hexagon for interaction with the lipid. The cholesterol-modified photoresponsive hexagon dimers were attached onto the lipid bilayer, and their diffusion was observed by HS-AFM. When these photoresponsive hexagon dimers were exposed to the UV light during AFM scanning, the dimer immediately dissociated into monomers. Then visible light was irradiated to the monomers; the monomers were associated to form dimer after the diffusion and collision. The reversible assembly and disassembly of hexagonal origami was directly imaged on the lipid bilayer surface. These photocontrolled manipulations for the molecular interactions are also expanded to the construction of the various switching devices.

5.2 Large Assembly of DNA Origami on the Lipid Bilayer

The lipid bilayer was further used to assembly the DNA origami structures into the large defined assemblies. We demonstrate a lipid bilayer-assisted assembly to assemble various DNA origami monomers into two-dimensional lattices (Fig. 10) (Suzuki et al. 2015a). DNA origami structures are electrostatically adsorbed onto the lipid bilayer in the presence of divalent cation. The origami structures were mobile on the lipid bilayer surface and assembled into large 2D lattices with micrometer size (Fig. 10a). Using the cross-shaped DNA origami monomer to form lattice by π–π interaction of the blunt ends (Liu et al. 2011), the dynamic processes including attachment and detachment of DNA origami monomers and reorganization of lattices were visualized using HS-AFM. In addition, the biotinylated lattices were modified with streptavidin; streptavidin binding can also be observed by HS-AFM. Using the cross-shaped origami monomer with ssDNA linkers at the ends to prevent the interaction, the monomers packed into 2D assemblies in a shape-fitted fashion (Fig. 10b). Other monomers including triangular and hexagonal monomers also assembled into packed form.

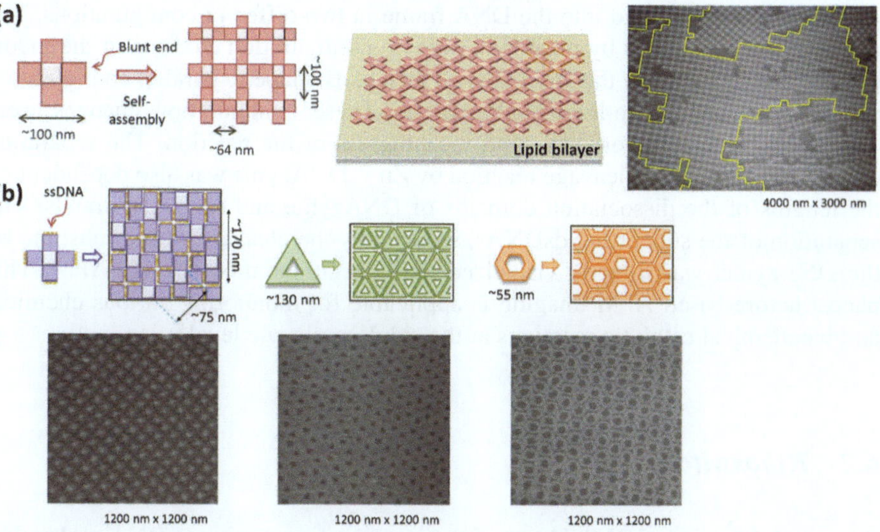

Fig. 10 Direct observation of assembly of DNA origami monomer into micrometer-sized structures on the lipid bilayer surface. (**a**) Assembly of cross-shaped origami monomers into lattice structures via π–π interaction of blunt ends on the lipid bilayer. AFM image of lattice structures. (**b**) Packed assembly of cross-shaped origami with ssDNA linkers at the ends (*left*), triangular origami (*middle*), and hexagonal origami in a shape-fitted fashion

Using the lipid bilayer for the assembly surface, DNA origami can be assembled into the predesigned 2D structures in micrometer scale. The method we developed here involving AFM imaging of dynamic assembly processes can be used for the precise design and preparation of the controlled micrometer-scale DNA architectures with various functions.

6 Direct Observation of the Catalytic Reactions and RNA Interactions in the DNA Nanostructures

6.1 DNAzyme

Catalytic nucleic acids (DNAzymes) attract an attention because these are applicable for various catalysts to be used for amplification and switching behaviors. One specific class of DNAzymes includes metal-ion-dependent DNAzymes that consist of various sequences of DNAzyme and substrate part. These metal-ion-dependent DNAzymes have been used for amplifying labels for sensing, functional components for logic gates, and stimuli-responsive DNA switches (Liu et al. 2014, Wang et al. 2015).

We demonstrate the single-molecule imaging of the catalytic reaction of the Zn^{2+}-dependent DNAzyme (Gu et al. 2013) in the DNA frame structure (Endo et al. 2015a). The DNAzyme and a substrate strands attached to two supporting dsDNAs were assembled into the DNA frame in two different configurations. The reaction was monitored by observing the two configuration changes of the incorporated DNA strands in the DNA frame, such as H-shape to parallel configuration and X-shape to double-loop configuration. These configuration changes were clearly observed in accordance with the progress of the reaction. The separation of the strands after the cleavage reaction by Zn^{2+}-DNAzyme was also dependent on the lengths of the dissociation domains of DNAzyme and substrate strands. The separation of the supporting dsDNAs, induced by the cleavage of the substrate by the DNAzyme, was directly visualized in two configurations by HS-AFM. This nanostructure-based AFM imaging is applicable for monitoring various chemical and biochemical catalytic reactions at the single-molecule level.

6.2 Riboswitch

Structural diversity of RNA is one of the important properties of RNA molecules, which exhibit unique functions such as specific complex formation and catalysis. In addition, RNA molecules uniquely form a complex through specific hairpin loops, called a kissing complex, which enables assembly of complementary RNA loops via Watson–Crick base pairing (Van Melckebeke et al. 2008). The kissing complex is widely investigated and used for the construction of RNA architectures such as polygonal structures and three-dimensional assemblies (Grabow and Jaeger 2014; Shukla et al. 2011). Molecular switches have also been created by combining a kissing loop and a ligand-binding aptamer to control the interactions of RNA molecules (Durand et al. 2014). We incorporated two kinds of RNA molecules into a DNA origami frame and used AFM to observe their ligand-responsive interactions at the single-molecule level (Takeuchi et al. 2015). A designed RNA aptamer called GTPswitch was used, which has a guanosine triphosphate (GTP)-responsive domain and can bind to the target RNA hairpin named aptakiss in the presence of GTP (Durand et al. 2014). Shape changes of the DNA/RNA strands were observed in the DNA origami, which are induced by the GTPswitch, into two different shapes in the absence and presence of GTP, respectively. We also found that the switching function in the nanospace could be improved by using a cover strand over the kissing loop of the GTPswitch and by deleting one base from this kissing loop. These newly designed ligand-responsive aptamers can be used for the controlled assembly of the various DNA and RNA nanostructures.

Protein binding-induced structural change of RNA was directly observed by HS-AFM. Ribosomal protein L7Ae binds to specific positions of three kink-turns in a double-stranded RNA (dsRNA) to form a triangular structure (Ohno et al. 2011). Using the HS-AFM system, the stepwise binding of three L7Ae proteins to the dsRNA to form triangular RNA–protein structure was clearly observed (Osada et al. 2014). Similar to the structural changes and reactions on DNA molecules, the detailed interactions and reactions of RNA–RNA and RNA–protein can be visualized in real time at nanoscale resolution using the nucleic acid scaffolds and HS-AFM system.

7 Conclusions

The precise placement and manipulation of target molecules are achieved in the designed nanospace constructed by DNA origami technology. The designed nanospace has been used to image and regulate biochemical reactions. Direct imaging of the mobile molecular systems on DNA origami structures has been demonstrated using the HS-AFM system. Using the advantage of DNA origami technology, single-molecule observation systems have been developed by creating target-oriented nanospaces and introducing target dsDNAs and RNAs for desired observations. The method we have developed can be expanded to the general observation of various biochemical reactions involved in gene expression and for other biological phenomena such as replication, transcription, and translation. Single chemical reactions such as complex formations, photoreactions, and bond formation/cleavage can also be imaged using this system. In addition, the movement of artificial mobile molecules was directly observed and manipulated in the designed systems, and the detailed mechanism for the movement of the molecules was investigated at the single-molecule level. The method can be applied to the development of nanomachines that can transport specific molecules and for the manipulation of more complicated movements on the DNA nanostructures. The lipid bilayer was also used to expand the HS-AFM imaging for direct observation of the dynamic formation of DNA origami assemblies. Using the designed DNA origami scaffolds and improving the HS-AFM imaging technique, novel nano-sized devices for single-molecule imaging and manipulation can be created to elucidate the physical properties of the molecules both in the biological and nonbiological environments.

References

Ando T, Kodera N (2012) Visualization of mobility by atomic force microscopy. Methods Mol Biol 896:57–69

Ando T, Kodera N, Takai E et al (2001) A high-speed atomic force microscope for studying biological macromolecules. Proc Natl Acad Sci USA 98:12468–12472

Asanuma H, Liang X, Nishioka H et al (2007) Synthesis of azobenzene-tethered DNA for reversible photo-regulation of DNA functions: hybridization and transcription. Nat Protoc 2:203–212

Bacolla A, Wells RD (2004) Non-B DNA conformations, genomic rearrangements, and human disease. J Biol Chem 279:47411–47414

Bacolla A, Wells RD (2009) Non-B DNA conformations as determinants of mutagenesis and human disease. Mol Carcinog 48:273–285

Bath J, Turberfield AJ (2007) DNA nanomachines. Nat Nanotechnol 2:275–284

Bath J, Green SJ, Turberfield AJ (2005) A free-running DNA motor powered by a nicking enzyme. Angew Chem Int Ed 44:4358–4361

Behe M, Felsenfeld G (1981) Effects of methylation on a synthetic polynucleotide: the B--Z transition in poly(dG-m5dC).poly(dG-m5dC). Proc Natl Acad Sci USA 78:1619–1623

Bruner SD, Norman DP, Verdine GL (2000) Structural basis for recognition and repair of the endogenous mutagen 8-oxoguanine in DNA. Nature 403:859–866

Crampton N, Yokokawa M, Dryden DT et al (2007) Fast-scan atomic force microscopy reveals that the type III restriction enzyme EcoP15I is capable of DNA translocation and looping. Proc Natl Acad Sci USA 104:12755–12760

David SS, O'Shea VL, Kundu S (2007) Base-excision repair of oxidative DNA damage. Nature 447:941–950

Dhakal S, Schonhoft JD, Koirala D et al (2010) Coexistence of an ILPR i-motif and a partially folded structure with comparable mechanical stability revealed at the single-molecule level. J Am Chem Soc 132:8991–8997

Dhakal S, Yu Z, Konik R et al (2012) G-quadruplex and i-motif are mutually exclusive in ILPR double-stranded DNA. Biophys J 102:2575–2584

Durand G, Lisi S, Ravelet C et al (2014) Riboswitches based on kissing complexes for the detection of small ligands. Angew Chem Int Ed 53:6942–6945

Endo M, Katsuda Y, Hidaka K et al (2010a) Regulation of DNA methylation using different tensions of double strands constructed in a defined DNA nanostructure. J Am Chem Soc 132:1592–1597

Endo M, Katsuda Y, Hidaka K et al (2010b) A versatile DNA nanochip for direct analysis of DNA base-excision repair. Angew Chem Int Ed 49:9412–9416

Endo M, Tatsumi K, Terushima K et al (2012a) Direct visualization of the movement of a single T7 RNA polymerase and transcription on a DNA nanostructure. Angew Chem Int Ed 51:8778–8782

Endo M, Yang Y, Suzuki Y et al (2012b) Single-molecule visualization of the hybridization and dissociation of photoresponsive oligonucleotides and their reversible switching behavior in a DNA nanostructure. Angew Chem Int Ed 51:10518–10522

Endo M, Yang Y, Sugiyama H (2013) DNA origami technology for biomaterials applications. Biomater Sci 1:347–360

Endo M, Takeuchi Y, Suzuki Y et al (2015a) Single-molecule visualization of the activity of a Zn (2+)-dependent DNAzyme. Angew Chem Int Ed 54:10550–10554

Endo M, Xing X, Zhou X et al (2015b) Single-molecule manipulation of the duplex formation and dissociation at the G-quadruplex/i-motif site in the DNA nanostructure. ACS Nano 9:9922–9929

Gehring K, Leroy JL, Gueron M (1993) A tetrameric DNA structure with protonated cytosine. cytosine base pairs. Nature 363:561–565

Gilmore JL, Suzuki Y, Tamulaitis G et al (2009) Single-molecule dynamics of the DNA-EcoRII protein complexes revealed with high-speed atomic force microscopy. Biochemistry 48:10492–10498

Gopaul DN, Guo F, Van Duyne GD (1998) Structure of the Holliday junction intermediate in Cre-loxP site-specific recombination. EMBO J 17:4175–4187

Grabow WW, Jaeger L (2014) RNA self-assembly and RNA nanotechnology. Acc Chem Res 47:1871–1880

Grindley ND, Whiteson KL, Rice PA (2006) Mechanisms of site-specific recombination. Annu Rev Biochem 75:567–605

Gu HZ, Chao J, Xiao SJ et al (2010) A proximity-based programmable DNA nanoscale assembly line. Nature 465:202–205

Gu H, Furukawa K, Weinberg Z et al (2013) Small, highly active DNAs that hydrolyze DNA. J Am Chem Soc 135:9121–9129

Guo F, Gopaul DN, Van Duyne GD (1997) Structure of Cre recombinase complexed with DNA in a site-specific recombination synapse. Nature 389:40–46

Guo F, Gopaul DN, Van Duyne GD (1999) Asymmetric DNA bending in the Cre-loxP site-specific recombination synapse. Proc Natl Acad Sci USA 96:7143–7148

Huppert JL, Balasubramanian S (2007) G-quadruplexes in promoters throughout the human genome. Nucleic Acids Res 35:406–413

Jovin TM, Soumpasis DM, Mcintosh LP (1987) The transition between B-DNA and Z-DNA. Annu Rev Phys Chem 38:521–560

Jungmann R, Steinhauer C, Scheible M et al (2010) Single-molecule kinetics and super-resolution microscopy by fluorescence imaging of transient binding on DNA origami. Nano Lett 10:4756–4761

Liang X, Mochizuki T, Asanuma H (2009) A supra-photoswitch involving sandwiched DNA base pairs and azobenzenes for light-driven nanostructures and nanodevices. Small 5:1761–1768

Liu W, Zhong H, Wang R et al (2011) Crystalline two-dimensional DNA-origami arrays. Angew Chem Int Ed 50:264–267

Liu X, Lu CH, Willner I (2014) Switchable reconfiguration of nucleic acid nanostructures by stimuli-responsive DNA machines. Acc Chem Res 47:1673–1680

Lund K, Manzo AJ, Dabby N et al (2010) Molecular robots guided by prescriptive landscapes. Nature 465:206–210

Mao CD, Sun WQ, Shen ZY et al (1999) A nanomechanical device based on the B-Z transition of DNA. Nature 397:144–146

Miyoshi D, Matsumura S, Nakano S et al (2004) Duplex dissociation of telomere DNAs induced by molecular crowding. J Am Chem Soc 126:165–169

Morikawa K, Matsumoto O, Tsujimoto M et al (1992) X-ray structure of T4 endonuclease V: an excision repair enzyme specific for a pyrimidine dimer. Science 256:523–526

Nash HM, Bruner SD, Scharer OD et al (1996) Cloning of a yeast 8-oxoguanine DNA glycosylase reveals the existence of a base-excision DNA-repair protein superfamily. Curr Biol 6:968–980

Ohno H, Kobayashi T, Kabata R et al (2011) Synthetic RNA-protein complex shaped like an equilateral triangle. Nat Nanotechnol 6:116–120

Osada E, Suzuki Y, Hidaka K et al (2014) Engineering RNA-protein complexes with different shapes for imaging and therapeutic applications. ACS Nano 8:8130–8140

Rajendran A, Endo M, Sugiyama H (2012) Single-molecule analysis using DNA origami. Angew Chem Int Ed 51:874–890

Rajendran A, Endo M, Hidaka K et al (2013a) Controlling the stoichiometry and strand polarity of a tetramolecular G-quadruplex structure by using a DNA origami frame. Nucleic Acids Res 41:8738–8747

Rajendran A, Endo M, Hidaka K et al (2013b) Direct and real-time observation of rotary movement of a DNA nanomechanical device. J Am Chem Soc 135:1117–1123

Rajendran A, Endo M, Hidaka K et al (2013c) HIV-1 nucleocapsid proteins as molecular chaperones for tetramolecular antiparallel G-quadruplex formation. J Am Chem Soc 135:18575–18585

Rajendran A, Endo M, Sugiyama H (2014) State-of-the-art high-speed atomic force microscopy for investigation of single-molecular dynamics of proteins. Chem Rev 114:1493–1520

Rothemund PW (2006) Folding DNA to create nanoscale shapes and patterns. Nature 440:297–302

Sannohe Y, Endo M, Katsuda Y et al (2010) Visualization of dynamic conformational switching of the G-quadruplex in a DNA nanostructure. J Am Chem Soc 132:16311–16313

Schitter G, Astrom KJ, DeMartini BE et al (2007) Design and modeling of a high-speed AFM-scanner. IEEE Trans Control Syst 15:906–915

Shirude PS, Okumus B, Ying L et al (2007) Single-molecule conformational analysis of G-quadruplex formation in the promoter DNA duplex of the proto-oncogene c-kit. J Am Chem Soc 129:7484–7485

Shukla GC, Haque F, Tor Y et al (2011) A boost for the emerging field of RNA nanotechnology. ACS Nano 5:3405–3418

Siddiqui-Jain A, Grand CL et al (2002) Direct evidence for a G-quadruplex in a promoter region and its targeting with a small molecule to repress c-MYC transcription. Proc Natl Acad Sci USA 99:11593–11598

Sun D, Hurley LH (2009) The importance of negative superhelicity in inducing the formation of G-quadruplex and i-motif structures in the c-Myc promoter: implications for drug targeting and control of gene expression. J Med Chem 52:2863–2874

Suzuki Y, Gilmore JL, Yoshimura SH et al (2011) Visual analysis of concerted cleavage by type IIF restriction enzyme SfiI in subsecond time region. Biophys J 101:2992–2998

Suzuki Y, Endo M, Canas C et al (2014a) Direct analysis of Holliday junction resolving enzyme in a DNA origami nanostructure. Nucleic Acids Res 42:7421–7428

Suzuki Y, Endo M, Katsuda Y et al (2014b) DNA origami based visualization system for studying site-specific recombination events. J Am Chem Soc 136:211–218

Suzuki Y, Endo M, Yang Y et al (2014c) Dynamic assembly/disassembly processes of photoresponsive DNA origami nanostructures directly visualized on a lipid membrane surface. J Am Chem Soc 136:1714–1717

Suzuki Y, Endo M, Sugiyama H (2015a) Lipid-bilayer-assisted two-dimensional self-assembly of DNA origami nanostructures. Nat Commun 6:8052

Suzuki Y, Endo M, Sugiyama H (2015b) Mimicking membrane-related biological events by DNA origami nanotechnology. ACS Nano 9:3418–3420

Takeuchi Y, Endo M, Suzuki Y et al (2015) Single-molecule observations of RNA-RNA kissing interactions in a DNA nanostructure. Biomater Sci 4(1):130–135

Torring T, Voigt NV, Nangreave J et al (2011) DNA origami: a quantum leap for self-assembly of complex structures. Chem Soc Rev 40:5636–5646

Uchihashi T, Kodera N, Ando T (2012) Guide to video recording of structure dynamics and dynamic processes of proteins by high-speed atomic force microscopy. Nat Protoc 7:1193–1206

Van Duyne GD (2001) A structural view of cre-loxp site-specific recombination. Annu Rev Biophys Biomol Struct 30:87–104

Van Melckebeke H, Devany M, Di Primo C et al (2008) Liquid-crystal NMR structure of HIV TAR RNA bound to its SELEX RNA aptamer reveals the origins of the high stability of the complex. Proc Natl Acad Sci USA 105:9210–9215

Vassylyev DG, Morikawa K (1997) DNA-repair enzymes. Curr Opin Struct Biol 7:103–109

Vassylyev DG, Kashiwagi T, Mikami Y et al (1995) Atomic model of a pyrimidine dimer excision repair enzyme complexed with a DNA substrate: structural basis for damaged DNA recognition. Cell 83:773–782

Walters DA, Cleveland JP, Thomson NH et al (1996) Short cantilevers for atomic force microscopy. Rev Sci Instrum 67:3583–3590

Wang F, Liu X, Willner I (2015) DNA switches: from principles to applications. Angew Chem Int Ed 54:1098–1129

Wickham SFJ, Endo M, Katsuda Y et al (2011) Direct observation of stepwise movement of a synthetic molecular transporter. Nat Nanotechnol 6:166–169

Wickham SFJ, Bath J, Katsuda Y et al (2012) A DNA-based molecular motor that can navigate a network of tracks. Nat Nanotechnol 7:169–173

Xu Y, Sugiyama H (2006) Formation of the G-quadruplex and i-motif structures in retinoblastoma susceptibility genes (Rb). Nucleic Acids Res 34:949–954

Xu Y, Sato H, Sannohe Y et al (2008) Stable lariat formation based on a G-quadruplex scaffold. J Am Chem Soc 130:16470–16471

Yang Y, Endo M, Hidaka K et al (2012) Photo-controllable DNA origami nanostructures assembling into predesigned multiorientational patterns. J Am Chem Soc 134:20645–20653

Yang Y, Endo M, Suzuki Y et al (2014) Direct observation of the dual-switching behaviors corresponding to the state transition in a DNA nanoframe. Chem Commun 50:4211–4213

You M, Chen Y, Zhang X et al (2012) An autonomous and controllable light-driven DNA walking device. Angew Chem Int Ed 51:2457–2460

Youngblood B, Reich NO (2006) Conformational transitions as determinants of specificity for the DNA methyltransferase EcoRI. J Biol Chem 281:26821–26831

Polymerase Reactions that Involve Modified Nucleotides

Masayasu Kuwahara, Kenta Hagiwara, and Hiroaki Ozaki

Contents

Abstract Polymerases are known to catalyze the synthesis of long DNA/RNA chains and play important roles in replication and translation in biological systems. Because their enzymatic activities are versatile, they have been widely employed for medical diagnoses and criminal investigations and also as research tools in biological studies. To date, various polymerases have been genetically engineered and are commercially available for selected applications that involve various chemically modified nucleoside triphosphate analogs. For example, dye-terminator sequencing, which made a great contribution to sequence determination in the human genome project, uses four kinds of 2′,3′-dideoxynucleoside-5′-triphosphate analogs that contain a fluorophore attached to the base moiety. Concomitantly, polymerase variants that can efficiently accept those analogs as substrates were developed. Various triphosphate analogs modified at their base/sugar/phosphate moieties have been designed and synthesized for the development of nucleic acid aptamers as therapeutic drugs, diagnostic agents, and molecular

M. Kuwahara (✉) • K. Hagiwara • H. Ozaki
Graduate School of Science and Technology, Gunma University, 1-5-1 Tenjin-cho, Kiryu, Gunma 376-8515, Japan
e-mail: mkuwa@gunma-u.ac.jp

S. Jurga et al. (eds.), *Modified Nucleic Acids in Biology and Medicine*, RNA Technologies, DOI 10.1007/978-3-319-34175-0_18

indicators. Because nucleoside analogs include xenonucleic acids (XNAs), which have unique sugar backbones, drastic modifications in polymerase engineering are being boldly pursued. In this chapter, we are focusing on polymerase reactions that involve chemically modified substrates and their applications to innovative life sciences and technologies.

Keywords Xenonucleic acid • Modified nucleoside triphosphate • Polymerase variant

1 Introduction

While adenine (A), guanine (G), cytosine (C), and thymine/uracil (T/U) are well known as the standard bases of DNA/RNA, modified bases can also be found in messenger, transfer, and ribosomal RNAs (Fig. 1a). To date, more than 100 kinds of modified bases, including 7-methyl guanine, pseudouracil, dihydrouracil, hypoxanthine, and large bases with amino acid adducts, have been reported (Limbach et al. 1994). Sugar modifications, such as 2′-methoxy (OMe) and 1″-,3-(5-″-phosphoryl)-ribosyl groups, have also been discovered. These modifications are considered to stabilize the three-dimensional structures of RNAs, allowing them to evade facile degradation by nucleases. Modified nucleotides are generally synthesized by modifying enzymes, such as methyltransferases and pseudouridine synthases, after transcription. For example, a methyltransferase initiates a nucleophilic attack by the thiolate (-S$^-$) of the cysteine residue at position 375 (Cys 375)

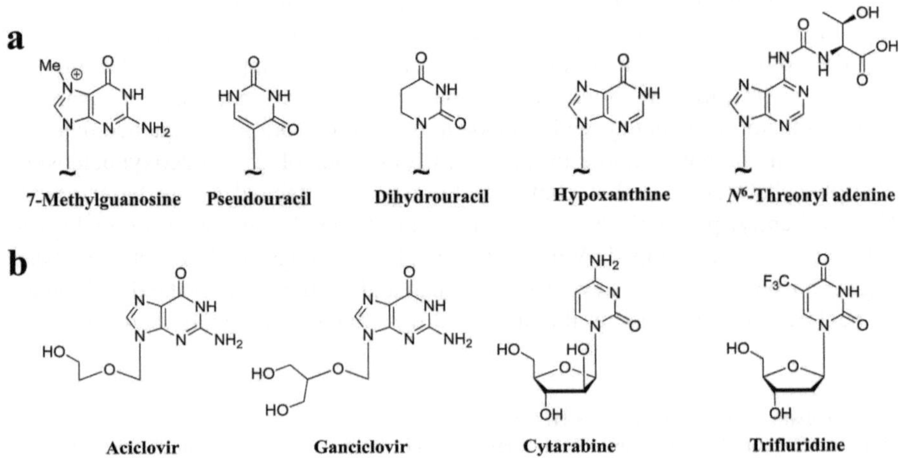

Fig. 1 (**a**) Modified bases found in messenger, transfer, and ribosomal RNAs; (**b**) polymerase inhibitors for antiviral drugs and antineoplastic agents

on the C6 position of cytosine, forming a Michael addition product that activates the C5 position of cytosine. Subsequent transfer of a methyl group from cofactor *S*-adenosyl-L-methionine yields 5-methylcytidine (Boschi-Muller and Motorin 2013).

Artificially synthesized nucleoside analogs, which were actively researched and developed from the 1960s to 1970s as polymerase inhibitors, are well known. These analogs, which include antiviral drugs (e.g., aciclovir and ganciclovir) and antineoplastic agents (e.g., cytarabine and trifluridine) (Fig. 1b), are phosphorylated through cellular metabolism and thereafter specifically inhibit polymerase activities to suppress virus proliferation and tumor growth (De Clercq 2001; Kufe et al. 1980). The other typical examples are fluorescence-labeled nucleoside triphosphate analogs, which are used in sequencing and microarray technologies for genetic research. Similar to fluorescent labeling, the introduction of foreign substituents to confer additional functionality, such as cell membrane permeability, nuclease resistance, and electroconductivity, has been studied for diverse applications.

2 Kinds of Polymerases

In general, DNA polymerases can extend a poly(oligo)nucleotide strand with a sequence complimentary to that of its template strand using nucleoside 5′-triphosphates as substrates. Some DNA polymerases have a proofreading 3′ to 5′ exonuclease activity. Since the time A. Kornburg discovered *E. coli* DNA polymerase I, various types of DNA polymerases have been isolated from prokaryotes, archaea, eukaryotes, and retroviruses. They are mainly classified into the following eight evolutionary families: A, B, C, D, E, X, Y, and reverse transcriptase (RT). For example, *E. coli* DNA polymerase I and mitochondrial DNA polymerase γ belong to family A; eukaryotic DNA polymerases α, δ, ε, and *E. coli* DNA polymerase II (pol II) are classified into family B; and *E. coli* DNA polymerase III (pol III) is a member of family C. Avian myeloblastosis virus (AMV) reverse transcriptase, which is an RNA-dependent DNA polymerase, can be mentioned as an example of an RT.

Highly thermostable DNA polymerases show optimal activity at temperatures around 75 °C. They are widely employed for polymerase chain reaction (PCR) to amplify/replicate DNA target sequences. To date, various engineered polymerases have been developed and supplied by manufacturers. One of the most popular thermostable DNA polymerases is *Taq* DNA polymerase, which belongs to family A and possesses 5′ to 3′ exonuclease activity, but not 3′ to 5′ exonuclease activity. It was isolated by A. Chien in 1976 (before the first report of the PCR technique in 1987) from an extremely thermophilic bacterium, *Thermus aquaticus* YT1, that was first discovered in the lower geyser basin of Yellowstone National Park, USA, in 1969 (Chien et al. 1976). The other examples of thermostable family A DNA polymerases are *AmpliTaq* and *Tth* DNA polymerases. The former, which was reported by F. C. Lawyer in 1989, is a modified form of *Taq* DNA polymerase obtained by expression of the gene in an *E. coli* host (Lawyer et al. 1989). The latter, which was isolated from the extremely thermophilic bacterium *Thermus thermophilus* HB8, is used for reverse transcription (RT)-PCR owing to its very efficient RT activity in the presence of Mg^{2+} ion (Rüttimann et al. 1985). Meanwhile, *rTth* DNA polymerase, a recombinant DNA polymerase, is known to exhibit efficient RT activity in the presence of Mn^{2+} ion.

Hyperthermostable DNA polymerases that belong to family B generally exhibit excellent proofreading ability during DNA chain extension, owing to their 3′ to 5′ exonuclease activity. These polymerases, such as *Vent* polymerase from *Thermococcus litoralis*, *KOD* from *Thermococcus kodakaraensis* KOD1, 9°N_m from *Thermococcus* species 9°N-7, *Tgo* from *Thermococcus gorgonarius* (Miroshnichenko et al. 1998), *Deep vent* from *Pyrococcus* species GB-D (Jannasch et al. 1992), *Pfu* from *Pyrococcus furiosus* (Fiala and Stetter 1986), and *Pwo* from *Pyrococcus woesei* (Rüdiger et al. 1995), were obtained from hyperthermophilic archaea found in deep ocean vents, volcanic marine mud, and solfataras on the seashore. Owing to their higher fidelity (i.e., lower error rate) and heat stability, compared with those of *Taq*, they and their variants have been well studied, and some of them are commercially available. *Vent* DNA polymerase, which was reported in 1991, was the first thermostable DNA polymerase having a 3′ to 5′ proofreading exonuclease activity (Mattila et al. 1991). Its D141A and E143A variant, which was engineered to eliminate the exonuclease activity and is known as*Vent(exo-)* DNA polymerase, was developed to improve the yield of PCR products and to be applied to dideoxy sequencing reactions (Kong et al. 1993). *KOD* DNA polymerase not only possesses excellent proofreading ability, with about 50-fold higher fidelity than *Taq* DNA polymerase, but can also elongate an oligonucleotide strand with about a fivefold higher reaction rate than the other family B DNA polymerases, such as *Pfu* and *Deep vent* DNA polymerases (Takagi et al. 1997). The N210D variant of *KOD*, known as *KOD(exo-)* DNA polymerase, which possesses one thousandth of the 3′ to 5′ exonuclease activity of *KOD* DNA polymerase, has also been developed (Nishioka et al. 2001). The blend of *KOD* and *KOD(exo-)* DNA polymerases, which is commercially available under the product names *KOD Dash* or *KOD XL*, enables the production of long double-stranded DNAs (~15 kbp) by PCR. *KOD FX* and *AccuPrimePfx* DNA polymerases were developed from *KOD* DNA polymerase for use in hot start PCR. This is an improved PCR technique that can evade nonspecific amplification of DNA at lower temperatures by inactivating the polymerase with its antibody. The 9°N_m DNA polymerase is the E143D variant of wild-type 9°N-7 DNA polymerase (Southworth et al. 1996). This 9°N_m variant exhibits reduced 3′ to 5′ exonuclease activity (0.4–5 % of wild-type exonuclease activity). Therminator DNA polymerase, the D141A, E143A, and A485L variant of 9°N-7 DNA polymerase, has no 3′ to 5′ exonuclease activity. This DNA polymerase exhibits enhanced incorporation of modified nucleotides, that is, efficient single-base incorporation of dideoxy and acyclonucleotides. Thus, various B family DNA polymerases have been discovered and genetically engineered to be applied in a number of applications.

In addition to the highly thermostable and hyperthermostable DNA polymerases, *E. coli* DNA polymerase I, Klenow fragment (KF), *Bst* DNA polymerase, φ29 DNA polymerase, and T7 DNA polymerase are well known. They display optimal activity at 30–37 °C, except for *Bst* DNA polymerase, which has an optimal temperature of 60–65 °C. KF is obtained as a large protein fragment by enzymatic cleavage of *E. coli* DNA polymerase I using the protease subtilisin (Klenow and Henningsen 1970). Although KF has lost its 5′ to 3′ exonuclease activity, it retains its 3′ to 5′ exonuclease activity; furthermore, its D355A and E357A variant, known as KF (3′–5′ exo-), lacks 3′ to 5′ exonuclease activity (Bebenek et al. 1990). *Bst* DNA polymerase is the large fragment of DNA polymerase I from the thermophilic bacterium *Bacillus stearothermophilus*, which is found in soil, hot springs, and ocean sediment and is generally unable to grow at temperatures below 35 °C (Kiefer et al. 1998). *Bst* DNA polymerase lacks 3′ to 5′ exonuclease activity, but possesses strong strand displacement activity, which allows isothermal DNA amplifications such as loop-mediated amplification (LAMP) and rolling circle amplification (RCA). Meanwhile, φ29 DNA polymerase, which is derived from the *Bacillus subtilis* phage phi29 (Φ29), is known to exhibit 3′ to 5′ proofreading exonuclease activity and extreme processivity, in addition to strong strand displacement activity (Blanco et al. 1989). Bacteriophage T7 DNA polymerase, which belongs to family A, also possesses 3′ to 5′ proofreading exonuclease activity (Grippo and Richardson 1971; Campbell et al. 1978; Adler and Modrich 1979). This DNA polymerase is known to be a complex comprising phage-encoded gene 5 protein and *E. coli* host thioredoxin, which enhances the processivity of the polymerase.

In eukaryotes, there are three main types of RNA polymerase: RNA polymerase I, which transcribes ribosomal RNA but not 5S rRNA; RNA polymerase II, which transcribes precursors of mRNA, snRNA, and microRNA; and RNA polymerase III, which transcribes ribosomal 5S rRNA, tRNA, and other small RNAs. Bacteriophage T7 (Davanloo et al. 1984), T3 (Majumder et al. 1979), and SP6 (Kotani et al. 1987) RNA polymerases can be mentioned as examples of commercially available DNA-dependent RNA polymerases. These RNA polymerases bind to their cognate promoters with very high sequence specificities and thereafter transcribe the DNA template downstream of the promoter to generate the complimentary single-stranded RNA. Some variants of T7 RNA polymerase have been developed to enhance the incorporation of modified nucleotides.

3 Enzymatic Synthesis of Modified Nucleotides

3.1 Base Modification

A variety of base-modified nucleotides, particularly C5-modified uridine analogs, have been reported to date (Fig. 2). Research results have demonstrated that nucleoside triphosphates bearing a C5-modified pyrimidine ring or a C7-modified 7-deazapurine ring can be more acceptable polymerase substrates than nucleoside

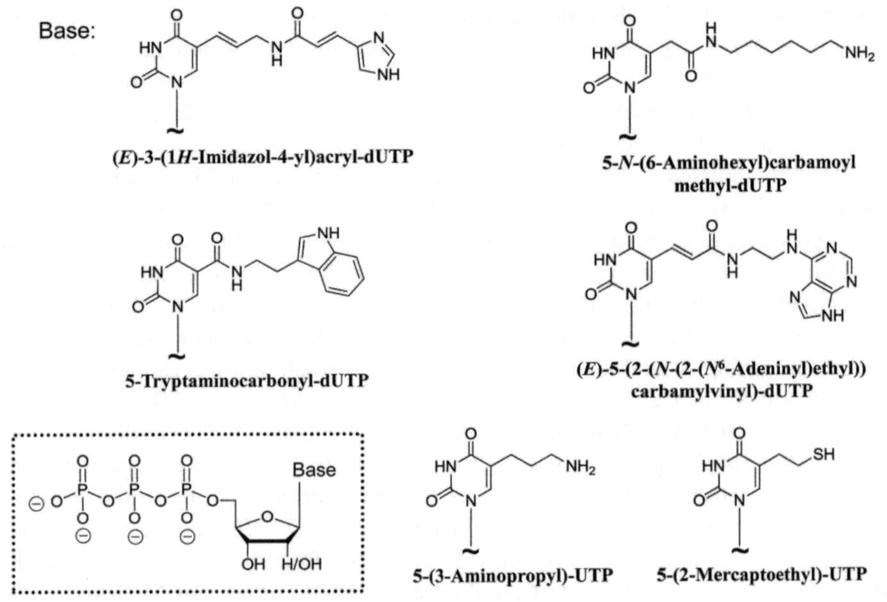

Fig. 2 Base-modified nucleoside triphosphates

triphosphates with substituents introduced at the other positions. Using these modified analogs in the presence of the four natural nucleoside triphosphates (dNTPs) allows foreign functionalities, such as fluorophores or biotin to be thinly incorporated into nucleic acid strands that can serve as probes or capture specific nucleotide targets. For example, in 1992, T. Ried et al. reported fluorescence in situ hybridization (FISH) probes, which were synthesized using *AmpliTaq* DNA polymerase-catalyzed PCR that incorporated dUTPs labeled with fluorescein, biotin, digoxigenin, and 2,4-dinitrophenol; these modified DNA probes were used to specifically visualize the centromeres of human chromosomes (Ried et al. 1992).

In 1998, K. Sakthivel et al. first reported the enzymatic synthesis of modified DNAs in which foreign functionalities were densely incorporated during PCR. They used ten different C5-modified 2′-deoxyuridine-5′-triphosphates (dUTPs) in the presence of the natural dNTPs, except for thymidine-5′-triphosphate (TTP) (Sakthivel and Barbas III 1998). Four different DNA polymerases (*Taq*, *Vent*, *Pfu*, and *rTth*) were examined. The C5-modified dUTP bearing an (*E*)-3-(1*H*-imidazol-4-yl)acryl group was accepted as a good PCR substrate for all of these DNA polymerases and provided the corresponding modified DNA very efficiently. The experimental results particularly emphasized that the rigid and extended α,-β-unsaturated arm adjacent to the imidazoyl group has a great influence on the incorporation efficiency of the modified dUs. Furthermore, modified UTPs bearing 5-(3-aminopropyl) or 5-(2-mercaptoethyl) groups were also synthesized and densely incorporated into RNAs with T7 RNA polymerase (Vaish et al. 2000).

Using these analogs instead of UTP, it should be possible to perform modified RNA aptamer/ribozyme selections.

In 2001, D. M. Williams and coworkers focused on the linker length and rigidity of substituents and systematically analyzed the substrate properties of ten different C5-modified dUTPs in the absence of TTP during PCR catalyzed by *Taq* DNA polymerase (Lee et al. 2001). The data showed that C5-modified dUTPs with linker arms containing rigid alkynyl and *trans*-alkenyl groups in the vicinity of uracil base were significantly superior to those with linker arms containing *cis*-alkenyl or alkyl groups. Furthermore, they synthesized C7-modified 7-deaza-dATP analogs with alkynyl, *cis*-alkenyl, and alkyl linker arms and found that the analog with the alkynyl linker arm acts as the best substrate for PCR catalyzed by *Taq* DNA polymerase (Gourlain et al. 2001). Intriguingly, using the preferred dUTP and 7-deaza-dATP analogs, modified dUs and 7-deaza-dAs could be simultaneously incorporated during PCR. In the same year, H. Sawai et al. first demonstrated, using PCR experiments incorporating dUTPs modified with a methylene linker at C5, that *KOD Dash* DNA polymerase is one of the most promising candidates as a catalyst for enzymatic syntheses of modified DNAs (Sawai et al. 2001). Indeed, *KOD* DNA polymerase-related products (e.g., *KOD XL*, *KOD FX*, *AccuPrimePfx*, and *KOD Dash*) are widely employed in applications involving enzymatic syntheses of modified DNAs.

In 2002, H. A. Held et al. demonstrated that *Pfu* and *Pwo* DNA polymerases, which are members of family B, as well as *KOD* DNA polymerase, can efficiently produce modified DNAs by primer extension (PEX) and PCR, using dUTPs modified at C5 with protected thiol groups (Held and Benner 2002). In successive incorporations of modified dUs, family B polymerases (*Pfu*, *Pwo*, *Vent*, and *Deep vent*) were found to be preferable to family A polymerases (*Taq*, *Tfl*, Hot *Tub*, and *Tth*). Thereafter, M. Kuwahara et al. systematically analyzed the PCR-based synthesis of modified DNA strands using C5-modified dUTPs and C5-modified dCTPs as substrates and family A (*Taq*, *Tth*) and B [*Vent*(*exo-*), *KOD Dash*, and *KOD*(*exo-*)] polymerases as enzymes and arrived at the same conclusion (Kuwahara et al. 2006). Furthermore, their kinetic studies using modified primers/templates/substrates revealed that modified group(s) adjacent to the extending terminus of the primer can greatly reduce catalytic efficiency, which resulted in low product yields with successive modified nucleotide incorporations.

D. M. Williams and coworkers applied the dual modification technique to modified DNAzyme selection, using modified dU and 7-deaza-dA analogs (Sidorov et al. 2004). Meanwhile, in 1999, D. M. Perrin et al. reported dual modification using C5-modified dUTP and C8-modified dATP in a PEX technique catalyzed using Sequenase Version 2.0 DNA polymerase, which is a genetically engineered form of T7 DNA polymerase with virtually no 3′ to 5′ exonuclease activity (Perrin et al. 1999; 2001). Using this technique, followed by in vitro selection, they were the first to produce a modified DNAzyme that acted as a metal-independent RNAase A mimic with two different functional groups ($k_{cat} = 0.044\ min^{-1}$) (Perrin et al. 2001). They also produced a modified DNAzyme with three different functional groups (i.e., amine, imidazole, and guanidine) which had an improved k_{cat}

value (0.134 min^{-1}) (Hollenstein et al. 2009). In 2003, T. Tasara et al. synthesized C5-modified dUTPs, C5-modified dCTPs, C7-modified 7-deaza-dATPs, and C7-modified 7-deaza-dGTPs with foreign functional groups, including biotin, Rhodamine Green, Cyanine 5, Evoblue 30, and Gnothis Blue 3 (Tasara et al. 2003). They examined successive incorporations of modified nucleotides in the absence of all four natural dNTPs using PEX. A reaction using *Vent(exo-)* polymerase and biotinyl dNTPs with four different bases provided a modified DNA with a 40-mer elongated strand, in which biotinyl dNs were successively incorporated. Similarly, in 2005, S. Jäger et al. synthesized modified dNTPs with four different bases possessing basic, acidic, and lipophilic substituents and examined their successive incorporations (Jäger et al. 2005). Full-length modified DNAs comprising 40-mer elongated strands were produced by successive modified dN incorporations using PEX catalyzed by *Pwo* DNA polymerase. Furthermore, these oligonucleotides, which were modified at high density, were found to be reverse-transcribed to natural cDNAs in a PEX procedure using *Pwo* DNA polymerase, four natural dNTPs, and a GC-rich reaction buffer containing 1.5 mM dimethyl sulfoxide. Eventually, the technical challenges in successive incorporations have encouraged researchers to explore various applications of enzymatically synthesized modified DNAs.

Postsynthetic derivatization is a convenient and alternative method for the preparation of long DNA strands with high density and/or bulky modifications (Fig. 3). In 2003, H. Sawai and coworkers reported the enzymatic synthesis of a 108-mer modified DNA containing 5-methoxycarbonylmethyl-dUs in place of natural T using PCR and its subsequent derivatizations via amide bond formation by ester–amide exchange reactions using amino functionalities such as tris (2-aminoethylamine), histamine, and hexamethylenediamine (Mehedi Masud et al. 2003). Conversion rates of the derivatizations were unfortunately not very high, i.e., 56 %, 72 %, and 76 %, respectively, owing to the hydrolysis of the methyl ester that occurred during the addition reactions. Thereafter, various postsynthetic derivatizations using different chemistries have been developed. For example, pyranosyl sugar-modified DNA derivatized from a 300-mer precursory modified DNA containing 5-ethynyl-dUs by Cu-catalyzed alkyne–azide cycloaddition, i.e., a "click" reaction (Gierlich et al. 2007); biotin-modified DNA derivatized from a 35-mer precursory modified DNA containing a 7-[5-{(4-azidobutyl)amino}-5-oxopent-1-yn-1-yl]-7-deaza-dA by Staudinger ligation using biotinylated phosphine (Weisbrod and Marx 2007); hydrazone-modified DNA derivatized from a 98-mer precursory modified DNA containing 5-(5-formylthiophene-2-yl)-dCs by condensation reactions of aldehydes with arylhydrazines (Raindlová et al. 2010); and biotin-modified DNA derivatized from a 414-mer precursory modified DNA containing 7-vinyl-7-deaza-dAs by inverse electron demand Diels–Alder reaction (Buβkamp et al. 2014). The precursory modified DNAs were prepared in good yields by PCR/PEX preferably using *KOD XL*, *Pwo*, and *KlenTaq* DNA polymerases; some postsynthetic derivatizations exhibited significantly improved conversion rates, i.e., >90 %.

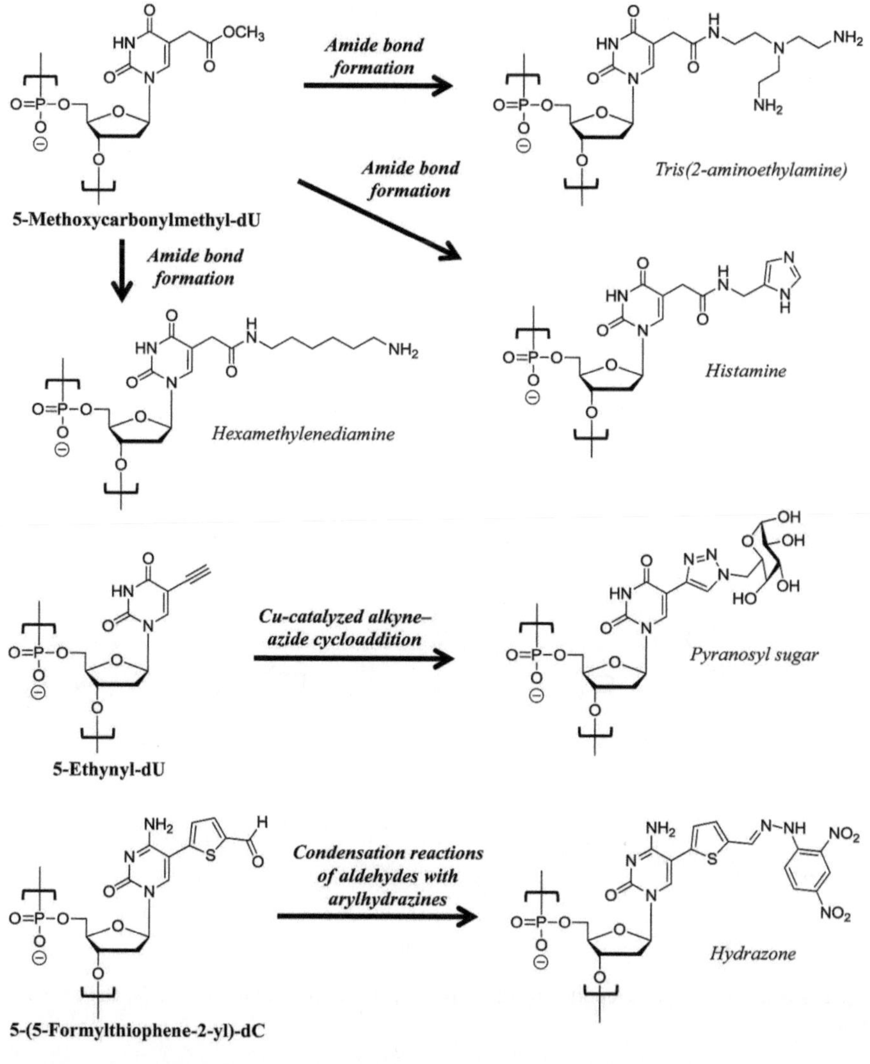

Fig. 3 Base modifications employed for postsynthetic derivatization

3.2 Special Base Modification (Artificial Base Pairs)

Artificial base pairs have great potential for expanding the genetic information system of life on earth. Such ambitious studies were considered to be able to pursue the basic principles of life phenomena and origin of life as well as technological applications.

In the late 1980s, S. A. Benner et al. first proposed expanded DNA alphabets (Switzer et al. 1989). In 1990, they demonstrated that artificial base pair formation,

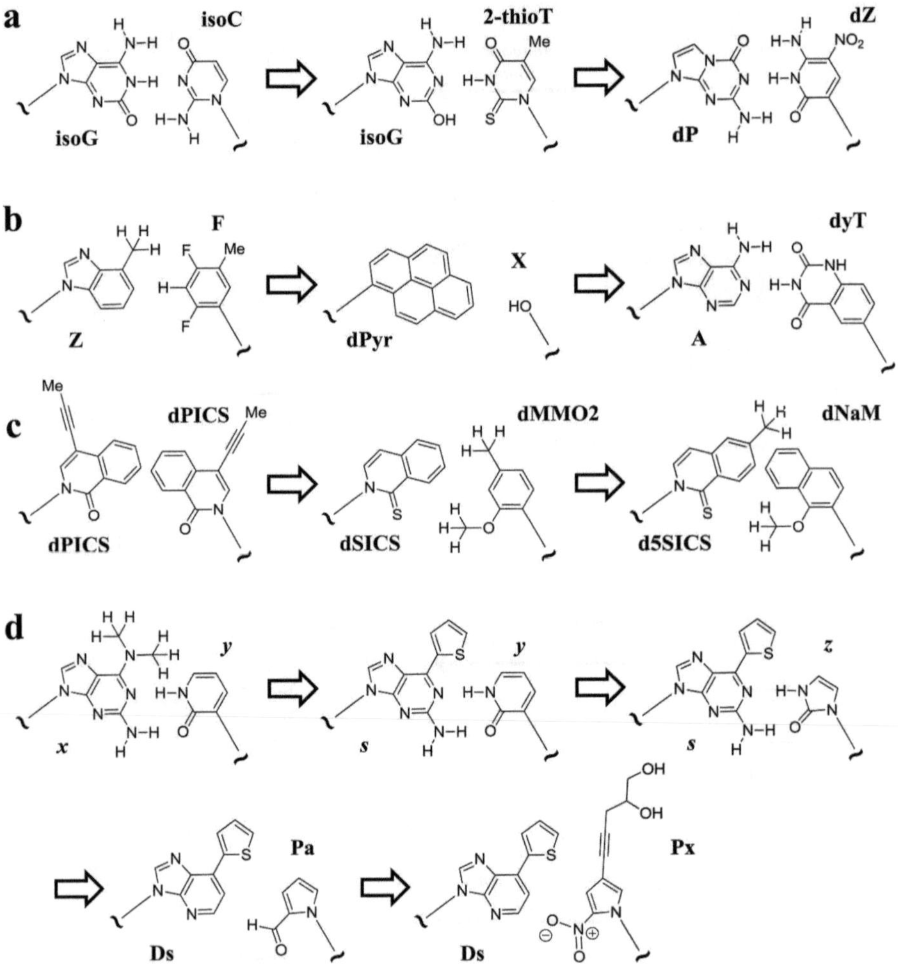

Fig. 4 Artificial base pairs reported by (**a**) Benner's, (**b**) Kool's, (**c**) Romesberg's, and (**d**) Hirao's groups

i.e., isoG–isoC (Fig. 4a), which differs from standard Watson–Crick base pairs, i.e., A–T(U) and G–C in the pattern of hydrogen bond formation, is possible during enzymatic DNA and RNA syntheses catalyzed by KF and T7 RNA polymerases, respectively (Piccirilli et al. 1990). Unfortunately, the incorporation of T opposite the enolic tautomer of isoG was also observed. To exclude the formation of the undesirable isoG–T pairing, they used 2-thioTTP instead of TTP, in addition to the other five triphosphate substrates, i.e., dATP, dGTP, dCTP, isoGTP, and isoCTP. As expected, the correct incorporation of isoC at a single isoG on a 51-mer template during PCR using *KlenTaq* DNA polymerase increased the value of fidelity-per-round in PCR from approximately 93 % to 98 % (Sismour and Benner 2005).

Furthermore, in 2006, they designed a novel third base pair comprising 2-amino-8-(2′-deoxy-β-D-erythro-pentofuranosyl)-imidazo[1,2-a]-1,3,5-triazin-4(8H)-one (dP) and 6-amino-5-nitro-3-(1′-β-D-2′-deoxyribofuranosyl)-2(1H)-pyridone (dZ) (Yang et al. 2006). PCR experiments using a 51-mer template containing a single dP residue showed high incorporation accuracies without using 2-thioTTP when dPTP and dZTP were used with the four natural dNTPs. The estimated values for fidelity-per-round were 94.4 %, 97.5 %, and 97.5 %, respectively, when *Taq*, *Vent* (*exo*-), and *Deep vent*(*exo*-) DNA polymerases were used (Yang et al. 2007). In the meantime, E. T. Kool, F. E. Romesberg, and I. Hirao were independently designing and developing their own candidates for the third base pair.

In 1998, E. T. Kool and coworkers reported a F(2,4-difluorotoluene)–Z (4-methylbenzimidazole) base pair (Fig. 4b), which does not form hydrogen bonds between the bases (Morales and Kool 1998). They demonstrated that efficiencies for a single incorporation of the Z residue opposite the F residue on a 28-mer DNA template were 130–1900-fold greater than those of the four natural dNs when KF (3′–5′ exo-) was used. Their results indicate that hydrogen bonding is not necessary for base pair formation, and, if anything, the size and shape of paired bases are more important to the adoption of artificial base pairs by an oligonucleotide duplex. Indeed, thereafter, it was demonstrated that pyrene deoxynucleoside (dPyr) as a nonpolar base analog can selectively be incorporated opposite an abasic nucleoside (X) or a tetrahydrofuran abasic analog (ϕ) to form a dPyr–X/ϕ pair during PEX (Matray and Kool 1999). Moreover, they developed "yDNA" (an abbreviation of "wide DNA"), which involves benzopyrimidine deoxynucleosides (dyT and dyC) bearing size-expanded pyrimidines, i.e., yT and yC, designed to form yT–A and yC–G pairs (Lee and Kool 2005). They then examined whether or not dyT and dyC can store and transfer genetic information in vitro and in bacterial cells (Chelliserrykattil et al. 2008). The results showed that the correct nucleotides could be inserted opposite yDNA residues in PEX using KF (3′–5′ exo-) and *Vent*(*exo*-) DNA polymerases. Furthermore, the first example of an encoding protein (GFP; green fluorescent protein) in a living organism, i.e., *E. coli*, by unnatural DNA base pair architecture was exhibited in 2008.

In 1999, F. E. Romesberg and coworkers reported that a stable 7-propynyl isocarbostyril nucleoside (dPICS) self-pair can be formed in duplex DNA (Fig. 4c), and dPICS triphosphate can be incorporated opposite dPICS on the template by KF (3′–5′ exo-) with reasonable efficiency (McMinn et al. 1999). However, after the dPICS incorporation, synthesis proceeded inefficiently. Thereafter, they determined the best pair from the 3600 (60 × 60) combinations of unnatural DNA base analogs, i.e., dSICS–dMMO2 (Leconte et al. 2008), and subsequently achieved d5SICS–dMMO2 and d5SICS–dNaM pairs, which exhibited the high values of 85.7–99.8 % for fidelity-per-round in PCR using *Taq*, *Deep vent*, and *Phusion high-fidelity* DNA polymerases (Seo et al. 2009; Malyshev et al. 2009). Recently, they finally managed to create a semisynthetic organism with an expanded genetic alphabet involving d5SICS–dNaM as the third base pair; the genetically engineered organism was *E. coli* that expresses an algal nucleotide triphosphate transporter, which has the efficient uptake of the triphosphates of d5SICS and dNaM, and thereby accurately replicates a plasmid containing d5SICS–dNaM (Malyshev et al. 2014).

In 2000, I. Hirao and coworkers designed and synthesized 2-amino-6-(*N*, *N*-dimethylamino)purine (denoted by *x*) and pyridin-2-one (denoted by *y*) deoxynucleoside analogs (Fig. 4d) (Ishikawa et al. 2000). They anticipated that the steric hindrance between the dimethyl at the N6 position of *x* and the 4-keto group of T would interfere with the formation of an *x*–T mismatch pair and, furthermore, that the unique pattern of hydrogen bonding between N1 and N2 on *x* and N1 and O2 on *y* would form a stable and specific *x*–*y* base pair. In PEX, using KF and KF (3′–5′ exo-), *y* was selectively incorporated opposite *x* on the template, which unfortunately was also erroneously incorporated opposites A and G. A ribonucleoside-5′-triphosphate analog of *y* was also synthesized, and the single incorporation of *y* opposite *x* in transcription was assessed using T7 RNA polymerase (Ohtsuki et al. 2001). As a result, *y* was incorporated opposite *x* with 95 % accuracy, while the erroneous incorporation of U opposite *x* was only occasionally observed (<5 % of instances). Thereafter, to improve incorporation efficiency and selectivity, they developed 2-amino-6-(2-thienyl)purine (denoted by *s*) and imidazolin-2-one (denoted by *z*) analogs to form *s*–*y* (Hirao et al. 2002)and *s*–*z* base pairs, respectively, which involve the formation of two hydrogen bonds (Hirao et al. 2004). Furthermore, the improvement of *s* provided the 7-(2-thienyl)-imidazo [4,5-b]pyridine (Ds) analog and of *z* yielded pyrrole-2-carboxaldehyde (Pa) (Hirao et al. 2006) and 2-nitro-4-propynylpyrrole (Px) analogs, which can form Ds–Pa and Ds–Px base pairs, respectively (Kimoto et al. 2009). The Ds–Px base pair in particular exhibited a high value of 99.9 % for fidelity-per-round in PCR using *Deep vent* DNA polymerase. Intriguingly, as models such as Kool's and Romesberg's possessing the F–Z and d5SICS–dNaM pairs, these base pairs do not involve hydrogen bonds. Using an analog bearing, i.e., a foreign functionality via the 4-propynyl group of Px, they have recently performed selections of modified DNA aptamers containing nucleoside analogs with the 5th base (discussed later).

3.3 Sugar Modification

In general, sugar-modified nucleic acids are called xenonucleic acids (XNAs). XNAs containing modified sugars with 2′-substituents such as methoxy (–OMe), fluoro (–F), amino (–NH_2), and azido (–N_3) groups are typical examples. Moreover, XNAs based on C2′-stereoisomers such as arabinonucleic acids (ANAs) and 2′-fluoroarabinonucleic acids (FANAs) have been extensively studied. Furthermore, XNAs containing unconventional sugars such as hexitol nucleic acids (HNAs), α-L-threofuranosyl nucleic acids (TNAs), cyclohexenyl nucleic acids (CeNAs), and 2′-*O*,4′-*C*-methylene-bridged/locked nucleic acids (2′,4′-BNAs/LNAs; hereinafter referred to as "LNA") have attracted the attention of researchers as informational biopolymer alternatives to DNA and RNA (Fig. 5).

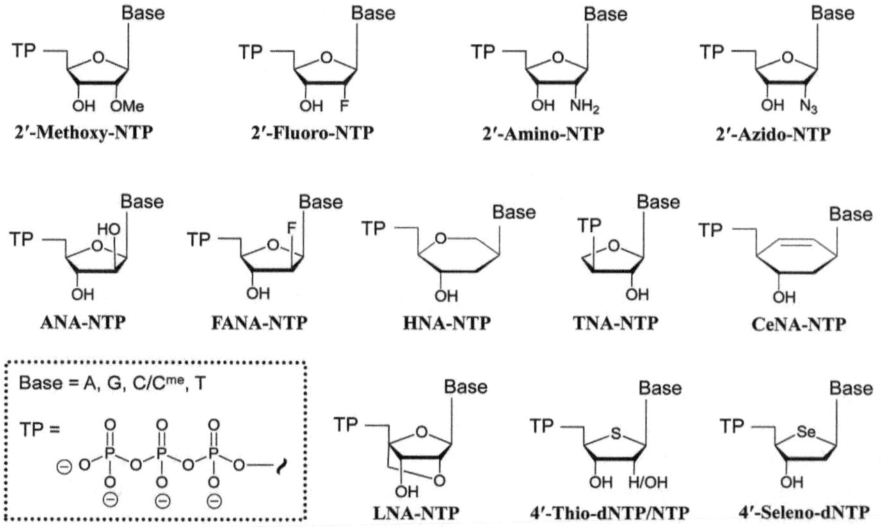

Fig. 5 Sugar-modified nucleoside triphosphates

In the late 1990s, to incorporate 2′-modified nucleotide analogs, screening of DNA polymerases that can incorporate 2′–F analogs was attempted, and *Vent(exo-)* and *Deep Vent(exo-)* DNA polymerases were found to be efficient catalysts (Ono et al. 1997). Moreover, the engineering of RNA polymerases was attempted. For example, T7 RNA polymerase variants, Y639F and Y639F/H784A, which enabled an efficient enzymatic incorporation of 2′–F or 2′–NH_2 analogs and those of the 2′–OMe or 2′–N_3 analogs, respectively, were developed (Padilla and Sousa 1999, 2002).

In 2000, K. Vastmans et al. demonstrated that *Vent(exo-)* DNA polymerase can elongate a 6-mer HNA strand on a DNA template using an HNA triphosphate (hATP) bearing an adenine base (Vastmans et al. 2000). In 2003, J. C. Chaput et al. examined DNA synthesis on a TNA template and TNA synthesis on a DNA template using various DNA polymerases (Chaput and Szostak 2003). Thereafter, they found that the Therminator DNA polymerase can polymerize TNA oligomers that are at least 80 nt in length using the following four TNA triphosphates, i.e., tTTP, tGTP, tCTP, and tDTP, which bear thymine, guanine, cytosine, and 2,6-diaminopurine, respectively, as a nucleobase (Ichida et al. 2005). In 2005, V. Kempeneers et al. reported that seven efficient successive CeNA incorporations were possible in the DNA-dependent CeNA polymerization using a CeNA triphosphate (CeATP) bearing an adenine base and *Vent(exo-)* DNA polymerase under conditions supplemented with 1 mM Mn^{2+} (Kempeneers et al. 2005). In 2007, R. N. Veedu et al. first examined the enzymatic incorporation of LNA nucleotides and observed that *Phusion High-Fidelity* DNA polymerase could accept LNA-triphosphates bearing thymine and adenine bases and catalyze primer extension reactions to yield DNA-based LNA strands (Veedu et al. 2007). In 2008,

M. Kuwahara et al. first demonstrated that *KOD Dash* DNA polymerase was superior to *Phusion High-Fidelity* DNA polymerase because of its reduced 3′,5′-exonuclease activity and could be applied for the synthesis of not only LNA but also other types of LNA, i.e., 2′,4′-BNA^{COC} and 2′,4′-BNA^{NC} (Kuwahara et al. 2008).

In 2012, P. Holliger et al. successfully created some XNA polymerases variants of *Tgo* DNA polymerase using compartmentalized self-tagging (CST) selection, which was performed on libraries of *TgoT* DNA polymerase that contained four amino acid mutations (V93Q, D141A, E143A, and A485L) (Pinheiro et al. 2012). For example, HNA polymerase was additionally mutated at V589A, E609K, I610M, K659Q, E664Q, Q665P, R668K, D669Q, K671H, K674R, T676R, A681S, L704P, and E730G. CeNA/LNA polymerase was additionally mutated at E654Q, E658Q, K659Q, V661A, E664Q, Q665P, D669A, K671Q, T676K, and R709K. ANA/FANA polymerase was additionally mutated atL403P, P657T, E658Q, K659H, Y663H, E664K, D669A, K671N, and T676I. These *TgoT* variants can produce the corresponding 72-mer XNA strands on a DNA template using four XNA triphosphates with different bases (A, G, C, and T).

As examples of other types of XNAs, 4′-modified analogs such as 4′-thioribonucleoside-5′-triphosphates, 2′-deoxy-4′-thionucleoside-5′-triphosphates, and 2′-deoxy-4′-selenonucleoside-5′-triphosphates, which have been developed by N. Minakawa and coworkers, are mentioned (Kato et al. 2005; Kojima et al. 2013; Tarashima et al. 2015). They have found that the first ones can be accepted by T7 RNA polymerase, whereas the second and third ones can act as good substrates for *KOD Dash* DNA polymerase.

Thus, continuous efforts in the screening and engineering of polymerases have enabled an efficient enzymatic production of various XNAs.

3.4 Phosphate Modification

Phosphorothioate nucleotides are well-studied phosphate-modified analogs in which one of the non-bridging α-phosphate oxygen is replaced by sulfur because introduction of phosphorothioate linkages to oligonucleotides can enhance nuclease resistance and antisense activity (Fig. 6). In 1968, F. Eckstein et al. first demonstrated enzymatic incorporation of a phosphorothioate using UTPαS (mixture of Sp- and Rp-isomers) and *Escherichia coli* DNA-dependent RNA polymerase (Matzura and Eckstein 1968). Thereafter, using optically pure Sp- or Rp-ATPαS, they discovered that the Sp-isomer could be more efficiently incorporated than the Rp-isomer during RNA strand elongation catalyzed by *E. coli* DNA-dependent RNA polymerase (Eckstein et al. 1976). In 1988, H. P. Vosberg et al. showed that the *Taq* DNA polymerase-catalyzed polymerase chain reaction (PCR) with Sp-dATPαS and dNTPs, except dATP, could efficiently amplify 310-mer double-stranded DNA fragments that contained multiple phosphorothioates (Nakamaye et al. 1988). The same results were also obtained when Sp-dNTPαS was used

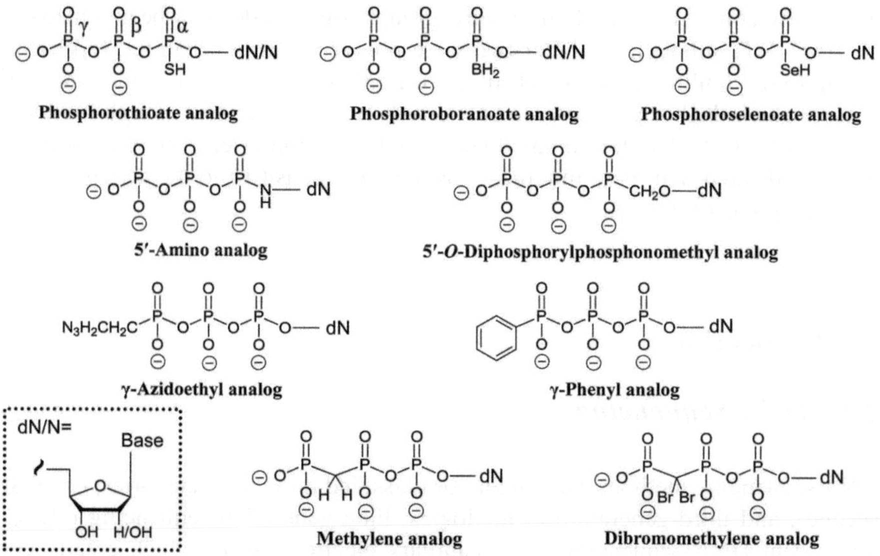

Fig. 6 Phosphate-modified nucleoside triphosphates

with a different base (G, C, or T). Other examples of non-bridging α-phosphate oxygen substituents, phosphoroboranoate and phosphoroselenoate analogs of 2′-deoxynucleoside-5′-triphosphates, have been reported. Unlike phosphorothioate analogs, the Rp-isomer of phosphoroboranoates was preferably incorporated using Sequenase DNA polymerase. One should note, however, that the Sp-isomer of phosphorothioates corresponds to the Rp-isomer of phosphoroboranoates in the configuration of the four substituents, i.e., $-P_2O_5{}^{3-}$, =O, nucleoside, and –SH/–BH_{3-}, that bond to the asymmetric phosphorus atom (Li et al. 1995; He et al. 1999). Meanwhile, both isomers of phosphoroselenoates were equally accepted as substrates by KF to produce corresponding DNA strands (Carrasco and Huang 2004).

Replacement of the bridging oxygen between α-phosphorus and the sugar 5′ carbon in a dNTP by other substituents is one of the means to enzymatically introduce chemical modifications into oligonucleotide backbones. J. L. Wolfe et al. reported that in the presence of natural dNTPs, KF (3′–5′ exo-) could accept 5′-amino-2′,5′-dideoxynucleoside-5′-*N*-triphosphates to produce the corresponding oligodeoxynucleotide strand in which multiple oxygen atoms in the 5′-position were replaced with imino (–NH–) groups (Wolfe et al. 2002). Similarly, P. Herdewijn et al. demonstrated the insertion of the methyleneoxy (–CH_2O–) group at this position using 5′-*O*-diphosphorylphosphonomethyl-2′-dA with Therminator DNA polymerase (Renders et al. 2007).

To analyze the effects of chemical modifications of the pyrophosphate moiety as a leaving group on polymerase activities, γ-phosphate oxygen substituents, i.e., γ-substituted dNTPs, were examined using AMT RT and DNA polymerase α (Alexandrova et al. 1998). Consequently, γ-substituted dNTPs with azidoethyl, aminoethyl, phenyl, or 2,4-dinitrophenyl groups could moderately be accepted only by AMT RT, although the latter two were poorer substrates than the former two owing to the bulkiness of the substituents. Furthermore, phosphate-modified analogs in which the bridging oxygen between β- and γ-phosphorus is replaced by the methylene ($-CH_2-$) or dibromomethylene ($-CBr_2-$) group were also examined. The results indicated that these modifications are more sensitive to the action of AMT RT than γ-substitutions.

4 Applications

4.1 DNA Sequencing

DNA sequencing methods can roughly be classified into three categories, i.e., first-, second-, and third-generation technologies. First-generation sequencing refers to fluorescent DNA sequencing using capillary electrophoresis, which greatly contributed to the Human Genome Project, 1990–2003. Fluorescent DNA sequencing methods, namely, the dye-primer method, which uses fluorophore-labeled primers, and the dye-terminator method, which uses fluorophore-labeled terminators, i.e., four 2′,3′-dideoxynucleoside-5′-triphosphates (ddNTPs) tagged with different fluorescent dyes, were devised in the 1980s (Fig. 7). Initially, the dye-primer method, which could provide more satisfactory long-read sequencing data, was mainly used, and KF (3′–5′ exo-) and Sequenase DNA polymerase were employed as enzymes for sequencing reactions. However, after *AmpliTaq* DNA polymerase FS, which exhibits a very weak 5′ to 3′ exonuclease activity and readily incorporates ddNTPs, had been developed, the dye-primer method was replaced with the dye-terminator method.

The second-generation sequencing technologies enable massively parallel sequencing. For example, Roche 454 sequencing technologies based on the pyrosequencing method employ dATPαS instead of dATP because luciferase accepts natural dATP as a substrate, giving rise to false-positive signals. As with dATPαS, some dATP analogs with base modifications can reduce false luciferase positives (Kajiyama et al. 2011). In pyrosequencing, KF (3′–5′ exo-) (Ronaghi et al. 1996), Sequenase Version 2.0 (Gharizadeh et al. 2004) DNA polymerase, and *Bst* DNA polymerase (large fragment) are used (Margulies et al. 2005). Lasergen developed other sequencing technologies based on the cyclic reversible termination method using 3′-OH unblocked nucleotides called Lightning Terminators. These triphosphates have a terminating 2-nitrobenzyl moiety attached to hydroxymethylated nucleobases and are efficiently incorporated by Therminator DNA polymerase (Gardner et al. 2012) (Fig. 7).

The third-generation sequencing technologies involve single-molecule real-time (SMRT) sequencing, which enables direct analysis of DNA/RNA extracted from

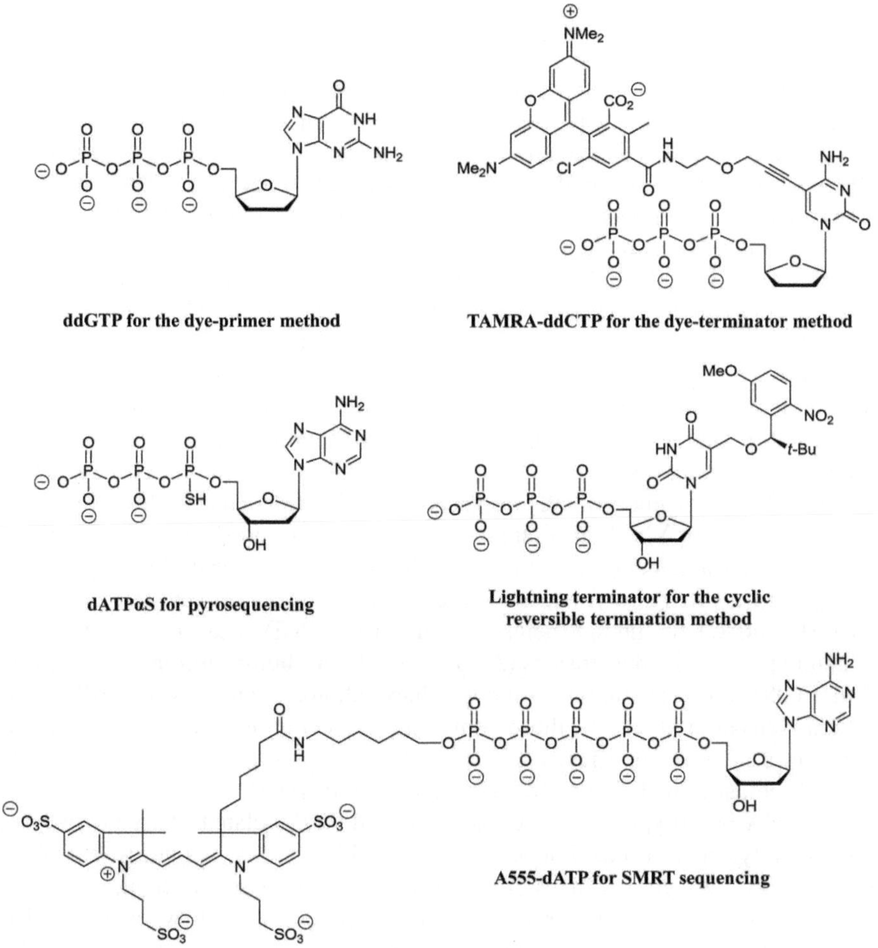

Fig. 7 Modified nucleoside triphosphates employed for DNA sequencing technologies

biological samples such as cells, tissues, and organs. For example, Pacific Biosciences developed an SMRT sequencing system using four modified dNTPs consisting of a γ-phosphate with different fluorophores and a complex of φ29 DNA polymerase and an analyte DNA template. The complex is solely immobilized at the bottom of a zero-mode waveguide, and attenuated light from the excitation beam penetrates only the lower 20–30 nm of each waveguide to provide a light microscope with a detection volume of only 20^{-21} L (Eid et al. 2009) (Fig. 7).

Thus, development of modified triphosphates and engineered polymerases continues to greatly contribute to the advancement of sequencing technologies.

4.2 Aptamer Development

To enhance nuclease resistance and improve target-binding affinities and specificities, various nucleoside triphosphates have been employed for the aptamer selection process, so-called systematic evolution of ligands by exponential enrichment (SELEX) (Tuerk and Gold 1990; Ellington and Szostak 1990). Macugen (pegaptanib sodium injection) for the treatment of wet age-related macular degeneration is a typical example of RNA-based aptamers. This modified RNA aptamer targeting vascular endothelial growth factor (VEGF) was selected from a modified RNA library. The library was enzymatically synthesized using T7 RNA polymerase and 2′-F pyrimidine nucleoside triphosphates (U and C) as well as natural ATP and GTP, followed by derivatizations, including post-SELEX modification, by which all natural purine nucleotides, except for two adenosine residues, could be replaced with 2′-OMe analogs (Ruckman et al. 1998). Furthermore, the use of a T7 RNA polymerase variant (Y639F, H784A, and K378R) enabled direct selection of modified RNA aptamers totally replaced with 2′-OMe nucleotides, which was demonstrated with VEGF as a target (Burmeister et al. 2005). Other examples of modified RNA aptamers created by SELEX using the corresponding triphosphate analogs and T7 RNA polymerase are those containing 2′-NH_2 pyrimidine nucleosides (U and C) for human neutrophil elastase (HNE) (Lin et al. 1994), those containing 4′-thioribonucleosides (A, G, C, and U) for human thrombin (Minakawa et al. 2008), those containing 5-iodouridines (photoaptamer) for the HIV-1 Rev protein (Jensen et al. 1995), those containing 5-(3-aminopropyl)uridines for ATP (Vaish et al. 2003), and those containing phosphorothioates (A, G, C, and U) for basic fibroblast growth factor (bFGF) (Jhaveri et al. 1998).

For DNA-based aptamer selection, *Vent* and *KOD*-related DNA polymerases have mainly been used. For example, modified DNA aptamers containing 5-(1-pentynyl)-2′-deoxyuridines for human thrombin (Latham et al. 1994) and those containing 5-(3-aminopropyl)-2′-deoxyuridines for ATP were obtained by SELEX using *Vent* DNA polymerase (Battersby et al. 1999). Meanwhile, modified DNA aptamers containing 5-*N*-(6-aminohexyl)carbamoylmethyl-2′-deoxyuridines for (R)-thalidomide(Fig. 2) (Shoji et al. 2007); those containing (*E*)-5-(2-(*N*-(2-(N^6-adeninyl)ethyl))carbamylvinyl)-2′-deoxyuridines for camptothecin derivatives (Fig. 2) (Imaizumi et al. 2013); those containing 5-tryptaminocarbonyl-2′-deoxyuridines (SOMAmers) for various protein targets such as fractalkine, osteoprotegerin, and cytotoxic T-lymphocyte-associated protein 4 (CTLA-4) (Fig. 2) (Gold et al. 2010); those containing the 2′-deoxynucleosides bearing Ds, i.e., one of the artificial bases from the Ds–Px base pair for VEGF and interferon-γ (Fig. 4d) (Kimoto et al. 2013); and those containing locked nucleic acid (LNA) nucleosides bearing thymine for human thrombin (Kasahara et al. 2013) were obtained by SELEX using *KOD*-related DNA polymerases such as *KOD Dash*, *KOD XL*, and *AccuPrimePfx* DNA polymerases (Fig. 5). In addition, SELEX using TaKaRa *Taq* Hot Start (HS) DNA polymerase was reported to have provided DNA-based aptamers comprised of six letters, i.e., A, G, C, T, Z (2(1H)-pyridone), and P (imidazo[1,2-a]-1,3,5-triazin-4(8H)-one), for HepG2 liver cancer cells (Zhang et al. 2015).

Furthermore, selection of the following XNA aptamers has been achieved using DNA polymerases: threose nucleic acid (TNA) aptamers for human thrombin (Therminator DNA polymerase) (Yu et al. 2012), hexitol nucleic acid (HNA) aptamers for hen egg lysozyme and HIV trans-activating response RNA (*TgoT* variant) (Pinheiro et al. 2012), and 2′-fluoroarabinonucleic acid (FANA) aptamers for HIV-1 reverse transcriptase (*TgoT* variant) (Alves Ferreira-Bravo et al. 2015).

5 Conclusions and Future Outlook

Since the 1980s, various nucleoside triphosphate analogs and polymerase variants have been developed. In the absence of the corresponding NTP(s)/dNTP(s), it became possible to produce oligonucleotides with single, double, triple, and quadruple substitutions with modified nucleotide(s) as research on screening and engineering of RNA/DNA polymerases progressed. Furthermore, oligodeoxynucleotides with expanded genetic alphabets and artificial biopolymers, i.e., XNAs with sugar structures quite different from β-D-ribofuranose in RNA or 2-deoxy-β-D-ribofuranose in DNA, can be enzymatically synthesized. Except for the abovementioned examples, various enzymatically synthesized polynucleotides that exhibit unique properties owing to introduced modifications, e.g., solvatochromic (Riedl et al. 2012), viscosity-sensitive (Dziuba et al. 2015), electroconductive (Patolsky et al. 2002), and amphiphilic functionalities (Fujita et al. 2015), have been reported. Such artificial biopolymers will serve as sensor materials for environmental and biological analyses and as programmable nanocapsules for drug carriers and gene deliveries in the near future.

Acknowledgments The authors are grateful to Dr. Hiroaki Sawai, Professor Emeritus of Gunma University, for his contribution to developing the research field and for the financial support by a Grant for Adaptable and Seamless Technology Transfer Program through Target-driven R & D, No. AS2525029M from Japan Science and Technology Agency (JST).

References

Adler S, Modrich P (1979) T7-induced DNA polymerase. Characterization of associated exonuclease activities and resolution into biologically active subunits. J Biol Chem 254:11605–11614

Alexandrova LA, Skoblov AY, Jasko MV et al (1998) 2′-Deoxynucleoside 5′-triphosphates modified at alpha-, beta- and gamma-phosphates as substrates for DNA polymerases. Nucleic Acids Res 26:778–786

Alves Ferreira-Bravo I, Cozens C, Holliger P et al (2015) Selection of 2′-deoxy-2′-fluoroarabinonucleotide (FANA) aptamers that bind HIV-1 reverse transcriptase with picomolar affinity. Nucleic Acids Res 43:9587–9599

Battersby TR, Ang DN, Burgstaller P et al (1999) Quantitative analysis of receptors for adenosine nucleotides obtained via in vitro selection from a library incorporating a cationic nucleotide analog. J Am Chem Soc 121:9781–9789
Bebenek K, Joyce CM, Fitzgerald MP et al (1990) The fidelity of DNA synthesis catalyzed by derivatives of *Escherichia coli* DNA polymerase I. J Biol Chem 265:13878–13887
Blanco L, Bernad A, Lázaro JM et al (1989) Highly efficient DNA synthesis by the phage phi 29 DNA polymerase. Symmetrical mode of DNA replication. J Biol Chem 264:8935–8940
Boschi-Muller S, Motorin Y (2013) Chemistry enters nucleic acids biology: enzymatic mechanisms of RNA modification. Biochemistry (Mosc) 78:1392–1404
Burmeister PE, Lewis SD, Silva RF et al (2005) Direct in vitro selection of a 2′-*O*-methyl aptamer to VEGF. Chem Biol 12:25–33
Buβkamp H, Batroff E, Niederwieser A et al (2014) Efficient labelling of enzymatically synthesized vinyl-modified DNA by an inverse-electron-demand Diels-Alder reaction. Chem Commun (Camb) 50:10827–10829
Campbell JL, Richardson CC, Studier FW (1978) Genetic recombination and complementation between bacteriophage T7 and cloned fragments of T7 DNA. Proc Natl Acad Sci USA 75:2276–2280
Carrasco N, Huang Z (2004) Enzymatic synthesis of phosphoroselenoate DNA using thymidine 5′-(alpha-P-seleno)triphosphate and DNA polymerase for X-ray crystallography via MAD. J Am Chem Soc 126:448–449
Chaput JC, Szostak JW (2003) TNA synthesis by DNA polymerases. J Am Chem Soc 125:9274–9275
Chelliserrykattil J, Lu H, Lee AH et al (2008) Polymerase amplification, cloning, and gene expression of benzo-homologous "yDNA" base pairs. Chembiochem 9:2976–2980
Chien A, Edgar DB, Trela JM (1976) Deoxyribonucleic acid polymerase from the extreme thermophile *Thermus aquaticus*. J Bacteriol 127:1550–1557
Davanloo P, Rosenberg AH, Dunn JJ et al (1984) Cloning and expression of the gene for bacteriophage T7 RNA polymerase. Proc Natl Acad Sci USA 81:2035–2039
De Clercq E (2001) Antiviral drugs: current state of the art. J Clin Virol 22:73–89
Dziuba D, Pohl R, Hocek M (2015) Polymerase synthesis of DNA labelled with benzylidene cyanoacetamide-based fluorescent molecular rotors: fluorescent light-up probes for DNA-binding proteins. Chem Commun 51:4880–4882
Eckstein F, Armstrong VW, Sternbach H (1976) Stereochemistry of polymerization by DNA-dependent RNA-polymerase from *Escherichia coli*: an investigation with a diastereomeric ATP-analogue. Proc Natl Acad Sci USA 73:2987–2990
Eid J, Fehr A, Gray J et al (2009) Real-time DNA sequencing from single polymerase molecules. Science 323:133–138
Ellington AD, Szostak JW (1990) In vitro selection of RNA molecules that bind specific ligands. Nature 346:818–822
Fiala G, Stetter KO (1986) *Pyrococcus furiosus* sp. nov. represents a novel genus of marine heterotrophic archaebacteria growing optimally at 100°C. Arch Microbiol 145:56–61
Fujita H, Nakajima K, Kasahara Y et al (2015) Polymerase-mediated high-density incorporation of amphiphilic functionalities into DNA: enhancement of nuclease resistance and stability in human serum. Bioorg Med Chem Lett 25:333–336
Gardner AF, Wang J, Wu W et al (2012) Rapid incorporation kinetics and improved fidelity of a novel class of 3′-OH unblocked reversible terminators. Nucleic Acids Res 40:7404–7415
Gharizadeh B, Eriksson J, Nourizad N et al (2004) Improvements in pyrosequencing technology by employing sequenase polymerase. Anal Biochem 330:272–280
Gierlich J, Gutsmiedl K, Gramlich PM et al (2007) Synthesis of highly modified DNA by a combination of PCR with alkyne-bearing triphosphates and click chemistry. Chemistry 13:9486–9494
Gold L, Ayers D, Bertino J et al (2010) Aptamer-based multiplexed proteomic technology for biomarker discovery. PLoS One 5:e15004

Gourlain T, Sidorov A, Mignet N et al (2001) Enhancing the catalytic repertoire of nucleic acids. II. Simultaneous incorporation of amino and imidazolyl functionalities by two modified triphosphates during PCR. Nucleic Acids Res 29:1898–1905

Grippo P, Richardson CC (1971) Deoxyribonucleic acid polymerase of bacteriophage T7. J Biol Chem 246:6867–6873

He K, Porter KW, Hasan A et al (1999) Synthesis of 5-substituted 2′-deoxycytidine 5′-(alpha-P-borano)triphosphates, their incorporation into DNA and effects on exonuclease. Nucleic Acids Res 27:1788–1794

Held HA, Benner SA (2002) Challenging artificial genetic systems: thymidine analogs with 5-position sulfur functionality. Nucleic Acids Res 30:3857–3869

Hirao I, Ohtsuki T, Fujiwara T et al (2002) An unnatural base pair for incorporating amino acid analogs into proteins. Nat Biotechnol 20:177–182

Hirao I, Harada Y, Kimoto M et al (2004) A two-unnatural-base-pair system toward the expansion of the genetic code. J Am Chem Soc 126:13298–13305

Hirao I, Kimoto M, Mitsui T et al (2006) An unnatural hydrophobic base pair system: site-specific incorporation of nucleotide analogs into DNA and RNA. Nat Methods 3:729–735

Hollenstein M, Hipolito CJ, Lam CH et al (2009) A self-cleaving DNA enzyme modified with amines, guanidines and imidazoles operates independently of divalent metal cations (M2+). Nucleic Acids Res 37:1638–1649

Ichida JK, Horhota A, Zou K et al (2005) High fidelity TNA synthesis by Therminator polymerase. Nucleic Acids Res 33:5219–5225

Imaizumi Y, Kasahara Y, Fujita H et al (2013) Efficacy of base-modification on target binding of small molecule DNA aptamers. J Am Chem Soc 135:9412–9419

Ishikawa M, Hirao I, Yokoyama S (2000) Synthesis of 3-(2-deoxy-β-d-ribofuranosyl)pyridin-2-one and 2-amino-6-(N, N-dimethylamino)-9-(2-deoxy-β-d-ribofuranosyl)purine derivatives for an unnatural base pair. Tetrahedron Lett 41:3931–3934

Jäger S, Rasched G, Kornreich-Leshem H et al (2005) A versatile toolbox for variable DNA functionalization at high density. J Am Chem Soc 127:15071–15082

Jannasch HW, Wirsen CO, Molyneaux SJ et al (1992) Comparative physiological studies on hyperthermophilic archaea isolated from deep-sea hot vents with emphasis on *Pyrococcus* strain GB-D. Appl Environ Microbiol 58:3472–3481

Jensen KB, Atkinson BL, Willis MC et al (1995) Using in vitro selection to direct the covalent attachment of human immunodeficiency virus type 1 Rev protein to high-affinity RNA ligands. Proc Natl Acad Sci USA 92:12220–12224

Jhaveri S, Olwin B, Ellington AD (1998) In vitro selection of phosphorothiolated aptamers. Bioorg Med Chem Lett 8:2285–2290

Kajiyama T, Kuwahara M, Goto M et al (2011) Optimization of pyrosequencing reads by superior successive incorporation efficiency of improved 2′-deoxyadenosine-5′-triphosphate analogs. Anal Biochem 416:8–17

Kasahara Y, Irisawa Y, Ozaki H et al (2013) 2′,4′-BNA/LNA aptamers: CE-SELEX using a DNA-based library of full-length 2′-O,4′-C-methylene-bridged/linked bicyclic ribonucleotides. Bioorg Med Chem Lett 23:1288–1292

Kato Y, Minakawa N, Komatsu Y et al (2005) New NTP analogs: the synthesis of 4′-thioUTP and 4′-thioCTP and their utility for SELEX. Nucleic Acids Res 33:2942–2951

Kempeneers V, Renders M, Froeyen M et al (2005) Investigation of the DNA-dependent cyclohexenyl nucleic acid polymerization and the cyclohexenyl nucleic acid-dependent DNA polymerization. Nucleic Acids Res 33:3828–3836

Kiefer JR, Mao C, Braman JC et al (1998) Visualizing DNA replication in a catalytically active *Bacillus* DNA polymerase crystal. Nature 391:304–307

Kimoto M, Kawai R, Mitsui T et al (2009) An unnatural base pair system for efficient PCR amplification and functionalization of DNA molecules. Nucleic Acids Res 37:e14

Kimoto M, Yamashige R, Matsunaga K et al (2013) Generation of high-affinity DNA aptamers using an expanded genetic alphabet. Nat Biotechnol 31:453–457

Klenow H, Henningsen I (1970) Selective elimination of the exonuclease activity of the deoxyribonucleic acid polymerase from *Escherichia coli* B by limited proteolysis. Proc Natl Acad Sci USA 65:168–175

Kojima T, Furukawa K, Maruyama H et al (2013) PCR amplification of 4′-thioDNA using 2′-deoxy-4′-thionucleoside 5′-triphosphates. ACS Synth Biol 2:529–536

Kong H, Kucera RB, Jack WE (1993) Characterization of a DNA polymerase from the hyperthermophile archaea *Thermococcus litoralis*. Vent DNA polymerase, steady state kinetics, thermal stability, processivity, strand displacement, and exonuclease activities. J Biol Chem 268:1965–1975

Kotani H, Ishizaki Y, Hiraoka N et al (1987) Nucleotide sequence and expression of the cloned gene of bacteriophage SP6 RNA polymerase. Nucleic Acids Res 15:2653–2664

Kufe DW, Major PP, Egan EM et al (1980) Correlation of cytotoxicity with incorporation of ara-C into DNA. J Biol Chem 255:8997–9000

Kuwahara M, Nagashima J, Hasegawa M et al (2006) Systematic characterization of 2′-deoxynucleoside- 5′-triphosphate analogs as substrates for DNA polymerases by polymerase chain reaction and kinetic studies on enzymatic production of modified DNA. Nucleic Acids Res 34:5383–5394

Kuwahara M, Obika S, Nagashima J et al (2008) Systematic analysis of enzymatic DNA polymerization using oligo-DNA templates and triphosphate analogs involving 2′,4′-bridged nucleosides. Nucleic Acids Res 36:4257–4265

Latham JA, Johnson R, Toole JJ (1994) The application of a modified nucleotide in aptamer selection: novel thrombin aptamers containing 5-(1-pentynyl)-2′-deoxyuridine. Nucleic Acids Res 22:2817–2822

Lawyer FC, Stoffel S, Saiki RK et al (1989) Isolation, characterization, and expression in *Escherichia coli* of the DNA polymerase gene from *Thermus aquaticus*. J Biol Chem 264:6427–6437

Leconte AM, Hwang GT, Matsuda S et al (2008) Discovery, characterization, and optimization of an unnatural base pair for expansion of the genetic alphabet. J Am Chem Soc 130:2336–2343

Lee AH, Kool ET (2005) A new four-base genetic helix, yDNA, composed of widened benzopyrimidine-purine pairs. J Am Chem Soc 127:3332–3338

Lee SE, Sidorov A, Gourlain T et al (2001) Enhancing the catalytic repertoire of nucleic acids: a systematic study of linker length and rigidity. Nucleic Acids Res 29:1565–1573

Li H, Porter K, Huang F et al (1995) Boron-containing oligodeoxyribonucleotide 14mer duplexes: enzymatic synthesis and melting studies. Nucleic Acids Res 23:4495–4501

Limbach PA, Crain PF, McCloskey JA (1994) Summary: the modified nucleosides of RNA. Nucleic Acids Res 22:2183–2196

Lin Y, Qiu Q, Gill SC et al (1994) Modified RNA sequence pools for in vitro selection. Nucleic Acids Res 22:5229–5234

Majumder HK, Maitra U, Rosenberg M (1979) Termination of transcription by bacteriophage T3 RNA polymerase: homogeneous 3′-terminal oligonucleotide sequence of in vitro T3 RNA polymerase transcripts. Proc Natl Acad Sci USA 76:5110–5113

Malyshev DA, Seo YJ, Ordoukhanian P et al (2009) PCR with an expanded genetic alphabet. J Am Chem Soc 131:14620–14621

Malyshev DA, Dhami K, Lavergne T et al (2014) A semi-synthetic organism with an expanded genetic alphabet. Nature 509:385–388

Margulies M, Egholm M, Altman WE et al (2005) Genome sequencing in microfabricated high-density picolitre reactors. Nature 437:376–380

Matray TJ, Kool ET (1999) A specific partner for abasic damage in DNA. Nature 399:704–708

Mattila P, Korpela J, Tenkanen T et al (1991) Fidelity of DNA synthesis by the *Thermococcus litoralis* DNA polymerase–an extremely heat stable enzyme with proofreading activity. Nucleic Acids Res 19:4967–4973

Matzura H, Eckstein F (1968) A polyribonucleotide containing alternation P=O and P=S linkages. Eur J Biochem 3:448–452

McMinn DL, Ogawa AK, Wu Y et al (1999) Efforts toward expansion of the genetic alphabet: DNA polymerase recognition of a highly stable, self-pairing hydrophobic base. J Am Chem Soc 121:11585–11586

Mehedi Masud M, Ozaki-Nakamura A, Kuwahara M et al (2003) Modified DNA bearing 5 (methoxycarbonylmethyl)-2′-deoxyuridine: preparation by PCR with thermophilic DNA polymerase and postsynthetic derivatization. Chembiochem 4:584–588

Minakawa N, Sanji M, Kato Y et al (2008) Investigations toward the selection of fully-modified 4′-thioRNA aptamers: optimization of in vitro transcription steps in the presence of 4-′-thioNTPs. Bioorg Med Chem 16:9450–9456

Miroshnichenko ML, Gongadze GM, Rainey FA et al (1998) *Thermococcus gorgonarius* sp. nov. and *Thermococcus pacificus* sp. nov.: heterotrophic extremely thermophilic archaea from New Zealand submarine hot vents. Int J Syst Bacteriol 48(Pt 1):23–29

Morales JC, Kool ET (1998) Efficient replication between non-hydrogen-bonded nucleoside shape analogs. Nat Struct Biol 5:950–954

Nakamaye KL, Gish G, Eckstein F et al (1988) Direct sequencing of polymerase chain reaction amplified DNA fragments through the incorporation of deoxynucleoside alpha-thiotriphosphates. Nucleic Acids Res 16:9947–9959

Nishioka M, Mizuguchi H, Fujiwara S et al (2001) Long and accurate PCR with a mixture of KOD DNA polymerase and its exonuclease deficient mutant enzyme. J Biotechnol 88:141–149

Ohtsuki T, Kimoto M, Ishikawa M et al (2001) Unnatural base pairs for specific transcription. Proc Natl Acad Sci USA 98:4922–4925

Ono T, Scalf M, Smith LM (1997) 2′-Fluoro modified nucleic acids: polymerase-directed synthesis, properties and stability to analysis by matrix-assisted laser desorption/ionization mass spectrometry. Nucleic Acids Res 25:4581–4588

Padilla R, Sousa R (1999) Efficient synthesis of nucleic acids heavily modified with non-canonical ribose 2′-groups using a mutant T7 RNA polymerase (RNAP). Nucleic Acids Res 27:1561–1563

Padilla R, Sousa R (2002) A Y639F/H784A T7 RNA polymerase double mutant displays superior properties for synthesizing RNAs with non-canonical NTPs. Nucleic Acids Res 30:e138

Patolsky F, Weizmann Y, Willner I (2002) Redox-active nucleic-acid replica for the amplified bioelectrocatalytic detection of viral DNA. J Am Chem Soc 124:770–772

Perrin DM, Garestier T, Hélène C (1999) Expanding the catalytic repertoire of nucleic acid catalysts: simultaneous incorporation of two modified deoxyribonucleoside triphosphates bearing ammonium and imidazolyl functionalities. Nucleosides Nucleotides 18:377–391

Perrin DM, Garestier T, Hélène C (2001) Bridging the gap between proteins and nucleic acids: a metal-independent RNAseA mimic with two protein-like functionalities. J Am Chem Soc 123:1556–1563

Piccirilli JA, Krauch T, Moroney SE et al (1990) Enzymatic incorporation of a new base pair into DNA and RNA extends the genetic alphabet. Nature 343:33–37

Pinheiro VB, Taylor AI, Cozens C et al (2012) Synthetic genetic polymers capable of heredity and evolution. Science 336:341–344

Raindlová V, Pohl R, Sanda M et al (2010) Direct polymerase synthesis of reactive aldehyde-functionalized DNA and its conjugation and staining with hydrazines. Angew Chem Int Ed Engl 49:1064–1066

Renders M, Emmerechts G, Rozenski J et al (2007) Enzymatic synthesis of phosphonomethyl oligonucleotides by therminator polymerase. Angew Chem Int Ed Engl 46:2501–2504

Ried T, Baldini A, Rand TC et al (1992) Simultaneous visualization of seven different DNA probes by in situ hybridization using combinatorial fluorescence and digital imaging microscopy. Proc Natl Acad Sci USA 89:1388–1392

Riedl J, Ménová P, Pohl R et al (2012) GFP-like fluorophores as DNA labels for studying DNA–protein interactions. J Org Chem 77:8287–8293

Ronaghi M, Karamohamed S, Pettersson B et al (1996) Real-time DNA sequencing using detection of pyrophosphate release. Anal Biochem 242:84–89

Ruckman J, Green LS, Beeson J et al (1998) 2′-Fluoropyrimidine RNA-based aptamers to the 165-amino acid form of vascular endothelial growth factor (VEGF165). Inhibition of receptor binding and VEGF-induced vascular permeability through interactions requiring the exon 7-encoded domain. J Biol Chem 273:20556–20567

Rüdiger A, Jorgensen PL, Antranikian G (1995) Isolation and characterization of a heat-stable pullulanase from the hyperthermophilic archaeon *Pyrococcus woesei* after cloning and expression of its gene in *Escherichia coli*. Appl Environ Microbiol 61:567–575

Rüttimann C, Cotoras M, Zaldivar J et al (1985) DNA polymerases from the extremely thermophilic bacterium *Thermus thermophilus* HB-8. Eur J Biochem 149:41–46

Sakthivel K, Barbas CF III (1998) Expanding the potential of DNA for binding and catalysis: highly functionalized dUTP derivatives that are substrates for thermostable DNA polymerases. Angew Chem Int Ed 37:2872–2875

Sawai H, Ozaki AN, Satoh F et al (2001) Expansion of structural and functional diversities of DNA using new 5-substituted deoxyuridine derivatives by PCR with superthermophilic KOD Dash DNA polymerase. Chem Commun 2604–2605

Seo YJ, Hwang GT, Ordoukhanian P et al (2009) Optimization of an unnatural base pair toward natural-like replication. J Am Chem Soc 131:3246–3252

Shoji A, Kuwahara M, Ozaki H et al (2007) Modified DNA aptamer that binds the (R)-isomer of a thalidomide derivative with high enantioselectivity. J Am Chem Soc 129:1456–1464

Sidorov AV, Grasby JA, Williams DM (2004) Sequence-specific cleavage of RNA in the absence of divalent metal ions by a DNAzyme incorporating imidazolyl and amino functionalities. Nucleic Acids Res 32:1591–1601

Sismour AM, Benner SA (2005) The use of thymidine analogs to improve the replication of an extra DNA base pair: a synthetic biological system. Nucleic Acids Res 33:5640–5646

Southworth MW, Kong H, Kucera RB et al (1996) Cloning of thermostable DNA polymerases from hyperthermophilic marine Archaea with emphasis on *Thermococcus* sp. 9 degrees N-7 and mutations affecting 3′-5′ exonuclease activity. Proc Natl Acad Sci USA 93:5281–5285

Switzer C, Moroney SE, Benner SA (1989) Enzymatic incorporation of a new base pair into DNA and RNA. J Am Chem Soc 111:8322–8323

Takagi M, Nishioka M, Kakihara H et al (1997) Characterization of DNA polymerase from *Pyrococcus* sp. strain KOD1 and its application to PCR. Appl Environ Microbiol 63:4504–4510

Tarashima N, Sumitomo T, Ando H et al (2015) Synthesis of DNA fragments containing 2′-deoxy-4′-selenonucleoside units using DNA polymerases: comparison of dNTPs with O, S and Se at the 4′-position in replication. Org Biomol Chem 13:6949–6952

Tasara T, Angerer B, Damond M et al (2003) Incorporation of reporter molecule-labeled nucleotides by DNA polymerases. II. High-density labeling of natural DNA. Nucleic Acids Res 31:2636–2646

Tuerk C, Gold L (1990) Systematic evolution of ligands by exponential enrichment: RNA ligands to bacteriophage T4 DNA polymerase. Science 249:505–510

Vaish NK, Fraley AW, Szostak JW et al (2000) Expanding the structural and functional diversity of RNA: analog uridine triphosphates as candidates for in vitro selection of nucleic acids. Nucleic Acids Res 28:3316–3322

Vaish NK, Larralde R, Fraley AW et al (2003) A novel, modification-dependent ATP-binding aptamer selected from an RNA library incorporating a cationic functionality. Biochemistry 42:8842–8851

Vastmans K, Pochet S, Peys A et al (2000) Enzymatic incorporation in DNA of 1,5-anhydrohexitol nucleotides. Biochemistry 39:12757–12765

Veedu RN, Vester B, Wengel J (2007) Enzymatic incorporation of LNA nucleotides into DNA strands. Chembiochem 8:490–492

Weisbrod SH, Marx A (2007) A nucleoside triphosphate for site-specific labelling of DNA by the Staudinger ligation. Chem Commun (Camb) 1828–1830

Wolfe JL, Kawate T, Belenky A et al (2002) Synthesis and polymerase incorporation of 5′-amino-2′,5′-dideoxy-5′-N-triphosphate nucleotides. Nucleic Acids Res 30:3739–3747
Yang Z, Hutter D, Sheng P et al (2006) Artificially expanded genetic information system: a new base pair with an alternative hydrogen bonding pattern. Nucleic Acids Res 34:6095–6101
Yang Z, Sismour AM, Sheng P et al (2007) Enzymatic incorporation of a third nucleobase pair. Nucleic Acids Res 35:4238–4249
Yu H, Zhang S, Chaput JC (2012) Darwinian evolution of an alternative genetic system provides support for TNA as an RNA progenitor. Nat Chem 4:183–187
Zhang L, Yang Z, Sefah K et al (2015) Evolution of functional six-nucleotide DNA. J Am Chem Soc 137:6734–6737

MIX
Papier aus verantwortungsvollen Quellen
Paper from responsible sources
FSC® C105338
www.fsc.org

If you have any concerns about our products,
you can contact us on
ProductSafety@springernature.com

In case Publisher is established outside the EU,
the EU authorized representative is:
Springer Nature Customer Service Center GmbH
Europaplatz 3, 69115 Heidelberg, Germany

Printed by Libri Plureos GmbH
in Hamburg, Germany